AF696277

# BIOPHYSICS OF CONSCIOUSNESS

## A Foundational Approach

# BIOPHYSICS OF CONSCIOUSNESS

## A Foundational Approach

Editors

**Roman R. Poznanski**
The Rockefeller University, USA
and
Universiti Teknologi Malaysia

**Jack A. Tuszynski**
University of Alberta, Canada

**Todd E. Feinberg**
Icahn School of Medicine at Mount Sinai
and
Mount Sinai Beth Israel Medical Center,
USA

World Scientific

NEW JERSEY • LONDON • SINGAPORE • BEIJING • SHANGHAI • HONG KONG • TAIPEI • CHENNAI • TOKYO

*Published by*

World Scientific Publishing Co. Pte. Ltd.
5 Toh Tuck Link, Singapore 596224
*USA office:* 27 Warren Street, Suite 401-402, Hackensack, NJ 07601
*UK office:* 57 Shelton Street, Covent Garden, London WC2H 9HE

**Library of Congress Cataloging-in-Publication Data**
Names: Poznanski, Roman R., 1962– | Tuszynski, J. A., 1956– | Feinberg, Todd E., 1952– .
Title: Biophysics of consciousness : a foundational approach / editors, Roman R. Poznanski, Jack A. Tuszynski, Todd E. Feinberg.
Description: New Jersey : World Scientific, [2016] | Includes bibliographical references and index.
Identifiers: LCCN 2015031941 | ISBN 9789814644259 (hardcover : alk. paper)
Subjects: LCSH: Consciousness. | Biophysics.
Classification: LCC QP411 .B56 2016 | DDC 612.8/233--dc23
LC record available at http://lccn.loc.gov/2015031941

**British Library Cataloguing-in-Publication Data**
A catalogue record for this book is available from the British Library.

Copyright © 2017 by World Scientific Publishing Co. Pte. Ltd.

*All rights reserved. This book, or parts thereof, may not be reproduced in any form or by any means, electronic or mechanical, including photocopying, recording or any information storage and retrieval system now known or to be invented, without written permission from the publisher.*

For photocopying of material in this volume, please pay a copying fee through the Copyright Clearance Center, Inc., 222 Rosewood Drive, Danvers, MA 01923, USA. In this case permission to photocopy is not required from the publisher.

Typeset by Stallion Press
Email: enquires@stallionpress

Printed in Singapore

*Dedicated to Gerald M. Edelman, Walter J. Freeman, and Karl H. Pribram*

Biophysics of Consciousness: A Foundational Approach
R. R. Poznanski, J. A. Tuszynski and T. E. Feinberg
Copyright © 2017 World Scientific, Singapore.

# CONTENTS

*"It is not consciousness that has been evolved — an elementary flash of consciousness may be a native property of matter — but a particular kind of integrated consciousness, a consciousness spun into a continuous thread by a faculty of memory, a consciousness embroidered as upon a canvas, whose function is to hold in place and in their proper relation the components of the picture."*

— Alfred J. Lotka

Biophysics of Consciousness: A Foundational Approach
R. R. Poznanski, J. A. Tuszynski and T. E. Feinberg
Copyright © 2017 World Scientific, Singapore.

# PREFACE

Over a century ago, Einstein developed his special theory of relativity that demonstrated electromagnetic waves, including light, could not be vibrations of aether. We now know that matter constitutes discrete atoms, that electricity is the flow of charged particles (electrons), that light waves are composed of particles of light energy called "photons" and that gravity is caused by mass warping spacetime. Likewise, at about the same time in the early 20th century founders of quantum mechanics introduced a mathematical abstraction, the quantum wave function, which allowed for the propagation of signals faster than the speed of light as "collapsing quantum entanglement" — a result forbidden by Einstein's relativity. Nor can quantum theory yet be reconciled with relativity's explanation of gravity. Thus, quantum mechanics and relativity are not yet compatible — a gap exists to this day in their unification. In neuroscience, there is also a gap between first-person ontology and third-person ontology called the explanatory gap. Our purpose is to bridge the explanatory gap through the development of specific theories of consciousness.

Placing ourselves by anology in those heady nineteenth-century days when the problem of the "aether" was so avidly being probed, a time at the brink of an earthshaking intellectual revolution, we set out on a similar journey of discovery, seeking the nature of consciousness. What adaptive and causal element in the natural world is consciousness? The current state of knowledge portrays either a reductionist view of consciousness, for example, the seat of consciousness is characterized as quantum gravity affects in microtubules (of the cytoskeleton) or a more recent viewpoint is that consciousness is the result of large-scale integration of information in the brain. The reductionist theories of consciousness claim that subjective experience is physical in nature and governed by physical laws. This is problematic since physical signals (that carry information) are accomplished objectively, in the third-person ontology, whereas subjective experience is first-person ontology and hence the explanatory gap remains. Answering how subjective experience can be encroached through emulation of the brain will undoubtedly have an enduring impact in biology similar to the effect Einstein's relativity had in physics at the turn of last century.

How our subjective experiences can be qualitatively and quantitatively explained by a theory of consciousness that is based on functioning of the brain and how these subjective experiences mediate human actions is a major enigma of modern neuroscience. Sentience refers to subjective experiences that cannot be algorithmically represented through reductionism even when the performance of all the relevant functions is explained through algorithms. How manifestations of consciousness are maintained in living beings, and how and from where experiences arise, remains a deep puzzle. Our approach encompassing ontological, evolutionary, neurobiological, and Freudian interpretations implies that consciousness must indeed involve ingredients that cannot adequately be simulated by mere computation alone. However, despite all these efforts, the physical basis of consciousness remains yet to be discovered. Some neuroscientists would say that it is the synaptic connections between neurons or the coherent firing patterns of large assemblies that harvest large-scale integration, but our approach is more foundational.

In this volume, we have assembled the best minds to forge an understanding on how it is possible that an objective brain produces subjective experience. The collection is a representative sample of contributions from various authors from diverse fields who provide insights into some of the specific theories of consciousness as we understand them today. We hope the approach we advocate will guide future studies and advances in the field.

The initiative for the production of this book originated from a special issue on the hard problem of consciousness published in the *Journal of Integrative Neuroscience*. We thank V. K. Sanjeed, Editor at World Scientific Publishing Co Pte Ltd for commissioning this book.

Roman R. Poznanski, Jack A. Tuszynski and<br>Todd E. Feinberg

Biophysics of Consciousness: A Foundational Approach
R. R. Poznanski, J. A. Tuszynski and T. E. Feinberg
Copyright © 2017 World Scientific, Singapore.

# FOREWORD: ADDRESSING THE HARD PROBLEM OF CONSCIOUSNESS

*John R. Searle*

*Department of Philosophy*
*314 Moses Hall 2390*
*University of California*
*Berkeley, CA 94720-2390, USA*

The hard problem of consciousness is concerned with what I take to be the leading problem in the biological sciences today. Indeed, the case could be made that it is the leading problem in science in general.

To put the problem at its most succinct: How exactly do neurobiological processes in the brain cause consciousness? This is sometimes called "the hard problem" to suggest that it is a different problem in kind from such other neurobiological problems as for example, how does the reflex arc work, what is the exact role of neurotransmitters in action potentials? I will now explain the problem in a way that will enable us to expand its formulation to include more aspects. First we will have to answer the question, what is consciousness? Consciousness is sometimes said to be hard to define but I think if we are just talking about a common sense definition that identifies the target of the discussion and not a scientific definition of the sort that typically comes at the end of the investigation, then consciousness is not hard to define. Here is the definition. By "consciousness" we mean all of our states of feeling or sentience or awareness. These typically begin in the morning when we wake from a dreamless sleep and they go on all day until we once again fall asleep, or otherwise become unconscious.

On this definition dreams are a form of consciousness. Consciousness so defined only exists insofar as experienced by human or animal subjects and we can say therefore it has a first person or subjective ontology (where "ontology" just means the mode of existence).

What exactly is the relation between consciousness so defined and the brain? Well opinions differ about that and indeed it is the traditional mind–body problem. But I think in general outline, we know enough about how the world works to state the general features of the answer to the question. All of our conscious states without exception are caused by lower level neuronal processes in the brain. The textbooks typically assume that the neuron is the basic functional unit but for all we know it may be something higher than the level of individual neurons, such as whole maps or clouds of neurons, or according to some theorists, it may be something lower such as microtubules within the neuron. But in any case the neuron is a distinctive feature of the brain and it seems reasonable to suppose that it plays a special role in the creation of consciousness. If consciousness is entirely caused by neuron firings then what exactly is the status of consciousness? What is its form of existence? And here again I think the answer is obvious. Consciousness *caused by* neuron firings is *realized in* the system of neurons. Just as, for example the liquid or solid behavior of the body of water in front of me is causally explained by the behavior of the $H_2O$ molecules, but liquidity and solidity are not some extra juice or substance squirted out by the molecules, they are just the conditions that the system of molecules is in; in exactly that same sense I want to say that consciousness is a condition of the neuronal system. The analogy only works so far because of course we are perfectly happy to reduce solidity and liquidity to features of molecular behavior in a way that we are reluctant to make a similar reduction of consciousness, precisely because consciousness has a subjective or first person ontology, unlike these other features which have an objective or third person ontology.

So that is in broad outline both a "solution" to the mind–body problem and a general description of the relationships between consciousness and the brain. I call this view *biological naturalism* because it is based on the assumption that the mind is a natural biological phenomenon that is as much a part of nature as photosynthesis, lactation or digestion, but at the same time the right level to discuss it is the level of biological processes and not for example, the level of sub-atomic physics, or the level of social structures. I think biological naturalism is exactly right but it raises a number of problems of its own. On this account, consciousness is a natural biological phenomenon like photosynthesis or digestion but it differs from these other biological phenomena in several respects, perhaps most importantly because it has a first person ontology.

The whole discussion of the relation of consciousness and the brain has such a sordid and confused history and it is very hard for us to overcome that history completely. Here in very brief outline are the two main elements of the history:

(1) The first called "dualism" says that consciousness is not a part of the physical world in the way that I have described it, but it belongs to some separate ontological realm. Consciousness on the traditional view is not a feature of the brain at all, it is a feature of the soul and the soul is not a part of the brain or any other phenomenon in the physical world. The soul has a completely immaterial, non-physical existence. This view has been incredibly powerful for millennia but I think it is hopelessly inadequate, and I do not have the space to expose its inadequacy here but I will assume that it is inadequate.

(2) The second view, which is almost as bad, takes itself to be a denial of dualism but in fact it accepts the worst mistake of dualism. It begins by saying that consciousness as I have defined it is not a part of the physical world but either it does not exist at all, or if it does exist it must be reducible to some feature of the physical world such as behavior, neuron firings, computer programs, etc. This view is sometimes called "scientific materialism" or "materialist monism". But it is almost as confused as the dualism it is designed to replace because it denies the obvious fact that consciousness as defined really does exist as part of the physical, biological world. Does anyone doubt this? We are all conscious pretty much all of our waking lives and furthermore, consciousness exists with this irreducible first person ontology. For any conscious state there is something that it feels like to be in that conscious state. And this *what it feels like* feature is always a feature of the individual who is conscious, who has the conscious feelings.

So given these twin mistakes, the whole history is strewn with the dead bodies of mistaken theories. I do not have much space but I will list five of the most outrageous mistaken theories.

(1) *Substance dualism.* This is the view that says we are identical with our souls and our souls are temporarily attached to our bodies, but the souls have no material existence. I find it hard to take this view seriously, but as I said earlier it has been incredibly influential.

(2) *Property dualism.* This view is not quite as bad but it is pretty bad. It says that there are not two different kinds of substances but two metaphysically different kinds of properties, the physical and the mental. Mental properties are not a part of the physical world, so this view immediately has the result that our mental life can have no causal effect. On this view called "epiphenomenalism", no human being in the history of the world ever drank because she was thirsty or ate because he was hungry, or raised his arm because he wanted to raise his arm. This view is again, I think not worth discussing but it survives to this very day in various confused philosophical theories.
(3) *Materialist reductionism.* On this view there really is no such thing as conscious phenomena as I have defined them. There are just third person phenomena. There is no such thing as irreducible ontological subjectivity. This view denies the very data that gets the subject going in the first place.
(4) *Behaviorism.* This view says consciousness does exist but it is really something else. It is really just behavior or dispositions to behavior. This view makes the same mistake as reductionist materialism, it just substitutes behavior for matter. But of course we all know from our own case that feeling of pain is one thing and emitting pain behavior, such as saying "Ouch!", is something else. You can have one without the other. You can have the pain without the pain behavior, and you can have the pain behavior without the pain. Behaviorism is so obviously false that it seems literally incredible that it should have been so influential for so long. The reason for its influence, is that it seems somehow more "scientific" than granting the obvious fact that we are all conscious and our consciousness is not reducible to something else.
(5) *Computationalism.* A view that has recently become influential is the idea that consciousness is really just a computer program or set of computer programs running in our brains. On this view, the mind is to the brain as the program is to the hardware. This view is obviously false. I refuted it over 30 years ago with the so-called "Chinese Room Argument" and I will not repeat the details here, just state the obvious point. The computer program is defined formally or syntactically. To put it slightly more technically, the notion "same implemented program" defines an equivalence class that is specified completely independently of the hardware realization. But we know independently that the syntax of the computer program is not by itself

> sufficient for the semantics, or the mental content, of an actual mental life whether conscious or unconscious. I proved this with the Chinese Room Argument where a person locked in a room could answer questions in Chinese simply by following the steps of the program even though he understood not a word of Chinese, either the questions or the answers. The syntax of the program was sufficient for his behavior but it did not produce the semantics of actual human cognition.

There are various other reductive theories out there also equally false. Currently there is an information theoretic view but it is, I believe, as confused as the views that I have refuted here.

Well if consciousness just is a biological process in the same sense as digestion or photosynthesis, then why do not we get busy and give a scientific explanation of consciousness? I think in fact this is the right approach and we are making progress but it is more difficult than many people ever expected it would be, and I want to explain why. In the standard method in sciences you go through three stages. First you find a correlate of the phenomenon that you are seeking to explain. The correlate of consciousness is called "the neuronal correlate of consciousness" or simply the NCC. Second, you perform the usual tests to figure out if the correlate is a causal correlation and you do this by withholding the NCC and seeing if you stop consciousness and resuming the NCC to see if you can restart consciousness. And third, you try to get a theory that explains why these phenomena have these effects. These are the stages by which we have for example come to understand heredity or come to understand the germ theory of disease. Why is it any different for consciousness? In fact I do not think it is any different. I think this is the right way to proceed but if progress has been slow, why? The main reason is that our existing research techniques for the most part are either matters of imaging, for example, where you image processes in the brain that are going on whenever a certain state of consciousness occurs, or they are single cell recordings, where you track the neuron firing of a single neuron in the brain. The results from these two methods are immensely impressive but they still have not given a solution to consciousness. Why not? I do not know, but here is one speculation. The problem with consciousness is not at the level of individual neurons and it is not even at the level of the sorts of patterns of neuron firings you find when the subject for example is trying to solve a certain problem. Rather, all of these things, the conscious perception or the attempt to solve a problem, only go on within a pre-existing *conscious field.* Almost all of the studies that I have seen to try to solve the problem of consciousness use subjects that are already conscious. But what we want to know is, how does the subject get to be conscious in the first place? What we want to know is, how does the brain create the unified *conscious field* within which the subject can then have a perception or a thought or an intentional action? It is much harder to get an account of how the brain creates the unified *conscious field* because our existing research techniques are not well suited to that discovery.

Though progress is slow, it is no slower than other scientific problems in the past. I think we are well on the way toward having an adequate account of how the brain causes consciousness and how consciousness is realized in the brain. That would be a solution to the hard problem.

Biophysics of Consciousness: A Foundational Approach
R. R. Poznanski, J. A. Tuszynski and T. E. Feinberg
Copyright © 2017 World Scientific, Singapore.

# CONTRIBUTING AUTHORS

**Michael A. Arbib, PhD**
The ABLE Project: Action, Brain, Language, Evolution
University of Southern California
Los Angeles, CA 90089-2520
USA

**Jalil Ali, PhD**
Laser Centre, IBNU SINA ISIR,
and Physics Department, Faculty of Science
Universiti Teknologi Malaysia
81310 Johor Bahru, Johor
Malaysia

**Dorian Aur, PhD**
Department of Comparative Medicine
Stanford University
Palo Alto, CA 94305
USA

**M. Bahadoran, PhD**
Laser Centre, IBNU SINA ISIR
Universiti Teknologi Malaysia,
81310 Johor Bahru, Johor
Malaysia

**Dani Bercovich, PhD**
Department of Molecular Genetics, Faculty of Life Sciences
Tel-Hai Academic College, Galilee 12210
Israel

**Gustav Bernroider, PhD**
Neurosignaling & Neuroecology Unit, Department of Ecology & Evolution
University of Salzburg
A-5020 Salzburg
Austria

**Francisco Di Biase, MD**
Department of Neurosurgery, Clínica Di Biase and Santa Casa Hospital
Barra do Piraí, Rio, 27120120
Brazil
and
Department of Post-Graduation UGB Universidade Geraldo Di Biase — Volta Redonda, Rio, Brazil

**L. A. Cacha, PhD**
Laser Centre, IBNU SINAR ISIR
Universiti Teknologi Malaysia
81310 Johor Bahru, Johor
Malaysia

**Antonio Chella, PhD**
Department of Chemical, Management, Computer, Mechanical Engineering
University of Palermo
Viale delle Scienze 90128, Palermo
Italy

**Travis J. A. Craddock, PhD**
Clinical Systems Biology Group
Institute for Neuro-Immune Medicine
College of Psychology
Nova Southeastern University
3424 South University Drive
Fort Lauderdale, FL 33328
USA

**Todd E. Feinberg, MD**
Mount Sinai Beth Israel
First Avenue at 16th Street
New York, NY 10003, USA

**Geoffrey Goodman, PhD**
Galil Genetic Analysis, Katzrin 12900
Israel

**Colin G. Hales, PhD**
Neuroengineering Laboratory, Department of Electrical and Electronic Engineering,
University of Melbourne, Victoria 3010
Australia

**Stuart R. Hameroff, MD**
Departments of Anesthesiology and Psychology
Center for Consciousness Studies Banner-University Medical Center
The University of Arizona
Tucson, AZ 85721
USA

**Huping Hu, PhD, JD**
Research & Development, Quantum Dream Inc.,
Stony Brook, NY 11790
USA

**Jon M. Mallatt, PhD**
School of Biological Sciences
Washington State University
Pullman, WA, 99164-4236
USA
and
University of Washington School of Medicine
4333 Brooklyn Ave NE, Seattle, WA 98105
USA

**Riccardo Manzotti, PhD**
Department of Marketing, Behavior and Communication
IULM University
via Carlo Bo, 8, 20143 Milano, Italy

**Sir Roger Penrose OM, FRS**
Mathematical Institute and Wadham College
University of Oxford, Oxford OX2 6GG
UK

**Alfredo Pereira Jr., PhD, Livre-Docente**
Department of Education
São Paulo State University (UNESP)
Rubião Jr. Campus, 18.618-970 – Botucatu, São Paulo
Brazil

**Roman R. Poznanski, PhD**
Department of Clinical Sciences
Faculty of Biosciences and Medical Engineering
Universiti Teknologi Malaysia
81310 Johor Bahru, Johor
Malaysia
and
Laboratory of Biological Modeling
The Rockefeller University,
10065 New York
USA

**Massimo Pregnolato, PhD**
Department of Drug Sciences
University of Pavia
12 Viale Taramelli, Pavia, 27100
Italy

**Mark Solms, PhD**
Department of Psychology
University of Cape Town
Rondebosch 7701
South Africa

**Jack A. Tuszynski, PhD, DSc**
Department of Oncology
University of Alberta
Edmonton, Alberta, T6G 1Z2
Canada

**J. L. Perez Velazquez, PhD**
Neuroscience and Mental Health Programme
University of Toronto,
Toronto, Ontario, M5G 1X8
Canada

**Rām Lakhan Pāndey Vimal, PhD**
Neuroscience & Consciousness Research Department
Vision Research Institute
25 Rita Street, Lowell, MA 01854
USA

**Maoxin Wu, MD, PhD**
Research & Development, Quantum Dream Inc.
Stony Brook, NY, 11790
USA

**Jasmy Yunus, PhD**
Faculty of Bioscience and Medical Engineering
Universiti Teknologi Malaysia
81310 Johor Bahru, Johor
Malaysia

**P. P. Yupapin, PhD**
Advanced Studies Center, Faculty of Science
King Mongkut's Institute of Technology
Bangkok 10520
Thailand

Biophysics of Consciousness: A Foundational Approach
R. R. Poznanski, J. A. Tuszynski and T. E. Feinberg
Copyright © 2017 World Scientific, Singapore.

# 1

# NEUROBIOLOGICAL NATURALISM

*Todd E. Feinberg* and Jon Mallatt*†*

*Neurology and Psychiatry, Icahn School of Medicine at Mount Sinai and Mount Sinai Beth Israel Medical Center, New York, USA;
†School of Biological Sciences, Washington State University, Pullman, WA, USA*

## Abstract

Neurobiological naturalism, an extension of John Searle's concept of biological naturalism, says primary (sensory) consciousness is a scientifically tractable problem based on natural laws although brains that possess consciousness display certain advanced neurobiological system-features. These features are complex, fast, hierarchical, system-wide, internal, and often topographically organized neuron–neuron interactions. We lay out the neurophilosophical problem of the ontological irreducibility of the subjective to the objective, describe the general and special neurobiological features of the conscious neural hierarchies in vertebrates, and then bring in evolutionary considerations to show how consciousness could have evolved in the first vertebrates. Our combined neurobiological, neuroevolutionary and neurophilosophical approach offers a solution to the hard problem of how and why physical brains can cause experiences and why consciousness and subjectivity are neurontologically irreducible.

*Keywords*: Consciousness; neurobiological naturalism; ontological subjectivity; hard problem; qualia; neural hierarchies; nestedness; emergence; reductionism; evolution.

## 1.1. Introduction

A primary philosophical obstacle to a scientific account of consciousness is the problem of ontological subjectivity; i.e., explaining how an objective and tangible brain creates experiences that have a first-person ontology. In this chapter, we review and expand a theory of *neurobiological naturalism* (Feinberg, 2012), that uses three domains of explanation — neurobiological, neuroevolutionary and neurophilosophical — to address different aspects of the "hard problem" of primary or sensory consciousness. We propose that by collectively considering these three overlapping approaches, sensory consciousness can be understood as the natural outcome of normal biological evolution coupled with several unique neurobiological and neurohierarchical functions.

## 1.2. What is Biological Naturalism?

The term "biological naturalism" was introduced by Searle (1984) as a philosophical solution to the so-called "mind–body" problem. He begins his book "The Rediscovery of the Mind" with the below explanation and definition:

> "The famous mind–body problem, the source of so much controversy over the past two millennia, has a simple solution. This solution has been available to any educated person since serious work began on the brain nearly a century ago, and, in a sense, we all know it to be true. Here it is: Mental phenomena are caused by neurophysiological processes in the brain and are themselves features of the brain. To distinguish this view from the many others in the field, I call it "biological naturalism." Mental events and processes are as much part of our biological natural history as digestion, mitosis, meiosis, or enzyme secretion" (Searle, 1992, p. 1).

According to this position, Searle (2004) asserts that conscious states are "real" (not illusory) but have a subjective first-person ontology. He argues that conscious feeling states cannot be "reduced" to their neural substrate because a third-person reduction by the observer omits this first-person ontology that needs to be explained in the first place:

> "....consciousness has a first-person or subjective ontology and so cannot be reduced to anything that has third-person or objective ontology. If you try to reduce or eliminate one in favor of the other you leave

> something out... biological brains have a remarkable biological capacity to produce experiences, and these experiences only exist when they are felt by some human or animal agent. You can't reduce these first-person subjective experiences to third-person phenomena for the same reason that you can't reduce third-person phenomena to subjective experiences. You can neither reduce the neuron firings to the feelings nor the feelings to the neuron firings, because in each case you would leave out the objectivity or subjectivity that is in question" (Searle, 1997, p. 212).

A critical neuroscientific limitation of Searle's theory of biological naturalism remains: *How can we produce subjective experience from objective neural processes yet this subjectivity cannot be objectively reduced to neural processes?* This apparent contradiction is the most problematic challenge to any theory that tries to "naturalize" consciousness. Indeed, were it not for this thorny issue, there would be no "famous mind–body problem," and no "hard problem" (Chalmers, 1995; 1996) of consciousness at all.

At the heart of this enigma is the undeniable fact that although we understand consciousness to be an "emergent feature" of complex and dynamic biological systems, it remains that *no other* complex biological or even non-biological system poses a similar reductionist challenge. For example, while the reduction of "life" to biology and chemistry presents no philosophical barriers, consciousness displays many emergent features that are resistant to scientific reduction.

### 1.2.1. *Reducing consciousness*

Searle argues that scientific reduction involves an identity relationship in which a property A is said to be reducible to property B, if it can be shown that A "is nothing but" B. If such a "nothing but" relationship exists between two entities, then we do not need to invoke any new property to explain A over and above those principles by which we explain B.

While Searle identifies several types of scientific reduction, for our understanding of consciousness, *ontological reduction* is the most pertinent:

> "The most important form of reduction is ontological reduction. It is the form in which objects of certain types can be shown to consist in nothing but objects of other types. For example, chairs are shown to be nothing but collections of molecules. This form is clearly important in the history of science. For example, material objects in general can be shown to be

> nothing but collections of molecules, genes can be shown to consist in nothing but DNA molecules" (Searle, 1992, p. 113).

Searle actually means "nothing but collections of objects *and their interactions*," allowing for the interactions between parts.

One impediment to successful ontological reduction is that in complex systems novel properties may emerge from the parts of the system and their interactions. Thus explaining the relationship between emergent properties in a complex system and their scientific reduction becomes critical. We must work out if, how and why the properties emerge within the confines of the system. This is especially true when one analyzes how consciousness — an emergent property of complex brains — can be reduced to neurons and their interactions.

There are *"weak"* and *"strong"* or *"radical"* versions of emergence theory, which differ in how they view an emergent feature in relation to its scientific reduction (Chalmers, 2006; Feinberg, 2001a, 2001b; Kim, 2006; Van Gulick, 2001). The weak version of emergence simply states that there are higher order properties of complex systems that are novel relative to the constituent parts that create them. A commonly cited example of a weakly emergent property is the fluidity of water relative to the non-fluid properties of single water molecules. Fluidity as a property of the aggregate of water molecules poses no scientific ontological "mystery". If one can explain all the principles of water molecules and fluid dynamics, the fluid behavior of water can be understood, making it a weakly emergent property of water molecules and their interactions.

In contrast to this weak version of emergence, the most important aspect of the strong version of emergence theory is the claim that there are emergent properties that cannot *in principle* ever be reduced to the component parts of a system. Dissecting such a theoretical system — if for example consciousness were a radically emergent process — would always leave something unexplained.

In biology, the most common reduction is an explanation of how whole systems are reduced to their parts and their interactions. Biological subsystems are conceived as the "material" parts of the reduction, and the interactions of these parts are processes with a time dimension. Thus, considering digestion for example, there are various constituent organs — stomach, colon, pancreas, etc. — that comprise the *digestive system*, and various *digestive processes* — enzymatic breakdown, secretion and peristalsis — that comprise the interaction of those parts.

The two features combined explain how digestion occurs, and we have no difficulty reducing digestion to its parts and their interactions. In a similar fashion, with a definition of life and observations of its properties and manifestations, Darwin deduced the interactions responsible for the process of evolution, and Watson and Crick elucidated the nucleotide bonds responsible for the structure of DNA. Ultimately, we find nothing that meets that definition of life that is not *reducible* to biological parts and processes that create it. By Searle's criteria, these systems display weak emergent features that can be ontologically reduced.

In neuroscience, we typically begin with a macroscopic scientific observation or a definable property (such as motor or sensory functions, an epileptic seizure, memory functions, emotions, etc.) and then attempt to explain its mechanism based upon more fundamental or known functional or anatomic properties. In this way, disorders of language or memory can be reduced to damage to particular brain regions and we understand the role these regions play in producing the function in question. Other successful ontological reductions in neuroscience and neurology include identifying how motor action is reducible to the physiology of the nerve and muscle, or determining that paralysis is reducible to injury of the motor system. There is no ontological "mystery of epilepsy" because we understand how the observable seizure can be reduced to abnormal electrical discharges of neurons in a particular brain region. In contrast, a theory that says consciousness is a radically emergent property of the brain would claim that consciousness in principle can never be reduced to the brain or explained by the laws of physics as scientists understand them.

A prototypical example of a radically emergent theory of consciousness was proposed by Nobel laureate Roger Sperry. He claimed that the mind is "more than the sum of the parts" of the material brain and that consciousness goes above and beyond the brain's physical processes. In Sperry's words: "...conscious phenomena are different from, more than, and not reducible to, neural events..." (Sperry, 1990, p. 383). This says consciousness does not fit Searle's definition of ontological reduction quoted above.

Sperry viewed the mind–brain relationship in hierarchical terms and supposed that the brain's neural elements combine in increasingly complex configurations until the mind emerges at the top of the hierarchy, and the mind cannot be reduced to the brain as the brain is currently understood by neurophysiology. According to his account, so many emergent properties and currently unknown principles must characterize the upper levels of the hierarchy that we are far from ever knowing this mysterious "brain code" (Sperry, 1983). While Sperry seems not to have embraced dualism nor said consciousness operates outside the brain or outside known laws of energy and matter, he seems to suggest that the advanced laws and properties of brains are so beyond our present understanding that we must for now treat consciousness as a "fundamental" entity. Sperry further proposed that this "central, emergent conscious force" has *causal* properties over the material brain:

> "...consciousness was conceived to be a dynamic emergent of brain activity, neither identical with, nor reducible to, the neural events of which it is mainly composed. Further, consciousness was not conceived as an epiphenomenon, inner aspect, or other passive correlate of brain processing, but rather to be an active integral part of the cerebral process itself, exerting potent causal effects in the interplay of cerebral operations. In a position of top command at the highest levels in the hierarchy of brain organization, the subjective properties were seen to exert control over the biophysical and chemical activities at subordinate levels. It was described initially as a brain model that puts "conscious mind back into the brain of objective science in a position of top command...a brain model in which conscious, mental, psychic forces are recognized to be the crowning achievement...of evolution" (Sperry, 1990, p. 382).

Sperry's idea can be criticized for attributing causal power to such a vaguely defined entity: just what might be those new spatiotemporal and mass–energy interactions at the top of the hierarchy that produce the conscious mind? Indeed, any claim for radical emergence will raise this issue of, "if the whole is so much more than the sum of the parts, then what *is* responsible for the whole?"

But such proposals of irreducibility are fully understandable, if not justified, by a recurring frustration in the effort to develop a neuroscientific reduction of consciousness. The seemingly insolvable problem is that when it comes to reducing consciousness to neural functions, there always appears to be something "left out" or to remain unreduced and unexplained, namely the subjective aspects of experience. Furthermore, we not only have been unable to *reduce* consciousness to neural systems, but we also appear unable to *derive* consciousness as an emergent feature of the nervous system. In other words we have difficulty explaining why should there be consciousness at all. As Chalmers claims, "Experience may *arise* from the physical, but is not *entailed* by the physical" (Chalmers,

1995, p. 208). With this we are back to the claim for aspects of subjective experience that cannot be reduced to objective neural functions and structures. In this chapter, we present a rebuttal and a solution that places consciousness squarely in the domain of the normal physical science.

### 1.2.2. *Neuroontologically irreducible features of consciousness*

We will begin addressing this problem by reviewing those unique features of consciousness that are most resistant to scientific reduction. These have been called *neuroontologically irreducible features* (NOIFs) (Feinberg, 2012) and include *referral, mental unity, qualia* and *mental causation* (Table 1).

The NOIFs of consciousness are those that are irreducibly subjective. This means that: (1) they have a first-person ontology and cannot be objectively detected but can only be subjectively experienced; (2) their ontological properties differ, depending upon whether they are externally described or subjectively embodied; and (3) therefore they are refractory to normal scientific reduction.

#### 1.2.2.1. *Referral of neural states*

The first fundamentally subjective NOIF is the "referral" or "projection" of neural states. A conscious experience does not refer to the material brain itself. That is, it is not experienced as the firing of neurons within the brain that produce it, nor is one aware of these firings *per se*. Rather the experience is referred either to something or somewhere in the world or to someplace on or in the body. Such conscious referral holds for *exteroceptive* experiences of the world that are projected out into the world as "mental images" as well as for *interoceptive* perceptions and *affective* feelings of the body that are experienced partly or wholly as internal bodily states (Feinberg, 2009, 2011; see *qualia* below).

**Table 1.** The unique, neurontologically irreducible features of consciousness.

| |
|---|
| *Referral* |
| *Mental unity* |
| *Qualia* |
| *Mental causation* |

For instance, consider the exteroceptive sensory experiences, which begin with stimuli from the external environment. In the case of the special senses (vision, hearing, taste and olfaction), Sir Charles Sherrington (1947) called one aspect of their referral *projicience*. Thus, one does not "feel" a visual stimulus on the retina or an auditory stimulus on the cochlea of the ear; nor does one experience the stimulus as originating in the brain. Instead, these sensations are "projected" outward toward the source of external stimulation, to the noisy object being viewed.

The property of projicience is obviously true for far-away distance reception, but some qualities are also "referred" to the environment from nearby stimuli that are directly on the body's surface. Thus, when we taste we do not experience "my tongue is sour," we experience the *lemon* on my tongue as sour. When holding an ice cube, we do not experience the innumerable local sensory stimuli on the hand (unless we try to), but the sensory qualities of cold, rectangular and hard are projected to the ice cube.

Some interoceptive feelings, like pain or itch, are referred or externalized *from* the neurons in the central nervous system *to* some other spatial location on or within the body. Affective states are also referred. These affects can be experienced entirely within the self (being happy or sad for instance), but not as if in the brain. In other words, when I am happy, I do not feel like my brain's neural circuits are happy. In all these cases, while the neurons as objectively observed are actually within the brain, the consciousness states as subjectively experienced are referred to some other part of the self or to the world.

#### 1.2.2.2. *Mental unity*

Though the nervous system as objectively analyzed consists of billions of individual neurons, consciousness as subjectively experienced appears as a unified field. For excellent summaries of the cognitive, philosophical, and neurological aspects of this unity, see Bayne (2010), Bayne & Chalmers (2003), Brook & Raymont (2010) and Cleeremans (2003). Sellars (1963) called this the "grain argument", meaning that while subjective experience is homogeneous, unified and "without grain" (see Teller, 1992), the brain is objectively a "gappy", heterogeneous, discontinuous "conglomerate of spatially discrete events" (Meehl, 1966). Thus, the nervous system as objectively observed and analyzed is material, divisible into parts, and extended, while consciousness is normally unified into a single overall experience. These differences create another impasse when attempting to reduce consciousness to neural events. How do the many produce the one?

1.2.2.3. *Qualia*

Qualia are individual, subjectively experienced features or qualities, such as the perceived color of an object, a recognized odor, or a negative affect. Many philosophers and neuroscientists regard the nature and origin of qualia as the central enigma of consciousness. Chalmers (1995) opines:

> "If any problem qualifies as *the* problem of consciousness, it is this one. In this central sense of "consciousness", an organism is conscious if there is something it is like to be that organism, and a mental state is conscious if there is something it is like to be in that state. Sometimes terms such as "phenomenal consciousness" and "qualia" are also used here, but I find it more natural to speak of "conscious experience" or simply "experience" (Chalmers, 1995, p. 201).

Crick and Koch agree that the neuroscience of qualia is perhaps the central issue we confront for a correct account of consciousness:

> "The most difficult aspect of consciousness is the so-called 'hard problem' of qualia — the redness of red, the painfulness of pain, and so on. No one has produced any plausible explanation as to how the experience of the redness of red could arise from the actions of the brain. It appears fruitless to approach this problem head-on" (Crick & Koch, 2003, p. 119).

Simply put, neurons as objectively observed are not "red," "painful," "hot," or "sweet," but the neural states that possess these subjective properties are. Thus, qualia are especially refractory to ontological reduction.

Related to qualia is another key aspect of sensory consciousness, selective *attention*. This is the ability to focus on one stimulus so that it becomes a quale, while filtering out or suppressing less-relevant stimuli. This attention is related to sensory alertness and arousal. To us, attention seems a far more tractable problem than the mystery of qualia. That is, investigators are starting to explain attention in terms of coordinated neuronal oscillations between brain regions that are tuned to the same stimulus, of neuromodulators that set arousal thresholds for certain sensory circuits, simple lateral inhibition of nearby neurons by the neurons that carry the strongest stimuli, and simple habituation of neurons that carry repetitive (unimportant) stimuli (Giurfa, 2013; Knudsen, 2011; Krauzlis *et al.*, 2013; Womelsdorf & Fries, 2007). The point is that selective attention seems reducible rather than irreducible, so we will not consider it further here (but see Chapter 2).

Returning to qualia, most philosophers and neuroscientists seeking to explain sensory consciousness view the nature of qualia as a single neurophilosophical issue. Crick and Koch in the quote above speak of the "redness of red" and the "painfulness of pain" as if the "mystery" behind both these kinds of qualia can be solved by one fundamental mechanism. While at some basic level this may ultimately turn out to be true, in fact *exteroceptive, interoceptive* and *affective* qualia differ — not only in the way they are subjectively experienced, but also in the underlying neural architectures that produce them.

Exteroceptive sensory consciousness, as described above, derives from external stimulation and depends on highly organized isomorphic (topographic) maps in the brain. The exteroceptive qualia are sensed features of external objects and do not inherently evoke emotions. This contrasts with affective consciousness, which refers to any negative or positive feeling which in humans includes our emotions. Affective sensory consciousness can derive either from the organism's reaction to external stimuli or from the "feeling" of internal body states. Unlike exteroceptive sensory processing, it is not strictly isomorphic and is self-referential rather than projicient. Thus, affective qualia such as the pleasantness or unpleasantness of suffering, joy, and anger lack the fine-tuned isomorphism of visual, discriminative-touch, or olfactory sensory representations. And in contrast to isomorphic representations and the mental images they create, these affective experiences are not projected outward but are more broadly experienced by the entire subjective organism: I see and hear the lion "out there" but I am afraid "in here."

Turning to interoceptive sensory consciousness, this has a "dual aspect" in that it shares features of both isomorphic sensory images and affects. Interoception is the sensing of inner body states, such as a full stomach, air hunger, or a heart palpitation. Also commonly included among interoceptive feelings are painful stimuli anywhere on the body such as a pinprick on the finger, skin temperature, itch, or inner bodily pains such as from a kidney stone (Craig, 2003). As an example of its dual aspect, a pin prick is topographically localized to the body surface, which is its exteroceptive aspect, but it also has an affective aspect because the pain and unpleasantness of the stimulus are experienced body-wide as emotional distress. The qualia associated with the pin prick are correspondingly dual (sharp and painful).

In the vertebrate animals these three types of sensory consciousness, and their associated qualia, localize to different parts of the brain. Exteroceptive qualia derive primarily from the cerebral cortex in mammals, and, as we recently proposed, from the optic tectum of the midbrain in fish and amphibians, which like the mammalian cortex has an isomorphic organization (Feinberg & Mallatt, 2013). Affective consciousness, on the other hand, stems from the subcortical limbic system (and then can be cognitively modulated by the cerebral cortex of mammals: Panksepp, 1998; Solms, 2013). Interoceptive consciousness, in line with its dual aspect, would arise from both major brain regions.

Therefore, although the exteroceptive, affective and interoceptive systems overlap and extensively interact, each has its own subjective attributes and associated brain regions. This points to an unexpected diversity in the types of consciousness and in the brain systems that associate with them. *Primary consciousness is not monolithic but multi-faceted.*

#### 1.2.2.4. *Mental causation*

The philosopher Jaegwon Kim defines the problem of mental causation as explaining "how it is possible for the mental to exercise causal influences in the physical world." (Kim, 1998, p. 29; also see Kim, 1993). With reference to neuroontological irreducibility, Revonsuo's definition of causation will serve nicely:

> "The idea that the mind or mental phenomena have causal powers to change some purely material (e.g., biological or neural) process in the brain" (Revonsuo, 2010, p. 298).

Or, how can consciousness — an apparently intangible, objectively unobservable, and wholly subjective entity — have causal properties upon objectively observable neurons, which in turn dictate behaviors that change the material world? How could an intangible process build the great pyramids or induce people to channel large rivers and thereby change the very surface of the earth?

## 1.3. Neurobiological Naturalism

In order to explain these enigmatic features of consciousness without invoking any new or "mysterious" forces or fundamental principles, one of us proposed a theory called *neurobiological naturalism* (Feinberg, 2012). Our current account of this theory (Table 2) rests on several major postulates: (1) Sensory consciousness and subjectivity are unique, emergent neurobiological, system-features that are based upon general and special biological functions; and (2) The origin of sensory consciousness can be traced back to when reflexes evolved into subjective mental images and affective feelings; (3) A complete explanation of subjectivity and the neuro-ontological irreducibility of sensory consciousness requires neurobiological, neuroevolutionary, and neurophilosophical approaches; and (4) Sensory consciousness is diverse, and different neural architectures can create it. We will first consider the unique neurobiological features that create ontological subjectivity. Then we will address evolution of the diverse types of consciousness, with particular reference to the transition from reflexes to subjectivity and to how a combination of neurobiological, neuroevolutionary and neurophilosophical perspectives is essential for a complete theory of neurobiological naturalism.

### 1.3.1. *Neurobiology: General and special brain features*

Biological systems demonstrate two broad categories of features that pertain to the neurobiology of consciousness (Table 3). The first are *general biological features* that apply to all biological systems (including nervous systems) and that all living things must possess. These are necessary precursors to consciousness but none of them taken separately or

**Table 2.** Postulates of neurobiological naturalism.

| |
|---|
| *Sensory consciousness and subjectivity are unique, emergent neurobiological-system features based upon general and special brain functions* |
| *The origin of sensory consciousness can be traced to when reflexes evolved into subjective mental images and affective feelings* |
| *A complete explanation of subjectivity and the neuroontological irreducibility of sensory consciousness requires neurobiological, neuroevolutionary and neurophilosophical approaches* |
| *Sensory consciousness is diverse, and different neural architectures can create it* |

**Table 3.** General and special neurobiological features of consciousness.

| |
|---|
| A. General biological features that apply to all living things |
| *Process* |
| *System and self-organization* |
| *Hierarchy, emergence, and constraint* |
| *Teleonomy* |
| B. Special neurobiological features that apply to conscious living things |
| *Reflexes and rates* |
| *Complexity* |
| *Representations that allow mental images, often isomorphically organized** |
| *Neural hierarchies and unique neural-neural interactions* |

**or, affective mental states*

collectively can account for consciousness. The second are *special neurobiological features,* the unique building blocks that make the emergence of consciousness possible. We propose that although consciousness is not "radically" different in any fundamental way from other emergent biological properties, nonetheless the special neurobiological features — in association with the general features — provide the advanced and unique nervous properties of consciousness.

### 1.3.2. *General biological features*

#### 1.3.2.1. *Process*

In the 19th century Charles Darwin observed the structure of living things and deduced the *process* of evolution. In the 20th century James Watson and Francis Crick observed the process of life and deduced the *structure* of DNA. Today, one of the missions of neuroscience is to explain both the structure of the brain and the process of consciousness. William James already understood that consciousness must be viewed as a process rather than a structure when he wrote:

> "To deny plumply that "consciousness" exists seems so absurd on the face of it — for undeniably "thoughts" do exist — that I fear some readers will follow me no farther. Let me then immediately explain that I mean only to deny that the word stands for an entity, but to insist most

> emphatically that it does stand for a function. There is, I mean, no aboriginal stuff or quality of being contrasted with that of which material objects are made, out of which our thoughts of them are made; but there is a function in experience which thoughts perform, and for the performance of which this quality of being is invoked. That function knows. "Consciousness" is supposed necessary to explain the fact that things not only are, but get reported, are known. Whoever blots out the notion of consciousness from his list of first principles must still provide in some way for that function's being carried on." (James, 1904, p. 478).

Alfred North Whitehead similarly argued that the mind should be identified with the process of the brain, not with any as yet undiscovered or more adequately defined form, or as he put it, "consciousness will be the function of knowing" (Whitehead, 1925, p. 171). What is essential here is that consciousness, like all biological functions, is a physiological process. Consciousness is an aspect of what the living brain *does.*

#### 1.3.2.2. *System and self-organization*

Living organisms in general and the nervous systems in particular are self-organizing, complex systems. Biological self-organization is defined by Camazine *et al.* (2001):

> "In biological systems self-organization is a process in which pattern at the global level of a system emerges solely from numerous interactions among the lower-level components of the system. Moreover, the rules specifying interactions among the system's components are executed using only local information, without reference to the global pattern. In short, the pattern is an emergent property of the system, rather than a property imposed on the system by an external ordering influence" (p. 8).

Indeed this is a feature of all *complex systems* or *complex adaptive systems* (Mitchell, 2009). Self-organization plays a role in the emergence of consciousness, and is a necessary but insufficient property.

#### 1.3.2.3. *Hierarchy, emergence, and constraint in biological systems*

Complex biological systems have a clear hierarchical structure (Salthe, 1985; Simon, 1962, 1973) in which evolutionary change tends to become concentrated upon the higher units (Buss, 1987, pp. 183–188). In general, the lower on the hierarchical scale an individual unit is located the more conservative are its biological features. For example, the genetic code of DNA is essentially universal across all living things, the cells of all animals have the same organelles, yet the tissues and organs differ more among the animal phyla. Then at the top, the kinds of animals are highly varied, from snails to lobsters to horses. This is an empirically observed trend, but its reason may be as follows. Although natural selection acts on every level of the hierarchy it is strongest at the top level, the individual organism, which has the most direct interaction with the challenges and changes of particular external environments. And the brain, at the highest level of the neuroaxis, has evolved great differences across the vertebrates (while always retaining a fundamental groundplan of forebrain, midbrain and hindbrain) (Butler & Hodos, 2005).

The complimentary principles of *emergence* and *constraint* are essential to all hierarchical systems. Kim (1992) and others (Simon, 1962, 1973; Salthe, 1985) have noted the intimate relationship between emergent features and hierarchical structure:

> "...although the fundamental entities of this world and their properties are material, when material processes reach a certain level of complexity, genuinely novel and unpredictable properties emerge, and that this process of emergence is cumulative, generating a hierarchy of increasingly more complex novel properties. Thus, emergentism presents the world not only as an evolutionary process but also as a *layered structure* — a hierarchically organized system of levels of properties, each level emergent from and dependent on the one below" (Kim, 1992, pp. 121–122).

An example from biology is the developing embryo, whose cells organize into tissues, which organize to form organs that start functioning one by one. The new heart beats, networks of neurons form and start to fire in the assembling brain, and soon all the organs are working together to support the complete body.

While the term emergence connotes the way that the parts in a hierarchy combine to form new properties at higher levels, constraint is the process by which higher levels impose control over the lower levels (Ahl & Allen, 1996; Allen & Starr, 1982; Salthe, 1985; Simon, 1962, 1973), a relationship called *downward causation* (Campbell, 1974). The biologist

Pattee wonderfully summarizes the importance of hierarchical constraint in the emergence of complexity in biological systems:

> "If there is to be any theory of general biology, it must explain the origin and operation (including the reliability and persistence) of the hierarchical constraints which harness matter to perform coherent functions. This is not just the problem of why certain amino acids are strung together to catalyze a specific reaction. The problem is universal and characteristic of all living matter. It occurs at every level of biological organization, from the molecule to the brain. It is the central problem of the origin of life, when aggregations of matter obeying only elementary physical laws first began to constrain individual molecules to a functional, collective behavior. It is the central problem of development where collections of cells control the growth or genetic expression of individual cells. It is the central problem of biological evolution in which groups of cells form larger and larger organizations by generating hierarchical constraints on subgroups. It is the central problem of the brain where there appears to be an unlimited possibility for new hierarchical levels of description. These are all problems of hierarchical organization. Theoretical biology must face this problem as fundamental, since hierarchical control is the essential and distinguishing characteristic of life" (Pattee, 1970, p. 119).

In biological systems such as a person, the individual cells of the body constrain their constituent organelles to perform cellular metabolism; the organs constrain their cells so that they secrete enzymes; the person's entire body constrains the organs to digest, breathe, and perform all the large-scale functions necessary for the life of the body. It is simple to see why these constraints persist. If a level fails to stay within these constraints, as when cellular metabolism or breathing stops, the system collapses in death. Only constrained organismal systems survive.

In order to analyze the neurohierarchical basis of consciousness we must consider two basic types of hierarchies (Fig. 1), *non-nested* or *control hierarchies* and *nested* or *compositional hierarchies* (Ahl & Allen, 1996; Allen & Starr, 1982; Feinberg, 2011; Pattee, 1970; Salthe, 1985). A non-nested hierarchy has a pyramidal structure in which the higher levels are not *physically composed* of the lower levels as, for example, in a military command with a general at the top and successively lower ranks of command below. The general is not physically composed of his lieutenants, nor are the lieutenants composed of the troops under their command. The reason non-nested

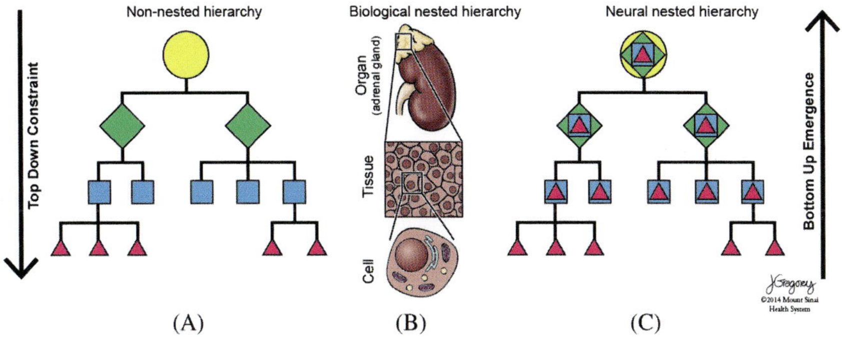


**Fig. 1.** Different types of hierarchies. All hierarchies display emergent properties that are created by the interaction of lower and higher levels and by constraint of higher upon lower levels. (A) In a non-nested hierarchy the entities or holons at higher levels of the hierarchy are physically independent from the entities at lower levels and there is strong constraint by higher upon lower levels. (B) In a biological nested hierarchy, higher levels are physically composed of lower levels, and there is no central control of the system, resulting in weak constraint of higher upon lower levels. (C) A neural nested hierarchy is a unique biological system that displays features of both non-nested and nested hierarchies: physically non-nested but functionally nested.

hierarchies are also called control hierarchies is because the constraints of higher upon lower levels are stronger (more controlling) than in nested hierarchies, which we will describe next.

In a *nested* or *compositional hierarchy*, any given level of the organization is entirely composed of the lower levels. Organs of a plant (flowers, leaves, xylem), and indeed the entire plant, are nested hierarchies in which the elements comprising the lower levels are physically combined or *nested* within higher levels to create increasingly complex wholes (also known as "holons:" Koestler, 1967, 1978). Allen & Starr (1982) explain this and contrast the two types of hierarchy:

> "A nested hierarchy is one where the holon at the apex of the hierarchy contains, and is composed of, all lower holons. The apical holon consists of the sum of the substance and the interactions of all its daughter holons and is, in that sense, derivable from them. Individuals are nested within populations, organs within organisms, tissues within organs, and tissues are composed of cells....Non-nested hierarchies relax the requirement for containment of lower by higher holons and also do not insist that higher holons are derivable from collected lower holons" (p. 38).

Later, we will see how the neural hierarchies that produce sensory consciousness display a unique combination of nested and non-nested hierarchical features.

#### 1.3.2.4. *Teleonomy*

Biological structures perform functional roles. This is not controversial, but stating it correctly can be difficult. We humans, being remarkably skilled builders, find it far too easy and natural to speak of every complex system as being designed for a purpose, just because all our own machines are. But invoking a "pre-designed" purpose in an explanation of a biological phenomenon is a form of *teleological thinking* — an effort to invoke the end result of a process as its cause, which is a logical impossibility. The evolution of biological traits proceeds step by adaptive step, with no targeted end-goal. If such success were forecast, the 99%-plus of all the species that ever evolved would not have become extinct — a soberingly high extinction rate revealed by the fossil record [en.wikipedia.org/wiki/Extinction]. Therefore, it is incorrect to say "the heart evolved to pump blood" as if by design.

To purge all implication of design, purists want to eliminate any mention of such purpose in evolution, but their corrected statements can be too long-winded and unwieldy: "Variant animals whose large artery by chance pumped blood more strongly were most likely to survive and they passed on more of their genes, until the species had evolved a true heart." Thus, even professional biologists who reject teleology and know that evolution never plans ahead, sometimes embrace brevity and figuratively say, "the heart evolved to pump blood."

On the other hand, the heart does have the adaptive *purpose* of pumping blood. Jacques Monod, a winner of the Nobel Prize for his contributions to molecular biology, said we should not reject the idea of purpose in biology, because it is clear that evolution produces organisms with biological functions that do serve purposes and are of adaptive advantage. He called this feature *teleonomy,* a word derived from the Greek word telos meaning "goal" or "endpoint." According to Monod, teleonomic processes are those that are end-directed, but not deliberately or intentionally so in any planned way:

> "Rather than reject this idea (as certain biologists have tried to do) it is indispensible to recognize that it is essential to the very definition of

> living beings. We shall maintain that the latter are distinct from all other structures or systems present in the universe through this characteristic property, which we shall call *teleonomy*" (Monod, 1971, p. 9).

Teleonomy allows us rightly to say, "The heart is an evolutionary adaptation to pump blood."

The same reasoning goes beyond a strictly evolutionary context. Theoretical biologist Mayr (1982) agreed that all genetically programmed biological processes and biological systems have teleonomic functions. Mayr wished to accept teleonomic language including phrases like "the function of the heart's beating is" or "the heart beats in order to" as valid as long as one realizes the metaphorical use of such language. An organism can perform a process that is teleonomic in that the process serves a survival function, yet the organism does not have to anticipate or "know" the end point of that process.

All living things are hierarchically constructed such that their parts are *constrained* to perform teleonomic goals — programed actions that serve an adaptive function for the organism. In a nested hierarchy, constraint is the means through which organisms achieve their goals: all levels are constrained toward attaining that goal. The critical role of teleonomy and evolution in the special nature of consciousness will be addressed later.

### 1.3.3. *Special neurobiological features*

We now turn to the neural features that are greatly enhanced in conscious nervous systems but are simpler or absent in other living systems (Table 3B). While these features derive from normal evolutionary processes, they make consciousness unique and explain neurobiological naturalism.

#### 1.3.3.1. *Reflexes and rates*

Neural reflexes are not in and of themselves conscious, but they are an essential ingredient of the neural substrate that makes consciousness possible. In a *monosynaptic reflex* like the knee jerk, two neurons, one sensory and one motor, are connected by a single synapse. The knee jerk is an involuntary neural response for maintaining balance. It does not require consciousness and will occur even if a person is asleep or in coma. The pupillary light reflex is a very complex, multi-synaptic yet involuntary reflex that also does not require consciousness. When a bright light is shined into the eye, the light reaches the retina and stimulates the retinal ganglion cells that project via the optic nerve to centers in the midbrain. These centers then activate a cranial nerve that signals a muscle in the iris to constrict, so less of the harmful light reaches the retina. As in the knee jerk, this reflex operates even if there is considerable impairment of higher brain functions as long as the midbrain centers and nerves involved in the light reflex are preserved. In other words, while these reflexes are examples of important neural functions that serve useful adaptive purposes (teleonomic functions), they do not require consciousness.

In evolution, reflexes can be expanded into complex neuronal hierarchies by adding more neurons, like links to a chain, and more hierarchical levels of processing. This makes reflexes an essential "royal road" to the creation of consciousness. One key characteristic is their speed. Some living hierarchical systems, such as a tree, have slow rates of operation, whereas others such as nervous systems have extremely rapid rates of operation. Reflexes operate at fast rates that higher neurohierarchical levels require for the emergence of consciousness. In the same way, the importance of tight (synaptic) connectivity between neurons in reflexes cannot be overemphasized. Tight and rapid connectivity made the neural hierarchy possible, allowing many signals to whiz back and forth through the system, and all the interesting and irreducible features of consciousness follow from this.

Reflexes also set up non-conscious but adaptive (teleonomic) external responsiveness. The eye-blink reflex automatically protects the eye from dust blowing in the air in the same way that a noxious pin-prick that in a fully conscious patient would be experienced as painful can provoke reflexive withdrawal in a patient in a coma. In these primitive reflexes, the world-organism, external–internal relationships are already evident.

#### 1.3.3.2. *Complexity*

There is an intimate relationship among adaptive and hierarchical systems, emergence and complexity. Herbert Simon, one of the godfathers of complexity and systems theory, was among the first to emphasize that nature builds complexity through hierarchical systems:

> "... complexity frequently takes the form of hierarchy, and that hierarchic systems have some common properties that are independent of their

specific content. Hierarchy, I shall argue, is one of the central structural schemes that the architect of complexity uses" (Simon, 1962, p. 468).

Hierarchies gain complexity in several ways: by adding more levels, by increasing the number and specialization of elements within each level, and by increasing the tight and fast interactions between and within levels. Then from this, new properties emerge. Paul Nunez emphasized the intrinsic relationship between complexity and emergence when he noted "Complex systems generally consist of multiple interacting parts such that a system taken as whole exhibits emergent properties not obvious in any single part" (Nunez, 2010, p. 260). Another good analysis that supports the role of neurohierarchical systems in the development of complexity is that of Sporns (2011). Tononi (2004, 2008; Tononi & Koch, 2008) offers an *integrated information theory* of the emergence of consciousness that fits Nunez's criteria for complexity. According to Tononi's account, consciousness corresponds to the amount of integrated information generated by a complex of elements and their interactions above and beyond that generated by the systems' individual parts. The more such information, the more the consciousness.

All these authors are suggesting that consciousness is an emergent property of the most complex neurohierarchical systems, those with the most parts and the most interactions among their parts, though science is not yet able to say exactly how the consciousness is produced. We restate the complexity solution as follows: Consciousness is most likely to exist in those neurohierarchies with the largest number of neuronal-response characteristics, the most synaptic connections, the greatest number and differentiation of levels, and the highest speed and density of interactions within and across levels.

Another major contributor to brain complexity, and presumably to consciousness, is cellular diversity. Many different kinds of neurons, with special roles, exist both across the animal kingdom and within a single brain. The different neurons have different numbers and branching patterns of dendrites (information-receiving processes) and of axons (information-sending processes). Neurons also vary by function, in how their axons carry signals: *tonically* by firing the whole time the neuron is stimulated, *phasically* by firing only briefly at the very start of a stimulus, firing seldomly (and only from the strongest stimuli), or firing in a *bursting* pattern with trains of impulses separated by quiescent intervals. Invertebrate neurons are a bit different from those of vertebrates, with their dendrites sometimes sending signals as well as receiving them, and axons sometimes receiving signals as well as sending them (Matheson, 2002).

Especially important for sensory consciousness, vertebrates have a great variety of sensory *receptors* associated with the sensory neurons; most of these receptors develop from embryonic tissues that exist only in vertebrates, called ectodermal placodes and neural crest (Sommer, 2013).

Such cellular variety is unique among the organ systems of the body. Mammal brain tissue has hundreds of neuron types, whereas the dominant tissues in kidney, muscle, ovary, liver and other organs have fewer than a dozen (Mescher, 2013). The great variety of neurons, as the communicating units in nervous systems, allows the complex information processing upon which consciousness depends.

#### 1.3.3.3. *Topographic and isomorphic representations, mental images*

One feature of brains that possess exteroceptive and interceptive sensory consciousness is the presence of topographic or isomorphic maps. A topographic map in the central nervous system is created when the spatial ordering of a lower-order sensory system is preserved at higher hierarchical levels. Classic topographic maps include *somatotopic* maps in the parietal lobe of the cerebral cortex where the body surface is represented in a distorted but still complete, point-by-point, topographic arrangement; and the *retinotopic* maps in the occipital cortex in which spatial relations of points in the visual field are preserved (Kaas, 1997; Kandel *et al.*, 2012; Ungerleider *et al.*, 2008).

This mapping function of the brain is one of its most obvious features, one that must necessarily be emphasized, and there is much speculation regarding its role in generating consciousness and the self (Edelman, 1989; Jbabdi *et al.*, 2013; Tinsley, 2008). One thing is certain: the vertebrate brain has very many neural maps, with some of the sensory maps encoding aspects of the external environment and others representing aspects of the body.

However, not all these brain maps are spatially organized in correspondence to a physical dimension of external space or the body, and in these cases the term *isomorphic* alone is sometimes used. Thus, auditory (*tonotopic*) representations are arranged by tones of the sounds, and smell (*chemotopic*) representations are arranged by odor combinations

(Gottfried, 2010; Kandel *et al.*, 2012; Shepherd, 2012). Yet ultimately, these reflect spatial maps of neuronal activity in the ear's cochlea and the brain's olfactory bulb, respectively, and are still strictly speaking topographical maps.

A critical point for ontological subjectivity is that when considered from the perspective of a conscious organism, the hierarchically organized and isomorphic neural-representations are experienced as *sensory mental images.* The term "sensory mental image" here describes those aspects of sensory consciousness that directly and immediately result from the brain's processing of sensory information in much the same way that Edelman (1992) defined *primary consciousness* as "the state of being mentally aware of things in the world — of having mental images in the present." Many other authors have suggested that isomorphic maps are critical to the creation of sensory consciousness (Damasio, 1999; Edelman, 1989; Feinberg, 2009, 2011; Feinberg & Mallatt, 2013). Thus, the phenomenon of isomorphic neuronal maps, which are physical things, forming subjective sensory images — a fully neuroontologically subjective experience — becomes the bridge between the material domain of neurobiology and the neurophilosophical domain of subjective experience and the hard problem (Feinberg & Mallatt, 2013).

#### 1.3.3.4. *Neural hierarchies and unique neural–neural interactions for sensory consciousness*

Many theories of the neurobiological basis of consciousness emphasize widespread *neural interactions* as the key to consciousness: from unity, to qualia, to selectively attending to the qualia. Crick & Koch (2003) proposed that unified consciousness is created by a "dominant coalition" or coordinated "assembly" of specific, interconnected, and interacting neurons; that is, by reciprocal interactions among the neurons in the back and front of the cerebral cortex. The dominant coalition was said to determine the contents of consciousness. Other theories bring in the thalamus, emphasizing interactions between the thalamus and cerebral cortex as the key determinant of mental and conscious unity (Beckers & Zeki, 1995; Engel *et al.*, 1999, 2001; Roskies, 1999; Singer, 1999; von der Malsburg, 1995; Zeki & Marini, 1998). Similarly, Tononi & Edelman (1998) proposed that a neural "dynamic core" based upon reciprocal thalamo-cortical interactions is responsible for the unification of consciousness.

More specifically, the neuronal interactions thought to be essential for consciousness are *coordinated oscillatory patterns* in the gamma frequency range (or in the slightly slower theta and alpha ranges). All the neurons in a sensory hierarchy that react to the same stimulus are said to be bound together in the one perception. The oscillations would also coordinate the actions of different hierarchies, to make the percept multisensory (Crick & Koch, 2003; Llinas *et al.*, 1998; Nunez & Srinivasan, 2010; Steriade, 2000; Uhlhaas *et al.*, 2009; Ursino *et al.*, 2009). Finally, as mentioned above, the oscillations seem to help direct selective attention.

All these theories that put hierarchic neural–neural interactions at the center of consciousness may be partly or entirely correct. The "neurobiological naturalism" theory we propose here says that whatever the exact mechanism of neural–neural interaction required for consciousness, the neural hierarchy creates unique biological properties and correspondingly unique subjective characteristics.

This hierarchical perspective helps explain how mental unity is achieved. A classic example of the puzzle of subjective mental unity is the phenomenon of "cyclopean perception": How is it that we experience a unified visual world that in consciousness appears to emanate from a single visual point of view but is based upon separate information coming from two eyes? In terms of our hierarchical systems approach, the explanation is that the lower order elements — in this case the information from each eye — are *constrained* to combine into a single unified visual percept, and it is via *top down constraint* that we see a single visual percept. Further, the brain creates an additional, emergent feature — depth perception — that is only possible by the integration of visual information emanating from both eyes.

With reference to ontological subjectivity, our critical point is that although multiple regions of the neural hierarchy are physically connected within the brain, *they are not a physically nested system,* as for example the way a cell is physically nested within an organ. In the visual hierarchy, the eye's retina is not contained in the thalamus and the thalamus is not contained in the cerebral cortex, so this is a non-nested system. But consciousness is *experienced as a functionally* nested and unified system, even though the brain is not a *physically* nested system. Therefore, mental unity can be explained as a function of particular and unique features of the neural hierarchy yet remain a neuroontologically unified and irreducible feature of consciousness.

Mental unity and consciousness actually require a unique blending of nested and non-nested hierarchical functions (Fig. 1). An example of the non-nested functions is "topical convergence" in sensory processing. Here, neurons at lower levels of the hierarchy with less-differentiated response characteristics project to neurons at higher levels with increasingly specific response characteristics until, in the temporal lobe at the "top" of this "neural pyramid", there are neurons that respond to highly integrated visual stimuli. Some such neurons respond only to a hand, or a face, or even the face of actress Julia Roberts or to specific buildings such as the Tower of Pisa (Barlow, 1995; Fried *et al.*, 1997; Gross, 2002, 2008; Gross *et al.*, 1969, 1972; Kreiman *et al.*, 2000; Quiroga *et al.*, 2005, 2008). Actually, though these high-level neurons do unify inputs like this, each neuron probably responds to several similar objects rather than just one (Quiroga *et al.*, 2005, 2008).

Here is how sensory consciousness also has the functional properties of a *nested* hierarchy, in which higher levels contain the lower ones. Consider the following multi-modal experience of a person observing an apple falling from a tree. The overall shape of the apple is composed of tens of thousands of individual line segments that are represented early in the processing hierarchy. Further up the hierarchy, in the visual association cortices, the short line segments are bound into longer line segments to create the outline of the apple, just as a small patch of red of the apple is bound to a larger red patch that is part of the redness of the entire image of the apple. Still farther up the visual hierarchy, a feature such as the stem emerges in awareness as "bound to" — or *nested* within — the overall shape of the apple. Next, via transmission from the primary visual cortex (V1) to color areas in V4 and the inferotemporal cortex, the overall shape is bound to the redness of the apple. Through input from area V5, these higher-order representations are bound to the apple's movement as it falls. Then the sound the apple makes as it hits the ground, which was initially processed within auditory regions of the temporal cortices, is integrated with visual information in higher-order association cortices. The end result of the process is that all these anatomically distributed and multimodal representations are bound together to create the subjectively unified experience of the apple's journey (Beckers & Zeki, 1995; Engel *et al.*, 1999, 2001; Roskies, 1999; Singer, 1999; Uhlhaas *et al.*, 2009; von der Malsburg, 1995; Zeki & Marini, 1998). Functional parts have made a functional whole.

In summary, consciousness is "united" by hierarchical organization and complex hierarchical interactions, synaptic connections, synchronized oscillations, upward emergence and downward constraint; and these processes could provide sufficient integration of neural activity to enable the binding across and within hierarchical levels to produce the unified consciousness. In this integrative manner, while the material brain is physically spread-out and piecemeal (with subparts, such as the different anatomic levels of the hierarchies), its functional, systemic and neurohierarchical *processes* are experienced by the subject as nested and unified. This makes the objective and subjective aspects of the neural system fully understandable while still being *ontologically irreducible.*

As the unique neurohierarchical features enumerated above create unity, they also create *qualia*. Exteroceptive qualia are the mapped mental images of the world that emerge from isomorphic projections at multiple levels of the neural hierarchy. The same is true for interoceptive qualia, at least for the isomorphically mapped and non-affective aspects of pain and of many inner-body sensations. Even the wholly affective qualia (the non-isomorphic affective states) depend on neural hierarchy, as sensory information is relayed up through several levels of the brain's "greater limbic system" to induce the affects (see Sec. 1.3.5). Thus, diverse kinds of conscious qualia — exteroceptive, interoceptive, and affective — obtain their subjective "feels" by the unique neural–neural interactions afforded by neurohierarchical processes. This is one reason why, as far as we know, only complex neurohierachically organized brains have qualia.

Finally, while the problem of mental causation (Table 1) generates a lot of philosophical discussion (for e.g., see Heil & Mele, 1993), this NOIF appears to be the easiest to understand as coming from a hierarchical system whose neurons interact extensively. The neurohierarchy responsible for conscousness is a *contained* neural system and all the NOIFs exist and operate only therein, so *the causal efficacy of consciousness is confined to the system*: I can only influence and fire the neurons *within my system.* I cannot affect events outside of my neural system and body "just by thinking of them." No telepathy exists. But from within my brain system, I can direct something outside of that system, namely my arm, to move and cause changes in the world outside my body. And I cannot move your arm strictly by intending to, without the intervention of my bodily actions signaled by my neural system. Mental causation is an example of a causally closed system property. It is within that system that all the special neurobiological features pertain, as they contribute to the ability of a brain to create consciousness. For this reason, we conclude that sensory consciousness is an emergent neurobiological, *system-feature*.

1.3.3.5. *Summary of the special neurobiological features*

Sensory consciousness arises in nervous systems that obtain a sufficient degree of neural complexity based upon rapid neuronal communication inherited from ancestral reflexes, activation of multi-level neural hierarchies, isomorphic representations that allow mental images, and unique neural–neural interactions including oscillatory binding for unity and directing attention. The emergence of all the irreducible features of consciousness including referral, mental unity, qualia and mental causation is linked to these special neurobiological features.

### 1.3.4. *Neuroevolution: The royal road in vertebrates from reflexes to consciousness*

Having treated the unique neurobiological features that contribute to primary consciousness, we will now address the neuroevolutionary issues. We propose that subjectivity and consciousness emerged during animal history and entered the realm of scientific theory over half a billion years ago when strictly reflexive responses transitioned to consciousness in the evolving lineage of chordate animals (Fig. 2) (Feinberg, 2012; Feinberg & Mallatt, 2013). This was in the first vertebrates. At this point the unique biological characteristics of the vertebrate nervous system can be related to its evolution, and then ultimately used to overcome the neurophilosophical obstacles that consciousness poses for scientific reduction.

Biologists often use the fishlike invertebrate animal, amphioxus, to approximate the neural capabilities of the invertebrates from which vertebrates evolved about 550 million years ago (Lacalli, 2008; Mallatt & Holland, 2013). The young larval stage, with its simple nervous system, seems especially informative of the ancestral state. Larval amphioxus is shown in Fig. 2. Although many of its actions are classically reflexive, some instead involve "central pattern generators" (CPGs) (Hirayama *et al.*, 2012; Lacalli, 2008; Nargeot & Simmers, 2011; Wicht & Lacalli, 2005). Its CPGs consist of short chains of neurons that signal the handful of simple behaviors the larval amphioxus can perform: fast swimming, slow swimming and staying still, all for escaping danger or passively

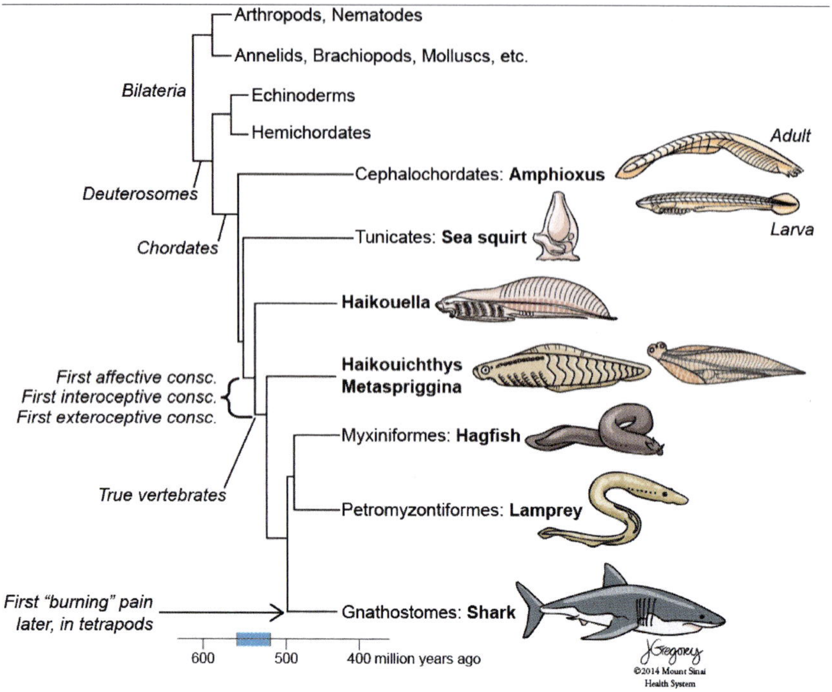


**Fig. 2.** Tree of animal relationships illustrating our thesis that consciousness evolved in the first true vertebrates around 550 million years ago. The larva of amphioxus (upper right) is used to approximate the ancestor of vertebrates, because living tunicates are highly specialized and *Haikouella*, *Haikouichthys* and *Metaspriggina* are only known from fossils.

filter-feeding in ocean waters. The central-pattern-generated movements are stereotyped and largely automatic, but can be modified by sensory input. Most of the sensory "hierarchies" from the body to the brain of larval amphioxus are chains of only *two* neurons (Lacalli, 2008), so short that they are not true hierarchies. The CPGs are actually just the middle parts of short, three- or four-neuron reflex arcs, too simple to produce sensory consciousness (recall the "Complexity" Sec. 1.3.3.2 above), so we deduce that larval amphioxus and early pre-vertebrates are/were not conscious.

In our model of the evolution of consciousness from a non-conscious amphioxus-like ancestor, the reflexes of this ancestor were adaptive, so teleonomy was already in place: larval amphioxus recoils reflexively from harmful touch, from a predator's shadow blocking the sunlight, and from bad food entering the mouth. And, in these simple reflexes, the world vs. organism and external vs. internal relationships already existed. Next, as more sensory stimuli came to be processed more thoroughly in the pre-vertebrates, the non-conscious reflexes gained more levels of intervening interneurons and more complex neural–neural interactions within this new, multilevel neurohierarchy. For efficiency of neuronal wiring and information processing (Kaas, 1997), the contained sensory pathways organized their pathways and neurons topographically. In this way, these evolving neural hierarchies eventually met the unique requirements necessary to produce sensory mental representations.

The evolution of visual images illustrates this sequence, and evidence suggests vision was the very first distance sense to elaborate in the vertebrate line (Feinberg & Mallatt, 2013; Lamb, 2011; Parker, 2003; Sestak *et al.*, 2013; Trestman, 2013; Vopalensky *et al.*, 2012). The neurons at each level of the visual hierarchy became arranged isomorphically (retinotopically) to create a visual map of the world. This can be attributed to the evolutionary transformation of simple light detectors (as in amphioxus) into camera eyes with a lens that formed sharp, detailed visual images (as in vertebrates), a process whose steps can be reconstructed with surprising ease by using the eyes of a variety of living invertebrates as analogues (Lamb, 2013; Nilsson, 2013). The strong adaptive advantages of extracting maximal visual information from the exquisite images of approaching predators, food items and mates drove further evolutionary changes. Additional hierarchically interacting levels of processing neurons were soon added, which improved visual discrimination by sharpening edges, increasing visual contrast, enhancing sensitivity to light, etc. Ultimately, the final creation of retinotopic sensory images occurred in the earliest vertebrates in a high level of the neural hierarchy, the optic tectum (Feinberg & Mallatt, 2013). These tectal images were the very first qualia to evolve.

Conscious smell, taste, touch and hearing soon followed vision in the evolving pre-vertebrates. The derivation of their sensory images followed a similar sequence, by establishing isomorphic neural hierarchies that mimicked the original, visual pattern (Butler, 2006; Feinberg & Mallatt, 2013). All these senses except smell likewise converged on the optic tectum, which produced their conscious images (as deduced from the tectum's role in perceiving these senses in modern fishes and amphibians: Dicke & Roth, 2009; Saidel, 2009). Smell and the conscious awareness of odors, by contrast, would have involved the cerebrum of the forebrain, as is the case in all living vertebrates known to be conscious (Baars, 2013).

The evolution of high-level isomorphic maps within the newly evolved sensory hierarchies also helps explain the unity and referral aspects of consciousness. The maps for all the different senses (except smell) are in register in the optic tectum of vertebrates (Stein & Meredith, 1993), providing a multisensory image of the world. Mammals similarly merge the different senses, but mainly in their cerebral cortex, which is said to lead to a conscious image there (Boly *et al.*, 2013). Merging all the senses into an image means, by definition, that the image is *unified*. And, any use of or access to this world image by the animal must entail consciously *referring* to the world. Although such full-blown referral characterized the first vertebrates, its roots were much older: the reflexive stage had already determined to where the responses were directed (referred), either to the environment or internally to maintain body homeostasis.

Any neuroevolutionary analysis must address whether consciousness is adaptive (of teleonomic value), and if so, then how? Philosophers have argued over whether consciousness is itself adaptive or just an epiphenomenon (by-product) spun off from some other adaptively selected trait, such as increased cognition. Nichols & Grantham (2000) review this debate and argue for a directly adaptive role. They point out that any biological system that is complex, such as consciousness, is likely adaptive and unlikely to have evolved by chance. This is because energy is needed for organisms to maintain complex structures and functions, and unless these are adaptive and valuable the energy is wasted and the features are selected against and lost. In other words, they would never evolve in the first place.

Indeed, all of the neurootologically irreducible features of consciousness — qualia, unity, and referral — can be shown to have evolutionary and adaptive aspects. Let us start with qualia, which discriminate different stimuli or qualities. The difference in experiencing certain wavelengths of light as "red" and others as "green" has a clear neurobiological explanation, as certain neural receptors and centers processing these two wavelengths differently. However, that differential responsiveness would not have evolved were it not important for the organism to distinguish a red herring from a green turtle. The same goes for discriminations in the other senses: telling the odor of a predator from the pheromones of a potential mate, a threatening roar from a mating call, or a nutritious meal from a poisonous one by taste. All these sensory discriminations promote survival and are adaptive. The same is true of *affective* qualia. Such "liking" and "disliking" aid survival by telling which harmful stimuli to avoid and which are beneficial to approach (Cabanac, 1996).

The mental unity aspect of consciousness is certainly adaptive, considering the difficulty an animal would have responding to a fragmented map of reality (e.g., failing to detect all the aspects of an approaching predator together). Even if the animal had some secondary, jerry-rigged way of compensating for this fragmentation it would be less efficient and waste energy. Finally, the referral aspect of consciousness also is adaptive, because it avoids wasting effort on attending to the internal workings of one's own neurons, which has no adaptive value (except perhaps to philosophers of consciousness!). Being conscious of the flashes at individual synapses would have no direct relevance to the real-world jobs of finding mates or food and sensing predators. And the creation of a mental image of the world would have little value to an organism if it were not experienced as emanating from the environment (see Sec. 1.3.6 on allo- and auto-ontological irreducibility).

A critical point is that the evolutionary factors *cannot be dissociated* from the neurobiological factors. All the general and special neurobiological features of consciousness (Table 3) are the product of evolution and its teleonomic aspects. Further, consciousness was a prime mover of vertebrate evolution by allowing vertebrates to go beyond mere reflexes and to map in detail, then assess, their external and internal environments in order to direct behavior (Feinberg & Mallatt, 2013).

### 1.3.5. *Sensory consciousness is diverse and different neural architectures can create it*

This section expands on our theme that primary consciousness is diverse and cannot be considered as one thing. We already covered the diversity of its exteroceptive vs. its affective vs. its interoceptive aspects, which differ in their qualia, referral location, anatomic pathways, brain regions and degree of isomorphism vs. affect. To emphasize these many differences, we list them in depth in Table 4.

Much diversity also exists within a single aspect, the isomorphic consciousness of the distance- and body senses. Yes, we said this type of

**Table 4.** Different aspects of sensory consciousness (qualia) and "something it is like to be."

**A. Exteroceptive**

*"Exteroceptive sensory consciousness"; "isomorphic sensory consciousness"*

*Projicient (refers to external environment) and is about external mapped objects*

*Experienced as local "mental images"*

*High sensory isomorphism*

*Low intrinsic affect/valence*

*Examples: all from distance receptors: vision, auditory, gustatory, olfactory, discriminative touch, sharp pain in skin*

*Anatomic pathways: visual, discriminatory touch, and auditory paths are exteroceptive, isomorphic. Pain and chemosenses (olfactory, gustatory) are both exteroceptive and interoceptives. Paths reach cerebral pallium or optic tectum.*

**B. Interoceptive awareness and pain**

*"Interoceptive sensory consciousness"*

*Has a "dual aspect": both projicient and inner-body/visceral (refers to both the inner body and the external environment). Often viewed as involving "homeostatic" processes*

*May be experienced as both local "mental images" and global inner-body/visceral and affective states*

*Moderate level of sensory isomorphism (combines features of isomorphic systems and non-isomorphic affective systems)*

*Sensations imbued with affect*

*Examples: skin and visceral pain, mechanical senses in body (e.g., stretch or irritation in organ walls), chemoreception (e.g., taste, olfaction)*

*Primary anatomic pathways and structures: spinothalamic and trigeminothalamic; insula, anterior cingulate cortex*

**C. Affective/limbic**

*"Sentience"; "affect"; "emotion"*

*Non-projicient*

*Experienced as global affective states (feelings, emotions); assigns values to internal and external sensory processes; referred to "self"*

*Low or no sensory isomorphism*

*High affect/valence: positive and negative affects lead to approach and avoidance behaviors*

*Examples: pleasure/displeasure, all the emotions*

*Anatomic structures: reticular formation, periaqueductal gray, mesolimbic reward system, basal forebrain, amygdala, insula, anterior cingulate cortex, parts of prefrontal cortex* consciousness always needs fast, complex, hierarchical chains of extensively interacting neurons (Table 3). But this requirement is met by different neuronal arrangements in different brain regions, from the classically emphasized stacks of laminae in the cerebral cortex of mammals to the different type of laminae in the optic tectum of anamniotes (fish and amphibians), to the olfactory part of the dorsal pallium of anamniotes that does not have laminae, to the likewise non-laminar nuclei in the dorsal ventricular ridge of the pallium of birds (see Chapter 2 of this volume).

Consciousness-related diversity even exists in the brain of a single animal. The following example involves smell vs. the other senses. In mammals, the conscious pathways for all senses relay through the thalamus — except for olfaction. Actually, the olfactory path does have a trans-thalamic branch but it is of secondary importance and olfactory consciousness does not depend on it (Gottfried, 2010; Mori *et al.*, 2013). In the dominant, non-thalamic path, neurons of the olfactory bulb send odor information onward to the piriform and primary olfactory cortex (of the lateral, then dorsal pallium), with this information transferred by gamma oscillations that are thought to encode the different odor combinations in map-like format, and lead to olfactory consciousness (Mori *et al.*, 2013). This is just like — but generated independently from — the thalamocortical oscillations that bind the other senses of vision, hearing, balance and touch in mammals, into conscious percepts for those senses. In mammals, the olfactory percepts would next be bound with those for the other senses over large areas of the prefrontal cortex, but the key point is until then they are two distinct systems of gamma-bound projections, as if evolved separately for 1) olfactory consciousness vs. 2) consciousness of the other senses.

From this, we can see that even within single brains and across species there are diverse paths to sensory consciousness and subjectivity, and thus no single anatomy can account for this diversity. So, the question is: What are the common neural factors across all types of sensory awareness? Or put another way: What does this heterogeneity tell us about the neural correlates of consciousness?

First, it reveals that the cerebral cortex is not a prerequisite for sensory consciousness as many have assumed. Second, it leads to the question whether *affective* consciousness, with its sub-cortical limbic basis (Solms & Panksepp, 2012) and without an isomorphic organization, fits all the other neurobiological features of consciousness

(Table 3). Yes, it seems to. That is, for the affects to emerge, sensory information travels in sensory hierarchies with complex neuron–neuron interactions (Craig, 2002; Critchley & Harrison, 2013; Saper, 2002); some reflexes lead quickly to affective reactions; affects have functional (teleonomic) roles, etc. Affective "mental states" are generated instead of isomorphically mapped "mental images," but all the other features of isomorphic conscioiusness seem to apply to affects. Thus, adding "or affective states" after "mental images" in Table 3 should make the list of features applicable to all three types of consciousness (extero-, intero-, affective).

### 1.3.6. *Neurophilosophy: Auto- and allo-ontological irreducibility*

Up to this point, we have considered the *positive* general features of life and special features of the nervous system that contribute to the emergence of consciousness (Table 3). To understand more fully the nature of neuroontological irreducibility, however, we must consider two *negative* features that are limiting factors to a scientific reduction of consciousness to the brain — auto-ontological and allo-ontological irreducibility. The nature of these negative features was well expressed by Globus, who spelled out the critical limitations:

> "The ontological claim that mental events are strictly identical with neural events unfortunately coalesces the perspectives of both subjective (S) and objective (O) observers. The term "mental events" implies the perspective of S who has the mental events immediately given by direct acquaintance (without inference), whereas the term "neural events" implies the perspective of O who is presumably the brain of S. Thus O cannot have S's mental events by direct acquaintance because they are private to S; for example, O cannot experience S's pain. For S, there are no neural events by direct acquaintance — that is, that his own mental events are physically embodied is not directly known by S, unless he observes his own brain in the manner of O using Feigl's imaginary "autocerebroscope" Mental events contain no information about any neural embodiments, for example, S's pain does not have the typical characteristics of physical objects in that S cannot see his pain or touch it. Nor is there anything about pain which seems at all like neurons... it does not appear that the brain in any way codes or represents in any way its own structure. (The nervous system has no sensory apparatus directed to its own structure.)" (Globus, 1973, p. 1129).

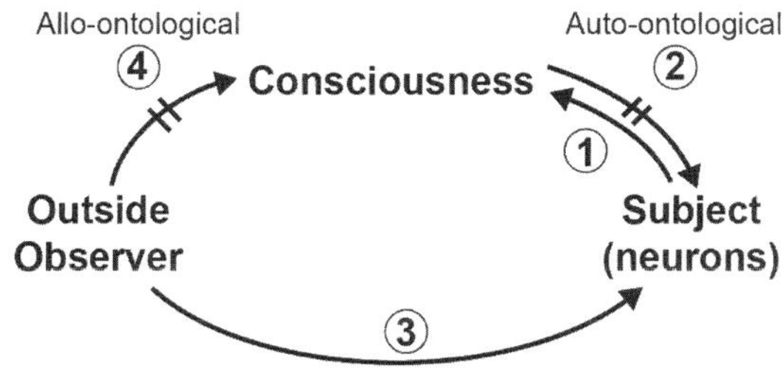


**Fig. 3.** Auto-ontological and allo-ontological reducibilities of consciousness. (1) Referral of consciousness; subject has access to his or her conscious experiences. (2) Auto-ontological irreducibility: subject lacks access to objective neurons. (3) Observer can access the subject's objective neurons. (4) Allo-ontological irreducibility; observer lacks access to subject's experience.

Figure 3 diagramatically relates this subjective-objective conundrum to auto-ontological and allo-ontological irreducibilities. These two negative features — what the brain cannot and does not do — are essential parts of what Schopenhauer (1967; cited in Globus, 1973), Globus (1973) called the "world knot," Searle called "ontological subjectivity" (Searle, 1992, 1997), and Chalmers called the "hard problem" (Chalmers, 1995, 1996). All the aforementioned authors have just arguments for their positions. The question is whether and how we can cross this impasse.

### 1.3.6.1. *Auto-ontological irreducibility*

The NOIFs of consciousness appear to create an unbridgeable gap between *observing* a brain in a feeling state and *being* a brain that it is in a feeling state. This dilemma is succinctly summarized by Searle:

> "…it is a general feature of such reductions that the phenomenon is defined in terms of the "reality" and not in terms of the "appearance." But we can't make that sort of appearance-reality distinction for consciousness because consciousness consists in the appearances themselves. *Where appearance is concerned we cannot make the appearance-reality distinction because the appearance is the reality*" (Searle, 1992, pp. 121–122).

Neural science and medical practice use all sorts of brain scanning machines for crudely "observing" the living nervous system. But the "gap" that these authors highlight philosophically is illustrated by the fictional device mentioned by Globus above that the philosopher Feigl (1967) called an "autocerebroscope." The autocerebroscope attaches to your head with a probe that goes through your skull and allows you to observe the firing neurons of your own brain:

> "This, for the time being, of course, must remain a piece of science fiction (conceived in analogy to the doctor's fluoroscope) with the help of which I would be able to ascertain the detailed configurations of my cortical nerve currents while introspectively noting other direct experiences, such as the auditory experiences of music; or my thoughts, emotions, or desires" (Feigl, 1967, p. 16).

The central problem raised by the possibilities of an autocerebroscope is that *even if one could objectively observe one's own neurons* in the act of creating experience — in the same fashion that they would be observed by an outside observer — there still remains an irreducible gap between the neurons as they are observed by you and the neurons themselves as they are experienced by you within your brain. As Globus (1973) puts it, the brain does not "represent in any way its own structure to the subject." There is no way that the subject can become aware of his or her own neurons from the embodied inside. They can be known objectively only from the "outside," third person point of view.

#### 1.3.6.2. *Allo-ontological irreducibility*

Allo-ontological irreducibility is the inverse of the auto-ontological. Just as the subject has no objective access to his or her brain that creates experience, the outside observer cannot know directly or directly observe features as they are experienced by the subject (see "(4)" in Fig. 3). Only the embodied neural system can experience these features (ontological subjectivity). It is a non-controversial fact that from the external viewpoint what is *observable* is the brain not the experience (neuroontological irreducibility) whereas from the inside what is *observable* is the subjective experience, not the way the neurons construct that experience. We accept these as **absolute** barriers.

#### 1.3.6.3. *Summary of neurophilosophical features*

Auto- and allo-perspectives on the brain are by their nature mutually irreducible. This is in fact the definition of the subjectivity and the origin of ontological subjectivity. We propose these dual perspectives arose in evolution at the critical branch point, over 500 million years ago (mya), when reflexes evolved into unified, referred, qualitative and subjective mental images.

## 1.4. Conclusions

We propose a theory of neurobiological naturalism that explains consciousness as the outcome of general and special adaptive neurobiological features (Table 3). We should not expect anything like qualia, mental unity, or referral to exist anywhere else in nature because the brain is unique in nature. Primary consciousness is restricted to those animals that have the general and special features.

The older term "biological naturalism" while conveying the idea that consciousness is a scientifically tractable problem, misses the special nature of consciousness in the natural world. Therefore, we use "*neuro*-biological naturalism" (Feinberg, 2012; and Table 2) to emphasize the role of the special features of the central nervous system. Once more, these features are the fast, complex, hierarchical, system-wide, representational, internal and unique neural–neural interactions (Table 3). Because they are internal and interactive, these neural features cause events that are irreducibly subjective. This raises the possibility of consciousness in invertebrates with complex brains, such as cephalopod molluscs and bees.

Evolutionary history elucidates the origins of the philosophical reductionistic impasse that we traced back to the ancient branch point between reflexes and sensory consciousness. Basal forms of sensory responsiveness were reflexive, innate, and they existed in the early, non-conscious invertebrate animals, but they were the building blocks of sensory consciousness. Even the primitive light reflex has within it the seeds of referral as the eye responds adaptively and appropriately to an external light source (by constricting the pupil). This can be fully, objectively, described by an outside observer because the reflex does not involve any consciousness: no allo-ontological (or auto-ontological) irreducibility exists in reflexes. But about 520 mya when a brain first turned reflexive responsiveness into a *mental image*, both kinds of irreducibility were abruptly created. Therefore, the specific manner in which consciousness evolved *created* the hard problem, and this occurred within an entirely "natural" biological framework. In this context, the two neurophilosophical "dilemmas" are resolved as a natural outcome of normal evolutionary processes. The irreducibilities are real, but fully explainable so the fundamental philosophical problem is solved.

A theory of neurobiological naturalism suggests, as does the older concept of "biological naturalism," that consciousness is a natural biological process that can be ontologically subjective and objectively irreducible but does not require any new physical properties nor violate any known ones. However, a theory of neurobiological naturalism emphasizes that in spite of consciousness being both biological and natural, *it differs from all other biological process in its unique irreducibility* that is based upon definable neurohierarchical processes.

In summary, in this chapter we have extended the theory of neurobiological naturalism by incorporating evolutionary ideas, using evolution to help bridge the gap between the known neurobiological features (Table 3) and the longstanding philosophical mysteries of conscious irreducibility (Fig. 3). We showed how an ancestral, reflex-based nervous system (exemplified by the fishlike invertebrate, amphioxus) could have evolved into a complex, hierarchically organized system that can produce the mental images and affects of sensory consciousness. We assign this event to the first vertebrates that appeared over half a billion years ago (Feinberg & Mallatt, 2013; Chapter 2 this volume). This change from reflexes to consciousness was adaptive because it enriched sensory access to the benefits and dangers of the environment. It created the subjective viewpoint, and by doing so generated the "hard problem" of allo- and auto-irreducibility, all within the physical brain.

Nor do the irreducibilities, or the gap between the subjective and the objective, harm or pose any problem to conscious organisms. Natural selection assures that they do not, by making an animal's subjective images and affects fit objective reality as accurately as possible, or as closely as is necessary for survival. The mad, the persistently hallucinating, or any other animals with distorted sensory perceptions and inappropriate emotional responses, quickly die in the "tooth and claw" state of nature. Evolution therefore sustains ("accurate") subjectivity. Natural selection keeps subjectivity grounded in reality.

The three domains of consciousness we have considered — neurobiological, neuroevolutionary and neurophilosophical — offer a solution to the "hard problem" of subjectivity and the nature of "ontological subjectivity". None can do so alone, but considered together it makes perfect sense why and how a brain can naturally evolve a "mind" through natural evolutionary processes. The evolutionary biologist Dobzhansky (1973) famously wrote, "Nothing in biology makes sense except in the light of evolution." With neurobiological naturalism, the evolutionary requirement can be applied to consciousness as well.

## References

Ahl, V. & Allen, T. F. H. (1996) *Hierarchy Theory: A Vision, Vocabulary, and Epistemology.* New York: Columbia University Press.

Allen, T. F. H. & Starr, T. B. (1982) *Hierarchy: Perspectives for Ecological Complexity.* Chicago: University of Chicago Press.

Baars, B. J. (2013) Multiple sources of conscious odor integration and propagation in olfactory cortex. *Front. Psychol.*, **4**, 930.

Barlow, H. (1995) The Neuron Doctrine in Perception. In: M. S. Gazzaniga, ed. *The Cognitive Neurosciences.* Cambridge, MA: MIT Press, pp. 417–435.

Bayne, T. (2010). *The Unity of Consciousness.* Oxford: Oxford University Press.

Bayne, T. & Chalmers, D. J. (2003) What is the Unity of Consciousness? In: A. Cleeremans, ed. *The Unity of Consciousness: Binding, Integration, and Dissociation.* Oxford: Oxford University Press, pp. 23–58.

Beckers, G. & Zeki, S. (1995) The consequences of inactivating areas V1 and V5 on visual motion perception. *Brain*, **118**, 49–60.

Boly, M., Seth, A. K., Wilke, M., Ingmundson, P., Baars, B., Laureys, S. & Tsuchiya, N. (2013) Consciousness in humans and non-human animals: Recent advances and future directions. *Front. Psychol.*, **4**, 625.

Brook, A. & Raymont, P. (2010) The Unity of Consciousness. In: E. N. Zalta, ed. *The Stanford Encyclopedia of Philosophy,* The Metaphysics Research Lab, Centre for the Study of Language & Information, Stanford University, CA 94305-4115.

Buss, L. W. (1987) *The Evolution of Individuality.* Princeton: Princeton University Press.

Butler, A. (2006) The serial transformation hypothesis of vertebrate origins: Comment on "The new head hypothesis revisited". *J. Exp. Zool. (Mol. Dev. Evol.)*, **306B**, 419–424.

Butler, A. & Hodos, W. (2005) *Comparative Vertebrate Neuroanatomy: Evolution and Adaptation*, 2nd edn. Hoboken, N. J.: John Wiley and Sons.

Cabanac, M. (1996) On the origin of consciousness, a postulate and its corollary. *Neurosci. Biobehav. Rev.*, **20**, 33–40.

Camazine, S., Deneubourg, J. L., Franks, N. R., Sneyd, J., Theraulaz, G. & Bonabeau, E. (2001) *Self-Organization in Biological Systems.* Princeton: Princeton University Press.

Campbell, D. T. (1974) Downward Causation in Hierarchically Organized Biological Systems. In: F. J. Ayala and T. Dobzhansky, eds. *Studies in the Philosophy of Biology.* Berkeley and Los Angeles: University of California Press, pp. 179–186.

Chalmers, D. J. (1995) Facing up to the problem of consciousness. *J. Conscious. Stud.*, **2**, 200–219.

Chalmers, D. J. (1996) *The Conscious Mind: In Search of a Fundamental Theory*. New York: Oxford University Press.

Chalmers, D. J. (2006) Strong and Weak Emergence. In: P. Clayton and P. Davies, eds. *The Re-emergence of Emergence*. Oxford: Oxford University Press, pp. 244–254.

Cleeremans, A., ed. (2003) *The Unity of Consciousness: Binding, Integration, and Dissociation*. New York: Oxford University Press.

Craig, A. D. (2002) How do you feel? Interoception: The sense of the physiological condition of the human body. *Nat. Rev. Neurosci.*, **3**, 655–666.

Craig, A. D. (2003) A new view of pain as a homeostatic emotion. *Trends Neurosci.*, **26**, 303–307.

Crick, F. & Koch, C. (2003) A framework for consciousness. *Nat. Neurosci.*, **6**, 119–126.

Critchley, H. D. & Harrison, N. A. (2013) Visceral influences on brain and behavior. *Neuron*, **77**, 624–638.

Damasio, A. R. (1999) *The Feeling of What Happens. Body and Emotion in the Making of Consciousness*. New York: Harcourt Brace & Company.

Dicke, U. & Roth, G. (2009) Evolution of the Amphibian Nervous System. In: J. Kaas, ed. *Evolutionary Neuroscience*. New York: Academic Press, 169–232.

Dobzhansky, T. (1973) Nothing in biology makes sense except in the light of evolution. *Amer. Biol. Teacher*, **35**, 125–129.

Edelman, G. M. (1989) *The Remembered Present: A Biological Theory of Consciousness*. New York: Basic Books.

Edelman, G. M. (1992) *Bright Air, Brilliant Fire: On the Matter of the Mind*. New York: Basic Books.

Engel, A. K., Fries, P., König, P., Brecht, M. & Singer, W. (1999) Temporal binding, binocular rivalry, and consciousness. *Conscious. Cogn.*, **8**, 128–171.

Engel, A. K., Fries, P. & Singer, W. (2001) Dynamic predictions: Oscillations and synchrony in top-down processing. *Nat. Rev. Neurosci.*, **2**, 704–716.

Feigl, H. (1967) *The "Mental" and the "Physical"*. Minneapolis: University of Minnesota Press.

Feinberg, T. E. (2001a) *Altered Egos: How the Brain Creates the Self*. New York: Oxford University Press.

Feinberg, T. E. (2001b) Why the mind is not a radically emergent feature of the brain. *J. Conscious. Stud.*, **8**, 123–145.

Feinberg, T. E. (2009) *From Axons to Identity: Neurological Explorations of the Nature of the Self*. New York: W. W. Norton.

Feinberg, T. E. (2011) The nested neural hierarchy and the self. *Conscious. Cogn.*, **20**, 4–17.

Feinberg, T. E. (2012) Neuroontology, neurobiological naturalism, and consciousness: A challenge to scientific reduction and a solution. *Phys. Life Rev.*, **9**, 13–34.

Feinberg, T. E. & Mallatt, J. (2013) The evolutionary and genetic origins of consciousness in the Cambrian Period over 500 million years ago. *Front. Psychol.*, **4**, 667.

Fried, I., MacDonald, K. A. & Wilson, C. L. (1997) Single neuron activity in human hippocampus and amygdala during recognition of faces and objects. *Neuron*, **18**, 753–765.

Giurfa, M. (2013) Cognition with few neurons: higher-order learning in insects. *Trends Neurosci.*, **36**, 285–294.

Globus, G. G. (1973) Unexpected symmetries in the "world knot". *Science*, **180**, 1129–1136.

Gottfried, J. A. (2010) Central mechanisms of odour object perception. *Nat. Rev. Neurosci.*, **11**, 628–641.

Gross, C. G. (2002) Genealogy of the "grandmother cell". *The Neuroscientist*, **8**, 512–518.

Gross, C. G. (2008) Single neuron studies of inferior temporal cortex. *Neuropsychologia*, **46**, 841–852.

Gross, C. G., Bender, D. B. & Rocha-Miranda, C. E. (1969) Visual receptive fields of neurons in inferotemporal cortex of the monkey. *Science*, **166**, 1303–1306.

Gross, C. G., Rocha-Miranda, C. E. & Bender, D. B. (1972) Visual properties of neurons in inferotemporal cortex of the macaque. *J. Neurophys.*, **35**, 96–111.

Heil, J. & Mele, A., eds. (1993) *Mental Causation*. Oxford: Clarendon Press.

Hirayama, K., Catanho, M., Brown, J. W. & Gillette, R. (2012) A core circuit module for cost/benefit decision. *Front. Neurosci.*, **6**, 123.

James, W. (1904) Does 'consciousness' exist? *J. Philos. Psychol. Sci. Meth.*, **1**, 477–491.

Jbabdi, J., Sotiropoulos, S. N. & Behrens, T. E. (2013) The topographic connectome. *Curr. Opin. Neurobiol.*, **23**, 207–215.

Kaas, J. H. (1997) Topographic maps are fundamental to sensory processing. *Brain Res. Bull.*, **44**, 107–112.

Kandel, E. R., Schwartz, J. H., Jessell, T. M., Siegelbaum, S. A. & Hudspeth, A. J. (2012) *Principles of Neural Science*, 5th edn. New York: McGraw Hill.

Kim, J. (1992) Downward Causation' in Emergentism and Nonreductive Physicalism. In: A. Beckermann, H. Flohr and J. Kim, eds. *Emergence or Reduction? Essays on the Prospects of Nonreductive Physicalism*. New York: Walter de Gruyter, pp. 119–138.

Kim, J. (1993) The Non-reductivist's Troubles with Mental Causation. In: J. Heil and A. Mele, eds. *Mental Causation*. Oxford: Clarendon Press, pp. 189–210.

Kim, J. (1998) *Mind in a Physical World. An Essay on the Mind-body Problem and Mental Causation*. Cambridge: MIT Press.

Kim, J. (2006) Being Realistic About Emergence. In: P. Clayton and P. Davies, eds. *The Re-emergence of Emergence*. Oxford: Oxford University Press, pp. 189–202.

Knudsen, E. I. (2011) Control from below: The role of a midbrain network in spatial attention. *Eur. J. Neurosci.*, **33**, 1961–1972.

Koestler, A. (1967) *The Ghost in the Machine*. Harmondsworth: Hutchinson Publishing Corp.

Koestler, A. (1978) *Janus: A Summing Up*. New York: Random House.

Krauzlis, R. J., Lovejoy, L. P. & Zénon, A. (2013) Superior colliculus and visual spatial attention. *Annu. Rev. Neurosci.*, **36**, 165–182.

Kreiman, G., Koch, C. & Fried, I. (2000) Category-specific visual responses of single neurons in the human medial temporal lobe. *Nat. Neurosci.*, **3**, 946–953.

Lacalli, T. C. (2008) Basic features of the ancestral chordate brain: A protochordate perspective. *Brain Res. Bull.*, **75**, 319–323.

Lamb, T. D. (2011) Evolution of the eye. *Sci. Am.*, **305**, 64–69.

Lamb, T. D. (2013) Evolution of phototransduction, vertebrate photoreceptors and retina. *Prog. Retinal Eye Res.*, **36**, 52–119.

Llinás, R., Ribary, U., Contreras, D. & Pedroarena, C. (1998) The neuronal basis for consciousness. *Philos. Trans. R. Soc. Lond. B Biol. Sci.*, **353**, 1841–1849.

Mallatt J. & Holland, N. (2013) *Pikaia gracilens* Walcott: Stem chordate, or already specialized in the Cambrian? *J. Exp. Zool. B Mol. Dev. Evol.*, **320**, 247–271.

Matheson, T. (2002) Invertebrate Nervous Systems. In: *Encyclopedia of Life Sciences*. Nature Publishing Group, pp. 1–5.

Mayr, E. (1982) *The Growth of Biological Thought: Diversity, Evolution, and Inheritance*. Massachusetts: Belknap Press.

Meehl, P. (1966) The Compleat Autocerebroscopist: A Thought Experiment on Professor Feigl's Mind/Body Identity Thesis. In: P. K. Feyerabend and G. Maxwell, eds. *Mind, Matter and Method*. Minneapolis: University of Minnesota Press, pp. 103–180.

Mescher, A. L. (2013) *Junqueira's Basic Histology: Text and Atlas*, 13th edn. New York: McGraw Hill.

Mitchell, M. (2009) *Complexity. A Guided Tour*. New York: Oxford Press.

Monod, J. (1971) *Chance and Necessity: Essay on the Natural Philosophy of Modern Biology*. New York: Vintage.

Mori, K., Manabe, H., Narikiyo, K. & Onisawa, N. (2013) Olfactory consciousness and gamma oscillation couplings across the olfactory bulb, olfactory cortex, and orbitofrontal cortex. *Front. Psychol.*, **4**, 743.

Nargeot, R. & Simmers, J. (2011) Neural mechanisms of operant conditioning and learning-induced behavioural plasticity in *Aplysia*. *Cell. Mol. Life Sci.*, **68**, 803–816.

Nichols, S. & Grantham, T. A. (2000) Adaptive complexity and phenomenal consciousness. *Philos. Sci.*, **67**, 648–670.

Nilsson, D. E. (2013) Eye evolution and its functional basis. *Vis. Neurosci.*, **30**, 5–20.

Nunez, P. (2010) *Brain, Mind and the Structure of Reality.* New York: Oxford University Press.

Nunez, P. L. & Srinivasan, R. (2010) Scale and frequency chauvinism in brain dynamics: too much emphasis on gamma band oscillations. *Brain Struct. Func.,* **215**, 67–71.

Panksepp, J. (1998) *Affective Neuroscience: The Foundations of Human and Animal Emotions.* New York: Oxford University Press.

Parker, A. (2003) *In The Blink Of An Eye: How Vision Sparked The Big Bang Of Evolution.* New York: Basic Books.

Pattee, H. H. (1970) The Problem of Biological Hierarchy. In: C. H. Waddington, ed. *Towards a Theoretical Biology 3.* Chicago: Aldine, pp. 117–136.

Quiroga, R. Q., Kreiman, G., Koch, C. & Fried, I. (2008) Sparse but not 'grandmother-cell' coding in the medial temporal lobe. *Trends Cogn. Sci.,* **12**, 87–91.

Quiroga, R. Q., Reddy, L., Kreiman, G., Koch C. & Fried I. (2005) Invariant visual representation by single neurons in the human brain. *Nature,* **435**, 1102–1107.

Revonsuo, A. (2010) *Consciousness: The Science of Subjectivity.* Hove: Psychology Press.

Roskies, A. L. (1999) The binding problem. *Neuron,* **24**, 7–9.

Saidel, W. M. (2009) Evolution of the Optic Tectum in Anamniotes. In: M. D. Binder, N. Hirokawa and U. Windhorst, eds. *Encyclopedia of Neurosciences.* Berlin: Springer, pp. 1380–1387.

Salthe, S. N. (1985) *Evolving Hierarchical Systems: Their Structure and Representation.* New York: Columbia University Press.

Saper, C. B. (2002) The central autonomic nervous system: conscious visceral perception and autonomic pattern generation. *Ann. Rev. Neurosci.,* **25**, 433–469.

Searle, J. R. (1984) *Minds, Brains and Science.* Cambridge, MA: Harvard University Press.

Searle, J. (1992) *The Rediscovery of the Mind.* Cambridge: The MIT Press.

Searle, J. (1997) *The Mystery of Consciousness.* New York: The New York Review of Books.

Searle, J. (2004) Biological Naturalism. In: M. Velmans and S. Schneider, eds. *The Blackwell Companion to Consciousness.* New York: Wiley-Blackwell, pp. 325–334.

Sellars, W. (1963) *Science, Perception, and Peality.* London: Routledge and Kegan Paul.

Šestak, M. S., Božičević, V., Bakarić, R., Dunjko, V. & Domazet-Lošo, T. (2013) Phylostratigraphic profiles reveal a deep evolutionary history of the vertebrate head sensory systems. *Front. Zool.,* **10**, 18.

Shepherd, G. M. (2012) *Neurogastronomy. How the Brain Creates Flavor and Why It Matters.* New York: Columbia University Press.

Sherrington, C. (1947) *The Integrative Action of the Nervous System.* New Haven: Yale University Press.

Simon, H. A. (1962) The architecture of complexity. *Proc. Am. Philos. Soc.*, **106**, 467–482.

Simon, H. A. (1973) The Organization of Complex Systems. In: H. H. Pattee, ed. *Hierarchy Theory. The Challenge of Complex Systems.* New York: George Braziller, pp. 71–108.

Singer, W. (1999) Neuronal synchrony: A versatile code for the definition of relations? *Neuron*, **24**, 49–65.

Solms, M. (2013) The conscious id. *Neuropsychoanalysis*, **15**, 5–19.

Solms, M. & Panksepp, J. (2012) The "Id" knows more than the "Ego" admits: Neuropsychoanalytic and primal consciousness perspectives on the interface between affective and cognitive neuroscience. *Brain Sci.*, **2**, 147–175.

Sommer, L. (2013) Patterning and Cell Type Specification in the Developing CNS and PNS. In: J. Rubenstein and P. Rakic, eds. *Specification of Neural Crest- and Placode-derived Neurons*. New York: Academic Press, pp. 385–400.

Sperry, R. W. (1983) *Science and Moral Priority: Merging Mind, Brain, and Human Values*. New York: Columbia University Press.

Sperry, R. W. (1990) Forebrain Commissurotomy and Conscious Awareness. In: C. Trevarthen, ed. *Brain Circuits and Functions of the Mind*. New York: Cambridge University Press, pp. 371–388.

Sporns, O. (2011) *Networks of the Brain*. Cambridge: MIT Press.

Stein, B. E. & Meredith, M. A. (1993) *The Merging of the Senses*. Cambridge, MA: The MIT Press.

Steriade, M. (2000) Corticothalamic resonance, states of vigilance and mentation. *Neuroscience*, **101**, 243–276.

Teller, P. (1992) Subjectivity and Knowing What It's Like. In: A. Berckermann, H. Flohr and J. Kim, eds. *Emergence or Reduction? Essays on the Prospects of Nonreductive Physicalism*. Berlin/New York: Walter de Gruyter, pp. 139–153.

Tinsley, C. J. (2008) Using topographic networks to build a representation of consciousness. *Bio Syst.*, **92**, 29–41.

Tononi, G. (2004) An information integration theory of consciousness. *BMC Neurosci.*, **5**, 42–72.

Tononi, G. (2008) Consciousness as integrated information: A provisional manifesto. *Biol. Bull.*, **217**, 216–242.

Tononi, G. & Edelman, G. M. (1998) Consciousness and complexity. *Science*, **282**, 1846–1851.

Tononi G. & Koch C. (2008) The neural correlates of consciousness: An update. *Ann. N.Y. Acad. Sci.*, **1124**, 239–261.

Trestman, M. (2013) The Cambrian explosion and the origins of embodied cognition. *Biol. Theor.*, **8**, 80–92.

Uhlhaas, P., Pipa, G., Lima, B., Melloni, L., Neuenschwander, S., Nikolić, D. & Singer, W. (2009) Neural synchrony in cortical networks: History, concept and current status. *Front. Integr. Neurosci.*, **3**, 17.

Ungerleider, L. G., Galkin, T. W., Desimone, R. & Gattass, R. (2008) Cortical connections of area V4 in the macaque. *Cereb. Cortex*, **18**, 477–499.

Ursino, M., Magosso, E. & Cuppini, C. (2009) Recognition of abstract objects via neural oscillators: interaction among topological organization, associative memory and gamma band synchronization. *IEEE Trans. Neural Netw.*, **20**, 316–335.

Van Gulick, R. (2001) Reduction, emergence, and other recent options on the mind-body problem: A philosophical overview. *J. Conscious. Stud.*, **8**, 1–34.

von der Malsburg, C. (1995) Binding in models of perception and brain function. *Curr. Opin. Neurobiol.* **5**, 520–526.

Vopalensky, P., Pergner, J., Liegertova, M., Benito-Gutierrez, E., Arendt, D. & Kozmik, Z. (2012) Molecular analysis of the amphioxus frontal eye unravels the evolutionary origin of the retina and pigment cells of the vertebrate eye. *Proc. NatL. Acad. Sci. USA*, **109**, 15383–15388.

Whitehead, A. N. (1925) *Science and the Modern World*. New York: The Free Press.

Wicht, H. & Lacalli, T. C. (2005) The nervous system of amphioxus: structure, development, and evolutionary significance. *Canad. J. Zoology*, **83**, 122–150.

Womelsdorf, T. & Fries, P. (2007) The role of neuronal synchronization in selective attention. *Curr. Opin. Neurobiol.*, **17**, 154–160.

Zeki, S. & Marini, L. (1998) Three cortical stages of colour processing in the human brain. *Brain*, **121**, 1669–1685.

Biophysics of Consciousness: A Foundational Approach
R. R. Poznanski, J. A. Tuszynski and T. E. Feinberg
Copyright © 2017 World Scientific, Singapore.

# 2

# THE EVOLUTIONARY ORIGINS OF CONSCIOUSNESS

*Todd E. Feinberg* and Jon Mallatt†*

** Neurology and Psychiatry, Icahn School of Medicine at Mount Sinai and Mount Sinai Beth Israel Medical Center, New York, NY, USA;*

*† School of Biological Sciences, Washington State University, Pullman, WA, USA*

## Abstract

In this chapter, we update our recent two-step model of the evolution of primary, phenomenal, sensory consciousness in vertebrate animals, a model based on neurobiological naturalism (see Chapter 1). The model proposes that this consciousness first appeared in the earliest vertebrates from mental reconstructions at the top of topographically organized neural hierarchies in the optic tectum of the midbrain (for vision and other senses) and in the pallium of the forebrain (for smell perception). Here, we add more information about the optic tectum that relates to such image reconstruction. Second, sensory consciousness leaped forward independently in the first mammals and then in the first birds, to be generated by an expanding dorsal pallium (cerebral cortex). We propose more reasons for this second step, all related to the favorable location of the dorsal pallium at the nexus of all kinds of sensory input, already highly processed, and near the hippocampus so that conscious experiences could be enriched by memories.

*Keywords*: Animal consciousness; vertebrate evolution; fish; lamprey; optic tectum; cerebral cortex; Cambrian period; topographic (isomorphic) mapping; mammal origin; bird origin.

## 2.1. Introduction

Here, we consider the evolution of the type of consciousness that experiences images of the external world. As a basic type that does not refer to self-awareness or the cognitive aspects of consciousness, it has been called primary consciousness, phenomenal consciousness and sensory consciousness (Boly & Seth, 2012; Edelman, 1989; Revonsuo, 2006). We will cover the basic sensory experiences without considering affects (affective consciousness: positive and negative feelings), which were discussed in Chapter 1 of this volume. This chapter updates our recently introduced hypothesis (Feinberg & Mallatt, 2013) which proposed that the most basally arising of all living vertebrates, the jawless fish represented by lampreys, are conscious, meaning all other fish and vertebrate animals are also conscious. But if consciousness originated in the very first vertebrates back in the Cambrian Period of earth's history (before 520 million years ago, or mya: Erwin & Valentine (2013)), then it advanced in a second great leap during the Mesozoic Era in the first mammals (about 220 mya) and first birds (about 165 mya). Our hypothesis contrasts with the currently dominant view that says consciousness requires a brain with an enlarged dorsal pallium, or cerebral cortex, thereby limiting its existence to humans or to mammals and birds (Baars *et al.*, 2013; Butler, 2008a; di Perri *et al.*, 2014; Edelman *et al.*, 2011; Min, 2010; Ribary, 2005; Rose *et al.*, 2014; Seth *et al.*, 2005; van Gaal & Lamme, 2012).

## 2.2. Evolutionary Step 1: Consciousness Appears

### 2.2.1. *Lampreys*

The relationships among the vertebrates, including the position of lampreys, are shown in Fig. 1. The jawless clade to which lampreys belong originated earlier than the jawed clade of gnathostomes (a *clade* is any natural group of organisms united by their common ancestry: Kardong, 2012). Before exploring if lampreys are conscious, we note that they are not the only living descendants of the ancient, jawless vertebrates. The others are the similarly eel-shaped hagfish. However, the blind and slimy hagfish are believed to be secondarily specialized for burrowing and living in the mud of the floor of the dark, deep sea (Mallatt, 1997; Northcutt, 1996), with some of their senses having regressed (vision and electroreception) and others having elaborated to an extreme degree (touch, smell, taste: Braun, 1996). These hags are not the best model of ancestral vertebrates.

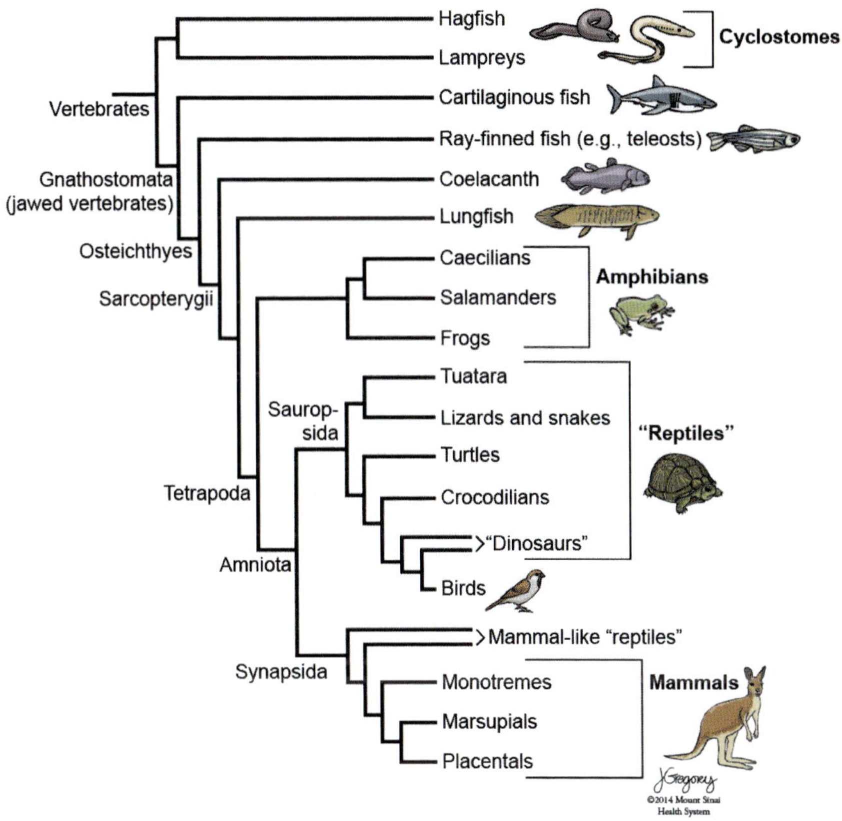


**Fig. 1.** Phylogenetic relations of the clades of vertebrates, from Benton & Donaghue (2007), Crawford *et al.* (2012) and others.

In contrast, the free-swimming lampreys, most species of which are parasites and prey on fish (Hardisty, 1979), seem to have few specializations beyond a sucker mouth, and to retain more of the sensory and brain characters of the first vertebrates. At least their neural characters are much more like those of the jawed vertebrates (Butler & Hodos, 2005; Collin, 2009; Glover & Fritzsch, 2009; Kardong, 2012; Nieuwenhuys & Nicholson, 1998; Nieuwenhuys *et al.*, 1998; Wicht, 1996). Lampreys have the vertebrate camera-eye and acute vision; the lateral line; olfactory system; neural-crest and placodal derivatives; peripheral nerves and trigeminal ganglion; and the three-part brain with a readily recognizable forebrain, midbrain, hindbrain and their subparts (Fig. 2 (A)). Even the

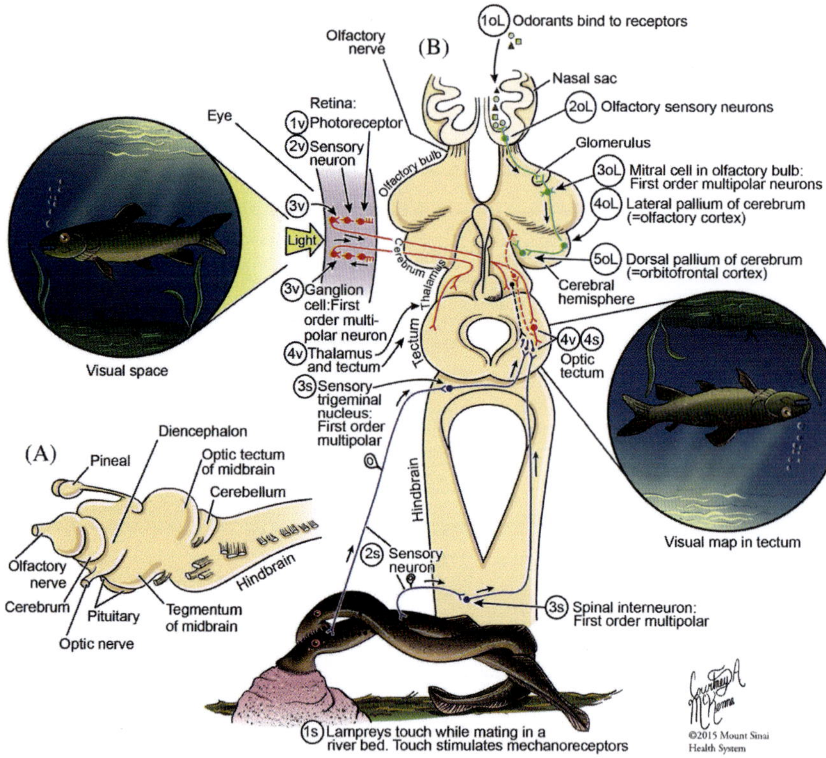


**Fig. 2.** Lamprey brain and sensory pathways. (A) Brain in side view, at left. (B) At center and right is brain in dorsal view to show sensory pathways for vision, touch/somatosensations and olfaction. Levels in the paths are labeled from 1 to 5, with 1oL meaning first level in the olfactory path, 3v meaning third level in the visual path, 2s meaning second level in the somatosensory path, etc.

sensory pathways to the lamprey brain (Fig. 2 (B)) are as typical and complete as in any other group of anamniotes (*anamniotes* = the fish and amphibians).

It must be emphasized, however, that the lamprey nervous system is not as fully developed as that of the jawed vertebrates including the well-known mammalian and human systems. For instance the cerebral pallium of most anamniotes is smaller and less elaborate than in the more encephalized *amniotes* (mammals, reptiles and birds) — and the pallium of lampreys is the smallest and simplest in any vertebrate (Northcutt, 2002). Lampreys and hagfish lack the myelin that surrounds axons and speeds impulse conduction in all the jawed vertebrates (Bullock *et al.*, 1984). They also have a simpler version of the autonomic nervous system for visceral motor control (Glover & Fritzsch, 2009; Haming *et al.*, 2011).

We will see that lampreys and the other anamniotes have some of the brain structures that associate with consciousness in humans, structures we can scrutinize to test whether the anamniotes are conscious. Knowledge of the *functions* of these brain structures in anamniotes, however, is plagued with gaps. This dearth of experimental data hinders investigations of whether any non-mammal or non-bird vertebrate is conscious, obstructing attempts to build a science of "comparative vertebrate consciousness." The scarcity of functional brain information in the anamniotes and reptiles is most acute for their telencephalon, the forebrain region that is most essential for consciousness in humans. Scientists do recognize the important evolutionary position of lampreys and have studied the lamprey brain better than the brain of any other anamniote (except zebrafish). Especially valuable contributions to lamprey neurobiology have come from Sten Grillner, Carl Rovainen, R. Glenn Northcutt, Rudolf Nieuwenhuys, and their colleagues (e.g., Cinelli *et al.*, 2013; Grillner *et al.*, 2013; Murakami & Kuratani, 2008; Nieuwenhuys & Nicholson, 1998; Northcutt, 1996; Northcutt & Wicht, 1997; Reperant *et al.*, 2009; Robertson *et al.*, 2014; Rovainen, 1979). Unfortunately, the functions of two especially important regions of the lamprey brain, the pallium and thalamus, are still poorly studied.

Despite the information gaps, enough has been learned about brain structure and function across vertebrates to let us investigate when consciousness began. Helping this quest are theoretical ideas about the kinds of neuronal arrangements that generate conscious mental images.

### 2.2.2. *Theory of isomorphic neural hierarchies*

Basic aspects of our hypothesis were presented in the previous chapter of this volume. The core concept is that chains of neurons extend through the vertebrate nervous system, conveying sensory information from one processing center to the next in a hierarchical series throughout which this information is mapped topographically (isomorphically). Sensory topography is a point-by-point mapping (Jbabdi *et al.*, 2013) that preserves the spatial arrangements of stimuli from the outer world (e.g., points in the visual field in the visual pathway) or of the receptors on or in the body (e.g., in the cochlea of the ear in the auditory pathway), or of mapped structures in a low-level processing center (e.g., the patterns of glomeruli of the olfactory bulb in the smell pathway). The hierarchy's different levels communicate back and forth with one another through reciprocal interactions, and, with other regions of the nervous system (Singer, 2001). The topographical mapping becomes increasingly complex or abstract toward the top of the processing hierarchy (Kaas, 1997) through successive re-representations that incorporate new information from other parts of the brain. Then, according to our hypothesis, the most extensively processed single-sensory or multi-sensory representations are experienced consciously as *mental images*. Other investigators have also proposed that topographic maps are necessary for sensory consciousness (Damasio, 2000; Edelman, 1989; Jbabdi *et al.*, 2013; Risi & Stanley, 2014; Tinsley, 2008; Ursino *et al.*, 2009). Our hypothesis says consciousness emerges at the fifth level and above in mammals, all in the cerebral cortex, but at the fourth level in anamniote vertebrates, namely in the optic tectum of the midbrain (see Table 5.3 in Feinberg & Mallatt, 2016).

Firing of neurons in coordinated, oscillatory patterns also seems essential for consciousness (Feinberg, 2012; Feldman, 2013; Melloni, 2007; Mori *et al.*, 2013; Singer, 2001; Uhlhaas *et al.*, 2009; Ursino *et al.*, 2009). These rhythmic oscillations are 'brain waves' in the gamma frequency-range of 30 to 80+ cycles per minute, or at somewhat lower frequencies (Bullock, 2002; Nunez & Srinivasan, 2010; Schomburg *et al.*, 2014). They occur in response to sensed stimuli (Cardin *et al.*, 2009), and the higher brain is what signals neurons at all the levels of a sensory hierarchy to fire in unison. Singer (2001) called this a transient synchronization of the discharges of neuronal assemblies of wide spatial distribution. This oscillatory synchronization is thought to integrate, or bind into a single perception, all the neurons in a hierarchy that are responding to the same object or scene.

The oscillations occur not only in single-sense hierarchies (e.g., for visual perception), but also work across senses. That is, they are thought to bind different modes of sensory information into one percept: to join the smell, texture, and visual appearance of an object being examined, for example. This cross-sensory binding would be responsible for the unified, multisensory, mental images of consciousness (Burck *et al.*, 2010; Feinberg, 2012; Magosso, 2010; Zmigrod & Hommel, 2013).

Besides unifying the conscious experience and making us aware of sensed objects, the synchronized oscillations could also sharpen and focus our attention on particular stimuli. This attention-directing function was emphasized by Goddard *et al.* (2012), Krauzlis *et al.* (2013), Miller & Buschman (2013), Schiff (2008) and Sridharan *et al.* (2011).

These concepts of topographic hierarchy and oscillatory binding in consciousness are usually applied to the cerebral cortex of humans and other mammals, where both the primary-sensory areas and the cortical-association areas are measurably active during conscious sensory experiences, as are gamma oscillations reverberating between the cortex and the thalamus at the top of the sensory hierarchies (Boly *et al.*, 2013; Dehaene & Changeux, 2011). But the cortex-centered consciousness of humans does not rule out the possibility that fish and amphibians, vertebrates without a cerebral cortex, generate consciousness in a more caudal part of their brain. That is, sensory consciousness might have shifted up the neural hierarchy when the mammalian cortex evolved. Anamniote consciousness should not be judged by the human condition but evaluated on its own evidence. In the anamniote vertebrates, most classes of sensory information converge and merge in the tectum of the midbrain, as will now be discussed.

## 2.3. The Optic Tectum

### 2.3.1. *Structure and function*

The optic tectum, called superior colliculus in mammals, forms most of the roof of the vertebrate midbrain (Butler & Hodos, 2005; Guirado & Davila, 2009; Saidel, 2009; Wullimann & Vernier, 2009a). It is relatively largest in the anamniotes, where its size sometimes exceeds that of the cerebrum (Figs. 2 and 3(A)). It is also big in reptiles and birds despite their larger cerebrums (Figs. 3(C) and 5(G)).

The tectum is the main visual-processing center in the anamniotes, receiving more visual input from the retina than does any other brain region. Fish whose tectum is destroyed behave as if blind (Northmore, 2011).

In all the clades of vertebrates, the tectum presents an extreme example of topographic mapping, not only of visual input but also of the input from other senses: hearing, touch, taste, lateral-line mechanoreception and electroreception — although not smell (Guirado & Davila, 2009; Knudsen, 2011; Stein & Meredith, 1993). This multi-sensory topography is reflected in the tectum's neuronal layering or lamination (Fig. 3(B)), where the different layers receive different classes of sensory input, and the mapped inputs from the different senses are in topographic register with one other. This provides a single coordinated and unified map of sensory space (Saidel, 2009). The sensory map in the tectum is aligned with a *motor* map. That is, the deep tectal layers send motor instructions to the brain's reticular formation and the striatum (basal ganglia), which signal body movements (Knudsen, 2011; Manger, 2009; Saitoh *et al.*, 2007). The studies from lampreys indicate that their tectum (Fig. 2) is not particularly primitive but is typical for a vertebrate (Iwahori *et al.*, 1998; Jones *et al.*, 2009; Nieuwenhuys & Nicholson, 1998; Pombal *et al.*, 2001; Reperant *et al.*, 2009; Robertson *et al.*, 2006; Sarvestani *et al.*, 2013).

The tectum performs important functions, which also seem largely constant across the vertebrates: (1) sensory = it processes the sensory information it receives; (2) motor = it controls behavioral actions; and (3) attentive = it participates in directing attention. For its sensory role, the tectum helps to recognize and discriminate among objects being viewed, by their size, shape, novelty, rate of change, and motion (Del Bene *et al.*, 2010; Dudkin & Gruberg, 2009; Graham & Northmore, 2007; Gruberg *et al.*, 2006; Gutfreund, 2012; Schuelert & Dicke, 2005). It also decides the salience (distinctiveness and importance) of viewed items.

For its motor functions, the tectum controls eye movements toward viewed objects and to track objects as they move. Tectal action orients the animal's body toward attended objects in the visual field and signals the animal to approach or avoid these objects (Cohen & Castro-Alamancos, 2010; Dutta & Gutfreund, 2014). It also guides vision-driven movements of the body through space, as when searching for food or avoiding obstacles.

For its role in attention, the tectum acts with adjacent centers in the brainstem called *isthmus nuclei* (Fig. 3(A)). Together, these structures serve 'selective attention' to sensed stimuli (Caudill *et al.*, 2010; Krauzlis *et al.*, 2013). That is, isthmus-tectal interactions determine which object in mapped, multisensory space is the most salient and then they attend to that object; they also prepare a possible response and select the targets of this response (Dutta & Gutfreund, 2014; Gruberg *et al.*, 2006; Knudsen, 2011).

The tectum is more for speed than for deep analysis. It performs the initial evaluation of sensory stimuli, or of a situation that requires immediate attention, fast decisions, and direct responses (Cohen & Castro-Alamancos, 2010; Sridharan *et al.*, 2014). Northmore (2011) described it as fast reactions to moment-to-moment changes in the environment. After this, the cerebral pallium may perform a further, slower, more extensive evaluation of sensory inputs — in those vertebrates whose cerebrum is

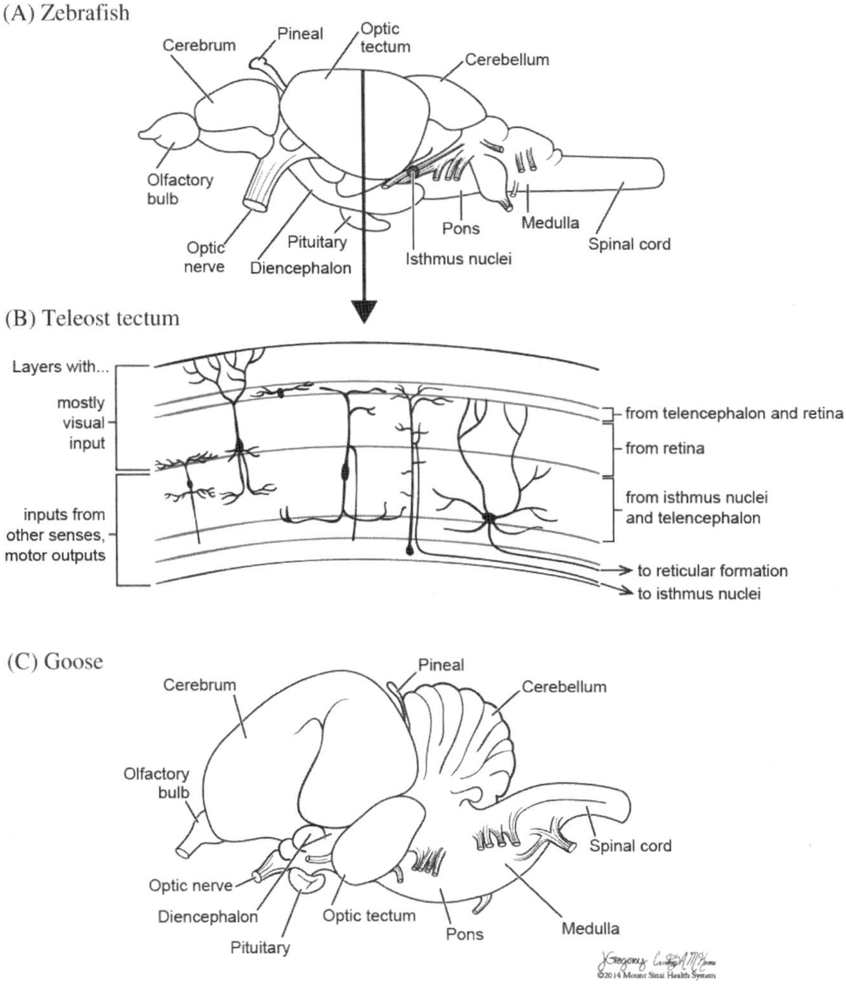


**Fig. 3.** Optic tectum, showing its size in a teleost zebrafish (A) and a bird (C), and its layered neuronal structure in a teleost (B). B is re-drawn from Northmore (2011).

elaborate enough and *if* more time is available to respond. The tectum, by contrast, has only one level of (multisensory) mapping and can go no further (Kaas, 1997). Its processing capacity may be less, as would be expected from its emphasis on speed and snap choices. By this reasoning, the tectum can sometimes make rushed and incorrect decisions in highly complex sensory environments (Krauzlis *et al.*, 2013).

Why do so many different senses project to the optic tectum, a fundamentally visual area? Besides allowing a more detailed and complete image of the world than vision alone can provide, the additional senses check for consistency (Kaas, 1997): do all the senses support the same, internally consistent reconstruction of the world? And the other senses act as a "spell checker" to correct any areas of weak or ambiguous visual input in the visual map (Burck *et al.*, 2010). Such image reinforcement improves sensory discrimination, object detection, and increases the speed of response to stimuli (Quinn *et al.*, 2014). The tectum's rigidly topographical, closely stacked, and aligned neuronal maps for the various senses are efficient in design because they keep the connections between the neurons short. As Kaas (1997) put it, they "effectively group neurons that most commonly interact, thus decreasing requirements for long, slow, and metabolically costly connections," this being the same advantage he assigned to topographic mapping in general. In summary, these are the many benefits of merging the senses into one, multimodal, and topographic image in the optic tectum. The benefits explain why, when vision and the tectum evolved in the very first vertebrates, topographic input from most other senses soon reached the visual tectum as well. This was the origin of mental **unity** of the exteroceptive senses (unity having been described in Chapter 1).

Among the above-listed roles of the anamniote tectum are types of sensory discrimination that seem sophisticated enough to be deemed conscious. At least, such tectal consciousness is what some scientists have implied. Dicke & Roth (2009) said:

> "In amphibians as in all other anamniote vertebrates, the tectum is the major brain center for integrating visual *perception* and visuomotor functions. In the amphibian tectum, localization and *recognition* of objects and depth perception takes place. Three separate retino-tectal subsystems for object recognition exist, which process information about (i) size and shape, (ii) velocity and movement pattern and (iii) changes in ambient illumination. These kinds of information are processed at the level of different retinal ganglion cells and tectal neurons in close interaction with neurons in other visual centers."

In this excerpt, we italicized the terms *perception* and *recogniton* as implying consciousness, which we also infer from Wullimann and Vernier (2009a) who stated the fish tectum is for "object identification and location." Others could argue that no, these tectal functions are entirely unconscious reflexes (albeit complex reflexes of the highest order). But we see *real consciousness* in this combination of the anamniote tectum building a multimodal sensory map of the environment, attending to the most salient objects in that map, and then, dictating motor behaviors that suggest the anamniotes are *accessing* the map.

### 2.3.2. *Tectal oscillations*

Electrophysiological recordings also suggest tectal consciousness. Several studies have detected gamma waves oscillating between the tectum and the isthmus nuclei (Goddard *et al.*, 2012; Knudsen, 2011; Marin *et al.*, 2007; Sridharan *et al.*, 2011, 2014). As mentioned above, the isthmus-tectal system participates in selective attention to salient stimuli, and, the authors of these studies explicitly stated how similar the tectal oscillations are to the gamma oscillations that reflect spatial attention and consciousness in the cerebral cortex of mammalzs. This could signify that the tectum is conscious, but a caveat is that the best evidence for the oscillations comes from birds, which have large and complex tecta and are not anamniotes. Sparser evidence for such gamma oscillations comes from the tectum complex of frogs (Caudill *et al.*, 2010) and teleost fish (Northmore & Gallagher, 2003); and all vertebrates have the required anatomic connections between the isthmus nuclei and the optic tectum (for lampreys, see Pombal *et al.*, 2001 and Robertson *et al.*, 2006). Additional electrical recordings of the isthmus-tectal signals in anamniotes could reveal definitively whether these tectal 'oscillations of consciousness' exist in basal vertebrates.

## 2.4. Telencephalon and Pallium

Could the cerebral-cortex equivalent in anamniotes, which is the dorsal pallium of their telencephalon (Medina, 2009) (Fig. 4), consciously perceive all the senses, as does the cerebral cortex of humans? The dominant view before about 1970 did not allow this because it claimed the anamniote telencephalon is only for olfactory processing, a "smell-brain" (Wullimann & Vernier, 2009b). Supporting this claim was the large size of the olfactory bulb in amniotes relative to the rest of the telencephalon of their brain (see Figs. 2 and 3(A)).

But in the 1970s this "smell-brain" idea was overturned, as new techniques for identifying brain structures and axonal projections showed that the telencephalons of fish and amphibians receive visual, auditory, touch, and other types of non-olfactory sensory input (Demski, 2013; Hofmann &

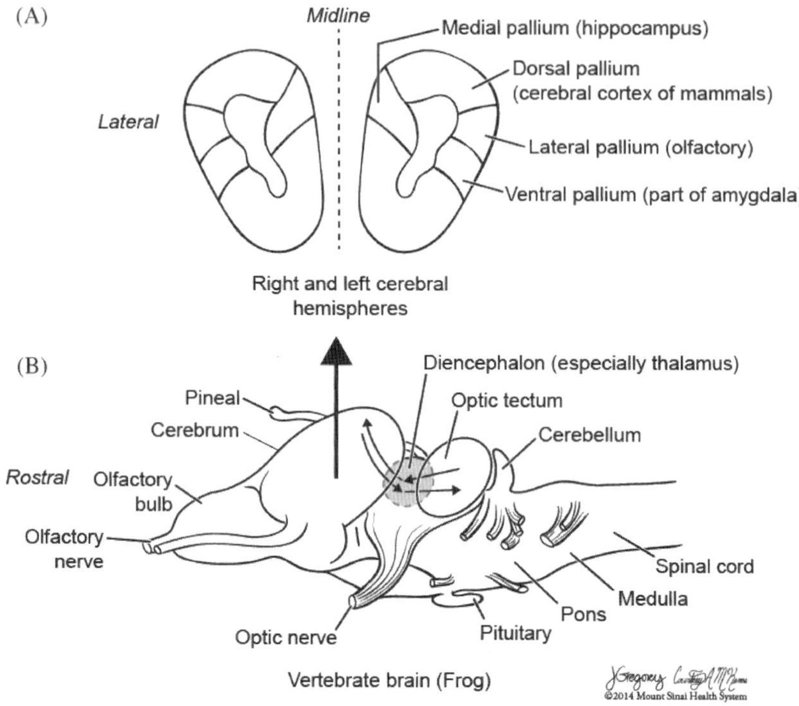


**Fig. 4.** Pallium of the cerebrum of the vertebrate brain. (A) Cross section through the right and left cerebral hemispheres of an embryo showing the four basic divisions of the pallium. (B) Orientation diagram of an adult vertebrate brain showing the plane of section through the cerebrum in part A. Pallial-tectal communications through the diencephalon are also shown (at the four arrows).

Northcutt, 2012; Reperant *et al.*, 2009; Wullimann & Vernier, 2009b). Anamniote telencephalons were also shown to have the same non-olfactory and non-sensory parts as in amniotes, including the hippocampus (medial pallium) for forming memories, the amygdala and other structures associated with affective "feelings," and the striatum (basal ganglia) for choosing which behavioral motor-programs to execute.

In recent years, however, the pendulum has swung back to the idea that olfactory processing was the main role of the telencephalic pallium in the first vertebrates. This is because the pallium of the most basally derived extant fish — lampreys, hagfish, and the cartilaginous fish — has been found to receive mostly olfactory projections, and to be far more smell-dominated than in the other vertebrates. And, upon entering the olfactory (lateral) pallium, the olfactory signals of these basal vertebrates project more extensively to the other pallial regions, such as the dorsal and medial pallium (Hofmann & Northcutt, 2012; Northcutt & Wicht, 1997; Wullimann & Vernier, 2009b).

However, it is still true that the anamniote pallium receives all classes of sensory input, not just smell. So we must still ask if the pallia of fish and amphibians map the non-olfactory senses into conscious images, as in mammals. Apparently not, because the pallial projections of these senses are not unimodal or topographically organized. Wilczynski (2009) wrote:

> "... all sensory systems not just olfaction reach large areas of the telencephalon. In this general way, amphibians are similar to other tetrapods... *The details, however, reveal a quite different functional organization from that which might be expected...* The telencephalic targets of ascending sensory pathways are all multimodal. There is no evidence for separate representations for each sensory system, *no indication of topographically preserved projection* from any thalamic nucleus to any telencephalic area and *no physiological evidence for a sensory (or for that matter motor) map. In essence, there is no evidence for the distinct, unimodal, mapped sensory representations that are so prominent in the mammalian cortex.* A possible exception may be the core olfactory-recipient regions of the lateral pallium. That is not to say there is or is not a homologue of mammalian neocortex within the amphibian telencephalon, but there are certainly no functional equivalents for the well-mapped, pure sensory zones that are so prominent in mammals..." (emphasis added).

Wilczynski was speaking of amphibians but his words also apply to the pallia of the *fish* that have been investigated (Northmore, 2011; Wullimann & Vernier, 2009b). As exceptions to the claim, however, certain teleost fish with especially keen senses show some separation by sensory modality in their pallial pathways (Demski, 2013; Prechtl *et al.*, 1998), and one group even shows pallial retinotopy (reef fish of the perch clade: Leo Demski, personal communication). But other evidence strengthens the original claim. In the carp clade, the visual and auditory pathways do not project to the dorsal pallium *at all* (Mueller, 2012). This means a very successful fish clade, which includes the often-studied zebrafish and goldfish, exists, sees, and hears without any possibility of a topographic, pallial consciousness of these senses. In conclusion, in anamniotes all the topographic sensory representations except the olfactory are in the optic tectum and not the pallium.

Where are the olfactory representations in the brains of fish and amphibians? The same olfactory pathway exists in all vertebrates including lampreys, and it targets the telencephalon (Fig. 2 (B), upper right) (Brodal, 2010; Eisthen, 1997; Green *et al.*, 2013; Northcutt & Puzdrowski, 1988; Wullimann & Vernier, 2009b). In this universal pathway, the olfactory neurons go to the olfactory bulb, which mainly projects to the lateral pallium (piriform cortex of mammals), which then projects to the dorsal pallium, which in mammals includes the orbitofrontal cortex for olfactory consciousness (Li *et al.*, 2010; Shepherd, 2007). Therefore, the site of any olfactory consciousness in the anamniotes must be the pallium (as in amniotes), although the optic tectum would be conscious site for all their other senses (unlike in mammals). But the olfactory pallia of anamniotes (lateral and dorsal) do not have multiple laminae like the tectum does (Northcutt & Wicht, 1997; Hofmann & Northcutt, 2012; Wulliman & Vernier, 2009b), so the pallium and tectum would have different anatomic substrates for consciousness.

The telencephalic pallium and midbrain tectum communicate with one another through the diencephalon, which lies between them (Fig. 4(B); also see Fig. 7 in Northmore, 2011; Butler 2008b; and Fig. 7 in Feinberg & Mallatt, 2013). Perhaps this pallial-tectal communication unifies the two types of consciousness in anamniotes, olfactory and non-olfactory sensory, although we cannot guess precisely where or how.

If the pallium of anamniotes is not involved with mapped multisensory consciousness, and thus is so unlike our own cortex, then what (non-smell) functions can it possibly perform? This is a challenging question, because a fish whose pallium has been experimentally removed can see and it appears to behave normally (Arhem *et al.*, 2008; Northmore, 2011). However, it is unable to learn from experience or from the consequences of its actions, nor can it learn the positions of objects in space (Broglio *et al.*, 2005; Fuss *et al.*, 2014). This is a memory deficit, and memory storage is a documented function of the pallia of vertebrates (Bingman *et al.*, 2009; Nadel *et al.*, 2012).

In addition, the dorsal and medial pallia are the main places in the vertebrate brain where olfactory input (from the lateral pallium) converges on input from all the other senses, from the optic tectum (Fig. 2 (B), upper right) (Butler & Hodos, 2005; Northcutt & Wicht, 1997). And all this sensory input is highly processed by the time it reaches the dorsal-medial pallium. Therefore, because it has access to more completely processed information than does the tectum, the dorsal-medial pallium could beneficially countermand the motor responses ordered by the tectum, which are based on quicker and rougher assessments and thus are more subject to error.

We reason that the dorsal-medial pallium of anamniotes, although small and not topographically conscious, has the distinct advantages of (1) storing memories and (2) being situated where it receives the widest variety of multisensory information. These advantages, we will argue, made it the favored site of brain elaboration in those vertebrates with greater mental capacities (see Sec. 3).

## 2.5. Memory and Consciousness

Memory is an important part of consciousness (Edelman, 1989, 1992; Edelman *et al.*, 2011). That is, the conscious experience involves images being called up from memory, and then modified by the newly arriving sensory information into an updated image of the world as it is experienced (Nadel *et al.*, 2012). In this way, the conscious act does not have to construct entirely new mental images from instant to instant but constantly modernizes the existing image. This permits speed, saves neurocomputational effort, and allows more detailed images to be built from a brain's existing computing capacity.

Memories are constructed by the medial pallium (hippocampus) of the vertebrate brain, not only *episodic* memories of experiences and objects but also memories of mapped spatial relationships for navigating the environment in which the animal lives (Allen & Fortin, 2013). For constructing these memories the hippocampus receives all the classes of sensory information, and it also has extensive, reciprocal connections to all major parts of the telencephalon and diencephalon (Bingham *et al.*, 2009; Brodal, 2010; Butler & Hodos, 2005; Northcutt & Ronan, 1992; Rolls, 2012). A hippocampus has been shown to exist in every clade of jawed vertebrates (Fuss *et al.*, 2014; O'Connell & Hofmann, 2011; Schluessel & Bleckmann, 2005), which is almost enough for us to conclude that memory served phenomenal consciousness in the first vertebrates. In lampreys, however, a hippocampus has not been demonstrated with certainty, the only evidence for one being expression of some hippocampus-defining genes in the developing lamprey brain (*Lfl.hx* 1/5 and 2/9: Watanabe *et al.*, 2008).

## 2.6. Arousal and Consciousness

We have established that besides qualia, the other critical component of phenomenal consciousness is selective attention or directed awareness. This component is related to vigilance and to general arousal (Schiff, 2008).

The brains of all vertebrates have structures for these attention-arousal functions, specifically the reticular activating system (RAS) of the brainstem's reticular formation and nuclei in the basal forebrain (BFN) (Butler & Hodos, 2005; Fuller *et al.*, 2011; Lee & Dan, 2012; Manger, 2009; Robertson *et al.*, 2007). In mammals, the RAS and BFN help to direct attention to salient stimuli by influencing the thalamo-cortical circuit of consciousness and attention. In anamniotes, however, some parts of that circuit are absent or undemonstrated: there is no reciprocal feedback from the pallium to the thalamus in anamniotes (Butler, 2008b). It is an incomplete or "broken" circuit. Thus, anamniotes could not use that circuit for any pallial consciousness nor for any pallial attention to specific stimuli.

Anamniotes do, however, have the arousal component. They have an RAS and BFN, and, as in the amniotes these two regions release neurotransmitters or neuromodulators that associate with general arousal, alertness, and vigilance. These neurochemicals include norepinephrine and acetylcholine (Lee & Dan, 2012; Manger, 2009). Arousal aids an animal's sensory awareness by increasing overall receptivity to sensory inputs. Therefore, the arousal system must contribute to sensory consciousness in fish and amphibians, as it does in amniotes.

In addition, lampreys have strong reciprocal links from the reticular formation to the dorsal thalamus to the optic tectum (Nieuwenhuys & Nicholson, 1998; Robertson *et al.*, 2006). These connections suggest lampreys have an RAS-arousal system that targets the *tectum*, analogous to the RAS-thalamic-cortical system of mammals. Again, this involves *arousal* alone. There is still no evidence that the reticular formation of anamniotes selectively directs attention to *specific* stimuli. But, as established above, the isthmus-tectal interactions play this attention-directing role in anamniotes (Caudill *et al.*, 2010; Dudkin & Gruberg, 2009; Northmore, 2011).

We have presented evidence that anamniote vertebrates have all the elements necessary for sensory consciousness. These are topographic maps for mental images (mostly in their optic tectum), memory storage (in their pallium), and mechanisms of selective attention and general arousal.

## 2.7. Amphioxus and Consciousness

Does amphioxus, the fish-like invertebrate whose larval brain is most likely to resemble that of the vertebrate ancestor (Fig. 2 in Chapter 1 of this volume), possess any of the hallmarks of sensory consciousness? For its sensory repertoire, amphioxus has photoreceptors but not image-forming eyes nor vision, and it has many mechanoreceptors and presumed chemoreceptors on its body surface (Lacalli, 2008; Vopalensky *et al.*, 2012; Wicht & Lacalli, 2005).

We find no hallmarks of sensory consciousness in amphioxus. First, its sensory neurons do not develop from embryonic neural crest or placodes, as do sensory neurons of vertebrates (Schlosser, 2005; Sommer, 2013), so these neurons are not even equivalent. Recall that sensory neurons are the essential first link in the conscious sensory hierarchies of vertebrates.

Second, most of the sensory pathways to the larval-amphioxus brain have just two neurons (Lacalli, 2008). This seems far too few to allow enough information-processing for any consciousness.

Third, there is the question of topographic organization. While it is not known for certain if amphioxus' sensory pathways are topographically constructed, Holland & Yu (2002) found hints of an unusual somatotopy of the mechanosensory axons in the larval spinal cord. The axons were arranged as (1) those from the 'dorsal and ventral' body together vs., (2) those from the mid-body. This differs from the mostly head-to-tail somatotopic arrangement in the central nervous system of vertebrates. It is not vertebrate-style topography.

Fourth, although amphioxus does have a simple reticular formation ("primary motor center" in its midbrain: Lacalli, 2008), it is missing all other components of the vertebrate arousal system. That is, no thalamus, tectum, or telencephalon is known to exist in amphioxus (Wicht & Lacalli, 2005). In theory amphioxus must have some sort of arousal system, if only to prime it for escape-swimming to avoid danger. But significantly, it lacks both of the arousal neuromodulators we mentioned above: Its forebrain has no acetylcholine and its central nervous system evidently lacks norepinephrine altogether (Candiani *et al.*, 2012). We conclude that the invertebrate amphioxus has none of the indicators of consciousness found in vertebrates.

## 2.8. Genes and Consciousness

In our recent paper, we explored sensory consciousness from the new perspective of participating genes (Feinberg & Mallatt, 2013). That is, we examined the literature for "genes of consciousness". We were able to document many genes that signal the embryonic development of the structures associated with consciousness throughout the vertebrates, from the three-part brain (e.g., the *Emx* gene) to the placodes and neural crest (e.g., *Hox* genes), to the eyes (e.g., *Pax 4/6* genes).

However, we could not find any universal genes for topographical organization, as our hypothesis predicted must exist. Recently, we discovered that such a gene family is known, the *Eph/ephrin* genes (Triplett & Feldheim, 2012). Their topography-signalling role is best documented in the visual system of vertebrates (Knoll & Drescher, 2002), but also in the auditory (Cramer & Gabriele, 2014) and somatosensory systems (North *et al.*, 2010). Neither *Eph/ephrin* nor the other genes associated with consciousness in vertebrates seem to be involved in nervous-system development in amphioxus (Bosne, 2010).

## 2.9. How Widespread is Vertebrate Consciousness?

Are all vertebrates conscious? Yes and no. We say "yes" because by our hypothesis and to our knowledge, no vertebrate line has ever lost the conscious adaptation. We conclude this because no vertebrate has lost all its distance senses or its mobility. That is, no species of fish, amphibian, or amniote is eyeless, earless, and sessile, as some invertebrates are (clams, corals). If hagfish, which are lampreys' sister clade among the earliest-evolving vertebrates, had a simple brain and sensory system, that might mean hagfish are not conscious. However, the hagfish brain is three times larger than that of lampreys (Northcutt, 2002) and it bulges with expansions that extensively process smell, touch, and taste stimuli (Andres & von During, 1993; Braun, 1996; Braun & Northcutt, 1998; von During & Andres, 1998). Therefore, by our criteria hagfish should have sensory consciousness.

We say "no," on the other hand, because every young vertebrate including humans must be unconscious in the early stages of its life cycle, before its brain develops and elaborates. Because a young zebrafish larva, with its simple brain and behaviors (Del Bene *et al.*, 2010; Portugues & Engert, 2009), can feed and survive in such a pre-conscious stage, then unconscious *adult* vertebrates are at least possible in theory. The evolutionary process by which larval or embryonic traits sometimes carry over into the adult stage is called paedomorphosis, and it has been demonstrated in many animal lines including salamanders and newts (Gould, 1977).

## 2.10. Summarizing Step 1, Conscious Origins

Between approximately 560 mya and 520 mya, the ancestors of vertebrates evolved image-forming eyes, which quickly led to detailed mental images of the world in their newly evolving optic tectum. Responsible for these mental images was a multi-neuron hierarchy of retinotopically arranged, visual-processing centers extending from the eye to the tectum. The visual images had the first qualia, which soon were enhanced by information from the other distance senses — hearing, touch, lateral-line, taste — as topographically organized and tectum-targeting hierarchies quickly evolved for those senses. Simultaneously, olfactory pathways were evolving and smell images appeared in the newly evolved telencephalic pallium.

All these well-organized and consciously experienced sensory images were adaptive. The reasons for their survival value were presented in depth in Chapter 1 of this volume. Here, we summarize why by saying the mental reconstructions were sufficiently broad and accurate approximations of the environment, good enough that they provided reliable behavioral access to the *actual* environment. By accessing the simulated reality in their minds, the vertebrates could direct their actions precisely and correctly toward any location in the world, for feeding, escaping attack, foraging or mating.

Furthermore, the conscious mental images were highly evolvable. That is, from the very start it was easy for these images to become more detailed, or more responsive to the most-informative and beneficial class of stimuli (vision-dominated in hawks, smell-dominated in bloodhounds), simply by adding more neurons and more neuronal processing to existing hierarchies. Such easily enhanced processing evolved again and again throughout vertebrate history, wherever selection pressures favored it.

Also essential to consciousness was the evolution of brain systems for directing attention to salient stimuli (the isthmus-tectal complex) and for general arousal that enhanced alertness to new stimuli (the RAS). Every vertebrate has all the requisite features of consciousness, but their invertebrate cousin, amphioxus, has none.

## 3.1. Evolutionary Step 2: Consciousness Leaps Forward in Mammals and Birds

### 3.1.1. *The leap described*

So far, we have argued that consciousness originated in the optic tectum of the earliest vertebrates some 560–520 mya. A second major step occurred several hundred million years later, independently in the two lines of amniote land-vertebrates. In both the synapsid (pre-mammal) and sauropsid (pre-bird) reptile lines, the dorsal pallium became the dominant brain region of sensory processing, and by our interpretation, of sensory consciousness (Figs. 1 and 5).

The most dramatic sign of this step was when the cerebrum underwent a big jump in size, first in the earliest mammals about 220 mya (Bi *et al.*, 2014) and then later when the birds evolved from dinosaurs about 165 mya (Lee *et al.*, 2014). Although the brains of living birds and mammals are around ten times larger than those of living reptiles when standardized to body weight, the first, Mesozoic members of these clades had brains only three to five times larger than those of their "reptile" ancestors, as far as can be judged by the fossil braincases (Alonso *et al.*, 2004; Jerison, 2009; Northcutt, 2002; Rowe *et al.*, 2011). Still, these were big increases of 300–500% in short geologic time spans of just tens of millions of years (Benton, 2014; Rowe *et al.*, 2011).

The increases were due to an enlargement of the sensory-processing pallium in both the birds and mammals, leading many experts to date the *initial, absolute,* origins of sensory consciousness to these parallel events (Arhem *et al.*, 2008; Boly *et al.*, 2013; Butler, 2008a; Butler & Cotterill, 2006; Butler *et al.*, 2005; Edelman & Seth, 2009). This view, which says only birds and mammals are conscious, has more adherents than our relatively new and opposing idea of consciousness in all the vertebrates.

It is possible that these marked advances in the first mammals and birds were just the largest steps in a gradual climb that began about 330 mya when the very first amniotes arose from "amphibian" anamniotes (Benton & Donoghue, 2007). However, a set of many gradual "baby steps" is unlikely. The braincases of the earliest, fossil amniotes are badly preserved and uninformative; but brain size could not have increased much when anamniotes evolved into amniotes. We can say this because the brains of living reptiles are less than twice as large as those of living amphibians (Northcutt, 2002; 2013). Reptiles' brains and behaviors are more complex than those of amphibians (Burghardt, 2013), but perhaps not enough so (Arhem *et al.*, 2008). No, the first mammals and first birds are the correct places to assign the revolutionary advances in brain size and consciousness.

Mammals and birds shifted from tectal to pallium-centered consciousness millions of years apart and each with a different dominant sense, smell in mammals and vision in birds. These differences make it difficult to discover commonalities that could explain why consciousness accelerated in the two lineages at roughly the same time.

### 3.2. The Not-So-Puzzling Mammals

The mammals' shift seems easier to explain. In the path from mammal-like reptiles to mammals ~220 mya, the pallial/cortical expansion has been tied to an evolutionary change from vision to smell (and touch) as the predominant senses (Rowe *et al.*, 2011; also see Aboitiz & Montiel, 2007; Bowmaker, 1998; Jacobs, 2012). This was interpreted as an adaptation to nocturnal living on the ground, where vision provides little information in the dark but the varied smells are highly informative and so are tactile stimuli mediated by a sensory, hairy pelt (Hall *et al.*, 2012). Fossil braincasts reveal that the olfactory bulbs and the olfactory lateral-pallium enlarged in the first mammals (Rowe *et al.*, 2011). So did the mechanosensory region of the cerebral cortex, in which, we assume, the tactile senses were now represented topographically. The apparent regression in vision in these proto-mammals would account for the relatively small size of the optic tectum in mammals as a clade (Fig. 5(E)).

We reason that when the mammalian tectum became less important, the enlarging cerebral cortex became the seat of consciousness for the senses that had previously been conscious in the tectum: hearing, touch, vision, and equilibrium. Logic says this was a shift from a declining region of the brain to a prospering one, to enrich the increasingly important smell images in the cerebrum with additional information from the other senses. This clear logic, in fact, is why we say the mammalian shift is rather easy to explain.

Not only do the fossilized braincases of the early mammals support this interpretation, so too do the brains of living mammals. The extant mammals are the monotremes that include the duck-billed platypus; marsupials such as kangaroos and opossums; and placentals, which comprise all others including humans, rats, cats, elephants, whales, etc. Along with their small tectum, all three clades share a large cerebral cortex with topographically organized primary-sensory areas for every conscious sense (Figs. 5(B) and 5(C)), plus adjacent sensory-association areas. Furthermore, in the basal members of each clade, the olfactory piriform cortex is the largest sensory area (Kaas, 2011). This common plan indicates that the cerebral cortex is the seat of consciousness in all living mammals. According to the fossil record, the latest common ancestor of the monotreme, marsupial, and placental mammals lived 190 mya or before (Benton & Donoghue, 2007; Bi *et al.*, 2014; Luo *et al.*, 2011). This is near the 220 mya date of the earliest documented fossil mammals and their mushrooming cerebrums, reaffirming that as the date when mammalian consciousness became completely pallial.

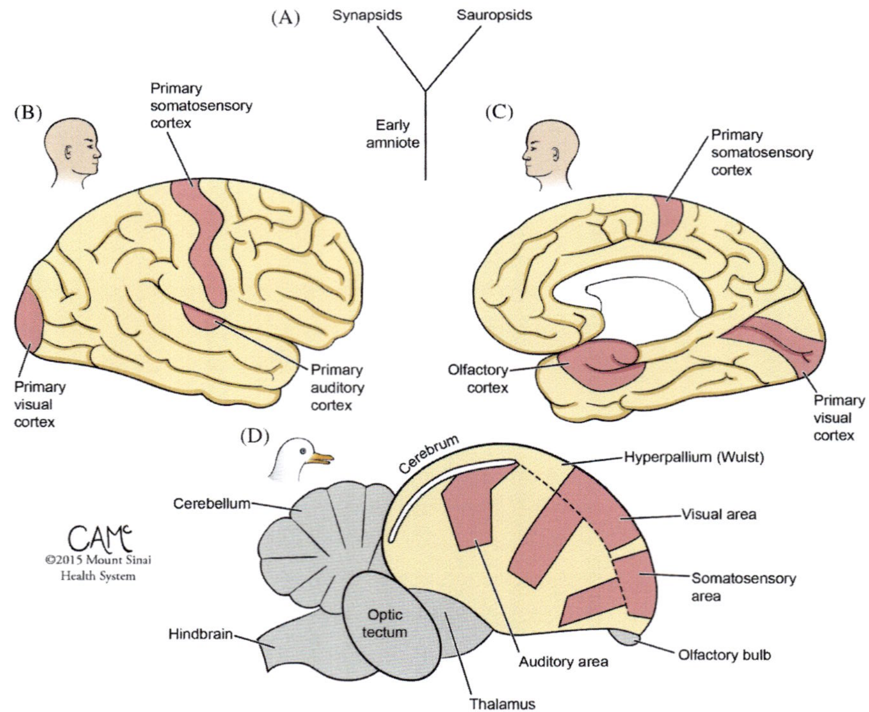


**Fig. 5.** Comparing the brains of different amniote vertebrates. (A) Simplified phylogenetic tree showing that early, reptile-like amniotes gave rise to synapsids (mammal-like reptiles and mammals) and sauropsids (the other reptiles and birds). This occurred about 330 mya. (B), (C) Primary sensory areas of the cerebral cortex of a mammal (human) in lateral (B) and medial (C) views. (D) The corresponding sensory areas in the bird pallium.

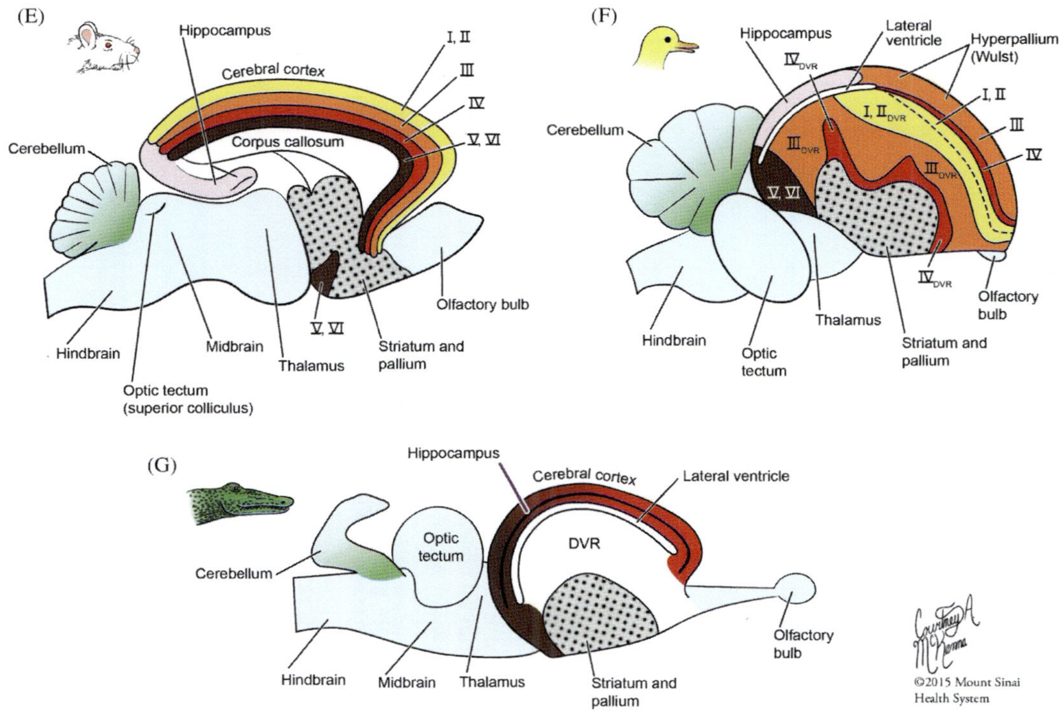


**Fig. 5.** (*Continued*) (E) Layers of the cerebral cortex in mammals, I–VI (left half of a cut rat brain). (F) Corresponding regions I–VI in dorsal pallium of a bird. (G) Dorsal pallium of a modern reptile. Parts (E–G) are simplifed sagittal sections of the brains. (D–G) are based on the work of Jarvis *et al.* (2013). The bird and reptile pictures are transparent views into the interior of the pallium.

## 3.3. The Puzzling Birds

For birds, the transfer to pallial consciousness is harder to grasp, because their associated sensory specializations are different than in mammals. As large-eyed dinosaurs evolved into birds the eyes enlarged even more, so an already excellent vision became even sharper but the sense of smell remained stable (Lee *et al.*, 2014; Zelenitsky *et al.*, 2011). This begs the following questions: If the optic tectum was the original site of visual consciousness in vertebrates, as we claim, then why did not the tectum *alone* enlarge when birds' vision improved? Why did their supposedly olfactory dorsal pallium enlarge even more than the tectum did and become the main region for visual processing and consciousness? Our hypothesis says the bird ancestors already had a good site for consciousness, in the tectum, so why make a new one?

## 3.4. Solving the Puzzle

The only possible explanation for why the dorsal pallium expanded abruptly in both vision-driven birds and smell-driven mammals is that the vertebrate pallium allows better, higher-order, multi-sensory processing than does the tectum. This line of reasoning holds promise because we already showed that all classes of sensory input converge in the dorsal pallium of all vertebrates, due to its advantageous location between the hippocampus for memory and the olfactory cortex for smell processing (Sec. 2.4). We also proposed a functional advantage of the dorsal pallium over the tectum: It more thoroughly processes the sensory inputs and compares these to memories retrieved by the adjacent hippocampus. This lets the pallium, when influencing behaviors, choose from a wider range of more-accurate decisions based on prior *learning*. But the extra processing is slow and energetically costly so it is not advantageous to all animals in all environments.

Because the dorsal pallium stores memories (Nadel *et al.*, 2012), its evolutionary enlargement must have allowed more memories to be stored: more memories of smells in mammals (Aboitiz & Montiel, 2007; Jacobs, 2012) and more visual memories in birds (Rattenborg & Martinez-Gonzalez, 2011, p. 242). With increased memory, conscious experiences became less ephemeral, less "only in the moment," less "the remembered present" (Edelman, 1989) and including more of the remembered past. More stored knowledge allowed better mental *prediction* of what events will happen next and of how best to respond (Gershman *et al.*, 2015). Increasing memory also provided mammals and (presumably) birds with a more mature type of pallium-based *selective attention*, an attention directed toward salient stimuli whose significance had been *learned* (Krauzlis *et al.*, 2013).

Now the question becomes whether the first birds and mammals inhabited environments that favored the evolution of advanced (pallial) sensory processing and of increased memory. Both groups are thought to have lived in forests, or for the proto-mammals, in other places that offered protective cover such as among rocks, in burrows, or under ground vegetation. We picture the nocturnal mammals as active on the forest floor amongst the litter, in a world filled with smells and things to touch, especially with their antenna-like vibrissae (whiskers); and the diurnal birds living up in the trees, a complex three-dimensional abode they had to master visually in order to fly from branch to branch and avoid collisions (Benton, 2014). Thus both lineages lived where they were flooded with a large variety of informative sensory stimuli. Both lineages had also experienced *miniaturization*, meaning they had markedly smaller bodies than their immediate ancestors, and both were probably preyed on by larger, carnivorous dinosaurs (Benton, 2014; Hall *et al.*, 2012; Kielan-Jawarowska, 2013; Kielan-Jawarowska *et al.*, 2013; Lee *et al.*, 2014).

The above scenario may have to be changed a little for the *bird* line. Balanoff *et al.* (2013) and Lee *et al.* (2014) showed that the brain was expanding throughout a whole, extensive radiation of small bird-related dinosaurs, whose bodies became successively smaller, from 10 kg down to 0.8 kg in the real birds. That is, the brain expansion did not all occur during the very last stage that produced the first true bird. Still, all the members of this dinosaur radiation had feathers and the "ability to fly in some form" (Balanoff *et al.*, 2013). Thus, even if consciousness advanced early in this radiation that led to birds, it still fits the flight-driven scenario we just presented.

Though tiny and relatively defenseless, both groups had good protection. Birds could fly from danger and hide in the forest canopy, and the nocturnal mammals were veiled by darkness and the ground cover under which they took refuge. In safety, they had time to weigh and choose among a variety of decisions based on the rich sensory inputs they continually received. So much sensory processing was slow, but this was no handicap in the secure habitats where they lived. Furthermore, both groups initially were insectivores (Gill *et al.*, 2014; Maynard, J. commenting on Godefroit *et al.*, 2013). Insects are an abundant and energy-dense food that could supply the energy required by more sophisticated brains and the high computational demands of increasing sensory consciousness. We conclude that both birds and mammals did originate in special environments that selected for mature, memory-augmented consciousness and pallial elaboration.

So how do we explain, in terms of our theory that tectal consciousness was ancestral, why pallial consciousness advanced so spectacularly, so similarly, yet independently in the first mammals and birds? We explain it as a case of convergent evolution, which can occur in different animal clades that are exposed to similar habitats, predators, and competitors.

The same evolutionary scenario could also be used to explain how the first mammals and birds each evolved consciousness *anew*, from unconscious reptilian ancestors, in the rival theory that says conscious is exclusively pallial and never existed in the optic tectum (Butler, 2008a; Butler *et al.*, 2005; Edelman *et al.*, 2005; Edelman & Seth, 2009). That rival theory, however, claims consciousness originated twice in vertebrates (in mammals then in birds), so it may be less parsimonious than our claim that it evolved just once, in the first fish. But, as pointed out by Arhem *et al.* (2008), this is not a serious weakness of the pallial theory given the above-mentioned likelihood of convergent evolution in the Mesozoic. And our own theory of tectal consciousness has a potential weakness also: It implies the brain of birds has two separate, dominant regions for conscious visual images, tectum and pallium. Is such duality even possible? Could the bird brain effectively bind two distantly generated mental images into one?

### 3.5. Pallial Diversity, Another Puzzle

No matter whether consciousness arose anew, or just advanced, in the first birds and mammals, another enigma emerges. These two clades seem to have achieved pallial consciousness by elaborating the dorsal pallium in different ways (Fig. 5). Granted, the pallia of both clades at first seem similar in having topographically arranged sensory regions for vision, hearing, touch, and olfaction (Figs. 5(B)–5(D)) (mammals: Kaas, 2011; Kandel & Schwartz, 2012; birds: Cunningham *et al.*, 2013; Kalman, 2009; Keary *et al.*, 2010; Muller & Leppelsack, 1985; Wild *et al.*, 2008). Also granted, the large pallia of some birds provide them with high cognitive abilities equaling those of large-brained mammals (Butler, 2008a; Edelman *et al.*, 2005; Pepperberg, 2009). But the neurons in the dorsal pallium of birds are grouped differently from those in the corresponding cerebral cortex of mammals, and for years neuroscientists labored to discover any commonalities or bases for comparison.

The mammalian neocortex has six layers (laminae) of neurons over its entire, caplike extent, from rostral to caudal and medial to lateral (Fig. 5(E)). But the dorsal pallium of birds has a large, layerless region named the dorsal ventricular ridge (DVR) and an outer, cortex-like "hyperpallium" or Wulst with three or four thick layers that do not really match the laminae of the mammalian cortex (Fig. 5(F)) (Jarvis *et al.*, 2013; Wild, 2009).

Several years ago, the enigma of these dissimilarities was solved by studies that found similarities in axonal connections, neuronal circuits, and gene-expression patterns in the pallia of mammals and birds (Dugas-Ford *et al.*, 2012; Karten, 2013; Jarvis *et al.*, 2013; Shanahan *et al.*, 2013). Neuroscientists successfully matched the six cortical laminae of mammals with the corresponding regions in birds. The results surprised many scientists. In birds, these regions are not stacked radially as in mammals but are spread "tangentially" through the pallium, which means from rostral to caudal and from medial to lateral (compare Figs. 5(E) and 5(F)). These bird regions take the form of cell clusters (nuclei) or thick bands, in contrast to the thin laminae of mammals. And, the visual area lies much farther rostrally in the bird pallium than in the mammalian (Figs. 5B–5D).

Yet many of the functional groups, microcircuits, and longer connections of neurons in the bird regions are the same as in mammals. This similarity applies to the topographically organized sensory areas for audition, vision, and touch (Karten, 2013). Thus, conscious sensory images could be generated in much the same ways in the pallia of birds and mammals. But the fact remains that in other respects the pallial structure has evolved quite differently in the two clades.

It is very difficult to learn how and when these bird-mammal differences evolved, or if either the bird- or the mammal condition is primitive. The reason for the difficulty is that the pallium of their last common ancestor, the early, extinct reptile-like amniote (Fig. 5(A)), is so hard to reconstruct. In the living reptiles (Fig. 5(G)), the dorsal pallium has a big DVR as in birds, and a cerebral cortex as in mammals, although the reptile cortex has only three laminae so it is much simpler. A problem with using living reptiles, however, is that they are all sauropsids (bird relatives: Figs. 1 and 5(A)), so one cannot assume a DVR ever existed in the pre-mammalian synapsids or in the early ancestral amniote (Butler *et al.*, 2011). Based on amniotes' closest living relatives among the anamniotes, namely the amphibians and lungfish, the dorsal pallium of the first amniote was a simple cortex with only one neuronal layer (Butler & Hodos, 2005). But we cannot know whether it had yet attained the multiple, topographically organized sensory zones we associate with consciousness. Discovering that these zones exist in the pallium of living reptiles would help, yet the necessary studies have not yet been performed (but see Sereno & Allman, 1991). The functions and neuronal homologies of the reptilian pallium are greatly understudied (for the current literature on reptiles, see Dugas-Ford *et al.*, 2012; Nomura *et al.*, 2013; Northcutt, 2013).

Despite all the unknowns, the documented differences between birds and mammals show conclusively that the neural substrate of pallial consciousness has different forms in different vertebrates. Such across-vertebrate diversity could be even greater. Hagfish, certain sharks, skates, rays, and teleost-fish groups also enlarged and elaborated their dorsal pallium, independently and not in the same way as each other but always by adding some new, specialized brain nuclei or stacked laminae (Demski, 2013; Hofmann & Northcutt, 2012; Huesa *et al.*, 2009; Wullimann & Vernier, 2009b; Yopak, 2012; Yopak *et al.*, 2012). These pallial elaborations, although unstudied at the circuit level, are always associated with advanced sensory processing in these particular fish. And all the above-listed clades of jawed fish, with their expanded pallia, originated in the Mesozoic Era (Maisey, 1996), the same era when the pallia and sensory consciousness were mushrooming in mammals and birds. Was the Mesozoic an age of repeated encephalization and widely increasing consciousness?

### 4.1. Conclusions

Sensory consciousness, or constructing and experiencing mapped images of the outer world (and inner body), went through two great advances in the evolutionary history of the vertebrates. In Step 1, when the vertebrates originated before 520 mya, consciousness also originated. It emerged from multimodal and topographical sensory representations in the brain, in the upper tiers of hierarchical chains of information-processing neurons. The first sites of consciousness were: (1) the optic tectum, which maps vision and the other non-olfactory senses, and is essential for recognizing sensed objects in fish and amphibians; together with (2) the pallium of the telencephalon for olfactory consciousness. Also necessary for consciousness and present in every vertebrate are a global arousal system in the reticular formation of the brainstem and thalamus, and mechanisms for selective attention to salient items, such as the isthmus-tectal complex of anamniotes. The basally arising jawless vertebrates, exemplified by lampreys, have all these characteristics.

From the very start, consciousness elevated vertebrate behavior beyond reflexes and simple, automatic motor responses. It contributed to the evolutionary success of this very sensitive and mobile subphylum of animals by allowing them to map and then assess the environment, leading to accurately targeted interactions with the world. Topographic organization is at the heart of such mapping, and *Eph/ephrin* genes direct the development of topographic connections.

In Step 2, when the first mammals and birds evolved in the Mesozoic Era, 220 and 165 mya respectively, the dorsal pallium of both clades enlarged with a new generation of topographically structured processing areas and became the brain's main center of sensory consciousness. No longer was the pallium just for *olfactory* consciousness. This change occurred independently in mammals and birds, by convergent evolution. In both clades, we attribute this change to the dorsal pallium's fortuituous location in the brain, at the crossroads where all types of sensory input converge, and near the hippocampus, the memory-recall center. Therefore, pallial consciousness built upon the dorsal pallium's ancestral function of slowly analyzing multi-sensory inputs to make complex decisions requiring access to memory and prior learning. This was an adaptive advantage, we reason, in the stimulus-rich but protected forests and ground habitats where the first birds and mammals lived.

In the mammals, the sensory-processing modules are stacked laminae in the pallial cerebral cortex, yet in birds, these modules are thicker and spread tangentially through the dorsal pallium (Fig. 5). This shows that pallial consciousness can be based on different neuronal arrangements of independant evolutionary origin.

Our considerations indicate that topographic sensory consciousness is more *widespread* than many investigators suspected (it exists in all vertebrates). It is also more *ancient* (over 520 million years old) and more *diverse* (tectal vs. pallial, and arising from the different pallial arrangements of birds vs. mammals). Our considerations of the affective consciousness of feelings (Chapter 1) also uphold this conclusion.

## References

Aboitiz, F. & Montiel, J. (2007) Origin and evolution of the vertebrate telencephalon, with special reference to the mammalian neocortex. *Adv. Anat. Embryol. Cell Biol.*, **193**, 1–116.

Allen, T. A. & Fortin, N. J. (2013) The evolution of episodic memory, *Proc. Nat. Acad. Sci. USA*, **110**, 10379–10386.

Alonso, P. D., Milner, A. C., Ketcham, R. A., Cookson, M. J. & Rowe, T. B. (2004) The avian nature of the brain and inner ear of *Archaeopteryx*. *Nature*, **430**, 666–669.

Andres, K. H. & von Düring, M. (1993) Cutaneous and subcutaneous sensory receptors of the hagfish *Myxine glutinosa* with special respect to the trigeminal system. *Cell Tissue Res.*, **274**, 353–366.

Århem, P., Lindahl, B. I. B., Manger, P. R. & Butler, A. B. (2008) On the Origin of Consciousness — Some Amniote Scenarios. In: H. Liljenstrom & P. Arhem, eds. *Consciousness Transitions: Phylogenetic, Ontogenetic, and Physiological Aspects*. San Francisco, CA: Elsevier, pp. 77–96.

Baars, B. J., Franklin, S. & Ramsoy, T. Z. (2013) Global workspace dynamics: Cortical "binding and propagation" enables conscious contents. *Front. Psychol.*, **4**, 200.

Balanoff, A. M., Bever, G. S., Rowe, T. B. & Norell, M. A. (2013) Evolutionary origins of the avian brain. *Nature*, **501**, 93–96.

Benton, M. J. (2014) How birds became birds. *Science*, **345**, 508–509.

Benton, M. J. & Donoghue, P. C. (2007) Paleontological evidence to date the tree of life. *Mol. Biol. Evol.*, **24**, 26–53.

Bi, S., Wang, Y., Guan, J., Sheng, X. & Meng, J. (2014) Three new Jurassic euharamiyidan species reinforce early divergence of mammals. *Nature*, **514**, 579–584.

Bingman, V. P., Salas, C. & Rodriguez, F. (2009) Evolution of the Hippocampus. In: M. D. Binder, N. Hirokawa & U. Windhorst, eds. *Encyclopedia of Neurosciences*. Berlin: Springer, pp. 1356–1361.

Boly, M. & Seth, A. K. (2012) Modes and models in disorders of consciousness science. *Arch. Ital. Biol.*, **150**, 172–184.

Boly, M., Seth, A. K., Wilke, M., Ingmundson, P., Baars, B., Laureys, S., Edelman, D. B. & Tsuchiya, N. (2013) Consciousness in humans and non-human animals: Recent advances and future directions. *Front. Psychol.*, **4**, 625.

Bosne, S. C. D. M. D. (2010) The Eph/ephrin gene family in the European amphioxus: an evo-devo approach, Master's Thesis in Biology. University of Lisbon, Department of Animal Biology.

Bowmaker, J. K. (1998) Evolution of colour vision in vertebrates. *Eye*, **12**, 541–547.

Braun, C. B. (1996) The sensory biology of the living jawless fishes: A phylogenetic assessment. *Brain Behav. Evol.*, **48**, 262–276.

Braun, C. B. & Northcutt, R. G. (1998) Cutaneous Exteroreceptors and Their Innervation in Hagfishes. In: J.M. Jorgensen, J.P. Lomholt, R.E. Weber & H. Malte, eds. *The Biology of Hagfishes*. London: Chapman and Hall, pp. 512–532.

Brodal, P. (2010) *The Central Nervous System*, 4th edn. New York: Oxford University Press.

Broglio, C., Gomez, A., Duran, E., Ocana, F. M., Jiménez-Moya, F., Rodríguez, F. & Salas, C. (2005) Hallmarks of a common forebrain vertebrate plan: specialized pallial areas for spatial, temporal and emotional memory in actinopterygian fish. *Brain Res. Bull.*, **66**, 277–281.

Bullock, T. H. (2002) Biology of brain waves: natural history and evolution of an information-rich sign of activity. *Adv. Electrophysiol. Clin. Pract. Res.*

Bullock, T. H., Moore, J. K. & Fields, R. D. (1984) Evolution of myelin sheaths: Both lamprey and hagfish lack myelin. *Neurosci. Lett.*, **48**, 145–148.

Bürck, M., Friedel, P., Sichert, A. B., Vossen, C. & van Hemmen, J. L. (2010) Optimality in mono-and multisensory map formation. *Biol. Cybern.*, **103**, 1–20.

Burghardt, G. M. (2013) Environmental enrichment and cognitive complexity in reptiles and amphibians: Concepts, review, and implications for captive populations. *Appl. Animal Behav. Sci.*, **147**, 286–298.

Butler, A. B. (2008a) Evolution of brains, cognition, and consciousness. *Brain Res. Bull.* **75**, 442–449.

Butler, A. B. (2008b) Evolution of the thalamus: A morphological and functional review. *Thalamus Relat. Syst.*, **4**, 35–58.

Butler, A. B. & Cotterill, R. M. (2006) Mammalian and avian neuroanatomy and the question of consciousness in birds. *Biol. Bull.*, **211**, 106–127.

Butler, A. B. & Hodos, W. (2005) *Comparative Vertebrate Neuroanatomy: Evolution and Adaptation*, 2nd edn. Hoboken, New Jersey: John Wiley & Sons.

Butler, A. B., Manger, P. R., Lindahl, B. I. B. & Århem, P. (2005) Evolution of the neural basis of consciousness: a bird–mammal comparison. *Bioessays*, **27**, 923–936.

Butler, A. B., Reiner, A. & Karten, H. J. (2011) Evolution of the amniote pallium and the origins of mammalian neocortex. *Ann. NY Acad. Sci.*, **1225**, 14–27.

Candiani, S., Moronti, L., Ramoino, P., Schubert, M. & Pestarino, M. (2012) A neurochemical map of the developing amphioxus nervous system. *BMC Neurosci.*, **13**, 59.

Cardin, J. A., Carlén, M., Meletis, K., Knoblich, U., Zhang, F., Deisseroth, K. ... & Moore, C. I. (2009). Driving fast-spiking cells induces gamma rhythm and controls sensory responses. *Nature*, **459**, 663–667.

Caudill, M. S., Eggebrecht, A. T., Gruberg, E. R. & Wessel, R. (2010) Electrophysiological properties of isthmic neurons in frogs revealed by *in vitro* and *in vivo* studies. *J. Compar. Physiol. A*, **196**, 249–262.

Cinelli, E., Robertson, B., Mutolo, D., Grillner, S., Pantaleo, T. & Bongianni, F. (2013) Neuronal mechanisms of respiratory pattern generation are evolutionary conserved. *J. Neurosci.*, **33**, 9104–9112.

Cohen, J. D. & Castro-Alamancos, M. A. (2010) Neural correlates of active avoidance behavior in superior colliculus. *J. Neurosci.*, **30**, 8502–8511.

Collin, S. P. (2009) Evolution of the Visual System in Fishes. In: M. D. Binder, N. Hirokawa & U. Windhorst, eds. *Encyclopedia of Neurosciences.* Berlin: Springer, pp. 1459–1466.

Cramer, K. S. & Gabriele, M. L. (2014) Axon guidance in the auditory system: Multiple functions of Eph receptors. *Neuroscience,* **277**, 152–162.

Crawford, N. G., Faircloth, B. C., McCormack, J. E., Brumfield, R. T., Winker, K. & Glenn, T. C. (2012) More than 1000 ultraconserved elements provide evidence that turtles are the sister group of archosaurs. *Biol. Lett.,* **8**, 783–786.

Cunningham, S. J., Corfield, J. R., Iwaniuk, A. N., Castro, I., Alley, M. R., Birkhead, T. R. & Parsons, S. (2013) The anatomy of the bill tip of kiwi and associated somatosensory regions of the brain: comparisons with shorebirds. *PloS one,* **8**, e80036.

Damasio, A. R. (2000) *The Feeling of What Happens: Body, Emotion and the Making of Consciousness.* New York: Random House.

Dehaene, S. & Changeux, J. P. (2011) Experimental and theoretical approaches to conscious processing. *Neuron,* **70**, 200–227.

Del Bene, F., Wyart, C., Robles, E., Tran, A., Looger, L., Scott, E. K. & Baier, H. (2010) Filtering of visual information in the tectum by an identified neural circuit. *Science,* **330**, 669–673.

Demski, L. S. (2013) The pallium and mind/behavior relationships in teleost fishes. *Brain Behav. Evol.,* **82**, 31–44.

Dicke, U. & Roth, G. (2009) Evolution of the Visual System in Amphibians. In: M.D. Binder, N. Hirokawa & U. Windhorst, eds. *Encyclopedia of Neurosciences.* Berlin: Springer, pp. 1455–1459.

Di Perri, C., Stender, J., Laureys, S. & Gosseries, O. (2014) Functional neuroanatomy of disorders of consciousness. *Epilepsy Behav.,* **30**, 28–32.

Dudkin, E. & Gruberg, F. (2009) Evolution of Nucleus Isthmi. In: M.D. Binder, N. Hirokawa & U. Windhorst, eds. *Encyclopedia of Neurosciences.* Berlin: Springer, pp. 1258–1262.

Dugas-Ford, J., Rowell, J. J. & Ragsdale, C. W. (2012) Cell-type homologies and the origins of the neocortex. *Proc. Nat. Acad. Sci. USA,* **109**, 16974–16979.

Dutta, A. & Gutfreund, Y. (2014) Saliency mapping in the optic tectum and its relationship to habituation. *Front. Integr. Neurosci.,* **8**, 1.

Edelman, G. M. (1989) *The Remembered Present: A Biological Theory of Consciousness.* New York: Basic Books.

Edelman, G. M. (1992) *Bright Air, Brilliant Fire: On the Matter of the Mind.* New York: Basic Books.

Edelman, D. B., Baars, B. J. & Seth, A. K. (2005) Identifying hallmarks of consciousness in non-mammalian species. *Conscious. Cogn.,* **14**, 169–187.

Edelman, G. M., Gally, J. A. & Baars, B. J. (2011) Biology of consciousness. *Front. Psychol.,* **2**.

Edelman, D. B. & Seth, A. K. (2009) Animal consciousness: a synthetic approach. *Trends Neurosci.,* **32**, 476–484.

Eisthen, H. L. (1997) Evolution of vertebrate olfactory systems. *Brain Behav. Evol.*, **50**, 222–233.

Erwin, D. H. & Valentine, J. W. (2013) *The Cambrian Explosion*. Greenwood Village, Colorado: Roberts & Company.

Feinberg, T. E. (2012). Neuroontology, neurobiological naturalism, & consciousness: a challenge to scientific reduction and a solution. *Phys. Life Rev.*, **9**, 13–34.

Feinberg, T. E. & Mallatt, J. (2013) The evolutionary and genetic origins of consciousness in the Cambrian Period over 500 million years ago. *Front. Psychol.*, **4**, 667.

Feldman, J. (2013) The neural binding problem(s). *Cogn. Neurodyn.*, **7**, 1–11.

Fuller, P., Sherman, D., Pedersen, N. P., Saper, C. B. & Lu, J. (2011) Reassessment of the structural basis of the ascending arousal system. *J. Compar. Neurol.*, **519**, 933–956.

Fuss, T., Bleckmann, H. & Schluessel, V. (2014) Place learning prior to and after telencephalon ablation in bamboo and coral cat sharks (*Chiloscyllium griseum* and *Atelomycterus marmoratus*). *J. Compar. Physiol. A*, **200**, 37–52.

Gershman, S. J., Horvitz, E. J., & Tenenbaum, J. B. (2015) Computational rationality: A converging paradigm for intelligence in brains, minds, and machines. *Science*, **349**, 273–278.

Gill, P. G., Purnell, M. A., Crumpton, N., Brown, K. R., Gostling, N. J., Stampanoni, M. & Rayfield, E. J. (2014) Dietary specializations and diversity in feeding ecology of the earliest stem mammals. *Nature*, **512**, 303–305.

Glover, J. C. & Fritzsch, B. (2009) Brains of Primitive Chordates. In: M.D. Binder, N. Hirokawa & U. Windhorst, eds. *Encyclopedia of Neurosciences*. Berlin: Springer, pp. 439–448.

Goddard, C. A., Sridharan, D., Huguenard, J. R. & Knudsen, E. I. (2012) Gamma oscillations are generated locally in an attention-related midbrain network. *Neuron*, **73**, 567–580.

Godefroit, P., Cau, A., Dong-Yu, H., Escuillié, F., Wenhao, W. & Dyke, G. (2013) A Jurassic avialan dinosaur from China resolves the early phylogenetic history of birds. *Nature*, **498**, 359–362.

Gould, S. J. (1977) *Ontogeny and Phylogeny*. Cambridge, MA, Belknap.

Graham, B. J. & Northmore, D. P. (2007) A spiking neural network model of midbrain visuomotor mechanisms that avoids objects by estimating size and distance monocularly. *Neurocomputing*, **70**, 1983–1987.

Green, W. W., Basilious, A., Dubuc, R. & Zielinski, B. S. (2013) The neuroanatomical organization of projection neurons associated with different olfactory bulb pathways in the sea lamprey. *Petromyzon marinus, PloS one*, **8**, e69525.

Grillner, S., Robertson, B. & Stephenson-Jones, M. (2013) The evolutionary origin of the vertebrate basal ganglia and its role in action selection. *J. Physiol. (Lond.)*, **591**, 5425–5431.

Gruberg, E., Dudkin, E., Wang, Y., Marín, G., Salas, C., Sentis, E. & Udin, S. (2006) Influencing and interpreting visual input: the role of a visual feedback system. *J. Neurosci.*, **26**, 10368–10371.

Guirado, S. & Davila, J. C. (2009) Evolution of the Optic Tectum in Amniotes. In: M.D. Binder, N. Hirokawa & U. Windhorst, eds. *Encyclopedia of Neurosciences*. Berlin: Springer, pp. 1375–1380.

Gutfreund, Y. (2012) Stimulus-specific adaptation, habituation and change detection in the gaze control system. *Biol. Cybern.*, **106**, 657–668.

Hall, M. I., Kamilar, J. M. & Kirk, E. C. (2012) Eye shape and the nocturnal bottleneck of mammals. *Proc. R. Soc. B: Biol. Sci.*, **279**, 4962–4968.

Häming, D., Simoes-Costa, M., Uy, B., Valencia, J., Sauka-Spengler, T. & Bronner-Fraser, M. (2011) Expression of sympathetic nervous system genes in lamprey suggests their recruitment for specification of a new vertebrate feature. *PloS one*, **6**, e26543.

Hardisty, M. W. (1979) *Biology of the Cyclostomes*. London: Chapman and Hall.

Hofmann, M. H. & Northcutt, R. G. (2012). Forebrain organization in elasmobranchs. *Brain, Behavior and Evolution*, **80**, 142–151.

Holland, N. D. & Yu Jr, K. (2002) Epidermal receptor development and sensory pathways in vitally stained amphioxus (*Branchiostoma floridae*). *Acta Zool.*, **83**, 309–319.

Huesa, G., Anadon, R., Folgueira, M. & Yanez, J. (2009) Evolution of the Pallium in Fishes. In: M.D. Binder, N. Hirokawa & U. Windhorst, eds. *Encyclopedia of Neurosciences*. Berlin: Springer, pp. 1400–1404.

Iwahori, N., Kawawaki, T. & Baba, J. (1998) Neuronal organization of the optic tectum in the river lamprey, *Lampetra japonica*: A Golgi study. *J. fur. Hirnforschung*, **39**, 409–424.

Jacobs, L. F. (2012) From chemotaxis to the cognitive map: the function of olfaction. *Proc. Nat. Acad. Sci. USA*, **109**, 10693–10700.

Jarvis, E. D. (2009) Evolution of the Pallium in Birds and Reptiles. In: M. D. Binder, N. Hirokawa & U. Windhorst, eds. *Encyclopedia of Neurosciences*. Berlin: Springer, pp. 1390–1400.

Jarvis, E. D., Yu, J., Rivas, M. V., Horita, H., Feenders, G., Whitney, O. & Wada, K. (2013) Global view of the functional molecular organization of the avian cerebrum: mirror images and functional columns. *J. Compar. Neurol.*, **521**, 3614–3665.

Jbabdi, S., Sotiropoulos, S. N. & Behrens, T. E. (2013) The topographic connectome. *Curr. Opin. Neurobiol.*, **23**, 207–215.

Jerison, H. J. (2009) Evolution and Brain-Body Allometry. In: M. D. Binder, N. Hirokawa & U. Windhorst, eds. *Encyclopedia of Neurosciences*. Berlin: Springer, pp. 1161–1165.

Jones, M. R., Grillner, S. & Robertson, B. (2009) Selective projection patterns from subtypes of retinal ganglion cells to tectum and pretectum: distribution and relation to behavior. *J. Compar. Neurol.*, **517**, 257–275.

Kaas, J. H. (1997) Topographic maps are fundamental to sensory processing. *Brain Res. Bull.*, **44**, 107–112.

Kaas, J. H. (2011) Neocortex in early mammals and its subsequent variations. *Ann. NY Acad. Sci.*, **1225**, 28–36.

Kalman, M. (2009) Evolution of the Brain at Reptile-bird Transition. In: M. D. Binder, N. Hirokawa & U. Windhorst, eds. *Encyclopedia of Neurosciences.* Berlin: Springer, pp. 1305–1312.

Kandel, E. R., Schwartz, J. H., Jessell, T. M., Siegelbaum, S. A. & Hudspeth, A. J. (2012) *Principles of Neural Science*, 5th edn. New York: McGraw Hill.

Kardong, K. (2012) *Vertebrate Comparative Anatomy, Function, Evolution.* New York: McGraw Hill.

Karten, H. J. (2013) Neocortical evolution: neuronal circuits arise independently of lamination. *Curr. Biol.* **23**, R12–R15.

Keary, N., Voss, J., Lehmann, K., Bischof, H. J. & Löwel, S. (2010) Optical imaging of retinotopic maps in a small songbird, the zebra finch. *PloS one*, **5**, e11912.

Kielan-Jaworowska, Z. (2013) *In Pursuit of Early Mammals.* Indiana University Press.

Kielan-Jaworowska, Z., Cifelli, R. L., Cifelli, R. & Luo, Z. X. (2013) *Mammals from the Age of Dinosaurs: Origins, Evolution, and Structure.* Columbia University Press.

Knöll, B. & Drescher, U. (2002) Ephrin-As as receptors in topographic projections. *Trends Neurosci.*, **25**, 145–149.

Knudsen, E. I. (2011) Control from below: the role of a midbrain network in spatial attention. *Eur. J. Neurosci.*, **33**, 1961–1972.

Krauzlis, R. J., Lovejoy, L. P. & Zénon, A. (2013) Superior colliculus and visual spatial attention. *Ann. Rev. Neurosci.*, **36**, 165–182.

Lacalli, T. C. (2008) Basic features of the ancestral chordate brain: A protochordate perspective. *Brain Res. Bull.*, **75**, 319–323.

Lee, M. S., Cau, A., Naish, D. & Dyke, G. J. (2014) Sustained miniaturization and anatomical innovation in the dinosaurian ancestors of birds. *Science*, **345**, 562–566.

Lee, S. H. & Dan, Y. (2012) Neuromodulation of brain states. *Neuron*, **76**, 209–222.

Li, W., Lopez, L., Osher, J., Howard, J. D., Parrish, T. B. & Gottfried, J. A. (2010) Right orbitofrontal cortex mediates conscious olfactory perception. *Psychol. Sci.*, **21**, 1454–1463.

Luo, Z. X., Yuan, C. X., Meng, Q. J. & Ji, Q. (2011) A Jurassic eutherian mammal and divergence of marsupials and placentals. *Nature*, **476**, 442–445.

Magosso, E. (2010). Integrating information from vision and touch: a neural network modeling study. *Inf. Technol. Biomed. IEEE Trans.*, **14**, 598–612.

Maisey, J. G. (1996) *Discovering Fossil Fishes*. New York: Holt.

Mallatt, J. (1997) Hagfish do not resemble ancestral vertebrates. *J. Morphol.*, **232**, 293.

Manger, P. R. (2009) Evolution of the Reticular Formation. In: M. D. Binder, N. Hirokawa & U. Windhorst, eds. *Encyclopedia of Neurosciences.* Berlin: Springer, pp. 1413–1416.

Marín, G., Salas, C., Sentis, E., Rojas, X., Letelier, J. C. & Mpodozis, J. (2007) A cholinergic gating mechanism controlled by competitive interactions in the optic tectum of the pigeon. *J. Neurosci.*, **27**, 8112–8121.

Maynard, J. (2013) www.itechpost.com/articles/9940/20130529/archaeopteryx-bird-first.htm.

Medina, L. (2009) Evolution and Embryological Development of Forebrain. In: M. D. Binder, N. Hirokawa & U. Windhorst, eds. *Encyclopedia of Neurosciences.* Berlin: Springer, pp. 1172–1192.

Melloni, L., Molina, C., Pena, M., Torres, D., Singer, W. & Rodriguez, E. (2007) Synchronization of neural activity across cortical areas correlates with conscious perception. *J. Neurosci.*, **27**, 2858–2865.

Miller, E. K. & Buschman, T. J. (2013) Cortical circuits for the control of attention. *Curr. Opin. Neurobiol.* **23**, 216–222.

Min, B. K. (2010) A thalamic reticular networking model of consciousness. *Theor. Biol. Med. Model,* **7**, 10.

Moret, F., Guilland, J. C., Coudouel, S., Rochette, L. & Vernier, P. (2004) Distribution of tyrosine hydroxylase, dopamine, and serotonin in the central nervous system of amphioxus (*Branchiostoma lanceolatum*): implications for the evolution of catecholamine systems in vertebrates. *J. Compar. Neurol.*, **468**, 135–150.

Mori, K., Manabe, H., Narikiyo, K. & Onisawa, N. (2013) Olfactory consciousness and gamma oscillation couplings across the olfactory bulb, olfactory cortex, and orbitofrontal cortex. *Front. Psychol.*, **4**, 743.

Mueller, T. (2012) What is the thalamus in zebrafish? *Front. Neurosci.*, **6**, 64.

Müller, C. M. & Leppelsack, H. J. (1985) Feature extraction and tonotopic organization in the avian auditory forebrain. *Exp. Brain Res.*, **59**, 587–599.

Murakami, Y. & Kuratani, S. (2008) Brain segmentation and trigeminal projections in the lamprey; with reference to vertebrate brain evolution. *Brain Res. Bull.*, **75**, 218–224.

Nadel, L., Hupbach, A., Gomez, R. & Newman-Smith, K. (2012) Memory formation, consolidation and transformation. *Neurosci. Biobehav. Rev.*, **36**, 1640–1645.

Nieuwenhuys, R. (1996) The greater limbic system, the emotional motor system and the brain. *Prog. Brain Res.*, **107**, 551–580.

Nieuwenhuys, R. & Nicholson, C. (1998) Lampreys, Petromyzontidae, Vol. 1. In: R. Nieuwenhuys, H. J. ten Donkelaar & C. Nicholson, eds. *The Central Nervous System of Vertebrates.* Berlin: Springer.

Nieuwenhuys, R., ten Donkelaar, H. J., Nicholson, C., Smeets, W. J. A. J. & Wicht, H. (1998) *The Central Nervous System of Vertebrates*, Vol. 1. Berlin: Springer.

Nomura, T., Kawaguchi, M., Ono, K. & Murakami, Y. (2013). Reptiles: A new model for brain evo-devo research. *J. Exp. Zool. Part B: Mol. Develop. Evol.*, **320**, 57–73.

North, H. A., Karim, A., Jacquin, M. F. & Donoghue, M. J. (2010) EphA4 is necessary for spatially selective peripheral somatosensory topography. *Develop. Dyn.*, **239**, 630–638.

Northcutt, R. G. (1996) The agnathan ark: The origin of craniate brains. *Brain Behav. Evol.*, **48**, 237–247.

Northcutt, R. G. (2002) Understanding vertebrate brain evolution. *Integr. Comp. Biol.*, **42**, 743–756.

Northcutt, R. G. (2013) Variation in reptilian brains and cognition. *Brain Behav. Evol.*, **82**, 45–54.

Northcutt, R. G. & Puzdrowski, R. L. (1988) Projections of the olfactory bulb and nervus terminalis in the silver lamprey. *Brain Behav. Evol.*, **32**, 96–107.

Northcutt, R. G. & Ronan, M. (1992) Afferent and efferent connections of the bullfrog medial pallium. *Brain Behav. Evol.*, **40**, 1–16.

Northcutt, R. G. & Wicht, H. (1997) Afferent and efferent connections of the lateral and medial pallia of the silver lamprey. *Brain Behav. Evol.*, **49**, 1–19.

Northmore, D. (2011) The Optic Tectum. In: A. Farrell, ed. *The Encyclopedia of Fish Physiology: From Genome to Environment*. San Diego: Academic Press, pp. 131–142.

Northmore, D. P. & Gallagher, S. P. (2003) Functional relationship between nucleus isthmi and tectum in teleosts: Synchrony but no topography. *Visual Neurosci.*, **20**, 335–348.

Nunez, P. L. & Srinivasan, R. (2010) Scale and frequency chauvinism in brain dynamics: Too much emphasis on gamma band oscillations. *Brain Struct. Funct.*, **215**, 67–71.

O'Connell, L. A. & Hofmann, H. A. (2011) The vertebrate mesolimbic reward system and social behavior network: A comparative synthesis. *J. Compar. Neurol.*, **519**, 3599–3639.

Pepperberg, I. (2009) *Alex and Me*. New York: HarperCollins.

Pombal, M. A., Marín, O. & González, A. (2001) Distribution of choline acetyltransferase-immunoreactive structures in the lamprey brain. *J. Compar. Neurol.*, **431**, 105–126.

Portugues, R. & Engert, F. (2009) The neural basis of visual behaviors in the larval zebrafish. *Curr. Opin. Neurobiol.*, **19**, 644–647.

Prechtl, J. C., von der Emde, G., Wolfart, J., Karamürsel, S., Akoev, G. N., Andrianov, Y. N. & Bullock, T. H. (1998) Sensory processing in the pallium of a mormyrid fish. *J. Neurosci.*, **18**, 7381–7393.

Quinn, B. T., Carlson, C., Doyle, W., Cash, S. S., Devinsky, O., Spence, C., Halgren, E. & Thesen, T. (2014) Intracranial cortical responses during visual–tactile integration in humans. *J. Neurosci.*, **34**, 171–181.

Rattenborg, N. C. & Martinez-Gonzalez, D. (2011) A bird-brain view of episodic memory. *Behav. Brain Res.*, **222**, 236–245.

Repérant, J., Ward, R., Médina, M., Kenigfest, N. B., Rio, J. P., Miceli, D. & Jay, B. (2009) Synaptic circuitry in the retinorecipient layers of the optic tectum of the lamprey (*Lampetra fluviatilis*). A combined hodological, GABA and glutamate immunocytochemical study. *Brain Struct. Funct.*, **213**, 395–422.

Revonsuo, A. (2006) *Inner Presence: Consciousness as a Biological Phenomenon*. MIT Press.

Ribary, U. (2005) Dynamics of thalamo-cortical network oscillations and human perception. *Progr. Brain Res.*, **150**, 127–142.

Risi, S. & Stanley, K. O. (2014) Guided self-organization in indirectly encoded and evolving topographic maps, In: *Proceedings of the Genetic and Evolutionary Computation Conference* (GECCO 2014). New York, NY: ACM.

Robertson, B., Saitoh, K., Ménard, A. & Grillner, S. (2006) Afferents of the lamprey optic tectum with special reference to the GABA input: combined tracing and immunohistochemical study. *J. Compar. Neurol.*, **499**, 106–119.

Robertson, B., Auclair, F., Ménard, A., Grillner, S. & Dubuc, R. (2007) GABA distribution in lamprey is phylogenetically conserved. *J. Compar. Neurol.*, **503**, 47–63.

Robertson, B., Kardamakis, A., Capantini, L., Perez-Fernandez, J., Suryanarayana, S. M., Wallen, P., Stephenson-Jones, M. & Grillner, S. (2014) The lamprey blueprint of the mammalian nervous system. In: G. Holstege, C. M. Beers & H. H. Subramanian, eds. *Breathing, Emotion and Evolution*, Oxford: Elsevier, pp. 337–350.

Rolls, E. T. (2012) Multisensory Neuronal Convergence of Taste, Somatosensory, Visual, Olfactory, and Auditory Inputs. In: B. E. Stein, ed. *New Handbook of Multisensory Processes*, Cambridge, MA: MIT Press, pp. 311–331.

Rose, J. D., Arlinghaus, R., Cooke, S. J., Diggles, B. K., Sawynok, W., Stevens, E. D. & Wynne, C. D. L. (2014) Can fish really feel pain? *Fish and Fisheries*, **15**, 97–133.

Rovainen, C. M. (1979) Neurobiology of lampreys. *Physiol. Rev.*, **59**, 1007–1077.

Rowe, T. B., Macrini, T. E. & Luo, Z. X. (2011) Fossil evidence on origin of the mammalian brain. *Science*, **332**, 955–957.

Saidel, W. M. (2009) Evolution of the Optic Tectum in Anamniotes. In: M. D. Binder, N. Hirokawa & U. Windhorst, eds. *Encyclopedia of Neurosciences*. Berlin: Springer, pp. 1380–1387.

Saitoh, K., Ménard, A. & Grillner, S. (2007) Tectal control of locomotion, steering, and eye movements in lamprey. *J. Neurophysiol.*, **97**, 3093–3108.

Sarvestani, I. K., Kozlov, A., Harischandra, N., Grillner, S. & Ekeberg, Ö. (2013) A computational model of visually guided locomotion in lamprey. *Biol. Cybernet.*, **107**, 497–512.

Schiff, N. D. (2008) Central thalamic contributions to arousal regulation and neurological disorders of consciousness. *Ann. NY Acad. Sci.*, **1129**, 105–118.

Schlosser, G. (2005) Evolutionary origins of vertebrate placodes: insights from developmental studies and from comparisons with other deuterostomes. *J. Exp. Zool. Part B: Mol. Develop. Evol.*, **304**, 347–399.

Schomburg, E. W., Fernández-Ruiz, A., Mizuseki, K., Berényi, A., Anastassiou, C. A., Koch, C., & Buzsáki, G. (2014) Theta phase segregation of input-specific gamma patterns in entorhinal-hippocampal networks. *Neuron*, **84**, 1–16.

Schluessel, V. & Bleckmann, H. (2005) Spatial memory and orientation strategies in the elasmobranch. *Potamotrygon motoro, J. Compar. Physiol. A*, **191**, 695–706.

Schuelert, N. & Dicke, U. (2005) Dynamic response properties of visual neurons and context-dependent surround effects on receptive fields in the tectum of the salamander *Plethodon shermani. Neuroscience*, **134**, 617–632.

Sereno, M. I. & Allman, J. M. (1991) Cortical Visual Areas in Mammals. In: A. G. Leventhal, ed. *The Neural Basis of Visual Function*. London: Macmillan, pp. 160–172.

Seth, A. K., Baars, B. J. & Edelman, D. B. (2005) Criteria for consciousness in humans and other mammals. *Conscious. Cogn.*, **14**, 119–139.

Shanahan, M., Bingman, V. P., Shimizu, T., Wild, M. & Güntürkün, O. (2013) Large-scale network organization in the avian forebrain: A connectivity matrix and theoretical analysis. *Front. Comput. Neurosci.*, **7**, 89.

Shepherd, G. M. (2007) Perspectives on olfactory processing, conscious perception, and orbitofrontal cortex. *Ann. NY Acad. Sci.*, **1121**, 87–101.

Singer, W. (2001) Consciousness and the binding problem. *Ann. NY Acad. Sci.*, **929**, 123–146.

Sommer, L. (2013) Patterning and Cell Type Specification in the Developing CNS and PNS. In: J. Rubenstein & P. Rakic, eds. *Specification of Neural Crest- and Placode-derived Neurons*. New York: Academic Press, pp. 385–400.

Sridharan, D., Boahen, K. & Knudsen, E. I. (2011) Space coding by gamma oscillations in the barn owl optic tectum, *J. Neurophysiol.*, **105**, 2005–2017.

Sridharan, D., Schwarz, J. S. & Knudsen, E. I. (2014) Selective attention in birds. *Curr. Biol.*, **24**, R510–R513.

Stein, B. E. & Meredith, M. A. (1993) *The Merging of the Senses*. Cambridge, MA: The MIT Press.

Stephenson-Jones, M., Kardamakis, A. A., Robertson, B. & Grillner, S. (2013) Independent circuits in the basal ganglia for the evaluation and selection of actions. *Proc. Nat. Acad. Sci. USA*, **110**, E3670–E3679.

Tinsley, C. J. (2008) Using topographic networks to build a representation of consciousness. *Biosystems*, **92**, 29–41.

Triplett, J. W. & Feldheim, D. A. (2012, February) Eph and ephrin signaling in the formation of topographic maps. In *Seminars in Cell and Developmental Biology*, Vol. 23. Waltham, MA: Academic Press, pp. 7–15.

Uhlhaas, P., Pipa, G., Lima, B., Melloni, L., Neuenschwander, S., Nikolić, D. & Singer, W. (2009) Neural synchrony in cortical networks: history, concept and current status. *Front. Integr. Neurosci.*, **3**, 17.

Ursino, M., Magosso, E. & Cuppini, C. (2009) Recognition of abstract objects via neural oscillators: interaction among topological organization, associative memory and gamma band synchronization. *IEEE Trans. Neural Netw.*, **20**, 316–335.

van Gaal, S. & Lamme, V. A. (2012) Unconscious high-level information processing implication for neurobiological theories of consciousness. *Neuroscientist*, **18**, 287–301.

von Düring, M. & Andres, K. H. (1998) Skin Sensory Organs in the Atlantic Hagfish *Myxine glutinosa*. In: J. M. Jorgensen, J. P. Lomholt, R. E. Weber & H. Malte, eds. *The Biology of Hagfishes*. London: Chapman and Hall, pp. 499–511.

Vopalensky, P., Pergner, J., Liegertova, M., Benito-Gutierrez, E., Arendt, D. & Kozmik, Z. (2012) Molecular analysis of the amphioxus frontal eye unravels the evolutionary origin of the retina and pigment cells of the vertebrate eye. *Proc. Nat. Acad. Sci. USA*, **109**, 15383–15388.

Wang, Y., Brzozowska-Prechtl, A. & Karten, H. J. (2010) Laminar and columnar auditory cortex in avian brain. *Proc. Natl. Acad. Sci. USA*, **107**, 12676–12681.

Watanabe, A., Hirano, S. & Murakami, Y. (2008) Development of the lamprey central nervous system, with reference to vertebrate evolution. *Zoolog. Sci.*, **25**, 1020–1027.

Wicht, H. (1996) The brains of lampreys and hagfishes: Characteristics, characters, and comparisons. *Brain Behav. Evol.*, **48**, 248–261.

Wicht, H. & Lacalli, T. C. (2005) The nervous system of amphioxus: Structure, development, and evolutionary significance. *Canad. J. Zool.*, **83**, 122–150.

Wilczynski, W. (2009) Evolution of the Brain in Amphibians. In: M. D. Binder, N. Hirokawa & U. Windhorst, eds. *Encyclopedia of Neurosciences*. Berlin: Springer, pp. 1301–1305.

Wild, J. M. (2009) Evolution of the Wulst. In: M. D. Binder, N. Hirokawa & U. Windhorst, eds. *Encyclopedia of Neurosciences*. Berlin: Springer, pp. 1475–1478.

Wild, J. M., Kubke, M. F. & Peña, J. L. (2008) A pathway for predation in the brain of the barn owl (*Tyto alba*): projections of the gracile nucleus to the "claw area" of the rostral wulst via the dorsal thalamus. *J. Compar. Neurol.*, **509**, 156–166.

Wulliman, M. F. & Vernier, P (2009a) Evolution of the Brain in Fishes. In: M. D. Binder, N. Hirokawa & U. Windhorst, eds. *Encyclopedia of Neurosciences*. Berlin: Springer, pp. 1318–1326.

Wulliman, M. F. & Vernier, P (2009b) Evolution of the Telencephalon in Anamniotes. In: M. D. Binder, N. Hirokawa & U. Windhorst, eds. *Encyclopedia of Neurosciences*. Berlin: Springer, pp. 1423–1431.

Yopak, K. E. (2012) Neuroecology of cartilaginous fishes: The functional implications of brain scaling. *J. Fish Biol.*, **80**, 1968–2023.

Yopak, K., Hofmann, M. H. & Northcutt, R. G. (2012) Forebrain organization in elasmobranchs. *Brain Behav. Evol.*, **80**, 142–151.

Zelenitsky, D. K., Therrien, F., Ridgely, R. C., McGee, A. R. & Witmer, L. M. (2011) Evolution of olfaction in non-avian theropod dinosaurs and birds. *Proc. R. Soc. B: Biol. Sci.*, **278**, 3625–3634.

Zmigrod, S. & Hommel, B. (2013) Feature integration across multimodal perception and action: A review. *Multisens. Res.*, **26**, 143–157.

Biophysics of Consciousness: A Foundational Approach
R. R. Poznanski, J. A. Tuszynski and T. E. Feinberg
Copyright © 2017 World Scientific, Singapore.

# 3

# HOW LANGUAGE EVOLUTION RESHAPED HUMAN CONSCIOUSNESS

*Michael A. Arbib*

*The ABLE Project: Action, Brain, Language, Evolution*
*University of Southern California*
*Los Angeles CA 90089-2520, USA*

**Abstract**

This chapter discusses the view that human consciousness may share aspects of "animal awareness" with other species, but has its unique form because humans possess language. Four ingredients of a theory on the evolution of human consciousness are offered: the view that a précis of intended activity is necessarily formed in the brain of a human that communicates in a human way; the notion that this defines a "communication plexus" essential to a particularly human form of consciousness; the mirror system hypothesis for the evolution of the human language-ready brain, and a view of the co-evolution of subtleties of emotion atop more basic systems for motivation such as thirst, hunger, fear and mating.

*Keywords*: Consciousness; animal awareness; language; evolution; brain; communication plexus; mirror system hypothesis; language-ready brain; emotion; motivation.

## 3.1. Introduction

The "core theory" here is what I will call the A & H theory — the account by Mary Hesse and myself (Arbib, 1985; Arbib & Hesse, 1986) of how human consciousness evolved in a way that gave a central but not exclusive role to language in our conscious experience. (When pain dominates our consciousness, the word "pain" is inessential.) A subsequent paper (Arbib, 2001) offered a first attempt to integrate this theory with the mirror system hypothesis of the evolution of the language-ready brain (Arbib & Rizzolatti, 1997; Rizzolatti & Arbib, 1998), a hypothesis which has been modified and expanded across the years. This paper reproduces, edits and re-organizes portions of the 2001 account, while also paying more attention to the role of emotion in consciousness, assessing the relevance of comparative accounts of monkeys, apes and humans (comparative neuroprimatology), and briefly relating our approach to other brain-related accounts of consciousness.

Section 2 contrasts brain states from mental states, stressing that the mental state is a gross abstraction of the brain state and is at *approximately* the same level as language in which a mental state may *or may not* be expressible in words, while the level of detail that we ascribe to a mental state is far coarser than a brain state. This serves as the basis for expressing our view of what an account of consciousness has to explain.

### 3.2. Neurology and Consciousness

One of the definitions of *consciousness* in the Oxford English Dictionary is: "The ... capacity from which awareness of thought, feeling, and volition and of the external world arises; the exercise of this. ... [Also] the aspect of the mind made up of operations which are known to the subject." We share, I suggest, some aspects at least of awareness of feelings concerning the external world with other species (perhaps especially with mammals) — I call this "animal awareness" — but humans are unique in that such awareness is so often accompanied by awareness of thought, volition and at least a folk theory of the mind. Of particular interest, here, is the way our conscious experience as humans is accompanied by a "running commentary" which makes language so much a part of our consciousness and of our ability to share at least some aspects of it with others. The etymological basis is the Latin *con + scio* — "to know *with*" and I suggest that it is the ability to share our experience *with others* that is the defining feature of consciousness, even though it remains active when we are alone. Dreaming would seem to be a further etiolation of consciousness — as some processes that support our conscious interaction with the physical and social world may continue even when our sensory and motor processes are in abeyance.

The prime aim of this paper is to place a theory of consciousness from the 1980s (Arbib, 1985; Arbib & Hesse, 1986) in the perspective offered by my recent thinking about language and human emotion in an evolutionary perspective strongly influenced by neuroscience. But first, I offer a few examples from the neurological literature to suggest some subtleties in linking consciousness to the brain. First is the phenomenon of *blindsight* which demonstrates that neural activity can effectively link perception and action without intervention of consciousness. Neurologists long held that a monkey (or human) without a visual cortex was blind. However, Humphrey (1970) argued that a monkey without visual cortex should have at least as much visual ability as a frog, since the role of tectum in directing whole body movements in frog (Lettvin *et al.*, 1959) is analogous to the role of superior colliculus (a mammalian midbrain [sub-cortical] homologue of tectum) in directing orienting movements in mammals. After two years of training, a monkey without visual cortex was able to use visual cues to grab at moving objects, and to use changes in luminance — such as an open door or obstacles with high luminance contrast from the background — for navigation, even though delicate processes of pattern recognition were never regained. Moreover, it was discovered that humans without visual cortex could also "see" in this action-oriented sense — but, remarkably, they were not conscious of the visual stimuli for their actions (Weiskrantz, 1974). As Stoerig and Cowey (1997) note: In human and monkey, destruction of the optic radiations and/or the primary visual cortex is characterized by an absence of any conscious vision, yet visual stimuli may nonetheless be processed. This unconscious vision includes neuroendocrine, reflexive, indirect, and forced-choice responses. In a provocative paper, Merker (2007a, 2007b) argues for "consciousness without a cerebral cortex," with a significant component of the neural underpinnings of consciousness located subcortically. However, the two statements are not equivalent. It is clear that cortical and subcortical structures work in concert in most behaviors, and that various sensory and motor functions survive limited cortical damage while others (dependent on the size and location of the lesion) are lost. Decerebration can still leave various forms of reactivity available, but there is no evidence that consciousness in the sense of concern here can survive this loss.

In Chapters 1 and 2 of this book, "Neurobiological Naturalism," and "The Evolutionary Origins of Consciousness," Todd Feinberg and Jon Mallatt focus on the evolution of "primary, phenomenal, sensory consciousness in vertebrate animals," offering a model based on neurobiological naturalism. They propose that such consciousness first appeared in the earliest vertebrates from mental reconstructions at the top of topographically organized neural hierarchies in the optic tectum of the midbrain (for vision and other senses) and in the pallium of the forebrain (for smell perception). They then argue that consciousness leaped forward independently in the first mammals and then in the first birds, to be generated by an expanding dorsal pallium (cerebral cortex). However, Humphrey's data on "What the Frog's Eye Tells the Monkey's Brain" and the data on blindsight in human and monkey leads me to take a different focus on consciousness (I will be more specific below) than any notion of consciousness that could be ascribed to an "early vertebrate," let alone a modern frog. Note, too, that a person with blindsight is indeed conscious — it is just that they have no conscious access to visual perception. The point, though, is that being able to react to stimuli in the environment must be distinguished from being conscious of such stimuli.

Further clinical studies of human patients with intact visual cortex showed a double dissociation between the "declarative" ability to communicate the size of an object, whether verbally or by pantomime, and the "procedural" ability to act upon objects. Goodale *et al.* (1991) studied a patient (DF) who had developed a profound visual form agnosia following a bilateral lesion of the so-called ventral pathway from primary visual cortex to temporal cortex, even though the dorsal pathway from occipital lobe toward the parietal lobe appeared to be intact. When the patient was asked to indicate the width of any one of a set of blocks either verbally or by means of her index finger and thumb, her response bore no relationship to the dimensions of the object and showed considerable trial to trial variability. Yet, when she was asked simply to reach out and pick up the block, the peak aperture between her index finger and thumb (prior to contact with the object) changed systematically with the width of the object, as in normal controls. A similar dissociation was seen in her responses to the orientation of stimuli. In other words, DF could preshape her hand accurately, even though she appeared to have no conscious appreciation (either verbal or by pantomime) of the visual parameters that guided the preshape.

Jeannerod *et al.* (1994) reported the case of a woman (AT) who was the "opposite" of DF. AT had a lesion of the occipito-parietal region that interrupted the dorsal route of visual processing, but left the inferotemporal lobe and the ventral pathway intact. AT could verbalize the diameter of a cylinder and could use her hand to pantomime its size, but could not preshape appropriately when asked to grasp it. Instead of an adaptive preshape, she would open her hand to its fullest, and only began to close her hand when the cylinder hit the "web" between index finger and thumb. But there was a surprise! When the stimulus used for the grasp was not a cylinder (for which the "semantics" contains no information about expected size), but rather a familiar object — such as a reel of thread, or a lipstick — for which the "usual" size is part of the subject's knowledge, AT showed a relatively adaptive preshape. This suggests that the ventral pathway provides the parietal areas with "default values" of action-related parameters for such objects, i.e., values which can serve in place of actual sensory data to, e.g., represent the approximate size of a known object to help the parieto-frontal system.

We thus see that consciousness is not a direct property of having neurons of a particular structure or complexity, because the visual data can be represented in two networks — the dorsal and ventral visual pathways — of comparable neural complexity, yet be accessible to consciousness only when the ventral network is intact. On this account, it is not the details of morphology or neurochemistry that distinguish the neurons of a "conscious" from an "unconscious" mode of operation so much as the larger circuits and systems of which they are a part. Some might argue that the blindsight patient or DF is perfectly conscious of the visual cues they use to guide action, but either cannot or will not give verbal or pantomimed expression to their consciousness, but I discount this possibility and suggest that the above examples support a critique of some of the other views of consciousness presented in this volume. Specifically, since there are no data whatsoever suggesting that the damaged portions of DF's brain contain microtubules different from those of the regions that support successful "non-declarative" hand movements, there seems to be no evidence for the role of microtubules in consciousness (see Chapter 14). In a related review paper, Hameroff & Penrose (2014) recall their proposal from the 1990's that consciousness depends on biologically orchestrated coherent quantum processes in collections of microtubules within brain neurons. Dramatically, they claim that their theory "establishes a connection between the brain's biomolecular processes and the basic structure of the universe" and that "consciousness plays an intrinsic role in the universe," but it is clear that their notion of consciousness is very far removed from that espoused by Feinberg and Mallatt, let alone the effort here to understand the sense in which human consciousness is far removed from that (if any) of the frog.

Natural selection can operate on the macromolecular building blocks of cells, on crucial cellular subsystems, and on the morphology of cells themselves, as well as the connectivity of these cells and their formation into diverse nuclei. What is selected about a subsystem, then, may be the impact of some change on a larger system or on a smaller detail, rather than the immediate change in the subsystem itself. Moreover, we understand that the genetic code may not specify adult forms so much as the processes of self-organization in cell-assemblies which can yield "normal" connectivity for the adult raised in a normal environment. Further, the environment which fosters adaptive self-organization may be as much social as physical in nature.

The manifold details here are beyond the scope of this chapter. Rather, my aim in this chapter is not to seek a magic transition from totally non-conscious other species to conscious humans, but rather to analyze how differing styles of behavior and social interaction may support an evolutionary path which renders plausible the emergence of consciousness in its human form. Given this, I consider consciousness to range from an almost reflex awareness of basic motor, sensory (which may include awareness of subtle social cues), and motivational states (so that even "simple" mammals may be aware of the difference between feeling maternal and feeling enraged) up to the subtle human consciousness that slips in and out of elaborate verbal arguments. I will give particular attention to how human development depended on diverse forms of language and communication, thus leaving room to appreciate the role of culture and bio-cultural events in the development of human consciousness.

## 3.3. Brain States, Mental States, and Consciousness

To move forward, we need to clarify the distinction between brain states and mental states. A human brain may be characterized in terms of hundreds of regions which may be distinguished, for example, by the types of neurons they contain, the chemical properties of those neurons, the arrangement of the neurons into layers or other structures, as well as by the patterns of connections they may have with receptors, effectors, and other regions of the brain. Each region in turn may contain millions, or even hundreds of millions of neurons. Current models of the brain focus, in particular, on two properties of neurons in a network:

(i) the "output" of each neuron (the pattern of electric signaling on its *axon*, the output fiber that may branch again and again to connect it to hundreds or thousands of other neurons).

(ii) the "strength" or "weight" (plus or minus, excitatory or inhibitory) of each *synapse*, i.e., each point where the axon of one neuron impinges upon another neuron. This *synaptic weight* measures the extent to which the signal in the first neuron's axon will affect the activity of the second neuron.

A great deal of work on learning and memory holds that the learning process changes the synaptic weights of the brain in such a way that on a later occasion the activity of the network will be changed to allow recall of a specific memory or exhibition of an acquired skill. The upshot of all this it that we can describe the state of the brain at two levels (and this is still only a partial approximation — for example, some synapses may play a neuromodulatory role, rather than being excitatory or inhibitory):

(i) *The activity state:* The current output of each individual neuron. The dynamic interaction of these "firing patterns" determine the brain's "computations" that underlie perception, thought and action, including those processes that in humans involve language.
(ii) *The synaptic state:* The current synaptic weight of each synapse. These "connection strengths" crucially determine the dynamic fate of a specific firing pattern — the same activity which might have faded without trace yesterday, might today trigger a strong emotional interaction because some intervening traumatic event has been encoded in the brain's synapses.

These states are of overwhelming complexity, with perhaps $10^{11}$ neurons and $10^{15}$ synapses in a single human brain. How do these then relate to mental state? Put simply, the mental state is a gross abstraction of the brain state which is at *approximately* the same level as language. The locution "at approximately the same level as language" is meant to convey two things: (i) that a mental state may *or may not* be expressible in words, while (ii) asserting that the level of detail that we ascribe to a mental state is far cruder than a brain state.

For example, I may see a face without being able to place it. The words "I've seen that person before — but where?" do not exhaust the mental state, for the latter includes the experience of facial features that we cannot put into words. If and when I recall the face, "Oh yes, during intermission at the opera last Saturday", the transition may be inexplicable at the "mental level", involving subtle neural processes which retrieved a memory of the scene involving that face, a representation rich enough to ground recognition of the context, with the subsequent return to the "mental level". The point here is that brain states are immensely complicated and causally complete. By this I mean that, brain states form a dynamic system: current state together with input determines (albeit with, perhaps, some noise) the next state and the output. By contrast, mental states are relatively simple and thus only sometimes causally efficacious — details of the brain state that we do not think about may be more important in determining our next thought than the preceding thought itself. In this regard, logic is not the essence of the mental, but is rather a crystallization of the limiting case where decisions and inferences can be made via inferential chains at the mental level with limited recourse to the brain's greater patternings.

Again, if we admire a sunset, our mental state involves our awareness and aesthetic appreciation of the rich patterning of red, orange and purple in the cloud formations banked above the horizon — the words "Look at that!" may then suffice to help a friend share aspects of that mental state without in any way reducing that state to those three neutral words. But the extent of that aesthetic appreciation does not begin to exhaust the complexity of the brain states which flash through the brain millisecond by millisecond as we enjoy the scene.

The relation of words to the above two vignettes is interesting. In the first case, the words are parts of our own thoughts, not tools for communication, and act as a précis of the current mental state that is itself a précis of the much more complex (and temporally extended) brain state. In the second case, the words we use for communication have nothing to do with the scene — but they convey the expectation that the friend too will appreciate the view of that sunset, an expectation that may be based either on one's knowledge of the friend's aesthetic preferences or on two more general propositions: "Most people enjoy watching a good sunset", and "It's good to increase the enjoyment of others." However, I again emphasize that logic is not "the essence of the mental," and it would in general not be the case that any explicit inference based on these two propositions was involved in your coming to say "Look at that!" I reiterate that causality is at the level of brain states, not mental states.

It should also be stressed that much of the brain state is irrelevant to one's mental state. For example, the state of the neural networks involved in the regulation of breathing do not affect our mental state save in times of crisis, and even then the effect on our mental state may be based on awareness of our overt pattern of breathing, rather than based on any monitoring of the relevant neural activity. But beware of that word "monitoring." We are *not* offering a dualist theory in which mind and brain are separate, with the mind monitoring the activity of the brain to extract highlights. Rather, we hold that the mental state is itself captured within the activity of the brain, forming some sort of précis of the broader neural activity *and* memory structures.

## 3.4. Consciousness, Language, and Distributed Schemas

Butler & Hodos (2005) show that the course of brain evolution among vertebrates has been determined in great part by

(a) Formation of multiple new nuclei through elaboration or duplication;
(b) Regionally specific increases in cell proliferation in different parts of the brain; and
(c) Gain of some new connections and loss of some established connections.

These changes can be accompanied by exquisite variations in neural morphology. These phenomena can be influenced by relatively simple mutational events that can thus become established in a population as the result of random variation. Selective pressures determine whether the behavioral phenotypes expressive of the central nervous system organization produced by these random mutations increase their proportional representation within the population and eventually become normal for a new species. What needs stressing here is that the genome does not come neatly packaged in terms of separate sets of genes for separate nuclei of the brain, nor does each nucleus control its own set of behaviors. At the lowest levels, the specification of a brain region can call on generic hierarchies of gene regulation and expression for the formation of boundaries amongst generic precursors of, e.g., neurons and glia. At a higher level, late structural genes that typify a taxon may elaborate on these in multifarious and marvelous ways. At the highest levels, an overt behavior may reflect a) the interactions of multiple brain regions and b) the social interaction of creatures with each other and their environment.

Moreover, and of great relevance to our later analysis of the evolution of language, *behavioral phenotypes* are not the result of "brain genes" alone; rather, they express both the brain's inherent organization, and the learning that has shaped it through the experiences of the individual organism, and these are determined in great part by the social milieu in which the organism is raised. For many species, this "social milieu" is hard to disentangle from the biology, but for primates we can discern a variety of "rituals", "practices" and "tribal customs" which constitute a body of culture constrained by, but in no sense determined by, the biological make-up of the social group. Thus, as we come to analyze the evolution of the hominids, culture comes to play an important role even in biological evolution, as well as being itself subject to change and selection.

Here, niche construction plays a crucial role (Arbib & Iriki, 2013; Iriki & Taoka, 2012). *Niche construction* refers to the process whereby creatures, by changing the physical environment, in turn change the adaptive pressures that will constrain the evolution of their own and other species (Odling-Smee *et al.*, 2003). What we add here is that by creating new patterns of behavior, creatures can alter the *cultural* niche in which they evolve (Arbib, 2011; Bickerton, 2009; Pinker, 2010). Importantly, such cultural niche construction can take place on both the biological time scale and on the much faster time scale on which culture evolves, with an increasing tempo in the history of *Homo sapiens*. Indeed, humans can induce such changes intentionally, to create a novel environmental niche, thus *intentional* niche-construction (Iriki & Sakura, 2008). Going further, Iriki & Taoka (2012) advanced the idea of *neural* niche construction, the notion that new behaviors may remodel the brain in such a way that creatures can take advantages of opportunities that would not otherwise be possible, thus rendering genes adaptive that can modulate the brain around this new stable state.

It should also be noted that certain genes seem to control the coordinated development of "brain systems" involving many brain regions, rather than being "specialists" for some local neural structures. From the work of Holmes (1939), we view the cerebellum's role as crucial to the graceful adaptation and coordination of movements, while conceding that the plan of movement is elaborated elsewhere. In mammals, cerebral cortex and cerebellum seem to be coupled in a genetically controlled system in such a way that they can co-evolve so that it now appears that lateral cerebellum can contribute to even the highly cognitive aspects of neural function (Ito, 2012).

Without wishing to equate consciousness with motor control — far from it — I would suggest that consciousness, too, may be a system function that involves networks including but not necessarily limited to cerebral cortex, and that as cerebral cortex evolves, so does consciousness. But just as cerebellar adaptation for eye movements is a far cry from cerebellar adaptation for, say, piano playing or speech, so we may expect the quality of consciousness to vary from species to species.

We have argued that we should not seek an all-or-none transition from other species viewed as totally non-conscious to conscious humans, but rather seek an evolutionary path which renders plausible the emergence of consciousness in its human form. I argue that we are conscious in a fully human sense only because we have language — i.e., that as awareness piggy-backs on all manner of neural functions, so too must it piggy-back on language, thus reaching a subtlety and complexity that would otherwise be impossible. However, I strongly deny that consciousness is merely a function of language. For example, as already noted in the previous section, one can be aware of the shape and shading and coloration of a face in great subtlety, and be totally unable to put one's vivid, conscious perception of that face into words. This point is well illustrated by Georgia O'Keefe's statement, painted on the foyer wall of the Georgia O'Keefe Museum in Santa Fe, that "The meaning of a word to me is not as exact as the meaning of a color. Colors and shapes make a more definite statement than words. I am often amazed at the spoken and written words telling me what I have painted." Nathalie Sarraute, in her book *Childhood*, says

> "For me... there is something prior to language: a sensation, a perception, something in search of its language, which cannot exist without language ... Scarcely does this formless [feeling], all timid, and trembling try to show its face than all powerful language, always ready to intervene so as to re-establish order — its own order — jumps on it and crushes it. Even the simple word "joy" cannot gather up what fills me, brims over in me, disperses, dissolves, melts into the pink bricks, the blossom-covered espaliers, the lawn, the pink and white petals, the air vibrating with barely perceptible tremors, with waves...waves of life, quite simply of life, what other word?...of life in its pure state..." (Sarraute, 1984).

What are we to make of that phrase, "which cannot exist without language?" Sarraute sees language as both expressive and destructive, and that tension between the verbal and non-verbal is surely the hallmark of our consciousness, setting it apart from whatever form of consciousness may be experienced by other creatures. The problem is that the writer, and of course the speaker, must try to express within words that which goes beyond words. I shall return to this topic in the section on motivation and emotion, offering some preliminary thoughts on the role of emotions and feelings in the development of consciousness.

Elsewhere (Arbib, 1978, 1992, 2013b), I have developed a theory of schemas as functional, as distinct from structural, units in a hierarchical analysis of the brain. The starting point was to describe perceptual structures and distributed motor control in terms of functional units called schemas which may be combined to form new schemas as coordinated control programs linking simpler (perceptual and motor) schemas, but these schemas provide the basis for more abstract schemas which underlie thought and language more generally. Thinking in terms of schemas reminds us that the behavioral phenotype of an organism is not necessarily linked to a localized structure of the brain, but may involve subtle patterns of "cooperative computation" between brain regions which support schemas in a distributed fashion. Selection may thus act as much on schemas and hence through neural systems as it does on localized neural structures. Developing this view, Arbib & Liaw (1995) argue that evolution not only yields new *schemas* connected to the old, but yields reciprocal connections which modify those older schemas. Three principles summarize and extend the key points of the above discussion:

**Principle 1: Cooperative computation.** The functions of perceptual-motor behavior and intelligent action of animals situated in the world can be expressed as a network of interacting schemas. Each schema itself involves the integrated activity of multiple brain regions. The method of interaction of schemas is "cooperative computation" (competition/cooperation) so that "computations" which are often seen as the province of traditional symbol-based processing are carried out by distributed "neuron-like" methods which do not involve explicit symbolic control. Cooperative computation not only serves as a basis for coordinated motor actions, but even for reactive planning, and intelligent behavior, including the use of language.

**Principle 2: Evolution and modulation.** New schemas and brain regions often arise as "modulators" of existing schemas or brain regions, rather than as new systems with independent functional roles.

**Principle 3: Interaction of partial representations.** A multiplicity of different representations — whether they be partial representations on a retinotopic basis, abstract representations of knowledge about types of object in the world, or more abstract "planning spaces" — must be linked into an integrated whole. Such linkage, however, may be mediated by distributed processes of competition and cooperation. There is no one place in the brain where an integrated representation of space plays the sole executive role in linking perception of the current environment to action.

In the next section, we will see the way in which Principle 2 was anticipated by the work of 19th century neurologist Hughlings Jackson, and suggest that this may offer a crucial clue as to the specific character of human consciousness. In the process, we may extend the observation of the last sentence of Principle 3 to state: "*There is no one place in the brain where an integrated representation of experience provides the essence of consciousness.*"

### 3.5. From Social Cooperation to Consciousness

Primitive communication subserves primitive coordination of the members of a social group, but processes which coordinate a group need not involve consciousness. Indeed, the social insects demonstrate a subtlety of group coordination that is in no way a precursor to consciousness. Here, we present the Arbib & Hesse (A & H) scenario for how group coordination may indeed have laid the basis for the co-evolution of consciousness and language (Arbib, 1985; Arbib & Hesse, 1986).

We each have only one body to act with and thus have a limited set of actions available to us at any one time. Thus, as we move towards the actual commitment of the organism to action, there would have to be a channeling from the richness of mental (and *a fortiori* neural) states to the well-focused choice, not necessarily conscious, of a course of action. Large ensembles of schemas in some sense interact, compete, cooperate to constitute a relatively well-focused plan of action that will commit the organism. This combination of many schemas within one body suggests a continuity of behavior by the one individual in similar situations, but also, as this repertoire builds up over time, the possibility that the schemas may eventually cohere in new ways, so that what had been an expected behavior in a certain set of situations may eventually give way, through new patterns of schema interaction, to new courses of behavior. Each individual has sets of schemas with some sort of coherence between them (this is not to claim that all of an individual's schemes are coherent); and the style of such a set of schemas can change over time to provide some sort of unity. I speculate that there are perhaps hundreds of thousands of schemas corresponding to the totality of what a human knows, with perhaps hundreds of these active at any time to subserve the current interaction of the organism with its environment. But what is the role of consciousness here? Would not these schemas do their jobs just as well if there were no such thing as consciousness? I cannot make a case for why we must *be* conscious. I think I can make a case for why we have patterns of schema activity that *correlate* with consciousness very well. But first, I do want to stress that our common sense about consciousness can be misleading, that there are many things that we think require our conscious thought but which do not — as we saw in our discussion of blindsight and DF.

The lesson is that even motor behaviors (reaching for a target, avoiding obstacles) that we think of as normally under conscious control can in fact proceed without our being conscious of the visual cues involved. A general point here is that what we may capture in correlates of a phenomenon within a limited account of one brain region pales before the richness of multi-level interactions playing out over tens, possibly hundreds of brain regions. The relation of visual cortex to superior colliculus in our exposition of blindsight can be related to the view of the British 19th century neurologist Hughlings Jackson of the brain in terms of levels of increasing evolutionary complexity (Jackson, 1874, 1878–1879). Jackson was very much influenced, as many people were at that time, by Darwinian concepts of evolution. In many cases, Jackson saw damage to the brain not so much in terms of the loss of function of the damaged region but rather in terms of the release of "older" brain regions from inhibition by the damaged area, which provided controls evolved later. In other words, such brain damage could often be seen as revealing behavior that is more primitive in evolutionary terms. As the brain evolves, new patterns of neural activity provide new "ecological niches" for the evolution of neural circuitry to exploit these patterns. Once such new circuitry evolves, there is a new "information environment" for earlier circuitry, which may thus evolve in turn to exploit the new patterns. (For an example of such a Jacksonian mode, see the account of optic flow in Arbib (1989a, pp. 342–344).)

Not only does the brain so evolve that new brain regions or schemas can exploit the prior richness of the brain but moreover — and this is the important point — *once these new systems are available they provide a richer environment for the older ones* which now have new possibilities for further evolution, whether evolution of brain regions over a biological time scale or the evolution of schemas over an individual time scale. If we compare, for example, the brain of a frog with the brain of a cat, we find that the corresponding regions of the visual mid-brain are much richer in the cat, not so much in terms of retinal input, but because there are so many descending pathways whereby the richness of cerebral cortex can influence what goes on in these more classical brain regions. In the terminology of schema theory, *evolution not only yields new schemas connected to the old, but yields reciprocal connections which modify those older schemas*.

Perhaps the development of animal and human communication may also be seen in terms of such evolutionary interaction to give us some insight into how consciousness might have evolved. A striking feature of human ability is that we come to incorporate tools into our body schema (Head & Holmes, 1911; Maravita & Iriki, 2004). Analogously, as creatures developed as social animals, their schemas came to represent their "bodies" as ending not at the extremities of the physical body but rather as extending to incorporate aspects of other members of the group. When we use a screwdriver our body ends at the tip of the screwdriver, not at the end of the hand; when we drive a car, our body ends at the rear bumper, not at our buttocks. Analogously, as creatures developed as social animals (and this account is not restricted to humans), the body might end not at the extremities of the physical body, but extend to incorporate aspects of other members of the group. However, coordinating others is a more subtle matter than just directing an arm or slightly adapting the hand to control a tool. The social animal has to find a way of expressing some précis of its mental state, of its richness of schema activity, so that it may then impinge upon others so that their behavior may be coordinated.

The A&H hypothesis of the evolution of human consciousness (Arbib & Hesse, 1986; Arbib, 1985) is that, with increasing richness of social interactions, though still at a prelinguistic stage in our evolutionary story, there would come the ability to form a *précis of schema activity* that is not necessarily relevant to deciding what to do next, but is relevant in terms of coordination with others. It is important to note that there is no claim that, at the earliest stages of its evolution, such a précis plays any role in deciding what to do next. However, in our "Jacksonian" terms, we can see that once the précis is available as part of the way the brain has evolved, older patterns of schema interaction may become modified in turn. There are occasions on which they can be better coordinated via the précis, and thus it would come about that the précis would develop into diverse states of a subnetwork (not necessarily localized) with bidirectional links into the schema network, and that this subnetwork would then evolve to have a role that is *sometimes* directive.

Primitive communication subserves primitive coordination of the members of a social group. As in blindsight, processes which coordinate group members need not involve consciousness. As communication evolves (by mechanisms we only begin to understand), the "instructions" that can be given to other members of the group increase in subtlety. Communication evolves at first purely as a way of coordinating the actions of a group. For this to succeed, the brain of each group member must be able not only to generate such signals, but also to integrate signals from other members of the group into its own ongoing motor planning. We suggest that the key transition in going from the limited set of vocalizations used in communication by, say, our common ancestors with vervet monkeys to the richness of human language came with a *migration in time* from:

(i) An execution/observation matching system enabling an individual to recognize the action (as distinct from the mere movement) that another individual is making, to
(ii) The individual becoming able to pantomime "this is the action I am about to take".

In the earliest stages of the evolution of this second ability, communication may have involved the accidental release of a motor plan from inhibition, thus allowing a brief prefix of the movement to be exhibited before the full action was released — but this "warning gesture" may have sufficed to alert others in time to bias their action, yielding benefits of adaptive value for groups that could both offer "signals of intention" and make use of them. This yields positive reinforcement to the individual accidentally releasing prefixes of actions, serving in turn as the basis for group selection, favoring the reproduction of those groups which can learn to emit and interpret such signals. This marks, at both the individual and species levels, the beginning of real communication as distinct from the release of signals. Note that we are not positing a form of Jungian *group consciousness* but only that *group selection* yields a population equipped to communicate more effectively and thus come to develop culture and cultural evolution which supports the sharing of ideas and ideologies in a manner that might be interpreted as group consciousness but which I would rather see as establishing memes and social schemas whose content is elaborated far beyond what is encoded genetically, which includes basic aspects of the motivational system.

However, Arbib and Hesse do not emphasize this external process of "group selection" in the population as a whole, but rather the changes within the individual brain made possible by the availability of a "précis" — a gesturable representation — of intended future movements (as distinct from current movements). They use the term *communication plexus* for the circuits involved in generating this representation. The Jacksonian element of their analysis is that the evolution of the communication plexus provides an environment for the further evolution of older systems. They suggest that once the brain has such a communication plexus, then a new process of evolution begins whereby the précis comes to serve not only as a basis for communication between the members of a group, but also as a resource for planning and coordination within the brain itself. This "communication plexus" thus evolves a crucial role in schema coordination. Elsewhere (Arbib, 2001), I asserted that "it is the activity of this co-evolved process that constitutes consciousness. As such it will progress in richness along with the increased richness of communication that culminates as language in the human line." I need to qualify that. I would rather say that "it is the activity of this co-evolved process that endows human consciousness with properties denied to other species," and add that the evolution of this novel brain system makes possible (Jackson again) the evolutionary enrichment of other systems in a way that makes it possible to enrich "animal awareness" with properties that other animals lack. What needs to be emphasized is the essentially distributed nature of the brain's representations (recall Principle 3 above). An anecdote may help with this. In 1961, when I was a novice PhD student at MIT, the eminent visual neurophysiologist Horace Barlow was discussing with me the research of Hubel & Wiesel (1959) who had discovered that cells in primary visual cortex of cats responded best to oriented edges crossing their receptive fields. Positing that such local data could be integrated to obtain a neural representation of distinctive contours in the visual field, Barlow said "And that's how you can recognize people from caricatures" to which I responded "But then how do I know you are not a caricature." This exchange left me with the understanding that any limited representation (such as that conveyed by words or communicative gestures) remains a pale approximation (caricature!) of what we can consciously experience. This reinforces the point: activity of the "communication complex" is not even approximately the extent of consciousness unless it is integrated into the activity of diverse systems (coordinating diverse schemas at multiple hierarchical levels) — just as activity in the visual system which provides contours that may aid the recognition of a familiar face in no way exhausts what we may experience on viewing that face in an image or when meeting the person in real life. We have already noted that a mental state is far more crude than a brain state is and, of course, the description of mental states by folk psychology's linguistic labels is even more limited. In the same spirit, we may note Elman's (2009) and Elman *et al.*'s (2004) discussion of words as "cues" for diverse mental entities, since a given word may have different meaning in different contexts and the earlier discussion of "Language, Metaphor and a New Epistemology" by Arbib & Hesse (1986) in which meaning emerges through the spread of activity in schema networks.

There are occasions on which "older" schemas can be better coordinated if they coordinate via the précis than if they coordinate by themselves. We then have a subnetwork of the schema network which provides a précis that may often have no directive role, and yet may evolve to have a role that is sometimes but not always directive. Since lower-level schema activity can often proceed successfully without this highest-level coordination, consciousness may sometimes be active, if active at all, as a monitor rather than as a director of action. In other cases, the précis of schema activity plays the crucial role in determining the future course of schema activity, and thus of action.

This evolutionary process, which occurs with subhuman species, then sets the stage, I would suggest, both for human consciousness and also for the evolution of language to express this rather coarse précis of the richness of the underlying schema activity. In summary, this theory sees human consciousness enriched by a précis of schema activity, evolving in such a way that it can elaborate certain mental processes at the level of language and logic and is related in part, but not entirely, to communication.

Arbib and Hesse's theory does not explain the phenomenology of consciousness — i.e., the way consciousness "feels" to each of us — but it does accord well with this phenomenology. The fact that lower-level schema activity can often proceed successfully without the high-level coordination afforded by the communication plexus explains why consciousness may sometimes be active as a monitor rather than as a director of action. In other cases, the formation of the précis of schema activity plays the crucial role in determining the future course of schema activity, and thus of action — and this accords with those occasions in which we experience a conscious effort in weighing a number of courses of action before we commit ourselves to behave in a specific way. O'Regan (2011, 2012) and O'Regan & Block (2012) develop a sensorimotor account of consciousness that is quite consistent with, and in some ways complementary to, the approach developed here. Of particular interest may be his paper "How to Build a Robot that is Conscious and Feels" (O'Regan, 2012), since this makes explicit the hypothesis that consciousness is to be understood at larger scales of organization (perhaps those addressed by schema theory) that can be shared by brains and machines, rather than depending on aspects specific to subneural physics.

So far, little has been said of emotions and feelings. Here, we may recall *The Feeling of What Happens* and *Self Comes to Mind* where Damasio (1999, 2012) claims that the first step in generating a "protoself" comes from primordial feelings, which humans share with other species. Similarly, Panksepp (1998) argues that primitive emotional feelings appear to lie at the core of our beings, and the neural mechanisms that generate such states may constitute an essential foundation for the evolution of higher, more rational, forms of consciousness. He further argues that a neural understanding of "the self" may arise from a study of how various biological value-coding systems (emotional circuits) converge and interact with representations of the body. I will devote a section to "Motivation and Emotion" below. Although I will not assess Damasio's and Panksepp's ideas explicitly I will offer at least a hint of how emotions and feelings relate to the A&H theory of the evolution of consciousness. The general thesis remains: just as pain, fear, hunger and so on may be a part of "animal awareness," so will they be part of human consciousness, but enriched by the interaction of primordial systems with the communication plexus — an interaction which has changed and enriched these systems across the course of primate evolution, building atop prior changes through mammalian evolution more generally. However, I first devote a section to the mirror system hypothesis (MSH) of the evolution of the human language-ready brain, a section which concludes with a very brief discussion of the notion of "theory of mind," which is not a scientific theory but rather the general understanding one has of the minds of oneself and others.

### 3.6. The Mirror System Approach to Language Evolution

More than a decade after working with Mary Hesse on the theory of the evolution of consciousness (the A&H theory) presented in the previous section, I developed with Giacomo Rizzolatti a preliminary version of MSH of the evolution of the language-ready brain. MSH fills in some of the details missing in the A&H theory as this section will now outline. But first I need to summarize the basic idea of "mirror neurons," as first discovered by Rizzolatti's group in the 1990s.

Premotor cortex of the macaque monkey contains an area F5 in which are located neurons specific for control of manual actions. Significantly, F5 contains a subclass of such neurons that discharge not only when the monkey grasps or manipulates objects, but also when the *monkey observes the experimenter* perform an action similar to the one that, when actively performed by the monkey, involves activity of the neuron (di Pellegrino *et al.*, 1992; Rizzolatti *et al.*, 1996). We distinguish such *mirror neurons*, which are active both when the monkey performs certain actions and when the monkey observes them performed by others, from *canonical neurons* which are active when the monkey performs certain actions but not when the monkey observes actions performed by others. We say that area F5 is endowed with an *observation/execution matching system*: When the monkey observes a motor act that resembles one in its movement repertoire, a neural code for this action is automatically retrieved. This code consists in the activation of a subset, the mirror neurons, of the F5 neurons which discharge when the observed act is executed by the monkey itself.

Two PET experiments (Grafton *et al.*, 1996; Rizzolatti *et al.*, 1996) were then designed to seek mirror systems in humans. The two experiments differed in many aspects, but both had a condition in which subjects observed the experimenter grasping a 3D object. Observation of the object alone was used as a control. (This condition also controlled for verbalization.) Grasp observation significantly activated the superior temporal sulcus (STS), the inferior parietal lobule, and the inferior frontal gyrus (area 45). All activations were in the left hemisphere. The last area is of special interest — areas 44 and 45 in left hemisphere of the human constitute Broca's area, a major component of the human brain's language mechanisms. Indeed, F5 is generally considered to be the homologue of Broca's area. But why should a mirror system for grasping be related to a system in the human brain for language production? For want of better data, we will assume that the common ancestor of humans and monkeys shared with monkeys a primate call system (a limited set of species-specific calls) and an oro-facial gesture system (a limited set of gestures expressive of emotion and related social indicators) — communication is inherently multi-modal, and body posture also plays a role in social communication, though I shall not emphasize this here. What is to be stressed here is that combinatorial properties for the openness of communication are virtually absent in basic primate calls and oro-facial communication, even though individual calls may be graded. Moreover, the neural substrate for primate calls is in a region of cingulate cortex distinct from F5, which we have seen to be the monkey homologue of human Broca's area. The *mirror system hypothesis for the evolution of the human language-ready brain* (Arbib, 2002, 2005b; Arbib & Rizzolatti, 1997; Rizzolatti & Arbib, 1998) suggests that human brain mechanisms supporting language evolved from a mirror system for manual actions in the last common ancestor of humans and macaques (LCA-m) in six stages:

***Pre-Hominid***

1. **A mirror system for grasping:** matching action observation and execution for grasping. Shared with common ancestor of human and monkey.
2. **A simple imitation system for grasping.** Shared with common ancestor of human and great apes.

***Hominid Evolution***

3. **A complex imitation system:** *complex imitation* combines the ability to recognize another's performance as a set of familiar movements with the ability to use this recognition to repeat the performance, and (more generally) to recognize that another's performance combines actions which can be imitated at least crudely by variants of actions already in the repertoire, with increasing practice yielding increasing skill.
4. **Protosign:** a manual-based communication system, breaking through the fixed repertoire of primate vocalizations to yield an open repertoire.

5. **Protospeech and multi-modal Protolanguage:** This rests on the "invasion" of the vocal apparatus by collaterals from communication system based on F5/Broca's area. Protosign and protospeech further co-evolve in an expanding spiral.

*Cultural Evolution in Homo Sapiens*

6. **Language:** The transition from action-object frames to verb-argument structures which express them; syntax and a compositional semantics: Co-evolution of cognitive & linguistic complexity.

In relation to Stages 4 and 5: Bickerton (1995) has spoken of *proto-language* as being language restricted, basically, to two word utterances (generally comprising a verb and a noun in some order), and suggests that it is common to chimps trained to exhibit some form of language-like use of symbols, two-year old humans, and people speaking pidgins. He then argues that proto-language was possessed by the hominid precursors of humans. However, I note that proto-language is only observed in creatures (chimps or humans) exposed to human language. Rather, I have argued that what early *Homo sapiens* possessed was the ability to name events with novel sequences of (manual and/or vocal) gestures, but that this capability does not imply the ability to separately name the objects and actions that comprised those events. I then claim that the latter ability was a momentous discovery made by humans perhaps 100,000 to 50,000 years ago, rather than a biological heritage from earlier hominids.

In relation to Stage 6: There is a danger, typified by Chomsky's notion of Universal Grammar (e.g., Chomsky, 1980), of thinking that there is some set of basic characteristics typical of all language and "hard-wired" into the human genome and thus the human infant's brain. I disagree, arguing that we can each know one or a few languages, but there is no such thing as Language-with-a-big-L that we all know, let alone "know in our genes". It is difficult to learn a foreign language because languages can be inherently different, but more-or-less adequate translations between languages are possible because languages have been shaped both by the need to express a range of basic human experiences, and by cultural diffusion.

The claim here is that biological evolution may not so much have yielded a human brain "equipped with" language so much as a human brain "ready for" language. The corollary to this is that even though an early *Homo sapiens* infant raised (after some miracle of time travel or cloning) in a human home in today's world would be completely human, an adult *Homo sapiens* of 200,000 years ago would have had a consciousness in many ways limited with respect to a modern human. Perhaps his awareness of the sights and sounds and smells around him would be more intense than most modern humans' experience, but it would lack the subtle overlays that the modern mind possesses precisely because of the recursive properties opened up by language.

Consciousness is shaped by the interaction of our biology (brain and body) with our environment (physical and social) and thus, I claim, has developed drastically since the human brain and body achieved their more-or-less modern genetic substrate. In this regard we may note the observation (Lee, 2000) that Whorf (1937) contrasts a "universe ... [of] conscious thinking abstracted from experience" with a "universe as unconscious thinking projected upon experience" and refers to "the organization of raw experience into a consistent and readily communicable universe of ideas though the medium of linguistic patterns" and suggests that in referring to "conscious thinking" what Whorf had in mind was the elements of language that allow that experience to be organized and reflected upon in shareable form.

The MSH is by now an oft-told tale, and I will not repeat it in any detail here. A book length exposition has been published (Arbib, 2012) and a summary of the book (Arbib, 2013c) was published a year later along with 12 critiques and a response which carried the hypothesis forward (Arbib, 2013a).

But what is the relationship of the MSH of the late 1990s, whose development still continues, and the earlier A&H hypothesis that a précis of intended action could have provided the basis for the emergence of a uniquely human form of consciousness? Does the temporal sequence of the formulation and publication of these ideas have any relation to the evolutionary development of what they describe in primates? I would suggest that the answer is "no" — in other words, the evolution of the language-ready brain and of a specifically human consciousness are intertwined. The original A&H story has two components:

(i) The notion that having an expressible précis of intended action could improve social interaction.

(ii) The further notion that such a précis could then create a new resource for brain function, in the style of Hughlings Jackson, that could create a system whose activity could be seen as providing the correlate of specifically human dimensions of consciousness.

The MSH complements this by suggesting how neural mechanisms supporting the expression of that précis could have evolved in a manner that, crucially, can link the production and perception of such expressions. However, as Dominey (2013) observes, *How the Brain Got Language: The Mirror System Hypothesis* (Arbib, 2012) says rather little about "the uniquely human drive to share psychological states with others and unique underlying forms of cognitive representation for doing so, as illustrated in the research of Tomasello *et al.* (2005). This includes both the behavior itself, i.e., the motivation to share mental states, as well as the underlying neurophysiology." Item (ii) of A&H might offer a start in this direction: that what starts as a mechanism for coordinating behavior evolves linkages with diverse brain systems beyond the premotor cortex, and that these come to incorporate more details of processing related not only to current action but also to motivational states and perceptual states that are currently being experienced, whether or not they have immediate links to practical action. However, the mechanisms that may have first evolved to coordinate actions with others can now be exapted to express aspects of the précis as enlarged by these new connections, and thus the transition from instrumental to declarative communication begins. We can enrich this account further by a brief discussion of motivation and emotion, since another dimension of language is indeed the expression of emotion, whether through the words that are chosen or through prosody. It is worth reiterating that all this talk of a précis and of consciousness in no way implies the localization of consciousness in any one region of the brain. Indeed different lesions may affect different dimensions of consciousness, just as lesions can differentially affect the syntax, semantics and prosody of language. Ramachandran (2011) also articulates different levels of representations corresponding to different levels of consciousness, emphasizes the role of language in the emergence of consciousness, and sees consciousness as emerging from a number of interacting brain processes.

It has been argued that fully developed conscious experience requires some degree of self-awareness, and hence a *theory of mind*, the human ability to ascribe mental states to oneself and others. Much has been written about the possible role of mirror systems in the development of such theory of mind, with particular attention to its partial breakdown in autism (Arbib, 2007; Dapretto *et al.*, 2006; Ramachandran & Oberman, 2006; Williams, 2008; Williams *et al.*, 2001).

One referee observed of an earlier draft that "the claim that language decisively contributes to the emergence of fully developed human consciousness leads to the conclusion that consciousness is [in great part] *narrative* or *discursive*. ... This is important in so far as it goes against a long tradition (especially in the continental philosophy) to speak of consciousness as a — more or less coherent — chain of mental images. Moreover, Arbib's theory seems also to undermine the idea of mentalese (language of thought). As I read Arbib's claims, he believes that there is more to consciousness than thought formulated in language (i.e., there are aspects of mental states that cannot be expressed in language), but at the same time he rejects the thesis that there exists a language of thought substantially different from the language we speak. These are just a couple of examples of philosophical doctrines which seem incompatible with Arbib's theory of consciousness; what they illustrate is that the conception proposed by Arbib is of a great significance for philosophy, as it is for cognitive science." Here I need to add two comments:

(a) The role of narrative in the evolution of language has been eloquently expressed by Barnard (2013) who argues that the evolution of narrative, and especially of myth, pushed the evolution of linguistic complexity, with language coevolving with mythology in symbolic frameworks which extended the capacity for verbal expression to the limits of cognition. My only demurrals are that narrative can perform valuable social functions without (or prior to) the emergence of myth, and that cognition co-evolved with means for verbal expression, rather than providing pre-existing limits.
(b) It is not that I "reject the thesis that there exists a language of thought substantially different from the language we speak" — I reject the notion that our mental processes are language-like save in limiting cases, with mental processing better defined at the level of schemas and as a mere abstraction of the even less language-like dynamics of brain states (Arbib, 1989b).

Given that the title of this chapter is "How Language Evolution Reshaped Human Consciousness," it may be surprising that my first paper on consciousness was entitled "Consciousness: The Secondary Role of Language" (Arbib, 1972). For our purposes, its key assertion was that "language arose as an elaboration of more fundamental functions of perception. This does not deny that the evolution of languages has greatly enriched human behavior, but does suggest the danger of trying to understand the human mind by regarding language as its primary attribute."

I suggested that the primary purpose of perceiving objects is not to name them, but rather to be prepared to interact with them — though you may perceive an object yet still leave it alone. In this spirit, the aim of the present chapter has been to suggest how the processes that marked the dramatic emergence of the human language-ready brain (or, more generally, the "culture-ready" brain (Whitehead, 2010) did not reduce human consciousness to thoughts expressible in language but, rather, have transformed more general forms of "animal awareness" to a uniquely human form of consciousness in which basic perception and awareness of action are enriched by a continuing "verbal commentary" on our awareness, and the ability to situate our current context beyond the here-and-now through what Suddendorf & Corballis (2007) refer to as "mental time travel."

### 3.7. Motivation and Emotion: The Motors of Behavior

We earlier quoted the emotive meaning of color to Georgia O'Keeffe and Nathalie Sarraute's description of her transcendent feelings which could not be reduced to language to make the point that our consciousness is by no means limited to what can be expressed in language. We experience nuances of awareness that go far beyond that which could be shared — let alone expressed — by an animal, and a skill of great writers is to find words that evoke these nuances of feeling and experience which most of us can seldom verbalize with any subtlety. In this section, I want to broach the issue of emotions and feelings in the development of consciousness. Simple mammals may be aware of the differences between "feeling maternal" and "feeling enraged," while the primate call system is a limited set of gestures expressive of emotionally charged behavior (e.g., fleeing a predator). However, MSH is at some pains to distinguish the medial brain systems implicated in primate calls from the lateral brain systems (like macaque F5) that seem homologous to the cortical areas most engaged in language processing.

This section is based on papers (Arbib, 2005a; Arbib & Fellous, 2004) that grew out of my work with Jean-Marc Fellous on the book *Who Needs Emotions: The Brain Meets the Robot*, (Fellous & Arbib, 2005), and was influenced by various contributions there. As suggested by the book's title, much of our concern there was in relating the neurobiology of animal and human emotion to the question of whether (future) robots could have emotions, and thus to the issue of what "emotion" would mean when cut loose from its biological moorings. Again we see the theme of trying to understand what aspects are specific to our biology, and which may in time be captured by advances in robotics and artificial intelligence, but here I will focus solely on the nature of human emotion and its roots in biological (and social) evolution. We analyzed emotion under two headings:

(1) *Emotional expression for communication and social coordination.*
(2) *Emotion for organization of behavior (action selection, attention, social coordination and learning).*

The former concerns "external" aspect of emotions; the latter "internal" aspects. In animals, these aspects have co-evolved. Animals need to survive and perform efficiently within their ecological niche and in each case patterns of coordination will greatly influence the suite of relevant emotions (if such are indeed needed) and the means whereby they are communicated. A key function of emotion is to communicate simplified but high impact information. A scream is extremely poor in information (saying nothing about the cause for alarm), but its impact on others is high. The similarity in facial expressions between certain animals and humans prompted classic evolutionary analyses (Darwin, 1872) which support the view that mammals (at least) have emotions, though not necessarily *human* emotions. Many researchers (e.g., Dalgleish, 2009; Damasio, 1994; Fellous & Ledoux, 2005; Jeannerod, 2005; LeDoux, 2012; Rolls, 2005) have proposed explicit functions for emotions: coordinating behavioral response systems; shifting behavioral hierarchies; communication and social bonding; short-cut cognitive processing; and facilitating storage and recall of memories. However, emotions are not always beneficial — if one is caught in a traffic jam, anger may easily set in, yet anger *in this case* has no apparent utility. How does the brain maximize the benefits of emotion while minimizing its occasional inappropriateness?

To echo something of the role of the précis in the A&H theory and the emergence of language in MSH, we now turn to general "cognitive architectures" in which the role of emotion can be situated at several levels. Ortony *et al.*, (2005) analyze the interplay of *affect* (value), *motivation* (action tendencies), *cognition* (meaning), and *behavior* at three levels of information processing:

*Reactive*: a hard-wired releaser of fixed action patterns and an interrupt generator. This level has only proto-affect.
*Routine*: the locus of unconscious well-learned automatized activity and primitive and unconscious emotions.

*Reflective*: the home of higher-order cognitive functions, including meta-cognition, consciousness, self-reflection, and full-fledged emotions.

They address the design of emotions in computational agents that must perform unanticipated tasks in unpredictable environments. They argue that such agents, if they are to function effectively, must be endowed with curiosity and expectations and a *sense of self* that reflects parameter settings that govern the agent's functioning. Sloman *et al.* (2005) also offer a three-level view of central processes:

*Reactive*: producing immediate actions. When inconsistent reactions are simultaneously activated one may be selected by a competitive mechanism.

*Deliberative*: using explicit hypothetical representations of alternative possible predictions or explanations, comparing them and selecting a preferred option.

*Meta-management*: allowing internal processes to be monitored, categorized, evaluated and controlled or modulated.

They further note the utility of an "alarm" system, a reactive component that gets inputs from and sends outputs to all the other components and detects situations where rapid global redirection of processing is required.

To reconcile the two schemes, Arbib & Fellous (2004) suggested four levels: reactive; routine; reflective-deliberative; and reflective-meta-management.

Emotion is not computed by a centralized neural system; emotions operate at many time scales and at many behavioral levels; and there is no easy separation between emotion and cognition. Each animal comes with a set of basic "drives" — for hunger, thirst, sex, self-preservation, etc. — and these provide the basic "motor", motivation, for behavior. Motivated behavior not only includes bodily behavior (as in feeding and fleeing, oro-facial responses, and defensive and mating activities), but also autonomic output (e.g., heart rate and blood pressure), and visceroendocrine output (e.g., adrenaline, release of sex hormones). These lie at the heart

[sic] of our emotional repertoire. However, the emotions that we talk about and perceive in others are both more restricted than this (how many people perceive another's cortisol level?) yet also more subtle, intertwining these basic motivations with our complex cognitions of social role and interactions, as in the cases of jealousy and pride. The core of the motivation system is provided by the nuclei of a deep-seated region of the brain called the *hypothalamus*. These nuclei are devoted to the elaboration and control of specific behaviors necessary for survival. Such behaviors include spontaneous locomotion, exploration, ingestive, defensive, and reproductive behaviors. Basically, the hypothalamus talks "downward" for basic behavioral control, and "upward" to involve the cortex in determining when particular behaviors are appropriate. The lateral hypothalamus plays a critical role in arousal, control of behavioral state, and reward-seeking behavior. It includes what Olds (1969) referred to as the "pleasure center" because rats will press a lever thousands of times per hour to deliver electrical stimulation to this region.

The role of the amygdala in the experience and expression of fear has received particular study. Stimuli that elicit fear reactions may be external (e.g., a loud noise) or internal, including those from memory structures such as hippocampus or prefrontal cortex. The amygdala can influence cortical areas via feedback from proprioceptive, visceral or hormonal signals, via projections to various "arousal" networks, and through interaction with the medial prefrontal cortex. Prefrontal cortex, in return, sends distinct connections to several regions of amygdala, allowing elaborate cognitive functions to regulate the amygdala's roles in emotion. Because of the tight interactions between amygdala and prefrontal cortex, it is likely that our ability to generalize and abstract is directed by (and influences, in return) some aspects of our emotional state. Connections between the amygdala, thalamus, and cortical areas allow for both fast emotion elicitation and more refined emotion control based on memory and high level sensory representations. This role of prefrontal cortex is exemplified by the observations that decision-making ability in emotional situations is also impaired in humans with damage to the medial prefrontal cortex, and abnormalities in prefrontal cortex may predispose people to develop fear and anxiety disorders (LeDoux, 2000). Moreover, activation of the human amygdala can be produced by observing facial expressions, and lesions of the human amygdala may cause difficulty in the identification of some such expressions (Rolls, 2000). The amygdala and prefrontal cortices are therefore involved in social as well as internal aspects of emotion, and together play a crucial role in the regulation of emotion, a key component of affective style, affective disorders, and social interactions (Adolphs, 2009).

The *medial prefrontal cortex* is adjacent to the *orbitofrontal cortex* and Rolls (2005) documents how damage to the orbitofrontal cortex produces emotional changes which includes a tendency to inappropriate responses. Orbitofrontal neurons also serve as part of a mechanism which evaluates whether a reward is expected, and different subregions of the prefrontal cortex are selectively involved during positive rewards or punishments (Schultz, 2000). Others have shown that different subregions of the medial temporal lobe memory system are selectively and differentially involved for emotional and neutral stimuli in human, and that this area was strongly correlated with amygdala activations during emotional stimuli. Many other brain areas have been involved in the experience and expression of emotions in humans, including the anterior cingulate cortex, insula, hippocampus, and fusiform gyrus.

LeDoux (2000) argues that emotion systems evolved as sensory-motor solutions to problems of survival. He distinguishes emotions from "feelings". Such "conscious emotions" are not, he suggests, the function that emotion systems evolved to perform. He has focused on fear-conditioning in the rat to define meaningful animal experiments and has been particularly concerned with the role of the amygdala in fear. In particular he has studied the role of the amygdala in conditioning of fearful behavior, as an animal comes to learn that certain situations may lead to danger. The laboratory equivalent might be for a rat to learn that it will get an electric shock if it approaches a particular place in its cage. LeDoux stresses that we cannot know what feelings the rat has if it is afraid, or indeed whether it has any feelings of fear. But it certainly can exhibit fearful behavior, trying to escape or, in many of LeDoux's studies, simply freezing in place. Such "freezing", a cessation of normal locomotion, is well known as a response to predators (if you do not move, he may not see you) and so is assumed to be a case of being too fearful to move.

Whether or not rats have emotional feelings as well as emotional behaviors, there is no doubt that humans do have feelings. How does "behavioral fear" relate to the feeling of fear or anxiety? The crucial element from an evolutionary point of view may be the set of reciprocal interactions between amygdala and cerebral cortex. Human emotions are strongly influenced by social milieu — what is embarrassing in one culture (e.g., nudity) may be regarded as perfectly normal in another. The amygdala can influence cortical areas by way of feedback either from proprioceptive or visceral signals or hormones, via projections to various "arousal" networks, and through the aforementioned interaction with medial prefrontal cortex. Fellous & LeDoux (2005) suggest that prefrontal-amygdala interactions may be involved in the conscious feelings of fear. However, this division between the cognitive cortex and emotional amygdala strikes me as too simple to do justice the brain's integration of widely distributed functions — both because not all parts of cortex give rise to conscious feelings and also because human emotions seems to be inextricably bound up with "cortical subtleties."

Going beyond the above hypothalamo–amygdala–cortical interactions, we note that language plays a unique role in shading and refining human emotions. It is thus interesting that recent research suggests that "mirror neurons" may provide a crucial part of the substrate for "empathy", the ability to recognize the emotional dispositions of others, as well as communication through language. We stress that the richness of human emotion is in part due to the linkage of emotion with cognitive processes of great subtlety. The neural substrate of emotion involves many interacting structures, from the hypothalamus to the orbitofrontal cortex to the diverse mirror systems, which have clear "cognitive" as well as emotional roles.

In tracing the macaque neurophysiology of the mirror system, we saw that neurons in the STS are active during movement observation though perhaps not during movement execution, and that STS and F5 may be indirectly connected via area PF which, like F5, contains mirror neurons. As Rizzolatti *et al.* (2001) observe, STS is also part of a circuit that includes the amygdala and the orbitofrontal cortex and so may be involved in the elaboration of affective aspects of social behavior. Human emotions are greatly influenced by our ability to empathize with the behavior of other people. We have all had moments when we felt a welling up of sadness on seeing the grief of another or smiled in response to another's smile. In *The Expression of the Emotions in Man and Animals*, Darwin (1872) observed that we share a range of physical expressions of emotion with many other animals. Our tendency to feel others' emotions, and the fact that certain emotions are so universal as to be shared with other mammals have led a number of researchers to suggest that in addition to a mirror system for manual actions, we have one for the generation and recognition of physical expressions of emotion. In fact, there seem to be multiple systems involving different classes of emotions.

The suggestion is that these systems enable us to empathize with others not simply by recognizing their emotions but by experiencing them. This is related to the *simulation theory* (Gallese & Goldman, 1998) which proposes that mirror neurons support our ability to read minds by "putting ourselves in the other's shoes", and using our own minds to simulate the mental processes that are likely to be operating in the other. Other findings suggest that the ability to recognize the auditory and visual details of a person's actions may be a factor in empathy (Gazzola *et al.*, 2006). The role of a brain region called the insula in a mirror system for disgust (Wicker *et al.*, 2003) suggests that emotional communication may involve diverse mirror systems. However, "emotion recognition" is not just a matter of empathy. If I were to see someone advancing toward me with a knife, I would recognize his anger yet feel fear; I would not empathize as I struggled to protect myself. And, indeed, much of the literature on mirror neurons does indeed distinguish between low-level visuomotor reaction and a more comprehensive understanding of the other's emotions. A crucial aspect of social interaction is our view of others. The term "Theory of Mind," briefly discussed at the end of Sec. 6, has become widely adopted for the ability of humans to understand that other people may be like themselves yet have different knowledge and points of view. Gallese and Goldman argue that observed action sequences are represented in the observer "off-line" to prevent automatic copying, as well as to facilitate further processing of this high-level social information. However, mirror neurons alone cannot mediate understanding. Instead, understanding rests on a network of associations distributed across multiple brain regions "beyond the mirror."

This cautions us to distrust talk of "the" mirror system but rather seek to understand the roles of, and interactions, between, multiple mirror systems such as those for hands and faces and — in humans — language. The previous section had its primary emphasis on intermediate stages in the evolution of mirror systems from the macaque-like mirror system for grasping to those that support language. I would suggest that, similarly, a number of stages would have to intervene in the evolution of the brain mechanisms that support emotion and empathy. This raises the challenge of connecting empathy with language evolution by assessing how our ability to experience others' emotions affected the way the language-ready brain evolved. But this challenge lies outside the scope of this chapter.

Appraisal theory (Ortony *et al.*, 1988) develops a catalogue of human emotions and seeks to provide a computational account of the "appraisals" which lead to invocation of one emotion over another, but without the "heat" provided by emotion's biological underpinnings in humans.

Appraisal theory is thus a good candidate for algorithms that enable a robot to *simulate* the appearance of human emotional behavior. However, in biology, the 4Fs (feeding, fighting, fleeing, and reproduction) (Pribram, 1960) are paramount. S-RETIC (Kilmer *et al.*, 1969) modeled the brain's reticular formation as a stack of modules, each with a slightly different selection of input, but each trying to decide to which 4F-like mode to commit the organism. They would communicate back and forth, competing and cooperating until reaching a consensus that would switch the organism's mode. Within any mode of behavior, many different acts are possible: if the cat should flee, will it take the mouse first or leave it, climb a tree or skirt it? The notion is that a hierarchical structure that computes modes, and then acts within modes, might in some sense be "better" than one that tries to determine successive acts directly. Consider an animal with a set of basic functions, each with appropriate perceptual schemas and access to various motor schemas. Each perceptual schema evaluates the current state to come up with an "urgency level" for activating various motor schemas, as well as determining appropriate motor parameters. A competition mechanism can adjudicate between these processes. A *motivational system* is a state-evaluation process that can adjust the relative weighting of the different functions, raising the urgency level for one system while lowering the motivation system for others depending on the context. One may then generalize "modes" to abstract groups of tasks, with many strategies gathered into a small number of modes such that it is in general more efficient, when faced with a problem, to first select an appropriate mode and then select a strategy from within that mode. The catch is that there may be cases in which rapid commitment to one group of strategies may preclude finding the most appropriate strategy — possibly at times with disastrous consequences. The brain has to balance this "fast commitment" process against more subtle evaluative processes that can check the suitability of a chosen strategy before committing to it completely (and thus we return to the varied levels of "Cognitive Architectures"). We might then liken "motivation" to biases that favor one strategy group over another, and "emotion" to the way in which these biases interact with more subtle computations. One way emotions may become maladaptive, then, is when biases favoring rapid commitment to one mode overwhelm more cautious analysis of available strategies.

Recall our earlier introduction of two "views" of emotion: (1) emotional expression for communication and social coordination; and (2) emotion for the organization of behavior. The very utility of "modes" in decision making (sense 2 of emotion) may well make fast communication of modal commitment an effective way to coordinate animals or humans.

Emotion changes the operating characteristics of cognition and action selection. There is neuroscientific evidence that affect is a prerequisite for establishing long-term memories. It also has important consequences for the allocation of attention. Fear tends to focus attention on local details; but under conditions of positive affect people tend to focus on the bigger picture (Gasper & Clore, 2002). Resource mobilization and prioritization of behaviors rely on compact signals that have high-impact on the functioning of an autonomous agent. The underlying hypothesis is that hierarchical assessment of task priorities can help animals cope with the complexity and unreliability of the (external or internal) environment — mirroring the rich interaction of cognitive and emotional processing in the mammalian brain. Adaptive mechanisms can adjust the hierarchy and grouping of tasks to match the constraints imposed by time, physical limits, and energy resources. In this regard, note that human emotion becomes maladaptive not only when rapid commitment processes overwhelm more cautious analysis, but also when worry or other related emotions that favor more cautious analysis overwhelm adaptive rapidity.

### 3.8. Concluding Claims About Consciousness

To conclude briefly, the A&H theory of consciousness has the following ingredients.

(i) Human consciousness as we normally experience it is a property of the brain, rather than some separate "mind stuff".
(ii) Consciousness is a distributed property that has access to many brain regions, and provides a précis based on the state of other brain regions that is only partially verbalizable.
(iii) It is possible that portions of our brain can support forms of "animal awareness" that may enrich human consciousness but seem qualitatively different in nature; but
(iv) What makes human consciousness so different is that it includes expression of our thoughts in language.
(v) The *communication plexus* underlying language has (by a process of "Jacksonian evolution") restructured the brain in such a way that consciousness may seem to be sometimes observer and sometimes controller.

The third acknowledges a possible evolutionary continuity in consciousness while the fourth stresses that the consciousness of humans who possess language is qualitatively different from that of other creatures; while the last ascribes a specific role to the evolution of a "communication plexus" for the communication of intended action in the emergence of human consciousness.

MSH complements the A&H theory by offering a more "brain-based" approach to the evolution of human language, tracing it through seven stages: (1) grasping, (2) a mirror system for grasping, (3) a "simple" imitation system, (4) a "complex" imitation system, (5) a manual-based communication system, (6) speech, and (7) language. Stages 1 and 2 are present in the monkey and posited to be shared by the last common ancestor of monkey and human; stage 3 is posited to be more fully developed in the last common ancestor of human and chimpanzee; stages 4, 5 and 6 are posited to occur along the hominid line. However, we do not argue that the "protospeech" of the first *Homo sapiens* was a human language in anything like the modern sense. Rather, the development of human languages and the attendant enrichment of human consciousness was, I argue, a process that took many tens of millennia.

We also outlined an account of the roles of empathy and emotion in the emergence of human consciousness. In this regard, it is worth noting the complementary nature of language and music (Arbib, 2013d; Arbib & Iriki, 2013; Levinson, 2013; Seifert *et al.*, 2013), the former more adapted to the expression of propositional content, the latter more directly linked to the evocation of emotion, suggesting that further understanding of their relationship will indeed enrich the approach to the emergence of consciousness sketched in these pages.

To conclude, I should mention that there is now a range of research linking theories of consciousness to data on the brain, but analysis of the recent material is outside the scope of this chapter. A fine example of such an approach is given by the review of "Experimental and Theoretical Approaches to Conscious Processing" by Stanislas Dehaene and Jean-Pierre Changeux (Dehaene & Changeux, 2011) which links data on macaque neurophysiology and human brain imaging with explicit modeling via computational neuroscience. However, it is worth noting that although the aim of their study is more ambitious in its attempt to review current neuroscience data it is also less ambitious in that it is limited in its delineation of consciousness: its main concern is with one question: "When a visual stimulus is presented to a human subject, they may or may not be able to declare (whether verbally, or by some motor response specified by the experimenter) what they have seen. What differences in neural activity characterize the difference between being aware vs. unaware of a visual stimulus?" The distinction is made more interesting by the fact that a visual stimulus may serve to prime a later response even though the subject remains unaware of the first stimulus.

In developing a theory to integrate the psychological and neuroscience data they review, Dehaene and Changeux situate themselves in agreement with a range of models of consciousness (see their paper for detailed citations) which emphasize four concepts:

- A supervision system,
- A serial processing system,
- A coherent assembly formed by re-entrant or top-down loops,
- A global workspace for information sharing. The theater metaphor

and indeed they call their framework the *global neuronal workspace* (GNW) hypothesis, and propose that associative perceptual, motor, attention, memory, and value areas interconnect to form a higher-level unified space where information is broadly shared and broadcast back to lower-level processors. My theory agrees to a limited extent (it sees supervision and serial processing as only partially in play, being side-effects of inherently distributed processes of competition and cooperation), but gives group coordination and the emergence of language (i.e., social rather than purely internal phenomena) a special evolutionary role. I suggest that language evolved for communication but as a corollary enriched our consciousness with an internal monolog. To simplify, I would claim that a frog has "no" consciousness (though it can adaptively react within its environment); a monkey has an extended animal awareness but no access to language; and a human has all this plus a "running commentary" that can structure or annotate experience but may only intermittently play a crucial role in committing the person to a course of action. Data are emerging that are relevant to all this, but their analysis to test my claims must await another day.

## Provenance

A conference was organized in Zaragoza, Spain, by Pedro Marijuan in 1999 to mark the centenary of the Spanish neuroanatomist Ramón y Cajal's *Textura del Sistema Nervioso del hombre.* Although Cajal wrote very little on consciousness in this great (and still relevant) classic of neuroanatomy, the theme of the conference was *Cajal and Consciousness: Scientific Approaches to Consciousness.* My article from the 1999 meeting was published as (Arbib, 2001), with an abridged version coming out recently (Arbib, 2014) in a special issue of the *Journal of Integrative Neuroscience* on the so-called "hard problem of consciousness." However, it must be noted that I do not claim to solve what some people call the *hard* problem of consciousness but instead focus on the *not at all easy* problem of how, given the presence of consciousness in ancestral species, humans came to have a consciousness with characteristics unshared by other creatures. The present paper combines some of the material from the previous papers with new material that should make it more relevant to the volume of which it is part, although it is not at the biophysical level suggested by the book's title, *Biophysics of Consciousness: A Foundational Approach.*

## Acknowledgments

The present chapter is based in part on work supported by the National Science Foundation under Grant No. BCS-1343544 "INSPIRE Track 1: Action, Vision and Language, and their Brain Mechanisms in Evolutionary Relationship" (Michael A. Arbib, Principal Investigator). My thanks to Marsha Kinder and three anonymous referees for their constructive responses to an earlier draft.

## References

Adolphs, R. (2009) The social brain: Neural basis of social knowledge. *Annu. Rev. Psychol.*, **60**, 693–716.

Arbib, M. A. (1972) Consciousness: The secondary role of language. *J. Philosophy*, **69**, 579–591.

Arbib, M. A. (1978) Segmentation, schemas, and cooperative computation. In: S. A. Levin, ed. *Studies in Mathematical Biology, Part I.* Providence, RI: Mathematical Association of America, pp. 118–155.

Arbib, M. A. (1985) *In Search of the Person: Philosophical Explorations in Cognitive Science.* Amherst, MA: University of Massachusetts Press.

Arbib, M. A. (1989a) *The Metaphorical Brain 2: Neural Networks and Beyond.* New York: Wiley-Interscience.

Arbib, M. A. (1989b) Modularity, schemas and neurons: A critique of Fodor. In: P. Slezak and W.R. Albury, eds. *Computers, Brains and Minds.* Netherlands: Springer, pp. 193–219.

Arbib, M. A. (1992) Schema theory. In: S. Shapiro, ed. *The Encyclopedia of Artificial Intelligence*. New York: Wiley-Interscience, pp. 1427–1443.

Arbib, M. A. (2001) Co-evolution of human consciousness and language. In: P.C. Marijuan, ed. *Cajal and Consciousness: Scientific Approaches to Consciousness on the Centennial of Ramon y Cajal's Textura*, pp. 195–220.

Arbib, M. A. (2002) The mirror system, imitation, and the evolution of language. In: K. Dautenhahn and C. L. Nehaniv, eds. *Imitation in Animals and Artifacts. Complex Adaptive Systems*. Cambridge, MA: The MIT Press, pp. 229–280.

Arbib, M. A. (2005a) Beware the passionate robot. In: J.-M. Fellous and M. A. Arbib, eds. *Who Needs Emotions? The Brain Meets the Robot*. New York: Oxford University Press.

Arbib, M. A. (2005b) From monkey-like action recognition to human language: An evolutionary framework for neurolinguistics. *The Behavioral and Brain Sciences*, **28**, 105–124.

Arbib, M. A. (2007) Autism — more than the mirror system. *Clin. Neuropsych.*, **4**, 208–222.

Arbib, M. A. (2011) Niche construction and the evolution of language: was territory scavenging the one key factor? *Interact. Stud.*, **12**, 162–193.

Arbib, M. A. (2012) *How the Brain Got Language: The Mirror System Hypothesis*. New York and Oxford: Oxford University Press.

Arbib, M. A. (2013a) Complex imitation and the language-ready brain. *Lang. Cogn.*, **5**, 273–312.

Arbib, M. A. (2013b) Neurons, schemas, persons and society — revisited. In: G. Auletta, I. Colagè and M. Jeannerod, eds. *Brains Top Down: Is Top-Down Causation Challenging Neuroscience?* Singapore: World Scientific, pp. 57–87.

Arbib, M. A. (2013c) Précis of how the brain got language: the mirror system hypothesis. *Lang. Cogn.* **5**, 107–131.

Arbib, M. A., ed. (2013d) *Language, Music and the Brain, a Mysterious Relationship. Strüngmann Forum Reports*, Vol. 10. Cambridge, MA: The MIT Press.

Arbib, M. A. (2014) Co-evolution of human consciousness and language (revisited). *J. Integra. Neurosci.*, **13**, 187–200.

Arbib, M. A. & Fellous, J. M. (2004) Emotions: From brain to robot. *Trends Cogn. Sci.*, **8**, 554–561.

Arbib, M. A. & Hesse, M. B. (1986) *The Construction of Reality*. Cambridge: Cambridge University Press.

Arbib, M. A. & Iriki, A. (2013) Evolving the language- and music-ready brain. In: M.A. Arbib, ed. *Language, Music, and the Brain: A Mysterious Relationship, Strüngmann Forum Reports*, Vol. 10. Cambridge, MA: MIT Press, pp. 359–375.

Arbib, M. A. & Liaw, J.-S. (1995) Sensorimotor transformations in the worlds of frogs and robots. *Artif. Intell.*, **72**, 53–79.

Arbib, M. A. & Rizzolatti, G. (1997) Neural expectations: A possible evolutionary path from manual skills to language. *Commun. Cogn.*, **29**, 393–424.

Barnard, A. (2013) Cognitive and social aspects of language origins. In: C. Lefebvre, B. Comrie and H. Cohen, eds. *New Perspectives on the Origins of Language* Amesterdam/Philadelphia: John Benjamins, pp. 53–71.

Bickerton, D. (1995) *Language and Human Behavior*. Seattle: University of Washington Press.

Bickerton, D. (2009) *Adam's Tongue. How Humans Made Language, How Language Made Humans*. New York: Hill & Wang.

Butler, A. B. & Hodos, W. (2005). *Comparative Vertebrate Neuroanatomy: Evolution and Adaptation*, Wiley & Sons, New York.

Chomsky, N. (1980) *Rules and Representations*. New York: Columbia University Press.

Dalgleish, T. (2009) The thinking person's emotion theorist: A comment on Bartlett's feeling, imaging, and thinking. *Br. J. Psychol.*, **100**, 199–201.

Damasio, A. (1994) *Descartes' Error: Emotion, Reason, and the Human Brain*. New York: Putnam Publishing.

Damasio, A. (1999) *The Feeling of What Happens: Body and Emotion in the Making of Consciousness*. San Diego, CA: Harvest/Harcourt.

Damasio, A. (2012) *Self Comes to Mind: Constructing the Conscious Brain*. New York: Vintage.

Dapretto, M., Davies, M.S., Pfeifer, J.H., Scott, A.A., Sigman, M., Bookheimer, S.Y. & Lacoboni, M. (2006) Understanding emotions in others: Mirror neuron dysfunction in children with autism spectrum disorders. *Nat. Neurosci.*, **9**, 28–30.

Darwin, C. (1872) *The Expression of the Emotions in Man and Animals (republished in* 1965). Chicago: University of Chicago Press.

Dehaene, S. & Changeux, J.-P. (2011) Experimental and theoretical approaches to conscious processing. *Neuron*, **70**, 200–227.

di Pellegrino, G., Fadiga, L., Fogassi, L., Gallese, V. & Rizzolatti, G. (1992) Understanding motor events: A neurophysiological study. *Exp. Brain Res.*, **91**, 176–180.

Dominey, P. F. (2013) The tip of the language iceberg. *Lang. Cogn.*, **5**, 177–187.

Elman, J. L. (2009) On the meaning of words and dinosaur bones: Lexical knowledge without a lexicon. *Cogn. Sci.*, **33**, 547–582.

Elman, J. L., Hare, M. & McRae, K. (2004) Cues, constraints, and competition in sentence processing. *Beyond nature-nurture: Essays in honor of Elizabeth Bates*, pp. 111–138.

Fellous, J.-M., & Arbib, M. A. (Eds.). (2005). *Who Needs Emotions: The Brain Meets the Robot*. Oxford, New York: Oxford University Press.

Fellous, J.-M. & Ledoux, J. E. (2005) Toward basic principles for emotional processing: what the fearful brain tells the robot. In: J.-M. Fellous and M. A. Arbib, eds. *Who Needs Emotions: The Brain Meets the Robot*. Oxford, New York: Oxford University Press, pp. 79–115.

Gallese, V. & Goldman, A. (1998) Mirror neurons and the simulation theory of mind-reading. *Trends Cognit. Sci.*, **2**, 493–501.

Gasper, K. & Clore, G. L. (2002) Attending to the big picture: Mood and global versus local processing of visual information. *Psychol. Sci.*, **13**, 34–40.

Gazzola, V., Aziz-Zadeh, L. & Keysers, C. (2006) Empathy and the somatotopic auditory mirror system in humans. *Curr. Biol.*, **16**, 1824–1829.

Goodale, M. A., Milner, A. D., Jakobson, L. S. & Carey, D. P. (1991) A neurological dissociation between perceiving objects and grasping them. *Nature,* **349**, 154–156.

Grafton, S. T., Arbib, M. A., Fadiga, L. & Rizzolatti, G. (1996) Localization of grasp representations in humans by positron emission tomography. 2. Observation compared with imagination. *Exp. Brain Res.*, **112**, 103–111.

Hameroff, S. & Penrose, R. (2014) Consciousness in the universe: A review of the 'Orch OR' theory. *Phys. Life Rev.*, **11**, 39–78.

Head, H. & Holmes, G. (1911) Sensory disturbances from cerebral lesions. *Brain,* **34**, 102–254.

Holmes, G. (1939) The cerebellum of man. *Brain,* **62**, 1–30.

Hubel, D. H. & Wiesel, T. N. (1959) Receptive fields of single neurones in the cat's striate cortex. *J. Physiol.*, **148**, 574–591.

Humphrey, N. K. (1970) What the frog's eye tells the monkey's brain. *Brain Behav. Evol.*, **3**, 324–337.

Iriki, A. & Sakura, O. (2008) The neuroscience of primate intellectual evolution: Natural selection and passive and intentional niche construction. *Philos. Trans. R. Soc. Lond. B. Biol. Sci.*, **363**, 2229–2241.

Iriki, A. & Taoka, M. (2012) Triadic (ecological, neural, cognitive) niche construction: A scenario of human brain evolution extrapolating tool use and language from the control of reaching actions. *Philos. Trans. R. Soc. B: Biol. Sci.*, **367**, 10–23.

Ito, M. (2012) *The Cerebellum: Brain for an Implicit Self.* Upper Saddle River, NJ: FT Press.

Jackson, J. H. (1874) On the nature of the duality of the brain. *Med. Press Circular,* **1,** 19–63.

Jackson, J. H. (1878–1879) On affections of speech from disease of the brain. *Brain,* **1**, 304–330.

Jeannerod, M. (2005) How do we decipher others' minds? In: J.-M. Fellous and M. A. Arbib, eds. *Who Needs Emotions: The Brain Meets the Robot*. Oxford, New York: Oxford University Press, pp. 147–169.

Jeannerod, M., Decety, J. & Michel, F. (1994) Impairment of grasping movements following a bilateral posterior parietal lesion. *Neuropsychologia,* **32**, 369–380.

Kilmer, W. L., McCulloch, W.S. & Blum, J. (1969) A model of the vertebrate central command system. *Int. J. Man-Machine Studies,* **1,** 279–309.

LeDoux, J. (2012) Rethinking the emotional brain. *Neuron,* **73**, 653–676.

Ledoux, J. E. (2000) Emotion circuits in the brain. *Annu. Rev. Neurosci.*, **23**, 155–184.

Lee, P. (2000) When is 'linguistic relativity' Whorf's linguistic relativity? In: M. Pütz and M. Verspoor, eds. *Explorations in Linguistic Relativity.* John Benjamins, pp. 45–68.

Lettvin, J. Y., Maturana, H., McCulloch, W. S. & Pitts, W. H. (1959) What the frog's eye tells the frog brain. *Proc. IRE*, **47**, 1940–1951.

Levinson, S. C. (2013) Cross-Cultural Universals and Communication Structures. In: M.A. Arbib, ed. *Language, Music, and the Brain: A Mysterious Relationship. Strüngmann Forum Reports, Vol.* 10. Cambridge, MA: MIT Press.

Maravita, A. & Iriki, A. (2004) Tools for the body (schema). *Trends Cogn. Sci.*, **8**, 79–86.

Merker, B. (2007a) Consciousness without a cerebral cortex: A challenge for neuroscience and medicine. *Behav. Brain Sci.* **30**, 63–81.

Merker, B. (2007b) Grounding consciousness: The mesodiencephalon as thalamocortical base. *Behav. Brain Sci.* **30**, 110–120.

O'Regan, J. K. (2011) *Why Red doesn't Sound like a Bell: Understanding the Feel of Consciousness*. Oxford: Oxford University Press.

O'Regan, J. K. (2012) How to build a robot that is conscious and feels. *Mind. Mach.*, **22**, 117–136.

O'Regan, J. K. & Block, N. (2012) Discussion of J. Kevin O'Regan's "why red doesn't sound like a bell: Understanding the feel of consciousness." *Rev. Philos. Psychol.*, **3**, 89–108.

Odling-Smee, F. J., Laland, K. N. & Feldman, M. W. (2003) *Niche Construction: The Neglected Process in Evolution*. Princeton, NJ: Princeton University Press.

Olds, J. (1969) The central nervous system and the reinforcement of behavior. *Am. Psychol.*, **24**, 114–132.

Ortony, A., Clore, G. L. & Collins, A. (1988) *The Cognitive Structure of Emotions*. New York: Cambridge University Press.

Ortony, A., Norman, D. & Revelle, W. (2005) Affect and proto-affect in effective functioning. In: J.-M. Fellous and M. A. Arbib, eds. *Who needs emotions* New York and Oxford: Oxford University Press, pp. 173–202.

Panksepp, J. (1998) The periconscious substrates of consciousness: Affective states and the evolutionary origins of the self. *J. Consciousness Stud.*, **5**, 566–582.

Pinker, S. (2010) The cognitive niche: Coevolution of intelligence, sociality, and language. *Proc. Natl. Acad. Sci. USA*, **107**(Suppl 2), 8993–8999.

Pribram, K. H. (1960) A review of theory in physiological psychology. *Annu. Rev. Psychol.*, **11**, 1–40.

Ramachandran, V. S. (2011) *The Tell-Tale Brain: A Neuroscientist's Quest for What Makes Us Human*. London: W. W. Norton & Company.

Ramachandran, V. S. & Oberman, L. M. (2006) Broken mirrors: A theory of autism. *Sci. Am.*, **295**, 62–69.

Rizzolatti, G. & Arbib, M. A. (1998) Language within our grasp. *Trends Neurosci.*, **21**, 188–194.

Rizzolatti, G., Fadiga, L., Gallese, V. & Fogassi, L. (1996a) Premotor cortex and the recognition of motor actions. *Cogn. Brain Res.*, **3**, 131–141.

Rizzolatti, G., Fadiga, L., Matelli, M., Bettinardi, V., Paulesu, E., Perani, D. & Fazio, F. (1996b) Localization of grasp representations in humans by PET: 1. Observation versus execution. *Exp. Brain Res.*, **111**, 246–252.

Rizzolatti, G., Fogassi, L. & Gallese, V. (2001) Neurophysiological mechanisms underlying the understanding and imitation of action. *Nat. Rev. Neurosci.*, **2**, 661–670.

Rolls, E. T. (2000) Neurophysiology and functions of the primate amygdala, and the neural basis of emotion. In: J.P. Aggleton, ed. *The Amygdala: A Functional Analysis*. 2nd edn. Oxford: Oxford University Press, pp. 447–478.

Rolls, E. T. (2005) What are emotions, why do we have emotions, and what is their computational role in the brain? In: J.-M. Fellous and M.A. Arbib, eds. *Who Needs Emotions: The Brain Meets the Robot*. Oxford, New York: Oxford University Press, pp. 117–146.

Sarraute, N. (1984) *Childhood (Translated by Barbara Wright in consultation with the author)*. George Braziller.

Schultz, W. (2000) Multiple reward signals in the brain. *Nat. Rev. Neurosci.*, **1**, 199–207.

Seifert, U., Verschure, P. F. M. J., Arbib, M. A., Cohen, A. J., Fogassi, L., Fritz, T. & Scherer, K. (2013) Semantics of internal and external worlds. In: M.A. Arbib, ed. *Language, Music, and the Brain: A Mysterious Relationship, Strüngmann Forum Reports*, Vol. 10. Cambridge, MA: MIT Press.

Sloman, A., Chrisley, R. & Scheutz, M. (2005) The architectural basis of affective states and processes. In: J. M. Fellous and M. A. Arbib, eds. *Who Needs Emotions*. New York & Oxford: Oxford University Press, pp. 203–244.

Stoerig, P. & Cowey, A. (1997) Blindsight in man and monkey. *Brain*, **120**, 535–559.

Suddendorf, T. & Corballis, M. C. (2007) The evolution of foresight: What is mental time travel, and is it unique to humans? *Behav. Brain Sci.*, **30**, 299–351.

Tomasello, M., Carpenter, M., Call, J., Behne, T. & Moll, H. (2005) Understanding and sharing intentions: The origins of cultural cognition. *Behav. Brain Sci.*, **28**, 675–691.

Weiskrantz, L. (1974) The interaction between occipital and temporal cortex in vision: An overview. In: F.O. Schmitt and F.G. Worden, eds. *The Neurosciences Third Study Program*. Cambridge, MA: The MIT Press, pp. 189–204.

Whitehead, C. (2010) The culture ready brain. *Soc. Cogn. Affect. Neurosci.*, **5**, 168–179.

Whorf, B. L. (1937) Discussion of Hopi linguistics. In: J.B. Carroll, ed. *Language, Thought, and Reality: Selected Writings of Benjamin Lee Whorf*. Cambridge, MA: MIT Press, pp. 102–111.

Wicker, B., Keysers, C., Plailly, J., Royet, J. P., Gallese, V. & Rizzolatti, G. (2003) Both of us disgusted in My insula: The common neural basis of seeing and feeling disgust. *Neuron*, **40**, 655–664.

Williams, J. H. G. (2008) Self-other relations in social development and autism: Multiple roles for mirror neurons and other brain bases. *Autism Res.*, **1**, 73–90.

Williams, J. H. G., Whiten, A., Suddendorf, T. & Perrett, D. I. (2001) Imitation, mirror neurons and autism. *Neurosci. Biobehav. Rev.*, **25**, 287–295.

Biophysics of Consciousness: A Foundational Approach
R. R. Poznanski, J. A. Tuszynski and T. E. Feinberg
Copyright © 2017 World Scientific, Singapore.

# 4

# CONSCIOUSNESS BY SURPRISE: A NEUROPSYCHOANALYTIC APPROACH TO THE HARD PROBLEM*

*Mark Solms*

*Department of Psychology, University of Cape Town*
*Rondebosch 7701, South Africa*

## Abstract

A neuropsychoanalytic approach to the hard problem of consciousness revolves around the distinction between the *subject* of consciousness and *objects* of consciousness. In contrast to the mainstream of cognitive science, neuropsychoanalysis prioritizes the subject. The subject of consciousness is the indispensable medium in which consciousness of objects is inscribed. This has implications for our conception of the mental in general. The subject of consciousness is not registered in the classical exteroceptive modalities; it is not a cognitive representation. On the contrary, the exteroceptive modalities are registered *by* the subject. Cognitive representations are mental solids (images of objects) embedded within subjectivity. The subject of consciousness is an endogenous state of the body. It is important to recognize that mental solids (including the body-as-object) are no more real than the subjective state in which

*This is a considerably revised version of an article first published in the *Journal of Integrative Neuroscience*, **13**: 173–185 (2014). I would like to thank my colleagues Andrea Clarici, Gyuri Fodor and Lois Oppenheim for helpful comments on earlier drafts of this revision, which greatly improved the chapter as published here.

they are inscribed (the body-as-subject). Also, the subjective state is not without quality. It is conventionally described quantitatively, as the level of consciousness or wakefulness. But it *feels* like something to be awake. The primary modality of subjective states is *affect*. Affect supplies sentience in general. Some implications of this approach are discussed here, in broad brush strokes, with particular emphasis on the need for us mental scientists to infer an abstract *mechanism* to explain both the objective and the subjective data of consciousness generation. (This is dual-aspect monism.) I propose that the abstraction we are looking for (i.e., the mechanism that lies behind both the labile neuronal networks that characterize consciousness physiologically and the unpredictability that characterizes it psychologically) is *uncertainty*. This mechanism (uncertainty) is equivalent to what Friston calls "surprise," which he defines mathematically in terms of "free energy."

*Keywords*: Neuropsychoanalysis; hard problem; dual-aspect monism; affect; prediction error; uncertainty; free energy.

## 4.1. Introduction

I welcome this opportunity to sketch a neuropsychoanalytic approach to the hard problem of consciousness. I call this approach meta-neuropsychological. I will explain later what I mean by the terms "neuropsychoanalytic" and "meta-neuropsychological."

## 4.2. Visual Consciousness vs. Consciousness Itself

I believe that the approach to consciousness taken by the mainstream of cognitive neuroscientists (following Crick, 1994) has been hampered by an excessive focus on exteroceptive, objectified forms of consciousness, especially *visual* consciousness. That approach gave rise to the formulation of a "hard problem" in the first place (Chalmers, 1995). I think we will find the going easier if we focus on consciousness itself.

My starting point is the distinction between the *subject* of consciousness and *objects* of consciousness. Visual consciousness, like all exteroception, pertains to objects. However consciousness of objects cannot exist without the state of *being conscious* — without a consciously perceiving subject. Objects of consciousness must always be perceived *by* a conscious subject. If the subject loses consciousness, the objects disappear. The converse does not apply. That is why I prefer to focus on the subject of consciousness; it is the nub of the matter. Notwithstanding the fact that no-one can point to them with their fingers, subjective states exist, they are part of nature; and they are the common ground of all experience. Wakefulness, for example, is a subjective state. We can only point with our fingers to the objective (physiological) markers of wakefulness, but these markers are not wakefulness itself. We cannot point with our fingers to wakefulness itself *because* it is a subjective state. Yet everyone knows from their most immediate experience[1] that wakefulness exists, and that it differs from other subjective states (like sleep, for example).

I am not saying that things in the world do not exist in themselves, independently of consciousness. All the evidences point in the opposite direction. I am only saying that in the absence of an experiencing subject there can be no consciousness of things.

### 4.3. Visual Consciousness of Objects vs. The Reality of Things

This is a good place to remind ourselves that things in themselves (as opposed to our perceptions of them) do not possess visual quality. Vision is a property of the brain (cf. Zeki 1993, p. 236, who made this point regarding color vision). We tend to think of objects as if they consist ontologically in the forms by which they are presented to us visually. We think of our visual image of the world as if it were, in a word, *reality*.

Please note: this is not a philosophical argument, it is a physiological one. The stimuli that impinge on the retina are *different in kind* from the signals propagated onwards from the retina to the brain. The external world is, as it were, *translated* by the nervous system. Moreover, in the brain the retinal translation of the world is merged with other physiological parameters, which are generated not by the external world but endogenously, by the perceiving subject (see in particular Sec. 4.7).

We picture reality as being synonymous with what we *see*, moreover, because human perception is dominated by vision. We do not typically think of reality as being made of smells, for example. This happens for good reason: vision (especially at the cortical level) generates "mental solids" — that is, images of objects — which yield more stable topological renditions of things than the other modalities of perception (Solms, 2013). The same

---

[1]In fact, following the logic of Descartes' *cogito ergo sum*, the existence of our subjective conscious state is the most valid and reliable inference that can ever be drawn from empirical data.

applies to a diminishing extent to the other perceptual modalities — after vision, come somatic sensation, hearing, taste and smell — proportionate to the amount of cortex allocated to them (see the end of Sec. 4.7).

By contrast, we do not think of subjective states as being real — notwithstanding the fact we have just acknowledged, namely that they do exist. Again, this happens for good reason: there is obvious biological value (increased chance of surviving to reproduce) in drawing a sharp distinction between subjective states and objective things — between what you desire and what you see, for example — in favor of the latter. This prejudice — the tendency to prioritize exteroceptive consciousness (and vision in particular) — is what has led consciousness studies astray. In short, it has led mental scientists to behave as if visual consciousness is synonymous with consciousness itself.

My general point (in this section) is that we think of reality — of the material world — as if it were defined by the qualities bestowed on it by our perceptual modalities. With our exteroceptive senses, we experience matter as being characterized by (visual) solidity, (tactile) tangibility, etc. We might experience reality differently and think about it differently if we were bats (cf. Nagel, 1974). We might likewise have a totally different conception of it if we possessed, say, only a sense of hearing; then we would presumably think that reality consists essentially in something like sound waves.

### 4.4. Perceptual Relativity vs. Abstractions of the Real

One conclusion that can be drawn from my argument regarding the distinction between conscious perception of reality and reality itself is the following: The quality you bestow on things depends on your observational perspective. For example, when you observe lightening visually, you see a flash; when you observe it aurally, you hear a clap. What, then, does lightning itself consist in? Actually it consists in neither of these perceptual objects; *in reality* it is a geo-electrical process that manifests differently in different sense modalities. Please note that this, the best description of the real state of affairs, requires *abstraction* from all the perceptual data; in this case it requires an abstraction called "electricity." *We are in need of an analogous abstraction to solve the hard problem of consciousness*, about which more below (Sec. 4.18).

If we extend the boundaries of human sensory capacity through artificial aids (like electron microscopes) we find that matter appears quite differently from how it looks to the naked eye. When we probe it to its core, we are driven to descriptions that transcend the properties of perception. Again, in the end, we are driven to abstractions, like "quarks," "weak force," "strong force," etc., and the very materiality of matter (its solidity, its tangibility) disappears. This is a significant fact for the hard problem of consciousness, and I will return to it.

For now, I want to conclude simply that the material attributes of objects are not absolute, they are relative. They can be shown to vary in relation to the observing subject (just as Einstein showed that the properties of space and time vary in relation to the velocity of the observer). The perceived materiality of reality varies in relation to the relative magnitude and position of the observer. If you expand the magnitude of the object by means of an electron microscope (thereby reducing the magnitude of the subject, relatively speaking) matter seems to lose its solidity and tangibility. Likewise if you change the position of the object relative to the sensory surfaces of the observer, it acquires different "material" qualities, for example, when you listen to an object rather than look at it, it loses its solidity.

## 4.5. Perception and Cognition vs. Reality

Since perception of reality must be distinguished from reality itself, from the thing being perceived, I also want to conclude that the distinction between "real" and "mental" does not coincide with the distinction between perception and idea. Perception of reality is part of the mental domain, no less than cognition is. Both entail translations of reality. This much is obvious, but it is not the way we commonly think. Common sense can be misleading — even (perhaps especially) to scientists.

Moreover, perceiving reality and cognizing it entail broadly the *same type* of translation. A defining feature of Descartes' *res extensa* — namely dimensionality — applies also to his *res cogitans,* to the extent that it consists in representations derived from the outer world (the 'impressions' of Hume and Locke). This conclusion is supported by recent mental imagery research: perception and imagination utilize the same topological representational structures. That is why, for example, the time taken to mentally rotate three-dimensional (3D) designs of increasing complexity tallies with the time taken in reality (Shepard & Metzler, 1971). Object representations are made of the same stuff, whether they be perceptions or ideas; both are *representations,* and both are dimensional and therefore measurable in principle. This is because both are derived from topological maps of the sensory receptor surfaces, which in turn literally map reality.

I believe that Descartes drew the distinction between mind and matter wrongly; he misconstrued the mind/body problem. It is a mistake to contrast ideas ("mind," i.e., cognized objects) with objects ("matter," i.e., perceived objects); both of these things are mental. Objects *as we know them* are perceptions. The subject of consciousness, not its objects, is the essence of the mental.

The mental, then, includes both perception and cognition. But that is by no means all it consists in. The prejudice that the mental consists only in perception and cognition is a major problem with mainstream cognitive accounts of consciousness, and I will return to it in Sec. 4.12.

### 4.6. The Conscious vs. the Mental

Before I shift focus from perception and cognition — which pertain to objects of consciousness — and address the subject of consciousness, the nub of the problem, I must draw another basic distinction that was overlooked by the philosophers who laid the foundations of modern psychology.

If the findings of recent cognitive science have taught us anything, they have taught us that it is necessary to distinguish between conscious and *unconscious* mental processes. Not all mental processes are conscious. This was, of course, Freud's great insight.

This has massive implications for psychology. If mental processes are not defined by consciousness, then what distinguishes mental from nonmental processes? What is the defining property of the mental? Freud's philosophy teacher, Brentano (1874) suggested that it was intentionality — or "aboutness." That is, he argued that mental processes always "intend toward" something — simply put, they are always *about* something (you cannot think without thinking about something; you cannot remember without remembering something; etc.). Freud's point was that the things in question need not be conscious. That is, according to Freud, the mental is intentional but what is "intended towards" is not necessarily conscious.

All representations are about something, including nonbiological representations like photographs (Strawson, 1994). The concept of aboutness, if limited to the exteroceptive conception of consciousness, therefore, is too bound up with the images of objects. I think the mental is more intrinsically bound up with subjectivity, as did Brentano, and therefore with intentionality in the colloquial sense — that is, with *volition* (see Secs.

4.11–4.16). Intentionality entails an active "reaching for" objects by the subject. The living subject is always in a state of desire.

But I am running ahead of myself. For now I want to add only this to my argument: the fact of unconscious mental processes applies equally to perception and cognition. It is possible to see without awareness[2] (blindsight), recognize without awareness, read without awareness, learn without awareness, remember without awareness, make decisions without awareness, and so on. In fact, just about every perceptual and cognitive task can be performed without awareness (Bargh, 2014).

Please note that such tasks pertain to the processing of *representations*. This implies that not only associative but also representational processing can operate unconsciously. (I point this out because so many people seem to think that unconscious "automatized" mental processes are exclusively associative, Pavlov–LeDoux type processes.) Reading of words and recognizing of faces, for example, can readily occur with zero-prime subliminal presentation of the stimuli, where the subject is completely unaware of having perceived any stimulus at all (i.e., where there is no *visual* consciousness of the stimulus (as opposed to emotional consciousness; see footnote 2). But such tasks require detailed and precise representational mapping and symbolic decoding of the spatial configuration of the stimulus; in other words they require cortex. I am unaware of any evidence for the notion that it is possible to read without cortex.

These facts have two implications. First, they lead to the conclusion that higher cortical functions are not intrinsically conscious. Second, they beg the question: what is consciousness for? What does consciousness add to the already impressive capacities of unconscious perception and cognition?

In later sections (mainly Secs. 4.11–4.16), I am going to argue that it adds something like intentionality in the colloquial sense.

## 4.7. Higher Cortical Functions vs. Consciousness

Moruzzi & Magoun (1949) and Magoun (1952) found decades ago — apparently to their own surprise — that cortical activity in cats was

---

[2]"Awareness" is an ambiguous term, which is often interpreted as knowing about experiencing, rather than experiencing itself. This has led to all sorts of confusion, especially in subliminal perception experiments where it is assumed that "lack of visual awareness" is synonymous with "no conscious response." (See Shevrin *et al.*, 2012.) Another sort of confusion about "awareness" is discussed in footnote 6.

mainly driven not by external sensory stimulation but endogenously, from the brainstem below. The same applies to all other animals, including humans. Penfield & Jasper (1954), for example, concluded that human consciousness depends upon the integrity, not of cortex, but rather of the upper brainstem — of what they called the "centrencephalic" region. The most striking recent evidence for this conclusion is provided by accounts of the behavior of hydranencephalic children, born without cortex (Shewmon *et al.*, 1999; Merker, 2007). These children — like all decorticated mammals — possess a normal sleep/waking cycle. They also show absence seizures, in which consciousness is lost and regained. They also display a full range of basic emotions, such as fear, joy and anger, elicited by appropriate stimuli. In fact, decorticate creatures are *excessively* emotional (Bard, 1934).

If the representational functions performed by cortex — by the so-called organ of the mind — are not the basis of consciousness, then what is cortex for? I think the intrinsic function of cortex is not consciousness but rather *memory* — specifically, memory of a kind that makes it possible for detailed and stable representations to be "held in mind." whether they be conscious or not (Solms & Panksepp, 2012).

### 4.8. The Hard Problem vs. The Easy Problem

Before turning to the question "what is consciousness for?" I want to address another issue that has just arisen. When I say that consciousness depends on the integrity, not of cortex, but rather of upper brainstem structures, I am locating the generation of consciousness in space (I am *correlating* mind with matter — the easy problem; Chalmers, 1995). I am also attributing the generating of consciousness to neuronal processes that occur in that location (matter *causing* mind — the hard problem; Chalmers, 1995).

I believe it is a mistake to do this, to say that neuronal processes "generate" consciousness. This mistake gives rise to the hard problem, which derives from our conflation of matter (solidity, tangibility, etc., neuron-like stuff) with reality, in the sense described in Secs. 4.3 and 4.4 — caused by our privileging of exteroceptive ways of thinking. This mistake is the essential difference between materialist monism and dual-aspect monism. For this reason, although my argument derives from physiological facts rather than philosophical premises, I prefer dual-aspect monism. Because we privilege vision, naïve materialists imagine that the object of the mind *as perceived visually* (i.e., the brain) causes or generates consciousness. It would be more correct to say that a *function* of the brain (i.e., something that has to be *abstracted* from the data of perception) causes or generates consciousness. This way of speaking makes it logically possible to solve the hard problem empirically. (See Secs. 4.18 and 4.20.)

## 4.9. The Body as An Object vs. The Subject of the Body

In exteroception, everything depends on the relative position and magnitude of the observer — on the observational perspective of the subject. In the case of the mind/body problem, that observational perspective is reflexive. The subject is looking at itself in the (mainly visual) form of an external object. The subject of consciousness thereby *represents* itself as an object of consciousness. The body-as-subject represents itself as body-as-object.

Subjective consciousness arises from *being* the portion of reality that generates consciousness. The relationship between *seeing* that portion of reality — representing it visually as, say, the innervations of the centrencephalic region — and simultaneously *feeling* it, is *not a causal relationship*. They are two manifestations of the same thing. It makes no sense to ask: how does an object of consciousness cause the subject of consciousness? *This is the traditional hard problem*. That question (the traditional hard problem) is even worse than asking: how does a flash of lightning cause a clap of thunder? I say "even worse" because the visual flash and auditory clap are at least *parallel forms* of consciousness — they are both *objects* of consciousness — whereas the functional anatomy of the centrencephalic region is a visually perceived *object* embedded within the subjective *state* of consciousness — which is, if anything, a hierarchical relationship opposite to the one dictated by common sense. The conscious state causes (or enables) conscious perception of a mental solid; not the other way around.

## 4.10. Observational Natural History vs. Inferential Science

Be that as it may, consciousness in all its manifestations must in reality be generated by processes that lie beyond consciousness. We (dual-aspect monists) can only *infer* those processes from the data of consciousness.

In doing so, we would of course be well advised to make use of the data provided by *all* the modalities of consciousness, including what it feels like to *be* conscious (i.e., including states of consciousness). Moreover, we would be well advised to describe the processes underlying consciousness in terms that transcend the qualities of the sense modalities; that is, we should describe this portion of reality — like all others — in *abstract* terms. We should describe the reality of the mind (the lawful relations that we infer from the data of consciousness) in nonperceptual language, ultimately using terms like "energy," etc., as the physicists do. I will do this shortly (Sec. 4.18.). The reality described by physicists and biologists and psychologists is, after all, the same old reality; and the laws inferred via our different perspectives on it must eventually be reconcilable with each other. The mind is not made of different stuff from the rest of the universe. The only difference about the mind is that we have an additional perspective on it; because we *are* the mind, we know what it feels like to be a mind.

This is the essence of the neuropsychoanalytical approach. It makes use of both objective and subjective data of consciousness, and where necessary it prioritizes the subject (the ground of experience). This is also why I use the term meta-neuropsychology. A meta-neuropsychological account of the mind goes beyond (meta) our material (neuro) and subjective (psychological) observations of it; it describes the real unconscious stuff of the mind in abstract terms, as a "mental apparatus" — to use Freud's (1900) felicitous term.

It is not easy to set out the neuropsychoanalytical approach, because we are not used to thinking like this. Common sense dictates that reality is identical with what we see. Seeing is indeed believing, but it is not reality. Seeing takes place in the mind.[3]

## 4.11. So, What is Consciousness for? (Part A: Perception and Cognition)

I will begin to answer the question from the exteroceptive (representational, mnemic) end of the mental apparatus, although I believe that consciousness is an endogenous property, not something intrinsically bound up with exteroception.

---

[3]Ironically, naïve materialism turns out to be a kind of crypto-idealism in which the mental nature of perceptual objects is overlooked, and the qualities of matter-as-it-looks are conflated with those of reality-in-itself.

What consciousness bestows on the objects of perception and cognition is *attention*. Consider the phenomenon of "attentional blindness." If you do not attend to something in your visual field (even a person in a gorilla suit) it does not enter your consciousness. The same applies to cognition. Remembering, recognizing, judging, navigating, thinking, etc., go on all the time, but these processes only becomes conscious when we pay attention to them. And we only pay attention to them when we *need* to. So the question, "what is consciousness for?" may be recast, insofar as it applies to exteroception: why do we pay attention to some perceptual/cognitive processes but not others?

The answer is that we pay attention to the salient ones, to the ones that matter to us. What matters to us living creatures — we infer, following Darwin — are at bottom the things that affect our chances of surviving to reproduce. But this cannot mean that we pay attention to *everything* that matters to us. That would be impossible. We know empirically that perceptual and cognitive consciousness is a very small place; we simply cannot "hold in mind" everything we need to (Bargh & Chartrand, 1999; see also Solms, in press [a]). The purpose of conscious exteroception, then, is to master the things which matter to us in such a way that we *no longer* need to pay attention to them — in such a way that we may render them unconscious again. Conscious processing of object relations is a temporary state. (Apparently the ultimate aim of all cognition is zombiedom; Solms, 2013.) In other words, we seek through perceptual and cognitive consciousness to establish predictive algorithms that render consciousness redundant. We seek to establish algorithms that work *automatically*: "when that happens, I do this." Freud (1911) called this achievement the "reality principle": through learning from experience, we establish internal models of reality so as to be equipped to meet our needs there. Realistic solutions to life's problems only require the conscious presence of the subject when the solutions fail or when the problems are novel; i.e., when things do not go according to plan or when there is no plan.[4]

Salient things, then, are the things we do not, or cannot, predict — things that unsettle our equilibrium — the things that generate *uncertainty*. Friston (2010) calls such things "surprise."

---

[4]Freudian "repression" consists in the *premature* automatization of such algorithms, before they meet the dictates of the reality principle. This leads to the constant threat of "surprise" (and therefore affect) and the return-to-consciousness of the repressed (see Solms, 2013, in press [b]).

## 4.12. So, What is Consciousness for? (Part B: Affect)

There are roughly two types of surprise. Things either go better or worse than expected. The determining of this distinction, it appears, is what consciousness is really for. Consciousness informs the subject, immediately, without learning, what to make of unexpected events (of their biological *meaning*). It enables the subject to *feel* its way through unknown quantities (deviations from homeostatic set-points) by answering the biological question: "is this good or bad for me?" The answer to the question — is this good or bad (for my survival and reproductive success)? — is initially provided not by memory but by feelings: the things that are good for us feel good (pleasurable) and the things that are bad feel bad (unpleasurable). The memory of what to do (prediction) flows from that.

It is no simple matter to calculate the biological meaning of unexpected events, as they happen, across the multiplicity of vital needs: nutritional, thermal, sexual, etc. This is what feelings provide. Or at least, this is what the simplest feelings provide (see Sec. 4.16). They provide an immediate global valuation of biological "state" (Mesulam, 2000). And they orientate the subject to what it should do in response to the unexpected events (e.g., approach vs. withdraw). They provide a global assessment along the lines: "this is what the current situation means to me."

It is important to recognize that this assessment is determined interoceptively: "what does this mean *to me*?" The assessment of the state of the subject is then *extended* (Damasio, 2010) onto the data of perception and cognition: "I feel like this (good or bad) *about that*." This is what consciousness adds to (intrinsically unconscious) exteroceptive perception and cognition. It adds subjective feeling and therefore valence to images of objects.

## 4.13. Qualia Belong to (and are bound by) the Subject

Valence is intrinsically qualitative, and in this instance intrinsically subjective too. That is why the objective properties of sensory stimuli and processes predict nothing about the associated qualia (cf. the qualia problem). Qualia belong to the subject. That is also why they all belong *to me* (cf. the binding problem).

The biological value of qualia is obvious. Feelings determine the subject's response to events that were not predicted by evolution (by reflex) or by prior individual experience (by learning). Most events, of course, cannot be predicted by evolution and responded to automatically. Our enhanced capacity to deal with unpredictability (learning guided by feelings) is what distinguishes us humans from, say, ants ... although I know nothing about the mental life of ants.

### 4.14. The Endogenous Sources of Consciousness

It is no surprise, in light of what I have said, to find that the neural correlates of consciousness (the so-called generators of consciousness) *are located in the upper brainstem, in close proximity and in tight connection with the structures that monitor the state of the visceral body (i.e., the subject) — the state of the internal milieu, of our vital functions* (Panksepp, 1998; Damasio & Carvalho, 2013). As Freud (1911) surmised even earlier, the elemental function of conscious feeling — of pleasure and unpleasure — is to detect oscillations in the tensions of our internal drive needs. The resultant feelings, in turn, have a compulsive influence on our actions. This is the "pleasure principle."

> "The id, cut off from the external world, has a world of perception of its own. It detects with extraordinary acuteness certain changes in its interior, especially oscillations in the tension of its instinctual needs, and these changes become conscious as feelings in the pleasure–unpleasure series. It is hard to say, to be sure, by what means and with the help of what sensory terminal organs these perceptions come about. But it is an established fact that self-perceptions — coenaesthetic feelings and feelings of pleasure–unpleasure — govern the passage of events in the id with despotic force. The id obeys the inexorable pleasure principle." (Freud, 1940, p. 198)

### 4.15. Levels vs. States of Consciousness

When Moruzzi & Magoun (1949) originally identified the anatomical location of the brainstem generators of consciousness, they made the mistake of describing their function in purely quantitative terms, in terms of the *level* of consciousness, or "wakefulness." They assumed that the *contents* of consciousness attached only to cortical processes. It fell to more recent investigators (Panksepp, 1998; Damasio, 2010) to recognize that *it feels like something to be awake*. These investigators recognized that the background *state* of consciousness — the subject of consciousness, *being* conscious — is intrinsically *affective*. Affective consciousness is secondarily extended upwards, to sample and test learnt representations of the outside world, and thereby to determine the meaning of ongoing perceptions and cognitions (see Solms & Panksepp, 2012; Solms, 2013).

> "Cathectic innervations are sent out and withdrawn in rapid periodic impulses from within into the completely pervious system (for perception). So long as that system is cathected in this manner it receives perceptions (which are accompanied by consciousness) and passes the excitation onwards to the unconscious mnemic systems; but as soon as the cathexis is withdrawn, consciousness is extinguished and the functioning of the system comes to a standstill. It is as though the [id] stretches out feelers, through the medium of the system (for perception), towards the external world and hastily withdraws them as soon as they have sampled the excitations coming from it." (Freud, 1925, p. 231; with acronyms spelled out in square brackets.)

It is nothing short of astonishing that cognitive science overlooks the affective states that color the whole of experience — and provide the global state of consciousness that is the precondition for all experience; the very medium in which object consciousness is written.

### 4.16. Instinctual Affects

It is important to recognize (contra the James–Lange theory) that the affective state of consciousness consists in more than read-outs of the current state of the body — although that is probably what the *dawn* of consciousness consisted in. It is also important to realize that conscious state consists in more than the pleasure–unpleasure series, generated (in the sense clarified above, in Sec. 4.8) at the level of the upper brainstem. The simplest form of consciousness — positive vs. negative hedonic valence — must have evolved more than 525 million years ago, as its functional anatomy (the reticular activating system) is shared by all vertebrates. It is not surprising that evolutionary pressures gave rise to more subtle affective distinctions; there are more than two meanings in the world (good vs. bad) demanding more than two types of response (approach vs. withdrawal). That is why we find, above the upper brainstem, a variety of instinctual–emotional circuits giving rise to a variety of affects and attendant stereotyped action tendencies. The general classes of affective circuits in mammals range from simple "homeostatic affects," like hunger and thirst, and "sensory affects" like pain, surprise, disgust, etc, to the so-called "basic emotions" which Panksepp (1998) capitalizes as SEEKING, FEAR, RAGE, GRIEF, etc. These latter circuits provide hard-wired psychobehavioral links between the inner and outer worlds. We cannot afford — biologically speaking — to learn from experience how to forage, flee, attack and attach.

## 4.17. Object Consciousness

Added to the precise and detailed representational capacities afforded by the vast memory space of cortex, is the extended consciousness of everyday perception and cognition — the mental solids of working memory, which threaten to hide everything else from view. (One is reminded of Plato's cave.) As I have said already, this form of consciousness (object consciousness) arises when consciousness itself (the subject of consciousness) samples representations of the external things affecting it: "I feel like this about that." Representations of the "thats" in question can be held in mind (in working memory) while we try to master the feelings associated with them (i.e., while we master their biological impact on us). Then, once we have mastered them, i.e., when we have established a predictive algorithm that works, and we have automatized the adequate response, we pay no further attention to them, and thus we render them unconscious once more. This important function — temporarily conscious perception/cognition — is not to be sneezed at, but as already stated: it tends to blind us as to the actually invisible, subjective, affective nature of consciousness itself.

The objects of consciousness are the stabilized mnemic imprints that meaningful perceptual constellations make on our affects. That is what I mean when I say that affect provides the medium in which perceptual consciousness is inscribed. I do not mean to imply by this that perceptual consciousness is the same as affective consciousness. When the upper brainstem activates learnt cortical representations,[5] a very different form of consciousness arises. This form of consciousness (images of objects, derived from learning) is characterized above all by its stability.

[5]Freud speaks of "cathexis" here (see Sec. 4.15).

It co-exists with affective consciousness. (There are incidentally no other forms — or contents — of consciousness.) That is, the stabilized form of consciousness coexists with the raw, untransformed, untamed consciousness of affect. (Friston speaks of "bound" vs. "free" energies here; see Sec. 4.18.) Since the purpose of the stabilized imprints (mental solids) is that we can hold them in mind while we learn how to meet our needs in reality, they are far from unimportant. That is why decorticate animals are excessively emotional.[6] An animal that cannot "hold things in mind" is in an extremely vulnerable state, unable to link its biological needs with an ever-changing and unpredictable external world.

### 4.18. What is Consciousness Really?

The hard problem now boils down to trying to *abstract* the processes underlying conscious experience. We have two types of access to these processes: one provided by exteroceptive perception (mainly vision) and the other by affect. That is, we have objective "solid" images of the relevant neural processes and we have subjective feeling states that correlate with them. The common denominator between these two terminal points of our knowledge of the mechanism of consciousness appears to be *uncertainty*. This, I believe, is the abstraction we are looking for.

In this connection, I would like to cite Freud's dictum: "consciousness arises instead of a memory-trace" (Freud, 1920, p. 25). When you know what to do (when you have already learnt what to do), there is no need for subjective (affective) *presence*, and therefore no need for consciousness. The memory trace does its job automatically (Solms, in press [b]). When memory (stable prediction) supervenes, consciousness (uncertainty) recedes.

It is only in the presence of uncertainty — of prediction error, or what Friston calls "surprise" — that the memory trace (the perceptual/cognitive representation) must be brought back to awareness, through hyperactivation by upper brainstem mechanisms (especially midbrain,

[6]In humans, there is the very important additional regulatory function of language — that is, symbolic re-representation — the bedrock of Freud's "secondary process" (see Freud, 1911). Panksepp points out that this level of representation would be better described as "tertiary process;" affective consciousness is primary (first person), objective consciousness is secondary (second person), reflexive consciousness is tertiary (third person). See Solms & Panksepp, 2012.

hypothalamus and nonspecific thalamus), and evaluated afresh. This amounts to a *temporary dissolution of the established LTM algorithm* (it amounts to consciousness arising instead of a memory trace). This is a regression to disorder — entropy — Friston's "free energy" (Friston, 2010). Physiologically speaking, this process coincides with the re-emergence of *lability* within the cells — or cellular networks (especially in basal ganglia (amygdala, nucleus accumbens, bed nuclei of the stria terminalis, septal nuclei) and hippocampus) — until the revised trace is "reconsolidated" (Misanin *et al.*, 1968; Tronson & Taylor, 2007).

This, I believe, is the mechanism of consciousness, and the mechanistic basis of the distinction between perceptual/cognitive and affective consciousness.

There is of course a great deal more to be explained, especially with regard to the physiological details. However, it is important to remember that what we must discern behind these details is an abstraction. The best abstraction will explain all the observable facts, and generate testable predictions. That is how science works.

### 4.19. Hallucinatory Wish Fulfillment

Since I am attempting here to sketch the neuropsychoanalytic approach only in broad outline, so as not to lose sight of the wood for the trees, I will limit myself — before closing — to one further point, and leave the details to other occasions.

Perception, according to Friston *et al.* (2012), following Helmholtz (1866), is an active process of hypothesis-testing by means of which we seek to confirm our predictive models of the world. I have argued that *conscious* perception occurs when our predictions are *disconfirmed* (prediction error) — or in conditions where prediction is impossible — that is, in conditions of *uncertainty*.

I would like to add that this means we constantly SEEK (in Panksepp's sense) to confirm our predictions. SEEKING (which manifests largely as foraging in lower animals) is expressed largely as saccadic eye movements in humans (Panksepp & Biven, 2012). Saccadic scanning may be considered a visual form of foraging (just as whisker twitching is a proxy for SEEKING in rodents), a form of hypothesis testing.

The role of *dopamine* in such behavior is well established. Of special interest therefore are the implications of this form of uncertainty — SEEKING to confirm predictions — for the mechanism of hallucination and delusion (for "wishful thinking") and other hyper-dopaminergic states. The analogy with REM sleep (a hyper-dopaminergic state characterized by rapid saccadic eye movements) and dreaming (a highly affective state of hallucinatory object consciousness) is obvious (Perogamvros & Schwartz, 2012). The relationship between dreaming and prediction error coding is an important topic for future research.

Of incidental interest in this connection is the fact that completely immobilized eyeballs rapidly become blind; the photosensitive cells give rise to no new stimuli (to a completely predictable state of affairs, as far as the central organ is concerned).

### 4.20. Conclusions

Consciousness registers *the state of the subject* within biological scales of values. The primal form of consciousness is affect. Pleasurable affect means "biologically good" and unpleasure means "biologically bad." There are many evolved varieties of the biological good and bad with subtler meanings (e.g., GRIEF means "loss of an attachment figure").

Perceptual/cognitive representations are intrinsically unconscious sensory-motor algorithms (learnt predictions as to how biological needs can be met in the world). Consciousness samples constellations of such representations in order to bestow meaning on them. Object representations are stable sensory-motor configurations which can be *held in mind*; they stabilize and shape raw consciousness into mental solids. Once meanings are established, the resultant predictions are automatized and rendered unconscious once more. However, the world is so uncertain, so full of surprises, that this "ideal" state is not easily achieved — and not easily maintained. Reality is unpredictable most of the time.

Prediction errors attract attention to the salient representations. This occurs through the generation of what Friston (2010), following Helmholtz (1866) and Freud (1911), calls "free energy:" Entropy of learnt representations causing uncertainty (loss of precision). This mechanism may be *visualized* as a cellular process — perhaps as decomposition of PKMz (Sacktor, 2008)[7] — but it is *felt* as consciousness. Consciousness persists, then, until a revised predictive model is reconsolidated (until the free

---

[7]I use this as an example of what I mean by "lability" in a hyperactivated memory-trace, prior to reconsolidation.

energy is "bound"). Consciousness of representations is predictive work in progress.

## References

Bard, P. (1934) On emotional expression after decortication with some remarks on certain theoretical views: Part 2. *Psychol. Rev.*, **41**, 424–449.

Bargh, J. (2014) Our unconscious mind. *Sci. Amer.*, **310**, 30–37.

Bargh, J. & Chartrand, T. (1999) The unbearable automaticity of being. *Amer. Psychol.*, **54**, 462–479.

Brentano, F. (1874) *Psychology from an Empirical Standpoint*. London: Routledge (trans: Rancurello, Terrell, D. & McAlister, L., 1973).

Chalmers, D. (1995) Facing up to the problem of consciousness. *J. Consciousness Stud.*, **2**, 200–219.

Crick, F. (1994) *The Astonishing Hypotheses*. New York: Scribner.

Damasio, A. (2010) *Self Comes to Mind*. London: Pantheon.

Damasio, A. & Carvalho, G. (2013) The nature of feelings: Evolutionary and neurobiological origins. *Nat. Rev. Neurosci.*, **14**, 143–152.

Freud, S. (1900) The interpretation of dreams. *Standard Edition*, **4–5**.

Freud, S. (1911) Formulations on the two principles of mental functioning. *Standard Edition*, **12**, 215–226.

Freud, S. (1920) Beyond the pleasure principle. *Standard Edition*, **18**, 7–64.

Freud, S. (1925) A note upon "the mystic writing-pad." *Standard Edition*, **16**, 227–232.

Freud, S. (1940) An outline of psychoanalysis. *Standard Edition*, **23**, 144–207.

Friston, K. (2010) The free-energy principle: A unified brain theory? *Nat. Rev. Neurosci.*, **11**, 127–138.

Friston, K., Adams, R., Perrinet, L. & Breakspear, M. (2012) Perceptions as hypotheses: saccades as experiments. *Front. Psychol.*, **3**, 151.

Helmholtz, H. (1866) Concerning the perceptions in general, in *Treatise on Physiological Optics*. 3rd edn. New York: Dover. (Southall, J. Trans.)

Magoun, H. (1952) An ascending reticular activating system in the brain stem. *AMA Arch. Neurol Psychiatry*, **67**, 145–154.

Merker, B. (2007) Consciousness without a cerebral cortex: A challenge for neuroscience and medicine. *Behav. Brain Sci.*, **30**, 63–134.

Mesulam, M.M. (2000) Behavioral neuroanatomy: Large-scale networks, association cortex, frontal syndromes, the limbic system and hemispheric lateralization. In: *Principles of Behavioral and Cognitive Neurology*. 2nd edn. New York: Oxford University Press, pp. 1–120.

Misanin, J., Miller, R. & Lewis, D. (1968) Retrograde amnesia produced by electroconvulsive shock after reactivation of a consolidated memory trace. *Science*, **160**, 554–558.

Moruzzi, G. & Magoun, H. (1949) Brain stem reticular formation and activation of the EEG. *Electroencephal. Clin. Neurol.*, **1**, 455–473.

Nagel, T. (1974) What is it like to be a bat? *Philos. Rev.*, **83**, 435–450.

Panksepp, J. (1998) *Affective Neuroscience*. New York: Oxford University Press.

Panksepp, J. & Biven, L. (2012) *The Archaeology of Mind*. New York: Norton.

Penfield, W. & Jasper, H. (1954) *Epilepsy and the Functional Anatomy of the Human Brain*. Oxford: Little & Brown.

Perogamvros, L. & Schwartz, S. (2012) The roles of the reward system in sleep and dreaming. *Neurosci. Biobehavioral Rev.*, **36**, 1934–1951.

Sacktor, T. (2008) PKMzeta, LTP maintenance, and the dynamic molecular biology of memory storage. *Prog. Brain Res.*,**169**, 27–40.

Shepard, R & Metzler, J. (1971) Mental rotation of three dimensional objects. *Science*, **171**, 701–703.

Shevrin, H., Panksepp, J., Brakel, L. & Snodgrass, M. (2012) Subliminal affect valence words change conscious mood potency but not valence: Is this evidence for unconscious valence affect? *Brain Sci.*, **2**, 504–522.

Shewmon, D., Holmse, D. & Byrne, P. (1999) Consciousness in congenitally decorticate children: developmental vegetative state as a self-fulfilling prophecy. *Dev. Med. Child Neurol.*, **41**, 364–374.

Solms, M. (2013) The conscious id. *Neuropsychoanal.*, **15**, 5–19.

Solms, M. (in press [a]) Affect theory today. *Psychoanalytic Psychology*.

Solms, M. (in press [b]) Reconsolidation: turning consciousness into memory. *Behav. Brain Sci.*

Solms, M. & Panksepp, J. (2012) The id knows more than the ego admits. *Brain Sci.*, **2**, 147–175.

Strawson, G. (1994) *Mental Reality*. Cambridge MA: MIT.

Tronson, N. & Taylor, J. (2007) Molecular mechanisms of memory reconsolidation. *Nat. Rev. Neurosci.*, **8**, 262–275.

Zeki, S. (1993) *A Vision of the Brain*. Oxford: Oxford University Press.

Biophysics of Consciousness: A Foundational Approach
R. R. Poznanski, J. A. Tuszynski and T. E. Feinberg
Copyright © 2017 World Scientific, Singapore.

# 5

# CAN QUALITATIVE BIOPHYSICS SOLVE THE HARD PROBLEM?

*Alfredo Pereira Jr.,* Ram Lakhan Pandey Vimal† and Massimo Pregnolato‡*

**São Paulo State University (UNESP), Institute of Biosciences, Campus Rubião Jr., 18618-970 — Botucatu-SP, Brazil*

*†Department of Neuroscience & Consciousness Research, Vision Research Institute 25 Rita Street, Lowell, MA 01854, USA*

*‡Department of Drug Sciences, University of Pavia, Viale Taramelli 12, Pavia, Italy*

## Abstract

A solution to the Hard Problem formulated by Chalmers should include a biophysical nonreductive explanation of the qualities of conscious experience ("subjective qualia"). This task is nearly impossible to be performed in the context of Modern Physics, since most influential scientists and philosophers conceived such qualities as being mental phenomena without a physical counterpart. We propose to extend the concept of multi-aspect state into the concept of *potential qualitative aspects* in fundamental physics, allowing for its expression in biophysical systems under adequate conditions. We conceptually model a structure, the N-dimensional dual-aspect state space of Nature, having *elementary waveforms* (EW) as the qualitative building blocks, and three dynamical phases: a) coherent wave functions, b) decoherent and c) recoherent macrostates. In coherent wave functions composing microstates, Elementary Waveforms (EW) are superposed and entangled. In decoherent macrostates, the Heisenberg matrix (HM) is reduced to the diagonal, generating — according to the principles of chemistry — one macro (not conscious) qualitative state. In living systems, chemical processes combine in a self-organizing manner, keeping the system at low entropy states and making room for the emergence of new structures and functions by means of an *order from fluctuation* mechanism. In the brain, the instantiation of macrostates is spatially distributed and unconscious. Brain recoherent macrostates are formed by means of the operation of quantum computing gates within ionic waves, and can be expressed by a hypermatrix (or hypertensor), corresponding to the binding of qualitative features in one integrated conscious episode. Once formed, the ionic waves feedback on brain activity, similarly to David Bohm's "pilot wave." In the theoretical context of systems theory, the global recoherent process would correspond to a phase in Scott Kelso's framework of *metastable dynamics*. In such a dynamics, each recoherent macrostate alternates in a time scale of nanoseconds with decoherent macrostates, but we consciously perceive only the sequence of nanosecond recoherent states, which appear to us in continuous chunks with the duration of — at least — hundreds of milliseconds.

*Keywords:* Quantum coherence; decoherence; recoherence; dual-aspect state; qualitative biophysics; Boltzmann entropy; Maxwell demon; synergy; metastability; consciousness.

## 5.1. Introduction

In this chapter, we propose to introduce the potentiality for a qualitative aspect of conscious experiences in fundamental physics, allowing for its expression in biophysical systems, under adequate conditions. In animals, especially in human subjects, we define the qualitative experiences as conscious processes containing at least four characteristics:

a) Sensations and feelings (such as hunger, thirst, happiness, sadness, fear, pain);
b) Perceptual qualities or 'qualia' (such as colors, sounds, smells, tastes, etc.);
c) Cognitive processes grounded on feelings and "qualia," such as attention, thinking, memory formation and recall, etc.
d) A spatiotemporal structure, composed of an egocentric spatial framework (Trehub, 2013) and a temporal duration of (conscious) episodes of 100 ms to 3 s (Poppel *et al.*, 1990).

Our proposal of a qualitative biophysics helping to solve the Hard Problem of Consciousness (Chalmers, 1995, 1996) is based on the idea that Elementary Waveforms (EW) that compose quantum microstates contain the potential for qualitative macrostates (QMSs), as those observed in the morphology and physiology of living systems. We further claim that the subjective qualities experienced in conscious episodes can be described by a hypermatrix/hypertensor composed of QMSs.

A biophysical nonreductive explanation of the qualities of conscious experience ("subjective qualia") is nearly impossible to be performed in the context of modern physics, since most influential scientists and philosophers conceived such "secondary" qualities as being mental phenomena without a physical counterpart. The ontology of modern physics does not include secondary qualities; these qualities are considered as existent in the mind of the observer (Martinez, 1974). The term "qualia" is assumed to cover both properties of conscious experiences and/or the respective properties/qualities of objects (Vimal, 2010c).

In the foundations of Modern physics, a decision was made to remove the conscious observer from the physical domain. The observer was brought back by Contemporary physics, to explain the collapse of the wave function in Quantum Theory, and to make explicit the dependence of physical knowledge on reference frames in Special Relativity. In both cases, the conscious observer with his/her perception of secondary qualities is introduced to solve physical problems, but is not considered as an object for physical explanation. With this remark, we are *not* suggesting that there is no interpretation of quantum mechanics (QM) without collapse and/or without observer, but assuming that a nonconscious detector can fulfil the explanatory needs of both QM and relativity.

David Hume stated that the main principle of Modern science and philosophy "is the opinion concerning colors, sounds, tastes, smells, heat and cold; which it asserts to be nothing but impressions of the mind, derived from the operation of external objects, and without a resemblance to the qualities of the objects" (Hume, 1992, p. 226–227). In this formulation, the mind appears as being a nonphysical system that adds subjective "secondary" qualities (impressions) to external objects in the process of perception. This kind of formulation is related to the famous Cartesian mind–body dualism (the theory that the conscious being is composed of two independent substances, the body and the thinking mind) and the dualist mind–body problem (how could an immaterial soul interact with the material body?), for which an acceptable solution (Descartes' appeal to the pineal gland notwithstanding) was not found yet and will probably never be, because of the way the issue was formulated.

A recent study (Buyse, 2013) has shown that Galileo, considered to be the original author of the distinction of primary and secondary qualities in modern times, was not a mind–body dualist and possibly made a formulation of the problem that is different from Descartes and Hume, one that may inspire contemporary researchers to find a solution. Buyse (2013) discovered a frequent mistake in the translation of Galileo's Italian originals. In most cases, Galileo's text is translated as meaning that secondary qualities exist only in the (immaterial) conscious mind of the observer, however "in the original text, Galileo nowhere writes that secondary qualities and emotions reside 'in consciousness.' On the contrary, he writes that they reside 'in the sensible body' (nel corpo sensitivo), or in other words, in the body of the perceiver, whether it be a human body or an animal body" (Buyse, 2013). As far as secondary qualities are instantiated in the body of animals, and living systems are considered to be physical systems, there must be a physics of the conscious mind. Galileo would probably agree with this challenge.

While recent developments of physical chemistry led to theories about living systems, such as Prigogine's Nobel Prize approach to dissipative structures, an understanding of the physics behind secondary qualities remains a challenge. Advances in neuroscience are not likely to provide this understanding, since a full physical model of brain function is still not available. Explanations of conscious activity by neuroscientists have been based on functional neuroanatomy together with innovative uses of information theory.

What is the aspect of the physical world that manifests itself as the subjective qualities experienced by the conscious observer, such as (in David Hume's words) "colours, sounds, tastes, smells, heat and cold"? This is a central issue raised by David Chalmers' "Hard Problem of Consciousness," in a now-classical paper that appeared in the second JCS issue (Chalmers, 1995).

One step towards the solution of this problem can be built from a philosophical current we call *Multi-Aspect Monism*, which includes Aristotles's *Hylomorphism* and Spinoza's *Substance Monism* (the latter, with two modes of existence: the body and the mind). In contemporary philosophy of science, this tradition was resumed in Max Velmans' *Reflexive Monism* (Velmans, 2008), and extensively developed by two of us (Pereira Jr., 2003, 2013, 2014; Pereira Jr. *et al.*, 2010; Pereira Jr. *et al.*, 2013; Vimal, 2008, 2009b, 2009c, 2010d, 2010e, 2010f, 2013).

The recognition of a qualitative aspect of nature finds echoes in other attempts of reconceptualization of physics, chemistry and molecular biology. The expression "Qualitative Physics" (Forbus, 1988) applies to qualities we perceive in everyday life. For example, there is a qualitative ontology implied in the sharing of our intuitions about the physical world ("tacit knowledge") with robots, using discrete systems of symbols. This knowledge is obtained from our experiences of external objects and processes, by means of the interaction of qualitative aspects of objects with qualitative features instantiated in our brains (Vimal, 2010b). Mitterauer claims for a qualitative approach in computational models and robotics, thus making room for the conceptualization of conscious experiences (Mitterauer, 2012).

In biology, qualitative features compose the *phenotype* of living systems. There is wide agreement that phenotypic structural and functional characteristics (including mental capabilities) derive from the interaction of a set of biological determinants (genes), among themselves and with the environment, composing the epigenetic process. Until recently, this process was conceived as being weakly emergent (there would be a specific gene responsible for each phenotypic feature), but the results of big data projects as the Human Genome suggest that it is strongly emergent: irreducible phenotypic features spontaneously result from the interaction of genes and environment (Hayden, 2010). Here, we conceive a similar process, starting from physical foundations and ending with biological and conscious phenomena.

We assume that the physical domain contains, besides the matter/energy aspect, patterns of distribution of matter/energy in space and time, which are pregnant of *potential qualitative aspects*. These patterns, according to the laws and principles of chemistry, become progressively actualized in both *micro* and *macrostates*. In living systems, chemical processes combine in a self-organizing manner, keeping the system at low entropy states and making room for the emergence of new structures and functions by means of an *order from fluctuation* mechanism. In the brain, the instantiation of macrostates is spatially distributed and unconscious, but a further mechanism of matching and selection of patterns — supported by quantum computational gates instantiated in ions and proteins — prompts the formation of recoherent macrostates, corresponding to the binding of spatially distributed patterns into conscious episodes experienced by the living organism. Nonconscious states include slow-wave dreamless deep-sleep, coma, vegetative and anesthetized states, implicit memory, long-term memory storage, paradoxical awareness as in the case of "blindsight" and subliminal perception (Rosenthal, 2009; Vimal, 2010a).

In our reconceptualization of physics, phenomenal experience — including the experience of "qualia" — is conceived as the actualization of complexes of natural potentialities in recoherent states, given adequate conditions. In our view, this conceptual move solves the Hard Problem, indicating how subjective conscious experiences emerge from nature. However, this strategy does not afford a reductionist approach to the conscious mind, since the emergence process is conceived as being of the strong modality, containing several evolutionary steps with results that cannot be deduced from antecedent states.

### 5.2. Elementary Waveforms (EW) and the N-Dimensional State Space of Nature

We depart from a concept of Nature as composed of EW that exist eternally, and combine to generate all natural processes. These waveforms are dual-state entities, having both matter/energy and a spatiotemporal pattern of activity that contain potential qualitative aspects. In this regard, they can be compared to the concept of fundamental reality advanced by Tesla: "To understand the true nature of the universe, one must think it terms of energy, frequency and vibration" (Morgan, 2014).

Tesla's concept is close to ancient Indian and Greek philosophy, considering the qualitative aspect as fundamental as the physical aspect. This is because matter (the physical aspect) at all levels (from the fundamental to microscopic quantum particles to macroscopic object levels) has an inseparable "form," which is a qualitative aspect and is consistent with Aristotle's concept of "form" (384–322 BC), Yājñavalkya's "rūpa"(form) (c. 1000–700 BC) in Bṛhadāraṇyaka Upaniṣad (Krishnananda, 1983), and Bādarāyaṇa's "rūpa" (form) (c. 500–400 BC) in Brahma Sūtra (Radhakrishnan, 1960).

Elementary Waveforms (EW) can also be compared to the fundamental strings proposed by Greene, considering that subatomic strings would be the elements of reality, which are reproduced in self-similar fashion in all spatial scales (Greene, 1999, 2004).

Greene seems to imply that string theory is more fundamental than QM. Mass, charge and spin of all elementary particles (six quarks: up, down, top, bottom, charm, strange; six leptons: electron, muon, tau, electron neutrino, muon neutrino, tau neutrino; five bosons: gluon, photon, z-boson, w-boson and Higgs boson) are associated with the pattern of vibration that a string executes (Greene, 1999). If string theory is correct, then we should be able to derive elementary *particles* from strings. The idea that the fundamental elements of reality are strings instead of particles is consistent with our proposal, since strings have both energy and patterns of activity with a degree of self-similarity across spatial dimensions (Flanagan, 2003), supporting the "brute fact" of existence of qualitative aspects in nature.

When introducing the qualitative aspect in physics, we address the issue of incompleteness of QM (Einstein *et al.*, 1935; Flanagan, 2003). It is well known that QM has many different and divergent interpretations, some of them purely epistemic, while others attempt to introduce ontological considerations to better understand the underlying reality. The debate famously began with Albert Einstein's dialogue with Niels Bohr (Einstein, 1971; Einstein *et al.*, 1935) about the incompleteness of QM, inspiring David Bohm's effort of postulating an *Implicate Order* (Bohm, 1952, 1980, 1985, 1987; Bohm & Hiley, 1993), a wavelike continuum informational domain to explain the observed, discrete *Explicate Order*. The philosophical and physical implications of quantum entanglement in terms of implicate order are elaborated in Caponigro *et al.* (2010).

There is a possibility of understanding EW as the components of Bohm's Implicate Order, being expressed in observable domains of reality, including the living brain. In neuroscience, Karl Pribram's work (Pribram, 1987, 1997, 1999, 2007) used Bohm's insight to figure how the brain could generate meaning and consciousness. Another theoretical approach to the reality underlying the quantum probabilistic picture is Quantum Field Theory, also used in the context of neuroscience and consciousness studies by Freeman and Vitiello (2006, 2011).

We begin our conceptual modeling with the proposal of a structure. Let *N* be a natural system. The ensemble of all possible states of *N* is given by the state space of *N*. The elements of *N* are EW. Each EW has a variety of possible states. Each state of an EW is a dual-aspect state, containing both the usual energy aspect and a potential qualitative aspect related to its pattern of activity.

An *N*-dimensional state space can be constructed when taking into consideration all potential qualitative states of EW. In a simplified approach, it may be convenient to model the state space of just one EW. This state space depicts all possible qualitative states assumed by the EW; different macroscopic qualities correspond to different regions of the state space. For example, consider qualitative states $q_{ri}$ ($i$ = 1 to nr) of a single EW underlying a macrostate *M*. This would imply that when all $q_{ri}$ of a single EW are combined in certain way a macrostate *M* would result. However, this model is not consistent with the idea of strong emergence based on interactions, because it assumes that a single EW could generate a macrostate.

In an interactionist approach (as proposed by Cacha & Poznanski, 2014), a macrostate should be conceived as the result of the interaction and temporal combination of a collection of EW. For example, the long wavelength light (related to the color red in our conscious experiences) is the result of the combination of several EW, because a wave is composed of several values within a range. This approach is consistent with the idea of strong emergence.

In a more realist — but overwhelming complex — modeling, it is necessary to conceive a system composed of many EW. This kind of model is more adequate to describe the temporal evolution of macrosystems, actualizing different potentialities of a collection of EW in time. This complex structure evolves into three phases of coherence–decoherence–recoherence (de Ponte *et al.*, 2010). The characteristics presented by macroscopic systems correspond to the region of the state space occupied by the system at each time of observation.

The dual state space of nature contains all its actual and potential states, each one with two aspects: energy/matter and qualitative aspects. In the construction of this conceptual state space, we assume that each quality ("primary" qualities as mass, extension, weight, pressure, charge, etc. as well as "secondary" qualities as color, sound, pain, pleasure, etc.) is described in a set of dimensions, e.g., three dimensions for fundamental colors, four dimensions for taste, four to seven dimensions for basic emotional feelings, etc. The result is a large set of dimensions that can accommodate every event that occurs in nature, be it a physical, chemical, biological, social, informational, computational and/or mental conscious event.

The actualization of the qualitative aspect of EW occurs in three dynamical phases, discussed in the following sections: coherent wave functions in microstates, decoherent (unconscious) and recoherent (conscious) macrostates. Qualitative conscious experience occurs only during the last phase.

### 5.3 Coherent Wave Functions and QMSs

Each coherent wave function (microstate) contains a collection of EW. This collection is described by the Heisenberg Matrix (HM), corresponding to Schrödinger's concept of a quantum wave function. While the collection is in the coherent kind of state, EW are superposed and entangled. Interaction with the environment, as well as the driving tendency to thermodynamic equilibrium described by the Second Law of Thermodynamics (stating that without external influences the entropy of a system always increases spontaneously) are related to the process of wave function decoherence.

The relation between entropic irreversibility and quantum decoherence has been demonstrated (Frasca, 2007), but it is still not clear what is the cause and what is the effect, of if both processes derive from a common cause (as Boltzmann`s "Principle of Molecular Disorder"; see discussion below).

In an attempt to recast the Copenhagen epistemic approach into an ontological picture, Zurek proposed to reconceive the idea of a "collapse" of the wave function by means of the *decoherence approach*, which is convenient also for our usage of the concept of *recoherence* (Zurek, 2005, 2007). Zurek (2005, 2007) uses HM to describe the (equivalent of) Schrödinger wave function and how it decoheres in time, when an interaction with the environment occurs. In this approach, the concept of decoherence is not exactly the same as the concept of a collapse, but takes its place.

In a decohered macrostate, only the diagonal of the HM survives (Zurek, 2005, 2007). The outcome of each decoherence process is a QMS, that corresponds to the combination of all EW of the diagonal; in other words, a QMS corresponds to the whole diagonal, not to a single outcome.

We have suggested that HM refer to *potential qualitative aspects* instantiated in material substrates. Considering the historical context when the concept of a "measurable" was formulated — dominated by philosophical Empiricism and Instrumentalism — it is comprehensible that reference was made to the process of measurement, instead of making reference to the *content* that appears in the observation act. As the purpose of our work is grounding qualitative conscious experiences in nature, we are disposed to argue in favor of the ontological thesis of "measurables" being *potential qualities*.

All scientific measurement ends with an observation that reveals qualitative features of nature. Flanagan reminds us that the relation between quantum fields and conscious "qualia" was anticipated by several approaches, including Herman Weyl's, quantum chromodynamics, gauge theory, and Kaluza–Klein (Flanagan, 2003, 2007). As argued by Flanagan (2003), the conscious qualities (qualia) that appear in our consciousness in the process of observation are supported by quantum fields, which can be identified with the so-called "observables." However, a compromise with Flanagan´s "Mind/Brain Identity" hypothesis is not necessary for our argument, since we propose a series of phases from the information carried by a photonic field to the final recoherent macrostate that instantiates a conscious episode.

In decohered macrostates, the HM is reduced to the diagonal, generating one macro (not conscious) qualitative state. There are two sub-phases in this process:

(1) *The formation of chemical elements by combinations of single atomic particles.*
The intrinsic qualities of each chemical elements are expressed by their atomic properties, such as the number of particles and the electronic configurations that determine their "reactivity" with other elements and their "affinities", which are comparable to human relationships (Goethe, 1980). Attempts to explain the periodic table by scientists like Bohr and Pauli have produced important concepts and have helped to shape QM itself. Although QM gives a perfectly good ab initio explanation of the lengths of periods of elements, it does not fully explain the precise order of shell filling, which is summarized by the "$n + 1$" or *Madelung Rule* (Scerri, 2006). It is well known that in living beings between 26 and 39 elements or more can be found.

Some elements are fundamental ions necessary to life: calcium, sodium, potassium and chlorine are the most abundant. Many others are considered oligoelements and occur mainly in enzymatic cofactors.

(2) *The formation of chemical substances by combinations of single elements.* Thanks to their peculiar characteristics, the elements can react together to form molecules with new emerging properties necessary for life. The most abundant elements in nature are: oxygen, carbon, hydrogen, nitrogen, sulfur and phosphorus, which reacting together form the main macromolecules of life. The combination of oxygen and hydrogen produces the ubiquitous water that is present in all life forms, from humans (with about 60–70%) to jellyfish (with about 99% of water). The combination of C, H, O produce carbohydrates and lipids; with two more elements, such as N and S, amino acids and proteins are obtained. The nucleic acids are a good example of macromolecules with huge informational properties deriving from the combinations of only five elements (C, H, N, O, P) forming four nucleotide building blocks. The DNA drives the process of production of proteins and the activity of proteins confers qualitative properties to the living cells. Metabolic reactions mediate the definition of the qualities of each cell.

We argue that all these materials and informational processes and their respective qualitative properties derive from quantum functions, which are ultimately derived from EW. However, at this phase (decoherent macrostates with chemical properties) the qualities occur unconsciously, until they are perceived by an adequate natural system.

## 5.4 Entropy Reduction and Life

Although the potentiality for life and consciousness is ubiquitous, these phenomena do not exist everywhere in nature. The actualization of conscious states requires the satisfaction of all necessary conditions of consciousness (such as formation of neural-networks, wakefulness, reentry, attention, information integration, working memory, stimulus contrast at or above threshold, and potential experiences embedded in neural-network: (Vimal, 2013)) the most salient of which possibly consists of distance from thermodynamic equilibrium (Schrödinger, 1944). This condition is provided by means of the action of local entropy reducers, without violating the Second Law of Thermodynamics.

When Thermodynamics and Quantum Theory are combined, a plausible interpretation suggests a connection between the processes of decoherence and irreversibility (Halliwell *et al.*, 1994). If quantum decoherence — understood as loss of correlation between particles that have interacted (Pereira Jr. & Rocha, 2000) — is related to entropy increase, a mechanism that locally decreases entropy — the Biological Maxwell Demon, as formulated by (Monod, 1970) — can allow microscopic potentialities to emerge at the macroscopic level, thus generating phenomena of life, affect, cognition and consciousness.

A balance of attraction and repulsion is found at the level of atomic particles (bosons and fermions), charged atoms and surfaces. This balance leads to the binding of elements forming aggregates. Under high entropy regimes, potential forms in entangled aggregates cannot reach the macro level because of the decoherence/entropy increase process. The operation of mechanisms that keep self-organizing systems at low entropy states allow some forms to be actualized at the macro level, leading to the emergence of life and consciousness.

Schrödinger (1944) argued that a local decrease of entropy in living systems occurs at the cost of increasing it in environment, thus implying that complex organization in living tissue could be generated by means of mechanisms that convert external into internal low entropy. The idea historically derives from conjectures made by Maxwell (1866) and resumed by Loschmidt (1876) in his objection to the H-Theorem (an attempt to mechanically demonstrate the tendency to entropy increase) presented by Boltzmann (1872). As per Narnhofer & Wreszinski (2014), "the (quantum) Boltzmann entropy increases if the initial (zero time) density matrix decoheres."

Boltzmann's work on physical irreversibility contains a concept of entropy based on the distinction of micro and macro states. A microstate refers to the movement values of particles (i.e., position, kinetic energy and direction of velocity), while a macrostate is conceived in terms of a coarse-grained description designed to match the measurable values of classical thermodynamics (temperature, volume and pressure). Each macrostate can be produced by a number of microstates. Boltzmann considered macrostates that could be produced by larger numbers of microstates as being more probable, and the increase of entropy as a spontaneous evolution from less to more probable macrostates.

His H-Theorem (Boltzmann, 1872) demonstrated that in closed ideal gas systems there is a spontaneous irreversible increase of entropy. It describes changes in the function of distribution of particles of a gas (conceived as the result of mechanical interactions of particles by means of elastic collisions depending on position, kinetic energy and angle of collision), obeying the principle of conservation of energy. Boltzmann used statistics to calculate the changes in the distribution function, making some simplifications to capture collective patterns of interactions of many particles in an ideal gas system. In this mechanistic context, the probability of the collision of two molecules would depend on their previous interactions. If they collided before, they could not be considered as statistically independent. However, in the derivation of the H-Theorem he considered the movements of the particles as statistically independent, even in the cases when they collided before. This assumption influenced how he calculated the number of collisions for each volume unit. As a consequence, the description of the evolution of the system was biased to go from less to more probable states. He called this unproved assumption "Stosszahlansatz" ("Principle for the calculus of the number of collisions"). In 1896, he recognized that this calculation was based on an ontological assumption of "molecular disorder" (the loss of correlation between microscopic particles, even those that interacted previously). In this interpretation, when considered in a contemporary perspective, it is possible to conjecture that molecular disorder would cause the increase of entropy and possibly be also related to the decoherence process, since superposition and entanglement are correlations between microscopic particles — the elimination of correlations would be also a loss of quantum coherence (for a contemporary view of the relation between decoherence and entropy increase, see Narnhofer & Wreszinski, 2014).

Could *molecular order* allow entropy *decrease*? From a historical perspective, the idea that coherent states supporting life and cognition can be generated from micro/mesoscopic mechanisms that locally decrease entropy was anticipated in a series of converging speculations by brilliant scientists, beginning with Maxwell (1860, 1866, 1871). Maxwell imagined a small agent (later called "Maxwell's Demon") placed in a wall separating two compartments of an isolated gas chamber in thermal equilibrium. The agent is able to separate the molecules according to their speed, causing the decrease of the entropy of the system. The imaginary entity, placed inside a recipient containing a gas, controls the small gate, separating faster from slower molecules. When a fast molecule approaches the gate the Demon leaves it open, but when a slow one approaches he closes it. The progressive result is the separation of both kinds of molecules: fast ones in the right and slow ones in the left sides of the recipient.

Boltzmann's work put forward the idea that entropy increase depends on a loss of microscopic correlations. This idea is coherent with Maxwell's conjecture, made more than 20 years before — and possibly an influence for Boltzmann's mature thought. In fact, when the small entity manipulates the gas particles to separate the slower from the faster molecules, he creates molecular order, a difference in concentration of molecules with different kinetic energy that can be used to produce work (e.g., moving a cylinder of an engine). The Demon uses information and also a small amount of energy to generate molecular order, thus decreasing the entropy of the system.

As long as there are Maxwell Demons operating in the universe, it will not reach thermodynamic "heat death" (thermodynamic free energy = 0, thermodynamic equilibrium). However, as the operation of Demons requires consumption of useful energy from the environment, it is not able to decrease the total entropy of the universe. The thermodynamic prediction would be that the universe as a whole is approaching heat death in the long run. One plausible possibility of escaping from this verdict lies in the fact that the Second Law does not apply to the microscopic level. Assuming (Boltzmann, 1896) the equiprobability of microstates and using the Ergodic Theorem (which states that in the long run isolated systems spend the same time at each available microstate), it is possible that in its eternal existence the universe alternates periods of entropy increase and decrease. An alternative view is a possibility allowed (but not implied) by Special Relativity equations: the existence of Final Causes that would direct the evolution of some kinds of systems towards low entropy states (Vannini, 2009). In this case, the reasoning would also require specification of mechanisms that reduce entropy.

## 5.5 Biological Self-Organization

Some approximations are needed to relate Maxwell's scheme with biological systems. Biological systems are not isolated, but depend on the assimilation of low-entropy parts of the environment (food) to stay alive. The work of Ilya Prigogine and his group in the field of NonEquilibrium Thermodynamics has showed that — under adequate conditions — dissipative processes could lead to macroscopic organization. Prigogine argued that an open system's internal production of entropy could be balanced by an external energy flux generating "dissipative structures", by means of an "order from fluctuation" mechanism (Prigogine & Stengers, 1984; Nicolis & Prigogine, 1987). In this explanation, the emerging organization is conceived as the result of the pressure towards equilibrium in open systems receiving a flux of free energy from the outside. Consumption and usage of external free energy makes possible the organization of a macro system into low entropy states, when some of the system's recessive potentialities (in our terminology, these possibilities correspond to off-diagonal terms) are amplified, making possible their emergence at the macro level.

A more recent explanation of entropy dynamics in nonequilibrium systems is given by the Maximum Entropy Production principle (Martyushev, 2010). However, it is still far from clear how this or the opposite (Minimum Entropy Production) principle would apply to biological systems (see discussion in Volk & Pauluis, 2010). A simpler explanation was advanced by Schrödinger (1944), in terms of a balance of internal lowering of entropy in living systems with entropy increase in the environment; the Second law is not violated because the total entropy of the whole universe is conserved.

Brillouin (1956), inspired by the similarity of the Boltzmann mathematical expression of entropy and the Weaver and Shannon (1949) expression for average information transmitted between a source and a receiver, proposed the identification of information and negentropy (the entropy that a system exports to keep its own entropy low). However, in order to undergo a negentropic trajectory, it is necessary for a system to possess adequate mechanisms to absorb low entropy from the environment. As indicated by Szilard's classical analysis of the Maxwell proposal, this kind of operation requires the use of information, which must exist previously to the operation of the Demon (Szilard, 1929). In biological systems, the DNA is the main source of structural information. Proteins operate as Maxwellian mechanisms using molecular and other available information to reduce entropy in metabolic systems that coordinate biological functions. Information is a requisite for, not the product of emerging organization; previous information patterns stored in the system are used to detect signals and to act upon the underlying processes to decrease the entropy of the system.

In an engine, the product of entropy reduction is movement (work), while for living beings movement is only a part of the result. The initial step that leads to several functions, including locomotion, is metabolism, a set of physio-chemical transformations. Two processes are central to understanding of the work of proteins in producing metabolic interactions: enzymatic reactions (catalysis: enzyme is not modified) and allosteric interactions (structure and activity of the enzyme are modified by the binding of a metabolic molecule). Since the pioneering work of Monod *et al.* (1963) on the allosteric model, we know proteins have a flexible structure limited to mutually exclusive states: relaxed and tense. The lock-and-key binding of effectors with one or more active microsites of the protein generates a transition of states, and then the protein can itself act as an effector to produce a change in another molecule. Allosteric proteins can form sequential chains to transport information, control gating mechanisms (like those present in membrane ion channels), and other functions. These processes, which are low-energy and cybernetic, coexist with biochemical processes that use energy from glucose to sustain such operations.

Brain QMS can be chemical and/or electric. The action of proteins depends on spatial immediacy of effectors and substrates. In overcoming this limitation, evolution developed physiological processes whereby some proteins encode messages in ionic populations, which broadcast and carry signals to distant places in the biological tissue. The evolution of perceptual, emotional and cognitive systems of the brain was based on the progressive development of neuro-glial connectivity, by means of bioelectric activity (electrical currents and corresponding magnetic fields generated by the movement of ions). In synapses, the bioelectric activity in a preceding neuronal fiber causes release of chemical neurotransmitters, which induces another wave of bioelectric activity in subsequent neurons throughout a signal-transduction pathway. In the neuronal axon, electric currents are generated by the movement of ions in and out the membrane, while in the astrocytes electric currents are produced by the movement of ions through cellular microdomains and gap junctions.

Signal-transduction pathways are complex mechanisms used not only for transmitting information between cells but also to connect two independent systems such as the environment and the genome. The neuronal ligand gated ion channel is a typical example of a biological Maxwell Demon, which uses information and low energy to control the movements of ions. The emergence of brain affective and cognitive functions can be explained by means of mechanisms of ionic control in neurons and astrocytes, inducing the formation of brain-wide coherent states (Freitas-da-Rocha *et al.*, 2001; Rocha *et al.*, 2005).

An example is the NMDA (N-Methyl-D-Aspartate) neuronal receptor, containing ion channels that cross the membrane. It is an allosteric protein with three or more binding sites, controlling $Ca^{++}$ entry in the post-synaptic neuron by means of a gate that, in its resting state, is blocked by magnesium ions ($Mg^{++}$). The opening of the NMDA gate depends on the activation of at least two sites in a hundred-millisecond temporal window. Such an opening depends on activation of recurrent neuronal networks by excitatory Glutamate (Glu) and Glycine (Gly) inputs to distal (NR2) and proximal (NR1) receptor sections, respectively, of apical dendrites of pyramidal neurons in neocortex.

Many neurons respond to environmental signals such as osmolarity, pH, and specific ion concentration through sensory receptors as components of the transduction pathway acting as a sensor. An example is rhodopsin, belonging to the G-protein-coupled receptor (GPCR) family, which is extremely sensitive to light and activate the visual phototransduction pathway. Other examples are odor receptors in olfactory receptor neurons, which are activated by interacting with molecular substructures on the odor molecules; taste receptors (gustatory receptors) in taste buds that interact with chemicals in food to produce an action potential, and mechanoreceptors that responds to mechanical pressure or distortion. The recently identified $H^+$-gated cation channels (ASICs, Acid Sensing Ion Channels) in sensory nerve endings are responsible for the activation of nociceptive afferents by acids (Waldmann, 2001).

The Second Law poses, for self-organizing individuals, the necessity of capturing low entropy from the environment to fuel an ensemble of Maxwell Demons that keep the system's organization and the corresponding capacity of actualization of forms. By means of such a capture, the individual stays alive and able to execute several operations that actualize morphological, physiological, affective and cognitive forms. Consciousness, as the final result of the process of actualization of elementary potential forms into complex aggregates, possibly evolved from ionic/molecular processes such as those involved in the signaling of hunger and thirst. For instance, Denton *et al.* (1999) propose that thirst sensations are generated by changes in sodium concentration detected by molecular sensors. The sodium-level detection system is an example of a signaling unit that generates conscious phenomena, making an early appearance in the phylogenetic scale.

Signal exchange with the entire body and external environment is necessary for normal brain development and functioning. Neuroplasticity means that dynamic patterns of brain activity are shaped by a history of interactions, the resulting patterns determining aspects of mature brain structures. For instance, an animal raised with sensory deprivation does not develop functional perceptual subsystems, whereas under normal conditions patterns of activation co-determine perceptual systems. In humans, sensory deprivation obtained by the use of an isolation tank is a practice of depriving a human being of the ability to perceive one or more types of sensory input (sight, hearing, touch, smell, taste). The sensory deprivation usually appears in a "deprivation tank" or flotation tank, an instrument invented by John Lilly in the late 50s. The subject is immersed in a tank containing water saturated with Epsom salts (magnesium sulfate), maintained at body temperature and fully soundproof in order to eliminate the tactile, visual and auditory sensations, and to isolate it from any kind of contact with the outside world (Lilly, 1956, 1977; Sireteanu *et al.*, 2008). Under these conditions, after reaching a certain state of relaxation, the person begins to feel so altered the boundaries of her own body, her arms seem lengthened or shortened and can manifest buzzing, flashing lights and other visions such as real hallucinations. In rare cases episodes of delirium reflecting the loss of contact with reality have been reported. According to a recent hypothesis the perception detachment from ordinary reality originates in molecular changes at the cellular level, where the transitional sequence: viscosity of membrane/Gs$\alpha$ protein/tubulin/microtubule nanowire network is somehow altered (Cocchi *et al.*, 2011).

Brain activity is located at the intersection of two cycles, the Epigenetic Cycle, a circular flux of information, composed of genetic expression into proteins and regulatory feedback from proteins to genes, and the Functional Cycle, a circular flux of efferent and afferent processes, originally proposed by Von Uexkull (1957). The entire body mediates the interaction of the brain with the environment. When the functional cycle is formed, the motor system, besides controlling behavior, also sends signals to the perceptual system (Gilbert, 2001). In the Epigenetic Cycle, genes are expressed into messenger RNA and proteins, which control brain activity. By means of both kinds of feedback, external and internal to the brain, an "inner world" (the Umwelt; Von Uexkull, 1957) emerges from an active functional cycle.

By means of such a series of interactive processes, the potential qualities of EW can progressively emerge in macroscopic systems. In biological systems, the interactive process of emergence of QMS includes a genetic phase. Genes combine and interact with the environment, resulting in phenotypic features. As a result, biological systems have two inseparable aspects — physical and mental — derived from the combination of elementary qualities and their interactions within themselves and with the environment. This proposal is compatible with a family of similar proposals: Dual-Aspect Monism (Vimal, 2008, 2010b, 2013), Triple Aspect Monism (Pereira Jr., 2013, 2014) and Triple-Aspect Monistic Model (Cacha & Poznanski, 2014).

## 5.6. Recoherent Macrostates and Qualitative Conscious Episodes

In open nonequilibrium systems, when entropy decreases the mechanism of Order through Fluctuations (Nicolis & Prigogine, 1977; Prigogine, 1977) allows the strong emergence of new QMS. Additionally, under the operation of adequate information integration mechanisms, the system can evolve to a "coherent macrostate", here called *recoherent*, following the use of the term by Chin *et al.* (2013), de Ponte *et al.* (2010) and Hsiang & Ford (2008). Recoherence may be also related to the temporal overlap implied by Heisenberg uncertainty (Chris Nunn, personal communication). Such overlaps may quantitatively map (if the "right" clock time duration is assumed) wavy classical events in the brain, resulting in recoherence in the sense of temporal binding.

In view of the diversity of interpretations of QM (Schreiber, 1995), the concept of recoherence is possibly not compatible with the Copenhagen view, because the latter is based on the problematic Interactive Substance Dualism and/or Idealism. However, in the Copenhagen interpretation the superposed eigen-states are conceived as all possible potential $n$ basis states, each with theoretical probability of $1/n$ and the sum of all probabilities is 1. This idea may be useful in the context of the decoherence/segregation process, for the analysis of a single dimensional attribute (such as redness) of a sub-modal attribute (such as color) of a modal attribute (such as vision) via matching and selection of a specific dimensional attribute (Vimal, 2010b). In our approach to recoherence, we use the concept of hypermatrix/hypertensor, where each diagonal element has probability equal to one, and their integrations/entanglements/integrations are represented in off-diagonal elements.

Brain recoherent macrostates are formed by means of the operation of quantum computing gates within ionic waves, and can be expressed by a hypermatrix (or hypertensor), corresponding to the binding of qualitative features in one integrated conscious episode. Once formed, the ionic waves feedback on brain activity, similarly to David Bohm's "pilot wave." One could argue that his *guiding wave* (implicate order) is not material; matter appears only when the continuous wave acts in discrete ways (explicate order). In addition, one could argue that an implicate order and an explicate order states are dual-aspect states, where the degree of the dominance of aspects may vary depending on the level of order. We can use Bohm's interpretation to conceptualize the recoherence process, although it has its own problems (Schreiber, 1995). Recently, a series of experiments have proven the possibility of a "pilot wave" producing quantum effects in macroscopic systems (Bush, 2014; Bush *et al.*, 2014; Oza *et al.*, 2014; Oza *et al.*, 2014).

Conscious episodes are not made of one single quality, as "redness." They are made of many qualities bound to each other. Each quality is determined by a decoherence process (all eigenstates being reduced to one in the act of "measurement"), but this product is a micro-consciousness (Zeki & Bartels, 1999) QMS located at some region of the primary sensory areas. In neuroscience, this is called segregation of attributes for analysis in specific brain areas, such as V4/V8/VO visual areas for color. A conscious episode requires the binding of this quality with many others.

In the nervous system of living individuals, including the human brain, the instantiation of macrostates is spatially distributed and unconscious. The formation of conscious episodes requires the formation of a recoherent collection of these macrostates. When all necessary conditions of consciousness such as activation of neural-networks, wakefulness, reentry, attention, activation integration, working memory, stimulus contrast at or above threshold, and potential experiences embedded in the neural-network: (Vimal, 2013) are satisfied, a recoherent state (corresponding to a conscious episode) emerges, from a collection of non-conscious QMS instantiated in spatially distributed neural circuits.

Brain recoherent macrostates result from the activity of entropy reducers, as ion channels and proteins composing intracellular signal transduction pathways. These mechanisms possibly instantiate quantum computing gates (Freitas-da-Rocha *et al.*, 2001; Rocha *et al.*, 2005). The operation of these gates form recoherent states, by means of informationally integrating a collection of QMS. Also Tononi proposed an information integration theory of consciousness, where "consciousness corresponds to the capacity of a system to integrate information" (Tononi, 2004). In addition, the experiential aspect of consciousness has two key properties (Tononi, 2004):

(i) The *differentiation*, which is the availability of a very large number of conscious subjective experiences (SEs);
(ii) The *integration*, which is the unity of such experiences. The materialistic identity theory-based Integrated Information Theory (IIT) (Balduzzi & Tononi, 2009; Tononi, 2004, 2008, 2012) explains only the objective Third Person Perspective (3pp)-physical aspect of a brain-mind state (corresponding to the physical aspect of reality) because the explanatory gap problem still remains. IIT can be developed in our Multi-Aspect Monist framework to explain the First Person Perspective mental aspect of brain–mind states (Vimal, 2009a, 2010c).

The recoherent macrostate as a whole would correspond to the binding of qualitative features in one integrated conscious episode available in the first person perspective for the living individual. It can be described by means of a hypermatrix or hypertensor with nonzero diagonal and off-diagonal terms, where each diagonal term corresponds to a macro quality.

Let us take the simple example of a "rightward moving red ball" to explain the coherence, decoherence and recoherence phases related to processes in physics (out there in real world), neurophysiology (in our brain) and reflected in psychology (in our mind). The three phases related to the "rightward moving red ball" and their processing are briefly elaborated as follows:

(i) The coherence phase consists of light reflected from the "rightward moving red ball" out there in real physical world, which has coherent physical information related to motion, color (long wavelength light) and spherical shape; the density matrix has three diagonal terms and six nonzero off-diagonal interference terms and hence information is in coherent and integrated form in physical domain out there in real world;
(ii) In the decoherence phase, the information encoded in wavelength and intensity of light is transduced by retinal cone photoreceptors into electrochemical signals and then segregated/decohered through processing further in retinal and LGN cells, and then specialized

cortical areas V5/MT for motion, V8/V4/VO for color, and multiple areas (such as STS-MT, FST, FEF, PFC and so on: (Sereno *et al.*, 2002) for shape. This is done in neurophysiological domains. In each specialized area, such as V8/V4/VO for color, the interference off-diagonal terms of diagonal terms of density matrix are destroyed to zero by anatomical/physiological segregation (we call it *decoherence*); only the diagonal terms representing the various potential color-experiences (such as redness, greenness, blueness, and so on) remain. Then, a specific QMS (such as redness of red ball) is selected through the matching and selection mechanisms (Vimal, 2010b). Similarly, a specific experience for rightward motion and roundness for shape are selected;

(iii) The three attributes, namely rightward motion, redness, and roundness, need to be integrated/bound for a unified experience/consciousness of "rightward moving red ball" in the recoherence phase. There are many models for this, such as 40 Hz synchrony (Aru *et al.*, 2012; Engel & Singer, 2001; Singer, 1999), microtubule-based quantum operators (Hameroff & Penrose, 1996, 1998; Hameroff & Powell, 2009) and quantum antennas in astroglial calcium waves (Pereira Jr., 2012). This recoherence phase constructs, by means of the binding/integration process, the 1pp-mental aspect of our brain-mind state and its inseparable 3pp-physical aspect. They appear as a conscious episode having the "rightward moving red ball" as the content. The diagonal and off-diagonal terms of hypermatrix/hypertensor are reconstructed by the recoherence process.

Here there are three QMS: redness, rightward motion and roundness of ball. An unified/integrated experience of these three qualities is obtained by inserting them along the diagonal of a hypermatrix; in this example, a $3 \times 3$ matrix. The off-diagonal six terms of this hypermatrix contain the interaction/entanglement of the two of the three diagonal terms; for example, the off-diagonal term (1, 2) is the interaction/entanglement of redness and rightward motion; the term (1, 3) is the interaction/entanglement of the redness and roundness of ball; and (2, 3) is the interaction/entanglement of the rightward motion and roundness of ball, and so on.

The probability of each of the terms of hypermatrix is $p$. Thus, a hypermatrix represents all possible experiences of "rightward moving red ball" with single or double attributes at a time, but not more than two, and does not represent (1, 2, 3)-interaction together simultaneously. This, however, can be represented using a hypertensor with 3rd order tensor with $p = 1$. If there are n attributes, then a *Nth* order tensor may be necessary to represent them.

The concept of recoherence is also different from Hameroff (2014). He believes that the collapse of the wave function generates a conscious state; however, collapse/decoherence, being a move towards higher entropy, would generate mostly unconscious states. The recoherence process after the decoherence phase generates the conscious state.

In sum, the conscious perception of a scene can be understood as containing the following steps. First, photonic fields in the environment decohere when they meet the eye or another sensor. Several photonic fields reach different sensory modalities at the same time. In each sensory cortex there is an unconscious qualitative pattern (color, smell, sound, etc.) being formed. These patterns are instantiated in dendritic fields and communicated to other neurons by means of action potentials, astroglial ionic waves and eventually electric synapses and dendro-dendritic connections. A conscious episode is experienced when all these features distributed in the brain bind, forming a "gestalt." The recoherence process corresponds to the formation of this "gestalt" and to the experience of a conscious episode. This expression depends on causal factors, such as physical forces, to occur, but the resulting patterns are not determined by these factors; they are determined by the mode of combination (e.g., matching, fusion, interference) of the elementary patterns.

Mechanisms of matching, resonance, selection and integration operate to define conscious episodes. In matching processes, unconscious qualities interact, interfere, cancel, inhibit or amplify each other. In resonant processes, when the matching is positive the amplification makes possible for a quality to be widespread in the system. In selection processes, the dominant qualities are selected to compose a conscious episode. In integration processes, the selected qualities are "entangled" to form a gestalt containing the qualities in an integrated multimodal scene. To sum up, the stimulus-dependent (or endogenous) feed forward (FF) signals interact with cognitive feedback (FB) signals (Vimal, 2010b); then resonance, selection, and integration of qualities/proto-experiences occur; and finally a unified/bound/integrated/recohered global whole is consciously experienced.

In the theoretical context of current systems theory, we use the concept of *metastability* (Kelso, 2013) to characterize the global dynamics of the brain:

(a) The local decoherence (through density matrix) would correspond to the segregation phase of metastability (an expression of brain-complexity that reach the maximum when the balance between segregation and its complement integration is achieved);
(b) Global recoherent state (described with the use of a hypertensor) would correspond to the synergic integration/binding phase of metastable dynamics, and
(c) Superposed potential experiences would correspond to attractors in a multistable landscape.

In such a dynamics, each recoherent macrostate alternates in a time scale of nanoseconds with decoherent macrostates, but we consciously perceive only the sequence of nanosecond recoherent states, which appear to us in continuous chunks with the duration of at least one second each.

## 5.7. Discussion

### 5.7.1. *On emergence*

There are many models for emergence (Bedeau, 1997; Broad, 1925; Chalmers, 2006; Fingelkurts *et al.*, 2010a, 2010b; Freeman, 1999; Freeman & Vitiello, 2011; Kim, 1999; McLaughlin, 1992; Seager, 2010, 2012). However, in all this literature the strong emergence of QMS is conceived as deriving only from the physical aspect of nature, or from the physiological aspect of brain activity. In our approach, the emergence of conscious SEs (and respective "qualia") is not properly from the material aspect of reality, but from the interaction of a collection of potential qualitative aspects of EW. This interaction process occurs in time, during the life history of complex macroscopic systems such as the living ones.

Materialist reductionist approaches make a category mistake in this regard, because the qualitative aspect belongs to an ontological category that is different from the material aspect. Our proposal of strong emergence of mental qualities in macro systems does not make the same category mistake, and is less mysterious than materialist emergentism. Materialists (Fingelkurts *et al.*, 2010a, 2010b) claim that mental qualities emerge irreducibly from nonexperiential nonmental matter, such as interacting neural signals; however, the same authors implicitly assume that matter does not have potentiality for mentality. This type of materialistic strong emergence (that still persists today) falls under "magic/miracle" theory and is ultimately mysterious.

We claim that mental qualitative states emerge irreducibly from interacting potential qualitative building blocks; this needs further research to clarify the "modus operandi," but it is less mysterious. For example, mixing appropriate amounts of redness related long wavelength and greenness related middle wavelength lights results in yellowness (Vimal *et al.*, 1987); however, our experience of yellowness is irreducible to redness and greenness. Strong emergence is claimed to halt scientific investigation (Seth, 2010), as if some vital life force acts to create novel attributes in a miraculous manner (Allen, 2005), or by being incompatible with empirical investigation and mechanistic explanation (Branch-Smith, 2014). Vimal (2013) claims for the possibility of unpacking strong emergence, assuming the potentiality of SEs in Nature, as well as the operation of brain matching and selection mechanisms (Vimal, 2010b) that select a specific conscious experience from the potential/possible experiences.

In the classical Kinetic Theory of Matter (Boltzmann, 1896), the probability of a macrostate (P(Mi)) is given by the number of microstates that can generate it. The Boltzmann probabilistic concept of entropy increase states that more probable macrostates are more stable and less probable macrostates are more unstable; therefore, once a system spontaneously reaches a probable macrostate it remains longer time in this state than when it reaches a less probable microstate. Considering that some regions of the universe as our planet are in a low entropy state, the tendency of natural systems in these regions is to evolve towards the most stable states.

Boltzmann´s approach was based on the methodological assumption of systemic closure. When taking into consideration open systems as living organisms and their brains, three complementary principles should be taken into consideration: *Interaction*, *Basic Emergence* and *Co-Evolution*.

(I) Principle of Interaction (PI): interactions between $N$ systems modulate the probability of their macrostates. PI makes possible for open systems receiving fluxes of useful energy from the environment to evolve towards low entropy states, when the second principle operates.

(II) Principle of Basic Emergence (PBE): when two or more $N$ systems interact regularly, instable macrostates may become stable, allowing the emergence of new macrostates. One possible explanation for PBE is that when two or more systems interact, the resulting state space expands more than the entropy of the resulting mega-system (the system formed by the union of the interacting systems).

(III) Principle of Coevolution (PC) of potential mental and physical aspects. Since a microstate is a dual-aspect state, both (potentially mental and physical) aspects strongly emerge into dual-aspect macrostates.

The idea of coevolution has similarities and differences in regard to the ancient Vedic account of reality. One common idea is that the universe (or multiverse) does not evolve from nothingness (Boyer, 2014a), and one difference is that we assume EW as potential states that evolve to actual states, while Boyer (2014b) assumes the existence of a full Universal Mind or Universal Consciousness at the foundations, from which all aspects of reality are derived.

Another theoretical possibility is combining Emergentism with Panpsychism (Seager, 2012). This idea seems to be counter-intuitive, since the idea of emergence implies that the conscious mind — as we know it — would depend on the actualization of an aspect of Nature that previously existed only in potential states. In other words, Emergentism seems to imply that before the necessary and sufficient conditions for consciousness appeared in some spatiotemporal regions of the universe there was no actual conscious mind. However, if a distinction of degrees of mentality is made, allowing nonbiological signaling to count as mental activity, Seager's claim would not be false.

Furthermore, combinatorial infusion (Seager, 2010) is a sort of diachronic strong emergence, because infusion mysteriously entails (i) the loss of constituents' properties, (ii) and then the emergence of a novel property. In other words, the parts/constituents infuse their properties into the whole and by so doing efface themselves (Seager, 2010) consistently with the diachronic strong emergence, which is a sort of jargon to hide our ignorance.

### 5.7.2. *Explanatory power of foundational theories*

One could argue that foundational approaches as in string theory lack predictive power. The spatial scale between strings and QM is too large, and it is unclear how elementary particles and QM can be derived from string theory.

Although the spatial scales are far distant, there is a degree of self-similarity (Flanagan, 2003). However, one could argue that the self-similarity between strings vibrating in complex six-dimensional compacted spaces and $Ca^{++}$ waves (or whatever) in 3D extended space as well as attributing some sort of physical reality to strings are debatable, as chains of physical causation would be needed for any meaningful mapping to occur. The disparities of scale and dimensionality make any such implications problematic at best (Nunn, personal communication in 2014).

We argue that the idea of our model is more complex. It is not a direct mapping. The becoming process of reality leading from strings to conscious states should have at least four stages:

(i) Vibrating strings in potential states;
(ii) Coherent quantum wave functions, where a collection of potential strings is superposed and entangled;
(iii) Decoherent macrostates, where the collection of strings is actualized into a system macrostate with material (interacting particles) and informational (forms that can be transmitted to other systems) aspects. Unconscious brain states are included in this category;
(iv) Recoherent macrostates that occur in systems as the living brain, where a collection of decoherent macrostates is bound into a conscious episode.

Furthermore, we note that a characteristic of the conceptual model is that each step transition involves strong emergence; therefore, one probably cannot be mathematically deduced from with other. This idea seems consistent with the failure of derivation of (a) elementary particles from string theory and (b) experiences from qualitative or physical aspects of strings, elementary particles, and classical objects.

However, one could still argue that "strong emergence" is just a jargon for hiding our ignorance or that we are using strong emergence because we do not know the underlying mechanism at present time.

On the other hand, one could also argue that strong emergence is not based on ignorance, but it occurs because the system is not fully deterministic. In an interacting system with degrees of freedom, each subsystem influences the others (as in Poincaré`s three-body problem); there is a meeting of independent causal lines, as in Cournot's concept of *chance* (Lungarzo & Pereira, 2009); as these lines are not previously coordinated, the outcome cannot be calculated. Another important fact is that the resulting state space of interacting subsystems is the *product* of their state spaces; there is a combinatorial explosion generating potentially infinite possibilities, as in human language (potentially infinite sentences from a finite alphabet).

### 5.7.3. *Quantum fields and perceptual states*

Do SEs (secondary qualities) reside (pre-exist) in fundamental physics? For instance, in one of the extra dimensions of string theory or in the "hidden" variables of QM, being projected in our 4D ($x, y, z, t$) conscious episodes? This is in analogy to the projection of a 3D sphere on 2D space.

This projection can be conceived:

(i) As SEs embedded in relevant neuro-networks as proto-experiences (pre-cursors of SEs) and then specific SE selected via matching and selection mechanisms (Vimal, 2010b) and/or
(ii) By means of an extension of the projection concept used in the "Quantum-like" paradigm proposed by Khrennikov & Basieva (2014). In other words, one of the possible answers of the question "where do SEs come from?" is that they can presumably come from one (or more) of the extra curled up dimensions of string theory as projections.

Moreover, the mass of a neuron is derived to be of the order of Planck mass and the brain vibration (~12 Hz) and Planck vibration can be described by the same second order Schrödinger equation (Marciak-Kozłowska & Kozlowski, 2014). In addition, (Chang, 2014) proposes that:

(i) the double helix DNA is analogue with an entangled string;
(ii) biological strings such as DNA, biological macromolecules, neurons, RNA, genes, proteins, and even sperms can be described by the superstring and the extensive string theory;
(iii) biological strings such as various loops and neural networks of living things may be described by closed string;
(iv) in biological self-organized systems, such as the permeable membrane, the enzyme, adenosine triphosphats (ATP), and so on, the complexity and bio-information increase and entropy decrease.

Furthermore, (a) the encoding of uncertainty in cognitive systems by superpositions of eigenstates, as a quantum(-like) approach to cognition (Khrennikov & Basieva, 2014), can be used in the decoherence or segregation phase for the analysis of specific attributes (such as color, motion, and shape) locally; such superpositions are resolved by a measurement that involves interactions between FF and FB signals using matching and selection mechanisms (Vimal, 2010b); and (b) the concept of superposition and the concept of generalized quantum observables given by positive operator-valued measures (POVMs) with interferences between localized diagonal terms representing segregated analyzed attributes (such as color, motion, and shape) (Khrennikov & Basieva, 2014) can be used in the recoherence or integration phase where localized micro-experiences are integrated using hypertensor for the generation of unified macro global conscious episode (such as the unified experience of "rightward moving red spherical ball" in our example).

Objects of our perception invariably include places, times, and secondary qualities in combination (5D: $x$, $y$, $z$, $t$, experience); here a primary quality is, for example, a wavelength of light (such as 650 nm) and the related secondary quality is its correlated experience (redness), which is the effect of the primary quality of things on the senses (Flanagan, 2003). This is a sort of consistent with Flanagan's (2003) proposal, which seems to imply that the secondary qualities (such as experience redness) may reside in one or more of the M-theory's curled up compactified (ultra-small) 6_spatial+1_brane extra-dimensions, along which strings latent/hidden/invisible to us from 3pp, this is in analogy to Kaluza–Klein 5D proposal.

According to Flanagan (2003), "we can make a natural mapping from the spectral colors to a color sphere…such a mapping could be made with red, green and blue for the axes of a unit sphere in Hilbert space. We could then easily map those color vectors to the photonic vectors with which they are associated, remembering that these "physical" vectors recapitulate the mathematics of colors under vector addition and multiplication. Then, any operation upon the photonic vector would naturally correspond to a rotation of the color vector, in a direct analogy with the mathematics of gauge theory and quantum theory generally."

Brian Flanagan continues his conjecture: "If such a color sphere were to 'sit over' every point in 4D space–time, that would seem to square with the facts of our experience of the visual field and also perhaps with the mathematics of string/M-theory, where a six-dimensional orbifold is often pictured 'sitting over' every point in 4D space–time...we might reflect on the fact that Calabi-Yau spaces are known to be projective spaces, and to include the SU(3) group of the standard model...as Weyl tells us, colors are isomorphic to the points of the projective plane... The theory is called 'chromo' dynamics because quarks and gluons recapitulate the behavior of spectral colors...photons and gluons are both gauge particles, and photons are continually associated with spectral color. So we might defy convention and ask: Is color a kind of gauge information?" (Flanagan, 2003).

Although possibly running against orthodox interpretations of quantum theory, Flanagan´s conjecture converges to our proposal of looking for potential qualia in the quantum domain.

### 5.7.4. *Conscious mental causation*

In the recoherent phase, once the pilot wave is formed it feeds back on the system that generated it. This action can be regarded as the physical aspect of the process of conscious mental causation, by which conscious experiences influences decision-making, memory formation, somatic processes and behavior.

How to physically describe this feedback? Here we do not attempt to give a full explanation of the problem, but to indicate how our model allows a treatment of this age-old issue.

In the calcium wave model of conscious processing (Pereira Jr. *et al.*, 2013), the integration-and-feeling operations that compose a conscious episode is related to the formation of a large calcium wave in the brain, which feeds back on the neural assemblies that generated it. This loop is illustrated by the "Carousel Effect" (Fig. 1).

The philosophical question that remains is about the kind of physical process involved in the feedback relation. Of course, there is an electromagnetic process, since moving ions are charged particles that generate electric currents and are affected by electromagnetic forces. However, the most relevant aspect would not be the matter or the energy, but the information patterns, or — in philosophical terms — it would be related to Aristotle's concept of a *formal cause*. It is the *form* of the waves that influence each other — the form of the spatially distributed unconscious brain waves influence the form of the unitary conscious recoherent wave, and then the form of the latter feeds back on the form of the local waves.

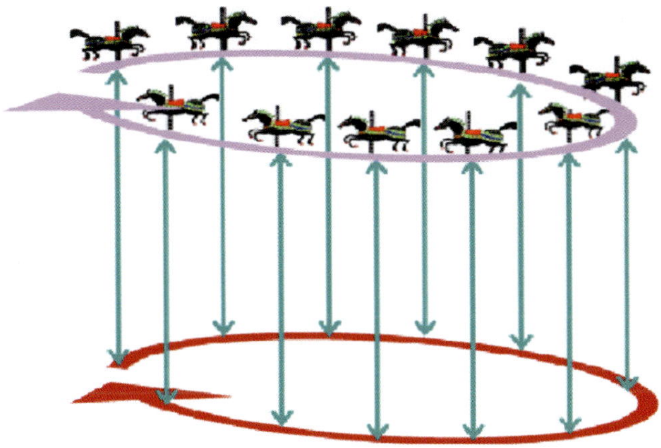

**Fig. 1.** The Carousel Effect (from (Pereira Jr. & Furlan, 2010)): The red circle represents a cycle of activity of a population of synchronized neurons, while the mauve circle represents one large calcium wave in neuro-astroglial domains. Each horse corresponds to a small local calcium wave. The coordination of smaller waves, generating the large wave, is orchestrated by synchronized neurons and feeds back on them (bidirectional arrows), independently of an actual causal relation between the smaller waves.

This inter-influencing of the waves occur because they are not identical or isomorphic; they are partially overlapping, but have important differences that make possible a dynamical "wave computing" process composed of constructive interferences of amplitude-modulated ionic waves combined with a degree of quantum entanglement (Pereira Jr., 2007, 2012).

## 5.8. Concluding Remarks: Has the Hard Problem Been Solved?

We argue that the Hard Problem can be convincingly solved by means of the above sketched *qualitative physics*. EW is conceived as the primitive building blocks that dynamically combine, in complex ways, to generate QMS. Qualities instantiated in unconscious brain macrostates are integrated into conscious experiences, when a set of conditions are fulfilled, as distance from thermodynamic equilibrium, operation of biological self-organizing mechanisms and information integration by quantum gates.

Using this explanatory strategy, we can explain why some natural systems have subjective conscious experiences, while others do not. The progressive interaction of EW generate a complex state space, of which some regions correspond to the first person conscious activity of living individuals. The existence of these regions is derived from the potentialities of EW, in a strongly emergent process. Other regions do not display conscious activity, because the necessary conditions of consciousness (such as formation of neural-networks, wakefulness, reentry, attention, information integration, working memory, stimulus contrast at or above threshold, and potential experiences embedded in neural-network: (Vimal, 2013)) for it are not satisfied.

The dynamical process above occurs in a temporal continuum. At one side of the continuum, there are forms in a potential state. When actualized, they compose physiochemical properties of substances and processes. In the middle, there are forms in an intermediary stage, such that they have mental but unconscious functions. At the other side, there are forms actualized in conscious episodes experienced by a living individual.

This proposal expresses a tendency of current approaches to consciousness to abandon both Materialism and Idealism, as well as Interactive Substance Dualism, moving towards *Multi-Aspect Monisms* (Pereira Jr. *et al.*, 2010; Pereira Jr., 2013, 2014; Vimal, 2008, 2010b, 2013).

## Acknowledgments

Thanks to FAPESP (Alfredo P. Jr.), Roman Poznanski, Chris Nunn and Sisir Roy for their valuable comments. RLP Vimal was supported by VP-Research Foundation Trust, and Vision Research Institute and Dristi Anusandhana Sansthana Research Funds. RLP Vimal is also affiliated with Dristi Anusandhana Sansthana, A-60 Umed Park, Sola Road, Ahmedabad-61, Gujrat, India; Dristi Anusandhana Sansthana, c/o NiceTech Computer Education Institute, Pendra, Bilaspur, C.G. 495119, India; and Dristi Anusandhana Sansthana, Sai Niwas, East of Hanuman Mandir, Betiahata, Gorakhpur, U.P. 273001 India.

## References

Allen, G. E. (2005) Mechanism, vitalism and organism in late nineteenth and twentieth century biology: The importance of historical context. *Stud. History Philos. Sci. C: Stud. History Philos. Biol. Biomed. Sci.*, **36**, 261–283.

Aru, J., Bachmann, T., Singer, W. & Melloni, L. (2012) Distilling the neural correlates of consciousness. *Neurosci. Biobehav. Rev.*, **36**, 737–746.

Balduzzi, D. & Tononi, G. (2009) Qualia: The geometry of integrated information. *PLoS Comput. Biol.*, **5**, e1000462.

Bedeau, M. A. (1997) Weak emergence. *Philos. Perspectives*, **11**, 375–399.

Bohm, D. (1952) A suggested interpretation of the quantum theory in terms of 'hidden' variables, I and II. *Phys. Rev.*, **85**, 166–179 and 180–193.

Bohm, D. (1980) *Wholeness and the Implicate Order*. Boston: Routledge & Kegan Paul.

Bohm, D. (1985) *Unfolding Meaning: A Weekend of Dialogue*. London, UK: Routledge.

Bohm, D. (1987) *Unfolding Meaning, A Weekend of Dialogue with David Bohm*. London, UK: Routledge & Kegan Paul.

Bohm, D. Hiley, B. J. (1993) *The Undivided Universe: An Ontological Interpretation of Quantum Mechanics*. London: Routledge.

Boltzmann, L. (1872) Further studies in the thermal equilibrium of gas molecules. In: S. Brush, ed. *Kinetic Theory*, Vol. 1. Oxford, London: Pergamon Press, pp. 88–175.

Boltzmann, L. (1896) *Lectures on Gas Theory* (S. Brush, Trans.). Berkeley, Los Angeles: University of California Press.

Branch-Smith, T. (2014) *Reducing the Emergence of the Gaps: Computation for Weak Emergence*. Ontario, Canada: University of Waterloo.

Boyer, R. W. (2014a) Did the universe emerge from nothing? Reductive vs. holistic cosmology. *NeuroQuantology*, **12**, 424–454.

Boyer, R. W. (2014b) Unless we are robots, classical and quantum theories are fundamentally inadequate. *NeuroQuantology*, **12**, 102–125.

Brillouin, L. (1956) *Science and Information Theory*. New York: Academic Press.

Broad, C. D. (1925) *The Mind and its Place in Nature*, 1st edn. London: Routledge & Kegan Paul.

Bush, J. W. M. (2014) Pilot-wave hydrodynamics. *Ann. Rev. Fluid Mech.*, **49**, 269–292.

Bush, J. W. M., Oza, A. & Molacek, J. (2014) The wave-induced added mass of walking droplets. *J. Fluid Mech.*, **755**, R7.

Buyse, F. (2013) *The Distinction between Primary Properties and Secondary Qualities in Galileo Galilei's Natural Philosophy*. Paper presented at the Talk given in September 28, 2012, at the Quebec Seminar in Early Modern Philosophy and in April 8, 2013, at the History of Science Collections in Bizzell Libraries at the University of Oklahoma.

Cacha, L. A. & Poznanski, R. R. (2014) Genomic instantiation of consciousness in neurons through a biophoton field theory. *J. Integr. Neurosci.*, **13**, 253–292.

Caponigro, M., Jiang, X., Prakash, R. & Vimal, R. L. P. (2010) Quantum entanglement: Can we "see" the implicate order? Philosophical speculations. *NeuroQuantology*, **8**, 378–389.

Chalmers, D. J. (1995) Facing up to the problem of consciousness. *J. Conscious. Stud.*, **2**, 200–219.

Chalmers, D. (1996) *The Conscious Mind: in Search of a Fundamental Theory*. Oxford/ New York: Oxford University Press.

Chalmers, D. J. (2006) Strong and weak emergence. In: P. Clayton and P. Davies, eds. *The Re-emergence of Emergence*. New York: Oxford University Press, pp. 244–256.

Chang, Y. F. (2014) Extensive quantum theory of DNA and biological string. *NeuroQuantology*, **12**, 356–363.

Chin, A. W., Prior, J., Rosenbach, R., Caycedo-Soler, F., Huelga, S. F. & Plenio, M. B. (2013) The role of non-equilibrium vibrational structures in electronic coherence and recoherence in pigment–protein complexes. *Nat. Phys.*, **9**, 113–118.

Cocchi, M., Tonello, L., Gabrielli, F. & Pregnolato, M. (2011) Consciousness and hallucination: Molecular considerations and theoretical questions. *NeuroQuantology*, **9**, 190–201.

de Ponte, M. A., Cacheffo, A., Villas-Boas, C. J., Mizrahi, S. S. & Moussa, M. H. Y. (2010) Spontaneous recoherence of quantum states after decoherence. *Eur. Phys. J. D*, **59**, 487–496.

Denton, D., Shade, R., Zamarippa, F., Egan, G., Blair, J., Mckinley, M., Lancaster, J. & Fox, P. (1999) Neuroimaging of genesis and satiation of thirst and an interoceptor-driven theory of origins of primary consciousness. *Proc. Natl. Acad. Sci. USA*, **96**, 5304–5309.

Einstein, A. (1971) Letter 84, 3 March 1947, The Born–Einstein letters: Correspondence between Albert Einstein and Max and Hedwig Born from 1916–1955, with commentaries by Max Born. London: Macmillan.

Einstein, A., Podolsky, B. & Rosen, N. (1935) Can quantum-mechanical description of physical reality be considered complete? *Phys. Rev. Lett.*, **47**, 777–780.

Engel, A. K. & Singer, W. (2001) Temporal binding and the neural correlates of sensory awareness. *Trends Cogn. Sci.*, **5**, 16–25.

Fingelkurts, A. A., Fingelkurts, A. A. & Neves, C. F. H. (2010a) Emergentist monism, biological realism, operations and brain-mind problem. *Phys. Life Rev.*, **7**, 264–268.

Fingelkurts, A. A., Fingelkurts, A. A. & Neves, C. F. H. (2010b) Natural world physical, brain operational, and mind phenomenal space-time. *Phys. Life Rev.*, **7**, 195–249.

Flanagan, B. J. (2003) Are perceptual fields quantum fields? *NeuroQuantology*, **1**, 334–364.

Flanagan, B. J. (2007) On the unification of mind and matter. *NeuroQuantology*, **5**, 331–345.

Forbus, K. D. (1988) Qualitative physics: Past, present, and future. Chapter 7, In: H. E. Shrobe, ed. *Exploring Artificial Intelligence*. San Francisco, CA, USA: Morgan Kaufmann Publishers, pp. 239–296.

Frasca, M. (2007) Thermodynamic limit and decoherence: Rigorous results. *J. Phys.: Conf. Ser.*, **67**, 012026.

Freeman, W. J. (1999) Consciousness, intentionality and causality. *J. Conscious. Stud.*, **6**, 143–172.

Freeman, W. J. & Vitiello, G. (2006) Nonlinear brain dynamics as macroscopic manifestation of the underlying many-body field dynamics. *Phys. Life Rev.*, **3**, 93–118.

Freeman, W. J. & Vitiello, G. (2011) The dissipative brain and non-equilibrium thermodynamics. *J Cosmology*, **14**, 4461–4468.

Freitas-da-Rocha, A., Pereira Jr., A. & Coutinho, F. A. B. (2001) N-methyl-D-aspartate channel and consciousness: From signal coincidence detection to quantum computing. *Prog. Neurobiol.*, **64**, 555–573.

Gilbert, P. F. C. (2001) An outline of brain function. *Cogn. Brain Res.*, **12**, 61–74.

Goethe, J. W. (1980) *Les Affinités Électives. Trad. Pierre du Colombier*. Paris: Gallimard.

Greene, B. (1999) *The Elegant Universe: Superstrings, Hidden Dimensions, and the Quest for the Ultimate Theory*. New York: W.W. Norton & Company.

Greene, B. (2004) *The Fabric of the Cosmos: Space, Time, and the Texture of Reality*. New York: Alfred A. Knopf.

Halliwell, J. J., Pérez-Mercader, J. & Zurek, W. H., eds. (1994) *Physical Origins of Time Asymmetry*. Cambridge (UK): Cambridge University Press.

Hameroff, S. (2014) Quantum walks in brain microtubules-a biomolecular basis for quantum cognition? *Top. Cogn. Sci.*, **6**, 91–97.

Hameroff, S. & Penrose, R. (1996) Orchestrated reduction of quantum coherence in brain microtubules: A model for consciousness? In: S. R. Hameroff, A. W. Kaszniak and A. C. Scott, eds. *Toward a Science of Consciousness — The First Tucson Discussions and Debates*. Cambridge, MA: MIT Press, pp. 507–540.

Hameroff, S. & Penrose, R. (1998) Quantum computation in brain microtubules? The Penrose–Hameroff 'Orch OR' model of consciousness. *Philos. Trans. R. Soc. London, Ser. A*, **356**, 1869–1896.

Hameroff, S. & Powell, J. (2009) The conscious connection: A psycho-physical bridge between brain and pan-experiential quantum geometry. In: D. Skrbina, ed. *Mind That Abides: Panpsychism in the New Millennium*. Chapter 5. Amsterdam: John Benjamins Publishing Company, pp. 109–127.

Hayden, E. C. (2010) Human genome at ten: Life is complicated. *Nature*, **464**, 664–667.

Hsiang, J.-T. & Ford, L. H. (2009) Decoherence and recoherence in model quantum systems. *Int. J. Mod. Phys. A*, **24**, 1705–1712.

Khrennikov, A. & Basieva, I. (2014) Quantum model for psychological measurements: From the projection postulate to interference of mental observables represented as positive operator valued measures. *NeuroQuantology,* **12**, 324–336.

Kim, J. (1999) Making sense of emergence. *Philos. Stud.*, **95**, 3–36.

Krishnananda, S. (1983) *The Brihadaranyaka Upanishad*. Rishikesh, Himalayas, India: The Divine Life Society.

Lilly, J. C. (1956) Mental effects of reduction of ordinary levels of physical stimuli on intact healthy persons. *Psyc. Res. Rep.*, **5**, 1–9.

Lilly, J. C. (1977) *The Deep Self: The Tank Method of Physical Isolation*. New York: Simon and Schuster.

Loschmidt, J. J. (1876) Sitzungsberichte der Akademie der Wissenschaften zu Wien. *Proc. Acad. Sci. Vienna*, **73**, 336.

Lungarzo, C. & Pereira, A. J. (2009) A cournotian approach to the emergence of relational collectives. *TECCOGS: Revista Digi. Tecnol. Cogn.*, **1**, 1–17.

Marciak-Kozłowska, J. & Kozlowski, M. (2014) Consciousness, brain vibrations and Planck mass. *NeuroQuantology,* **12**, 382–390.

Martinez, J. A. (1974) Galileo on primary and secondary qualities. *J. History Behav. Sci.*, **10**, 160–169.

Martyushev, L. M. (2010) The maximum entropy production principle: Two basic questions. *Philos. Trans. R. Soc. Lond. B. Biol. Sci.*, **365**, 1333–1334.

Maxwell, J. C. (1860/1965) Illustration of the dynamical theory of gases. In: S. Brush, ed. *Kinetic Theory,* Vol. 1. Oxford, London: Pergamon Press, pp. 148–171.

Maxwell, J. C. (1866/1965) On the dynamical theory of gases. In: S. Brush, ed. *Kinetic Theory*. Vol. 2. Oxford, London: Pergamon Press, pp. 23–87.

Maxwell, J. C. (1871) *Theory of Heat (reprinted in 2001)*. New York: Dover.

McLaughlin, B. P. (1992) The rise and fall of British emergentism. In: A. Beckermann, H. Flohr and J. Kim, eds. *Emergence or Reduction? Essays on the Prospects of Nonreductive Physicalism (Foundations of Communication)*. Berlin: De Gruyter, pp. 49–93.

Mitterauer, B. (2012) Qualitative information processing in tripartite synapses: A hypothetical model. *Cogn. Comput.*, **4**, 181–194.

Monod, J. (1970) Le Hasard et la Nécessité: Essai sur la Philosophie Naturelle de la Biologie Moderne (Chance and Necessity: An Essay on the Natural Philosophy of Modern Biology). Paris: Editions du Seuil.

Monod, J., Changeux, J. P. & Jacob, F. (1963) Allosteric proteins and cellular control systems. *J. Mol. Biol.*, **6**, 306–329.

Morgan, J., ed. (2014) *Quintessence: Three Visions* (Contributions by Kim Stanley Robinson, John Sims, Amina Eagle, Michael Schneider, and David Rubin): CreateSpace Independent Publishing Platform.

Narnhofer, H. & Wreszinski, W. F. (2014) On reduction of the wave-packet, decoherence, irreversibility and the second law of thermodynamics. *Physics Reports*, **541**, 249–278.

Nicolis, G. & Prigogine, I. (1977) *Self-Organization in Nonequilibrium Systems — From Dissipative Structures to Order through Fluctuations*. New York: Wiley.

Oza, A., Harris, D. M., Rosales, R. R. & Bush, J. W. M. (2014a) Pilot-wave dynamics in a rotating frame: On the emergence of orbital quantization. *J. Fluid Mech.*, **744**, 404–429.

Oza, A., Wind-Willassen, O., Harris, D. M., Rosales, R. R. & Bush, J. W. M. (2014a) Pilot-wave dynamics in a rotating frame. *Exotic orbits. Phys. Fluids*, **26**, 082101.

Pereira Jr., A. (2003) The quantum mind / classical brain problem. *NeuroQuantology*, **1**, 94–118.

Pereira Jr., A. (2007) Astrocyte-trapped calcium ions: The hypothesis of a quantum-like conscious protectorate. *Quant. Biosyst.*, **2**, 80–92.

Pereira Jr., A. (2012) Perceptual information integration: Hypothetical role of astrocytes. *Cogn. Comput.*, **4**, 51–62.

Pereira Jr., A. (2013) Triple-aspect monism: A conceptual framework for the science of human consciousness. In: A. Pereira Jr. and D. Lehmann, eds. *The Unity of Mind, Brain and World: Current Perspectives on a Science of Consciousness*. Cambridge, UK: Cambridge University Press, pp. 299–337.

Pereira Jr., A. (2014) Triple-aspect monism: Physiological, mental unconscious and conscious aspects of brain activity. *J. Integr. Neurosci.*, **13**, 201–227.

Pereira Jr., A., Barros, R. F. & Santos, R. P. (2013) The calcium wave model of the perception-action cycle: Evidence from semantic relevance in memory experiments. *Front. Psychol.*, **4**, 252.

Pereira Jr., A., Edwards, J. C. W., Lehmann, D., Nunn, C., Trehub, A. & Velmans, M. (2010) Understanding consciousness: A collaborative attempt to elucidate contemporary theories. *J. Conscious. Stud.* **17**, 213–219.

Pereira Jr., A. & Furlan, F. A. (2010) Astrocytes and human cognition: Modeling information integration and modulation of neuronal activity. *Prog. Neurobiol.*, **92**, 405–420.

Pereira Jr., A. & Rocha, A. F. (2000) Auto-Organização Físico-Biológica e a Origem da Consciência. In M. E. Gonzales and I. e D'Ottaviano, eds. *Auto-Organização: Estudos Interdisciplinares*. Campinas: CLE / UNICAMP.

Poppel, E., Ruhnau, E., Schill, K. & Steinbuchel, N. V. (1990) A hypothesis concerning time in the brain. In: H. Haken and J. Stadler, eds. *Synergetics of Cognition*. Heidelberg: Springer.

Pribram, K. H. (1987) The Implicate Brain. In B. J. Hiley and F. David Peat, eds., *Quantum Implications: Essays in Honour of David Bohm*. New York: Routledge & Kegan Paul, pp. 365–371.

Pribram, K. H. (1997) Holonomic brain theory and motor gestalts: Recent experimental results. Abstract of the keynote lecture, 10th Scientific Convention, Society for Gestalt Theory and its Applications. Vienna, Austria.

Pribram, K. H. (1999) Quantum holography: Is it relevant to brain function? *Inform. Sci.*, **115**, 97–102.

Pribram, K. H. (2007) Holonomic brain theory. *Scholarpedia*, **2**, 2735.

Prigogine, I. (1977) Time, structure and fluctuations. Nobel lecture.

Prigogine, I. & Stengers, I. (1984) *Order Out of Chaos. Man's Dialogue with Nature.* New York: Bantam Books.

Radhakrishnan, S. (1960) *Brahma Sutra: The Philosophy of Spiritual Life.* London: Ruskin House, George Allen & Unwin.

Rocha, A. F., Massad, E. & Pereira Jr., A. (2005) The brain: From fuzzy arithmetic to quantum computing. In: J. Kacprzyk, ed. *Studies in Fuzziness and Soft Computing,* Vol. 165. Berlin, Heidelberg, Germany: Springer-Verlag, pp. 1–225.

Rosenthal, D. (2009) Concepts and definitions of consciousness. In: P. W. Banks, ed. *Encyclopedia of Consciousness.* Amsterdam: Elsevier, pp. 157–169.

Scerri, E. R. (2006) *The Periodic Table: Its Story and Its Significance.* 1st edn. Oxford University Press.

Schreiber, Z. (1995) The nine lives of Schrödinger's cat: On the interpretation of non-relativistic quantum mechanics. Ph.D Thesis, Department of Physics, University of London, Imperial College of Science, Technology and Medicine.

Schrödinger, E. (1944) *What is Life? The Physical Aspect of the Living Cell.* Cambridge: Cambridge University Press.

Seager, W. (2010) Panpsychism, aggregation and combinatorial infusion. *Mind Matter,* **8**, 167–184.

Seager, W. (2012) Emergentist panpsychism. *J. Conscious. Stud.,* **19**, 19–39.

Sereno, M. E., Trinath, T., Augath, M. & Logothetis, N. K. (2002) Three-dimensional shape representation in monkey cortex. *Neuron,* **33**, 635–652.

Sireteanu, R., Oertel, V., Mohr, H., Linden, D. & Singer, W. (2008) Graphical illustration and functional neuroimaging of visual hallucinations during prolonged blindfolding: A comparison to visual imagery. *Perception,* **37**, 1805–1821.

Seth, A. K. (2010) Measuring autonomy and emergence via Granger causality. *Artif. Life,* **16**, 179–196.

Singer, W. (1999) Neuronal synchrony: A versatile code for the definition of relations? *Neuron,* **24**, 49–65, 111–125.

Szilard, L. (1929) Uber die Entropieverminderung in einem thermodynamischen System bei Eingriffen intelligenter Wesen [About the Decrease of Entropy in a Thermodynamic System by the Intervention of intelligent beings]. *Z. Physik,* **53**, 840–856.

Tononi, G. (2004) An information integration theory of consciousness. *BMC Neurosci.,* **5**, 42.

Tononi, G. (2008) Consciousness as integrated information: A provisional manifesto. *Biol. Bull.,* **215**, 216–242.

Tononi, G. (2012) Integrated information theory of consciousness: An updated account. *Arch. Ital. Biol.,* **150**, 293–329.

Trehub, A. (2013) What am i? redux. *J. Conscious. Stud.,* **20**, 207–225.

Vannini, A. (2009) A syntropic model of consciousness. *Syntropy,* **1**, 109–138.

Velmans, M. (2008) Reflexive monism. *J. Conscious. Stud.,* **15**, 5–50.

Vimal, R. L. P. (2008) Proto-experiences and subjective experiences: Classical and quantum concepts. *J. Integr. Neurosci.*, **7**, 49–73.

Vimal, R. L. P. (2009a) Meanings attributed to the term 'consciousness': An overview. *J. Conscious. Stud.*, **16**, 9–27.

Vimal, R. L. P. (2009b) Subjective experience aspect of consciousness Part I — Integration of classical, quantum, and subquantum concepts. *NeuroQuantology*, **7**, 390–410.

Vimal, R. L. P. (2009c) Subjective experience aspect of consciousness Part II: Integration of classical and quantum concepts for emergence hypothesis. *NeuroQuantology*, **7**, 411–434.

Vimal, R. L. P. (2010a) Consciousness, non-conscious experiences and functions, proto-experiences and proto-functions, and subjective experiences. *J. Conscious. Exploration Res.*, **1**, 383–389.

Vimal, R. L. P. (2010b) Matching and selection of a specific subjective experience: Conjugate matching and subjective experience. *J. Integr. Neurosci.*, **9**, 193–251.

Vimal, R. L. P. (2010c) On the quest of defining consciousness. *Mind Matter*, **8**, 93–121.

Vimal, R. L. P. (2010d) Towards a theory of everything part I — Introduction of consciousness in electromagnetic theory, special and general theory of relativity. *NeuroQuantology*, **8**, 206–230.

Vimal, R. L. P. (2010e) Towards a theory of everything part II — Introduction of consciousness in Schrödinger equation and standard model using quantum physics. *NeuroQuantology*, **8**, 304–313.

Vimal, R. L. P. (2010f) Towards a theory of everything part III — Introduction of consciousness in loop quantum gravity and string theory and unification of experiences with fundamental forces. *NeuroQuantology*, **8**, 571–599.

Vimal, R. L. P. (2013) Emergence in dual-aspect monism. In: A. Pereira Jr. and D. Lehmann, eds. *The Unity of Mind, Brain and World: Current Perspectives on a Science of Consciousness*. Cambridge, UK: Cambridge University Press, pp. 149–181.

Vimal, R. L. P. & Davia, C. J. (2008) How long is a piece of time? — Phenomenal time and quantum coherence — Toward a solution. *Quant. Biosyst.*, **2**, 102–151.

Vimal, R. L. P. & Davia, C. J. (2010) Phenomenal time and its biological correlates. *J. Conscious. Exploration Res.*, **1**, 560–572.

Vimal, R. L. P., Pokorny, J. M. & Smith, V. C. (1987) Appearance of steadily viewed light. *Vision Res.*, **27**, 1309–1318.

Volk, T. & Pauluis, O. (2010) It is not the entropy you produce, rather, how you produce it. *Philos. Trans. R. Soc. Lond. B Biol. Sci.*, **365**, 1317–1322.

Von Uexkull, J. (1957) A stroll through the worlds of animals and men. In: C. H. Schiller, ed. *Instinctive Behavior*. New York: International Universities Press.

Waldmann, R. (2001) *Adv. Exp. Med. Biol.*, **502**, 293–304.

Weaver, W. & Shannon, C. E. (1949) *The Mathematical Theory of Communication.* Urbana: University of Illinois Press.

Zeki, S. & Bartels, A. (1999). Toward a theory of visual consciousness. *Consciousness and Cognition*, **8**, 225–259.

Zurek, W. H. (2005) Probabilities from entanglement, Born's rule $p_k = |\psi_k|^2$ from envariance. *Phys. Rev. A*, **71**, 052105.

Zurek, W. H. (2007) Relative states and the environment: Einselection, invariance, quantum Darwinism, and the existential interpretation. arXiv 0707.2832v.

Biophysics of Consciousness: A Foundational Approach
R. R. Poznanski, J. A. Tuszynski and T. E. Feinberg
Copyright © 2017 World Scientific, Singapore.

# 6

# THE CAUSAL ROOTS OF INTEGRATION AND THE UNITY OF CONSCIOUSNESS

*Riccardo Manzotti*
*Institute "GP Fabris', IULM University, via Carlo Bo, 8, 20143 Milano, Italy*

*Antonio Chella*
*Department of Chemical, Management, Computer, Mechanical Engineering, University of Palermo, Viale delle Scienze, 90128, Palermo, Italy*

**Abstract**

A fundamental feature of consciousness is unity. The problem is whether unity is compatible both with the physical underpinnings of conscious experience and with the fabric of the physical world in general.

*Keywords*: Consciousness; unity; causality; overdetermination; ontology; integration; binding problem; wholes; computation; information integration; Alexander's dictum; neural networks.

## 6.1. Introduction

On one side, the mind is characterized by an overall unity encompassing the whole of one's experiences — an enduring whole partially overlapping with the notion of the self (Feinberg, 2009). On the other side, one's experience appears to be organized into perceptual and phenomenal unities such as those at the center of Gestalt psychology. In other words, phenomenal experience is integrated. The resulting phenomenal integration is one of the features of consciousness that contrasts with the standard description of the physical world (Feinberg, 2012; Revonsuo, 2010; Tononi, 2004). Conscious subjects perceive the world as though it were made of a hierarchy of macro-entities — are these macro-entities real? If they are not real in terms of macro-objects in the external world, do they correspond to unities made of integrated neural processes? In this regard, the physicist Max Tegmark recently commented 'One of the most striking features of our physical world is that we perceive it as an object hierarchy [...] If you are enjoying a cold drink, you perceive ice cubes in your glass as separate objects' (Tegmark, 2015). Such a perception is surprising from a physicist's perspective since the existence of macro-objects is something that does not figure out anywhere in physics. There is a sharp contrast between the world one experiences consciously and the physical world described by physics. This gap is in addition to other traditional gaps between consciousness and the physical world — such as phenomenal quality, intentionality, first-person perspective, and mental causation (Feinberg, 2012, p. 18).

The issue of integration is paramount if the system under scrutiny is considered to be the physical realizer of the mind. Significantly, from Golgi's and Cajal's time, neurosciences have had to tackle with the notion of integration. In fact, consciousness appears to be a highly integrated phenomenon (Manzotti, 2008; Bayne, 2004; Revonsuo, 1999; Cleeremans, 2003). It is still unclear whether this integration is the result of information processing (Tononi, 2004; Aleksander & Gamez, 2010), cognitive modularity (Baars & Ave, 1997; Shanahan & Connor, 2008), sensori-motor coordination (Hurley, 2003; Gangopadhyay & Kiverstein, 2008), teleological integration (Mirolli & Baldassarre, 2013; Manzotti *et al.*, 2012).

By and large, four approaches have been devised to address the issue of unities. First, one may deny it, endorsing some version of eliminativism of the mind and unities as well — if there were no consciousness, the problem of unity could be set aside. Second, one may embrace some version of dualism — the gap is taken to be real and thus it must be endorsed by a real ontological gap. Third, one may advocate some form of emergentism that would explain how the creation of unities is possible. Fourth, one may consider whether nature has the resources to generate unities.

The first option is not feasible because it explains away an indisputable fact — namely that conscious experience is organized in wholes. The second option amounts to dualism and thus it can be discarded because it lies outside of the scope of scientific enquiry. The last two options are the ones on which this chapter will focus — either some version of neurobiological emergentism or some account of unities as part of physics. We will favor the latter option. The culprit with any form of strong emergentism is that they endorse downward causation and downward causation runs afoul the argument of causal exclusion (Kim, 1999; Kim, 2006; Kim, 1998). In this regard, the philosopher Kim wrote that:

> "If there is systematic causal overdetermination in all cases of downward causation, emergents cannot fulfill their causal promise; anything they causally contribute can be, and is, contributed by a physical cause. This result, unless it is successfully rebutted, threatens to bankrupt one of the central claims of emergentism. If downward causation goes, so goes emergentism." (Kim, 2006, p. 598)

In all his works, Kim has argued against the actual possibility of downward causation and, in many respects, we find his skepticism well-grounded as we will argue in the following. Yet, irrespective of downward causation and emergentism, wholes and unities are a fact of the world we experience. Thus, in this chapter, we will consider an alternative that strives to ground the ontology of integration (and thus of wholes) at a fundamental physical level rather than being the upshot of higher levels of organization such as those found in complex dynamic biological systems. In a physical perspective, there are only material parts and their causal processes. Such an approach does not entail that consciousness is an entity causing integration but rather that the processes leading to consciousness exploits integration — if an agent achieves consciousness, its biological structures must exploit physical processes of the kind that produces unities.

In the following section, it is questioned whether integration occurs at a physical or different level — for instance, neural or biological. The possibility to outline a fundamental causal account of physical integration as the case of consciousness is taken into consideration. Then, in Sec. 6.3, it is explained why a causal and physical notion of integration is taken into account rather than on a biological, neural, or cognitive one. Because of the argument of causal overdermination and causal exclusion, there are worries about the ontology of any extra level but the causal and the physical ones. In Sec. 6.4, integration as a methodological approach and integration as natural phenomenon are contrasted. A few requirements about a theory of integrative function of consciousness are laid out. The punch line is that a criterion of physical integration ought to satisfy a series of mandatory requirement such as physical grounding, causal efficacy, circularity, and scaling. In Sec. 6.5, after setting the theoretical landscape, some attempts to address the unity of neural processes underlying consciousness are outlined and discussed. These approaches suffer from a common shortcoming — namely that they either give integration for granted or they hypothesize some emergent version of integration. In Sec. 6.6, a causal model of integration based on joint causation is put forward. Integration is seen as the result of causal interaction. The model, presented in causal and physical terms, suggests a kind of process that may be the building block of more complex hierarchical processes in biological and neural systems. Moreover, it should be able to scale-up to more complex neural, cognitive, and phenomenal systems. In Sec. 6.7, the relation between integration and time is discussed — the link is mandatory since experience takes place in time. The basic claim is that the moment present is a form of unity that arises from the same kind of joint causal processes.

## 6.2. Is Integration a Physical or a Biological Feature?

It is fair to state that, so far, there is no final evidence that the problem of integration in consciousness is exclusively a problem either of neural/biological integration or of functional/computational integration, although both levels of explanation may be revealing. More generally, the fact that a conscious being exhibits integration — either at the level of what one experiences or at the level of its cognitive and neural underpinnings — poses the question as to the physical underpinnings of unity.

Consider entropy. Biological systems increase entropy. Computational systems do too. Biological organisms grow olds and dies. Yet, entropy is not the upshot of anything specifically related either on the biological or on the computational level. It would be misleading to deal with entropy *only* because of biological processes. Entropy is the result of the fundamental structure of reality. This chapter addresses the possibility that unity — and thus integration — is something that takes place at a level comparable to that of entropy. According to this hypothesis, integration is exploited by biology and not created by it.

From a phenomenal perspective, conscious experience unfolds into unities hierarchically organized through time. Starting from the most elementary percepts up to the highest-level cognitive thoughts, everything takes place in form of discrete chunks of meaning that tend to coalesce into integrated unities. Subjects do not perceive pixels — they perceive objects. Subjects do not perceive waves of pressure — they perceive words and sounds. William James spoke of *buds* of experience to refer to the discrete blocks into which experience seems to organize itself — "acquaintance with reality grows literally by buds or drops of perception" (James, 1911, p. 34). These "buds or drops" correspond to the stuff one's experience is made of. What do they correspond in physical terms, then? Are they just a delusion, or do they correspond to physical structures? Is the physical world organized into real discrete macro/physical entities? In this regard, James stated that "The unity — into which the present mental state binds the individual past facts with each other and with itself — does not exist until the thought is there" (James, 1890, p. 182) — thereby suggesting that such wholes are the outcome of perceptual processes rather than being a fact of nature. Yet, such a view is coherent only with a dualistic ontology. In contrast, if perceptual processes are a part of the physical world, their structural properties must be physical too. If perceptual processes exhibit unity, something into the physical world must exhibit unity. Later, James changed his opinion eventually and considered the possibility that such unities had a neutral ontological foundation (James, 1904a, b).

Contemporary neuroscience and cognitive science do not restrain from emphasizing the unifying role of cognition and consciousness. The self, in this respect, if often modeled as the unity that integrates all experience, past and present, into one discrete and unified meaningful dynamic whole (Tononi, 2004, 2008; Shanahan, 2010; Baars, 1997; Baars *et al.*, 2011; Feinberg, 2009). This unification is often taken to be a functional/computational/cognitive operation rather than a natural phenomenon like, say, digestion, or growth. The quest for an integrative functional architecture of the brain mirrors that for an integrative theory of cognition (Barrett & Satpute, 2013; Poznanski, 2002). By the same token, despite the huge complexity of underlying neural structures, a unified will acting as a whole in the world is often assumed to play a role in one's behavior (Zschorlich & Köhling, 2013; Libet, 1985; Haggard *et al.*, 2002). Even if such beliefs were just a massive delusion — as it has been sometimes claimed (Metzinger, 2003) — it would remain to be explained why such beliefs should arise and to which end.

This familiar relation between one's experience and the occurrence of unities is likely to be the expression of physical structures underpinning conscious experience. Consciousness is an opportunity to peer into the circumstances that, in the physical world, lead to unities. As a working hypothesis, here consciousness is taken to be a physical phenomenon and not a massive delusion, therefore it is fair to look for fundamental physical conditions capable of endorsing such a property. By the same token, it is fair to assume that the physical underpinnings of consciousness have an integrative role to flesh out the world one experiences.

The connection between consciousness and unity is intimate. In facts, without consciousness, there would not be any need to bother with the ontology of unities — integration could just be a methodological approach. However, the world one experiences is made of objects and one is a unity of experience. Thus, there is something that needs to be explained in terms of its physical underpinnings (Koch, 2004; Revonsuo, 2010) and not just explained away. In the following, for the sake of discussion, consciousness is taken to be a physical process that is implemented by the neural and bodily structures during their everyday causal entanglement with the environment — "the stream of experience that appears when we wake up in their morning and disappears when we fall into deep sleep at night" (Gamez, 2014, p. 2).

Bottom line: Consciousness exhibits unity as a natural feature that, although may be addressed from a functional/computational perspective, is likely to be the expression of some more fundamental structure of the physical world. The gist of the chapter is whether it is possible to provide an elementary causal account of integration that is exploited by biological structures.

## 6.3. Causal Overdetermination and Levels

A crucial and controversial issue is the connection between integration and levels such as the chemical, the biological, the neural, the functional and, possibly, the computational one. It is an open issue whether such levels are of epistemic or ontological nature. In the former case, they would be nothing but levels of description. In the latter case, they would be levels of reality with novel properties and causal powers.

To cut short a long story, while such levels have a crucial epistemic value, it is unclear whether they have any ontological solidity. Such a worry stems from the causal closure of the physical world and the connected conservation principles (Dowe, 2000; Walter, 2008; Kim, 1998, 1999, 2006). The ontological status of any additional level besides the physical one (for instance the biological level) is thorny and problematic.

Although today many scholars consider the possibility that there might be levels of reality not reducible to the physical, it is difficult to unveil definitive empirical evidence. The main problem is that, if a level had any concrete ontology, its elements should have non-trivial causal powers. For instance, if one is a realist about the biological level, one ought to demonstrate that such level causes events that cannot be explained in terms of the underlying physical processes. This is questionable. It is a thesis that conflicts with various traditional problems — ranging from a violation of several principles of conservation to issues of causal overdetermination. Worryingly, it entails that biological processes — qua biological ones — do something more than the underlying physical processes. This is very puzzling, to say the least. All the more since it is widespread the acceptance of the reducibility of life to its physical underpinnings — "the reduction of 'life' to biology and chemistry poses no philosophical barriers" (Feinberg, 2011, p. 432), "Ultimately, we find nothing that meets that definition of life that is not reducible to biological parts and processes that create it." (Feinberg, 2012, p. 16). In fact, claiming otherwise entails to endorse some version of animism according to which some stuff, in virtue of being alive, is capable of doing something additional to what it would be if it were not. The notion of any additional causal power is problematic since such a causal power would conflict with that of the original constituents (Kim, 2006).

The notion that new causal properties and new ontological levels result from complex hierarchical organizations is puzzling for several reasons that are worth of being recapped:

(1) A genuinely new level of reality should have genuinely new causal properties that would conflict both with conservation principles and with the causal over determination argument.

(2) The emergence of a new level of reality must be the result of some natural law and not a pure miracle. Yet, to the extent that a new level is the result of some law, it is not genuinely emergent. In contrast, to

the extent that is not the result of some law, it is tantamount to dualist ontology.

(3) The appeal to the existence of additional levels of reality with their own specific laws and properties does not explain their constitution — it merely describes their juxtaposition.

Of course, these concerns hold only if one aims to use such levels as an explanation for additional phenomena (for instance unities, integration, mental causation and the like). However, postulating that a certain systemic organization leads to a level with integration does not constitute a satisfactory explanation since it does what Occam advises not to — it multiplies the explanatory entities. While it can be argued that a new level does not presupposes new entities but only new causal powers, this is not totally since it is assumed that a sufficient condition of identity between events is represented by the identity of causes and effects (Davidson, 1969, p. 304). Thus, introducing new levels is no different from multiplying one's ontological framework. In this regard, Kim stresses that any theory that suggests a certain level is presumptively committed to the causal efficacy of such a level: "for any phenomenon to have an explanatory role, its presence or absence in a given situation must make a difference — a causal difference" (Kim, 1998, p. 31).

For instance, consider a center of mass. Such a notion, albeit very useful in dynamics, has no real causal and ontological existence. A center of mass is not a real thing with real causal powers but an epistemic shortcut devised to refer to complex states of things. While some scholars enjoy arguing that, since such a notion must be real since is used in physics, it is conceivable to disagree on the ground of causal considerations. Take a galaxy and a starship. It seems *as though* the starship is pulled towards the center of mass of the galaxy. In truth though, the starship is pulled towards the particles of mass that compose the galaxy wherever they are. Two entities — namely the collection of particles and the center of mass — compete for the same causal role. Which one of them is really pulling the starship — the collection of particles or their center of mass? Of course, the former. Center of masses are not real, they are just epistemic shortcuts.

Likewise, the notion of a different level of reality (for instance, biology vs. physics, neural vs. physics, or mental vs. physics) ought to be causally grounded. As mentioned above, Kim stressed that the notion of levels of reality require some positive account of downward causation

(Marras & Yli-Vakkuri, 2008; Walter, 2008; Kim, 2005, 2006). In short, if the upward level (for instance, the functional level) were real, it should be able to produce effects on the underlying level (for instance the physical structures that realize the upward level). Moreover, these effects should not be reducible to those that the underlying levels would produce by themselves. Otherwise, the upward level is nothing but an epiphenomenal one because all the causal powers are drained by the lower level. A level must have its own unique and not reducible causal powers. If a level does not have any special causal power, such a level is nothing but an explanatory shortcut akin to a center of mass (Kim, 2003, 2005; Block, 2003).

The causal overdetermination argument suggests how to judge whether something is real. Suppose that you have a physical system that exhibits certain phenomena. For instance, the growth of a cell (C). You may have two explanatory principles — chemical reactions that replicate certain molecules (A) or a living force (B). There is no reason *a priori* to choose among the two. The point is that they cannot be both true since they cause the same outcome. The outcome C is causally overdetermined because it cannot be the result of both A and B. Thus, either A or B is superfluous. Since chemical replication is a more general phenomenon, the presence of a living force is usually set aside. Based on such a criterion, in order to be real, a level should show some form of not overdetermined causal power, which is another way to address downward causation. Therefore, before attributing any ontological status to a level, one should be able to have an explanation as to why biological systems should be required to add anything to their underpinning physical processes (Chauvet, 2004, p. 28). One way to put the issue is the following. Suppose that it were possible to provide a detailed description of what takes place in an organism at an atomic level. Would there still be any need to introduce any biological fact or law? Would not the physical description suffice to describe everything that happens? Alas, so far, there has been no definitive evidence supporting downward causation — "how emergent properties can have distinctive causal powers of their own — in particular, powers to influence events and processes at the basal level" (Kim, 2006, p. 559). In the lack of such evidence, it is preferable to stick to fundamental causal and physical processes rather than appeal to higher level of description and explanation. Thus, the claim that biology has special laws (Chauvet, 2004, p. 36) is controversial since it hinges on whether there is any actual proof of downward causation. So far, it is not obvious whether this is the case. The supporters of emergence have not been entirely convincing (Bishop, 2006; Kelso, 1995; Megill, 2012; Thompson, 2004; Maturana & Varela, 1980).

This is why here the focus is on a causal and physical notion of integration rather than on a biological or neural one. Since the issue of the ontological status of biological and neural levels is still open, it is safer to see whether it is possible to provide an account of integration that does not depend on yet-to-be-settled ontological commitments about biological processes. Of course, the adopted stance does not entail that the study of these levels has no explanatory power. However, it suggests that the mental, cognitive, computational, functional, biological, chemical and physical levels are primarily useful epistemic constructs. They are levels of description rather than levels of reality. So far, the ontological commitments and empirical ground of several related notions are still to be ascertained — such as stigmergy, swarm intelligence, distributed cognition, downward cognition and functional organization of the living (Dorigo *et al.*, 2000; Dorigo & Blum, 2005; Theraulaz & Bonabeau, 1999; Passino *et al.*, 2014; Santana & Correia, 2010; Chauvet, 2004; Maturana & Varela, 1980). Is it possible to keep an ontologically neutral approach and to consider whether, at an elementary level, a set of causal conditions that may underpin integration and unity can be singled out?

## 6.4. The Integrative Function of Consciousness

As anticipated at the onset, if it were not because of conscious experience, the world could be described just as a whirlwind of quarks and electrons. This is in contrast with everyday experience of the world made of macrophysical objects. How can such a contradiction be explained? The working hypothesis is thus that one perceives the world as made of unities because, in some ontological — and thus physical — sense, wholes are real. If wholes are real there must be some physical processes that, in virtue of some nomological natural principle, brings wholes into existence. Whether such wholes are internal or external to one's body is a matter of discussion. In both cases, though, wholes are taken to be real and thus to be physical. Thus, here a realist stance as to the existence of objects and wholes is adopted.

Before proceeding further, a terminological caveat is useful — in this chapter, the expressions "to bring a whole into existence" and "to instantiate integrative function/processes" are taken to be largely synonymous.

The first expression will often be preferred because it is less loaded with implicit assumptions as to the whereabouts of the integrative process. Furthermore, it emphasizes the crux of the matter — namely that the outcome of integrative activities has to be something additional with respect to unintegrated sets of phenomena.

If wholes are taken to be real physical instances, the key empirical issues to be solved are the location where integration takes place, in what conditions it occurs, and what is the resulting integrated whole.

What is integration? By and large, integration is a physical process that brings together parts of nature what would have otherwise been isolated. Integration is then a physical process characterized by specific causal conditions. In which conditions, are separate physical phenomena integrated in a strong physical sense? A legitimate concern is that the notion integration is invoked at different levels from that in which it would have its real explanatory power.

Moreover, it ought to be possible to put forward the minimum criteria for integration at a neutral physical level in which no reference to complex neural systems were necessary. Irrespective of whether any two events A and B are taken to be either as biological or as neural, it should be possible to distinguish when A and B are physically integrated. A minimal notion of integration should be available in simple cases. Concisely, the issue of integration is twofold. On one hand, it addresses the issue of integrating together different perspectives, tools, and evidences on certain phenomena that, to a certain extent, manifest integration — for instance, behavior, cognition, and perception. On the other hand, it addresses the issue as to whether there are phenomena that are not genuinely divisible into their components. To recap, there are three approaches to integration:

- *Methodological*: Merging different approaches that have been held historically separate.
- *Epistemological*: Phenomena whose understanding cannot be achieved unless they are considerate in an integrated perspective.
- *Physical*: Phenomena consisting in something physically more than the parts that contribute to their occurrence. The mind appears to be one of those, but is it physically explainable?

Methodological approach is the less controversial case. In fact, here the methodological approaches aspects of integration are not addressed. In this regard, an integrative approach is the development of an interdisciplinary stance — it suggests that a set of phenomena, which have been historically considered in isolation, are to be considered conjointly. For instance, such methodological approach to integration is given by Chauvet (1996) (Williams *et al.*, 2008; Williams, 2006; Poznanski, 2002; Fass, 2006; Amari *et al.*, 2002). In this chapter, the importance of these three integrative approaches to neuroscience is by no means reduced — an integrative approach is key to provide the necessary breakthroughs to the study of neural systems. However, it should be noticed that, in many cases, the notion of integration does not refer to a specific natural phenomenon but rather to an integrative methodology (i.e., methodological approach). Hence, one thing is to discuss integrative neuroscience as a methodology. Another thing is to look for a notion of integration capable of causing a physical difference in the physical phenomena under scrutiny. In the former case, "integrative" refers to the method and not to the object of study. Obviously, the application of an integrative method of study to a certain phenomenon (for instance perception) allows singling out the relation between different levels of analysis or different empirical outcomes that would otherwise remain undetected and understudied. However, a different case is to consider whether the integration between levels is a fact of nature or, even more controversially, whether it has a causal role in shaping the phenomenon under scrutiny.

Epistemic integration is a notion that is parasitical on physical integration. If there were no physical integration, it would be questionable whether there might be genuine epistemic integration, which would then be nothing but a kind of conceptual reshuffling (Fodor, 1974; Kim, 1992). Some authors provided a coherent account of epistemic irreducibility of the mind. Daniel Dennett claimed that the intentional stance allows us to get a grip on certain intentional features of behavior that, although not identical to any corresponding physical underpinnings, are nonetheless epistemically autonomous (Dennett, 1987). Dennett argued that, although mental states are not physical things, they explain people's behavior and thus have an explanatory role. Yet, it is difficult to resist thinking that this notion of intentionality depends for its existence to the mind.

Finally, consider physical integration. Is integration a physical fact? Is integration able to make a physical difference? In the following, physical integration will be addressed in detail since it may underpin both cognitive and phenomenal integration. To recap, considering that the mind and its constituents — such as intentions, goals, percepts, feelings — appear to be integrated wholes, are there any physical phenomena showing some kind of physical integration able to be causally efficacious?

If the answer is negative, it follows that the mind has no real unity and that all cognitive and phenomenal phenomena, which seem so convincingly unified, are just scattered collections of separate atomic physical events. Nevertheless, before drawing such a drastic conclusion, a convincing explanation as to why consciousness and percepts manifest such a degree of phenomenal integration ought to be provided.

While it is convenient to use the term "integration" to address phenomena that are integrated, such a notion should be questioned in terms of physical elementary phenomena. Is integration a property of nature or a representational structure? Is a physical principle or an epistemic stance? In the available literature in neuroscience and other disciplines, there is some confusion in this regard. For instance, does a bee colony exist as an integrated system or is it just convenient to study it as such? This is a typical example that depends on the scale at which the observer is going to consider the beehive. One may choose to consider the behavior of the swarm as a unified entity. In such a case, the behavior of the individual bees is integrated. On the other hand, one may focus on what happens at the level of the single bee. At that level, each interaction between bees may be satisfactorily explained as the result of the state of every single bee. When two bees meet, the outcome of their interaction is decided by local factors such as the bees' nervous systems, the presence and density of local neurotransmitters on their bodies and in the space occupied by the insects, and so forth. Although these local factors are the result of past interactions between bees that belong to the colony, at the precise time of their encounter, the existence of the colony *as an integrated whole* may be set aside. The fact is that the heralded emergence of some kind of distributed intelligence whereas "group behavior is more than the sum of its parts" (Klarreich, 2006) is highly controversial. Empirical evidences shows that the group intelligence is explained thanks to local interactions among individual organisms (Couzin, 2009, 2007; Goldstone & Gureckis, 2009).

Similar considerations can be made for neural systems whereas there are multiple levels of interactions and a hierarchy of functions as well as of systems. In this regard, biological systems may require to be analyzed at multiple levels (Chauvet, 2006, p. 31). However, the necessity to structure our explanations of a biological organism in a hierarchy of levels of explanation does not necessarily imply that such levels are anything but epistemic tools. Moreover, the interplay between local interactions and global processes is key is the mechanism that brings into existence an integrated system out of a multiplicity of interacting processes and functions. Consider a neural system. According to a hypothetical molecular neuroscientist, the notion of integration does not make a difference in the behavior synapses, neural firings, cell membranes, and the like. Neurons behave locally based on the local properties of cellular components and their proximal environment (some authors suggested a different interpretation, for instance Chauvet (2006)). At the best of current knowledge, if stimulated by the same chemicals the same neuron would not behave differently inside a large network or in an isolated condition. If there are any differences in its behavior, they occur because so far scientists have been unable to recreate exactly the same chemical conditions that would obtain inside a brain. Nevertheless, is fair to maintain that neurons have their output based on non-local states of other neurons. Of course, there are also long-range connections. To allow causal influence between any two neurons in one's brain, a chain of local interactions must hold. On one hand, the neural system appears different because of its humongous complexity that conjoins neuroscientists to adopt specific strategies. On the other hand, it is assumed that neural system have a role in making possible the integration of our mind. Key requirements for integration.

If the mind and its components have some kind of genuine integration, integration must be constitutive of physical reality. Integration — if physical — must satisfy a series of criteria:

(a) It must be grounded on some physical structural principle,
(b) It must be able to cause effects,
(c) It must be definable in no circular terms,
(d) It must be scalable.

First, integration must correspond to some physical underpinnings. There has to be something that ground integration — causal relations, energy transference, or some conservations principle. In other words, integration cannot be just in the model that one uses to describe a phenomenon. It must be realized into some physical aspect of the phenomenon at hand. Integration must be based on some observable physical difference.

Second, it must make a causal difference. For instance, if the fact that a system is considered integrated had no discernible difference, then it would follow that the notion of integration at hand would be epiphenomenal. Hence, integration would not correspond to any physical property and could be set aside.

Third, one ought to be able to provide an account of integration that does not hold any circularity to the notion of integration itself. That is, one ought to be able to explain what integration is without referring to the notion of integration.

Fourth, the notion of integration should be scalable from simple cases up to the most complex ones. Consider a brain. It looks like a complex highly integrated system. While studying integration at the level of consciousness is surely important, it should be possible to provide an account for integration at a much more elementary level and then to scale it up to more complex cases. Therefore, the notion of integration should not be invoked only at the level of highly complex neural networks.

Consider a simple physical system such as the movement of two particles affected by mutual gravitational attraction. Are they integrated? Somehow they are. On the other hand, it looks as if their movements are just the sum of elementary interactions. In which conditions, what is going on is more than the sum of the interactions between the two particles? In physics, the standard case is offered by gravitational interaction between *n* bodies. And, although the available mathematical solutions do not allow any algebraic shortcut (Diacu, 1996) and the mathematical solution are simply too demanding to avoid approximations (Wang, 1991; Sundman, 1912), the movements of *n*-bodies brings into existence nothing but gravitational attraction. Thus, the *n*-body problems is a case of irreducibility insofar the complexity of the phenomena at hand is such that it is impossible to provide a more concise description of it than by using a complete simulation. However, the fact that any prediction of the future of the system requires passing through a complete simulation does not entail that the interaction between *n*-bodies brings into existence any special additional property. The future of the system is still the result of the sum of the interactions between all particles, period. Is the mind a different kind of phenomena? Namely, does the brain bring into existence something that is somewhat intrinsically more than the interactions between its parts?

To recap, a working notion of integration should be able to address the simplest physical cases and then to scale up to more complex circumstances like those offered by neural systems. Moreover, it should be grounded onto some discernible physical difference producing actual effects.

### 6.5. Kinds of Integration

As aforementioned, while conscious experience is characterized by an intrinsic sense of unity (Bayne, 2004; Bayne & Chalmers, 2003; Bayne, 2008; Baars, 1997; Koffka, 1935; Köhler, 1947), the physical world seems immune to it. Thus, a series of approaches have been envisaged to fit unity — or integration — into a physical framework.

In mereology and logic, the problem of wholes is known as the special composition problem. Inwagen formulated the problem as follows "in what circumstances do things add up or compose something? When does unity arise out of plurality?" (van Inwagen, 1990b, p. 31). A more quantitative version of the problem sounds as follow "Given *n* elements (or initial entities of any aforementioned kind) how many wholes are there? If any?"

There are three possible categories of answers. First, the principle of *Unrestricted Composition* holds that for any group of elements, there is a whole that they constitute (Lewis, 1986; Bigelow & Pargetter, 2006) — an option which had also been labeled as Universalism (van Inwagen, 1990a; Lewis, 1991). Succinctly, it states that "whenever two things exist, then there is also a third thing that contains those two as parts" (Bigelow & Pargetter, 2006, p. 486). Such view admits the largest possible number of wholes (for $n$ parts, it accepts $2^n - 1$ wholes). Such a view is counterintuitive since it admits gerrymandered objects like Daniel Nolan's left ear plus the Alpha Centauri system. The second category — *No Composition* at all — maintains that there are no wholes. Inwagen, which calls this alternative Nihilism. There are only simple entities, "the physical world consists entirely of quarks, leptons, and bosons — there is just nothing else in it, for these particles have no parts and they never add up to anything bigger." (van Inwagen, 1990b, p. 72). Finally, the third category — sometimes referred to as *Restricted Composition* — limits the total number of wholes. Roughly speaking, sometimes a group of simples gives somewhat rise to a whole and sometimes it does not. However, no accepted criteria have been proposed so far. Here, a version of restricted composition is put forward.

What distinguishes an integrated system from a juxtaposition of scattered events? Consciousness is the case in which integration seems to be a mandatory property the debate Golgi vs. Cajal debate. The quest for integrative neural underpinnings is motivated by the unity that is empirically experienced in one's consciousness. Is it possible to pinpoint some version of integration that is: (i) causally effective, (ii) physically grounded in concrete processes, (iii) not circular and genuinely explanatory and (iv) scalable?

A promising account of physical integration is offered by Tononi (2004, 2008). He suggests that certain properties of how information is processed might lead to the emergence of integration. Tononi's approach provides a mathematical formulation of the degree of integrated information inside a system. By dwelling on the notion of mutual information, Tononi introduces a certain property, dubbed Integrated Information, which ought to allow a quantitative measure of the extent of integration. The main idea is to consider all possible bipartitions A and B in which a system may be divided. For each bipartition, measure the degree of causal coupling by means of the maximum Efficient Information[1] is measured. For any system S, it is defined a quantity defined information integration **Φ**.[2] Roughly, Tononi suggests considering the worst possible case in terms of connections between units and then, for that case, to inject the maximum available information (condition of maximum entropy) to see how much the information in the other half is causally coupled. **Φ** measures the degree of causal coupling internal to a given system.

The model has inspired attempts to estimate the degree of awareness either in patients with severe conditions (Rosanova *et al.*, 2012) or in normal subjects during sleep and anesthesia (Alkire *et al.*, 2008; Tononi & Massimini, 2008). Nonetheless, there are still various worries concerning the relation between information integration and the occurrence of a unified integrated mind. A few may be listed here. First, the relation between

---

[1]That is $EI(A \leftrightharpoons B) = EI(A \rightarrow B) + EI(B \rightarrow A)$. More precisely, Tononi considers the minimum partition for efficient information normalized by the maximum entropy $EI\,(A \leftrightharpoons B)/H^{max}(A \leftrightharpoons B)$. In turn, the notion of maximum efficient information is derived from that of mutual information given condition of maximum entropy $EI(A \rightarrow B) = MI(A^{hmax}; B)$, where, finally, $MI(A; B) = H(A) + H(B) - H(AB)$.

[2]**Φ** is defined as $EI(^{MIB}A \leftrightharpoons B)$ whereas MIB is the bipartition {A, B} of S "for which $EI(A \leftrightharpoons B)$ reaches a minimum (the informational 'weakest link')" (Tononi, 2004).

potential and actual information integration. Second, it does not seem easier to explain the emergence of qualitative states from information integration than from, say, neural firings. The theory postulates that information integration is phenomenal consciousness, rather than explaining why information integration should have a phenomenal character. Third, information integration, as it is defined, has no causal role. In fact, in order to compute $\Phi$, one has to simulate all the possible states and causal interactions between unities at a local level. One may ask whether information integration makes any difference at the local level. In addition, the impression is that it does not make any difference.

Moreover, Tononi's approach suggests a questionable identity between a suggested global parameter and a real physical property of the system. To use a well-known example, consider again the notion of center of mass. It surely is a useful global parameter. It surely has physical meaning and predictive power. Unfortunately, it does not correspond to any actual physical entity or property. It is a pure epistemic, albeit physically meaningful, notion.

Nevertheless, it is worthwhile to stress that integrated information attempts to single out, at a neural level, a property whose existence is suggested by phenomenal experience. Tononi writes that his view "is motivated by two key phenomenological properties of consciousness: differentiation — the availability of a very large number of conscious experiences; and integration — the unity of each such experience." (Tononi, 2004). Tononi seems to suggest that the causal interactions between units in a neural system may result into an emergent property whose properties are no less than those usually attributed to consciousness. Tononi focuses on the unified aspect of the mind rather than onto other features such as a qualitative aspects or intentionality. As a result, he does not provide any justification about why information integration should realize quality.

In AI and cognitive science, many models have tried to tackle the issue of integration. Although there have been many differences, the prevailing strategy considers some kind of centralized control in which various parallel modules compete. Eventually, the upshot of this activity is broadcasted and imposed to peripheral modules. It should be obvious that, in this description, labeling modules either as *central* or *peripheral* is not explanatory innocent. By and large, there have been three related approaches: (1) higher order control, (2) centralized dashboard, and (3) fair competition. In one way or another, all these models posit a privileged module and put a stress on it. However, it is not enough to stipulate that a module integrate information.

As a paradigmatic example of these approaches, consider Shanahan's extension of Baars' global workspace (Baars, 1997; Baars & Franklin, 2003; Shanahan, 2005, 2010, 2012). Once again, it is worth stressing that the interest for integration is motivated by the relation with either cognition or consciousness. In fact, both Baars and Shanahan are explicitly trying to devise a cognitive model that may cope with consciousness.

In essence, Shanahan's premise is that consciousness is the outcome of some kind of cognitive integration whose main goal is to provide support for cognition (Shanahan, 2010, p. 69). Consciousness is identified with a style of information integration that allows behavior flexibility in order to handle unexpected circumstances (Shanahan, 2010, p. 85). Flexibility and autonomy are possible thanks to integration (Shanahan, 2010, p. 117). Apart from the fact that he does not explain why integration and phenomenal experience should be correlated, he does not address the issue of integration in any strong physical sense.

If consciousness were nothing but a way to integrate information and to generate flexible behavior, a model of cognitive integration would be an acceptable solution. However, preliminarily, one ought to show why information integration and flexible behavior are a satisfactory account for consciousness.

What seems to be a weakness of most of these approaches is that they address the issue of integration only at a cognitive level. However, are cognitive processes real entities capable of endorsing physical properties? Is cognition a real physical phenomenon? According to many authors, a cognitive process is just a handy way to describe what an agent is doing. Surprisingly, Shanahan himself adopts a nominalist attitude. Like many cognitive scientists, Shanahan seems to divide reality in two separate domains: The former is made of substantive entities (mass, charge, tables and trees) which are taken to be real, the latter is made of fictitious entities that are taken to have a somewhat instrumental existence. The problem arises when one deals with cognitive processes. Are they just conceptual tools or real physical entities? For instance, is computation a short-hand to complex causal chains or is it something more than that (Manzotti, 2012)?

The issue at stake is not merely conceptual. If the cognitive level is only a conceptual level, it is not wise to look for integration at the level of cognitive "boxes." Of course, the supporters of a cognitive approach to integration may retort that recent approaches have stressed the importance of embodiment and situatedness. Hence, a cognitive process would no longer be a symbolic abstract entity, but a concrete physical process. Crossing embodiment and cognitive theory, Shanahan claims that "the inner life of a human being arises from the combination of a global neuronal workspace with such an internal sensorimotor loop" (Shanahan, 2010, p. 5). And that "cognition is to be understood in terms of its role in embodied interaction with the environment, and its proper locale is the sensorimotor loop by means of which the embodied agent is coupled to that environment' (Shanahan, 2010, p. 100). However, even so two problems arise. First, it is not clear whether the notion of cognition should provide any advantage. If what it matters is the loop with the environment, one may simply set aside the notion of cognition. Second, it still seems that the role played by internal integration is paramount. This is the legacy of Baars' global workspace theory. Both Baars and Shanahan insist that consciousness is produced by choosing one among many competing processes and giving it center stage (Shanahan, 2010, p. 91). Hence, the environment has no real role as far as integration is concerned. The notion of global workspace focuses on internal processing flows rather than on embodiment. In fact, according to Shanahan, "global communications infrastructure in the brain that underlies the inner life of a human being. This communications infrastructure, [...] receives information from, and disseminates information to, numerous parallel processes operating on multiple levels, and thereby integrates their otherwise segregated activity" (Shanahan, 2010, p. 69).

In the neuroscientific literature, a well-known problem related with integration is the binding problem (Wolfe, 2012; Treisman, 1998; Feldman, 2013; Di Lollo, 2012; Singer, 2001; Roskies, 1999) that addresses how different perceptual features and different modalities end up as if they constitute some kind of perceptual unity. Once again, it is relevant that if one would confine his research to purely neural — and thus physical — phenomena, there would be no need to look for any binding. The binding problem arises as an attempt to find an explanation of a experienced unity. If one had access purely to neural data, there would be no reason to suppose any binding. The fact that information is split in different areas and then processed in time and modules variously connected is not an issue unless one is faced with the unity of conscious experience.

The binding problem stems out of the difference between neural activity, which is scattered in time and space through the cortex, and phenomenal experience, which has perceptual unity. Color, movement, form, edges, features are processed in different parts of the brain (Bayne & Chalmers, 2003; Cleeremans, 2003; Zeki *et al.*, 1999; Zeki, 2003; Bartels & Zeki, 2005). Why perception is constituted by unified percepts such as color, movement, and form? Neuroscience looks for a binding mechanism that — by means of synchronization, broadcasting, or information integration — could get together all these otherwise separate processes (Engel & Singer, 2001; Roskies, 1999; Treisman, 1998; Cleeremans, 2003). Up to now, though, no convincing answer has been provided.

## 6.6. Physical Integration: A Causal Account

The issue of integration cuts across disciplines as diverse as ontology and neuroscience, cognitive science and information processing, psychology and phenomenology. It does so because it originates from the contrast between experience and the atomistic description of physical reality — key is the tension between consciousness (or experience) and the reductionist view of the physical world. According to the former, there are natural wholes of some kind. Yet, as we have seen, it is still unclear what such wholes are (Simons, 1987; van Inwagen, 1987; Hudson, 2003). Here, *integration is the process by means of which a whole is brought into existence* and the objective of an account of integration is to define the causal requirements of such a process.

Here, a causal account is put forward. Concisely, the key hypothesis is that *two events are integrated whenever they cause a joint effect*. This hypothesis entails that to be integrated is tantamount to bring into existence something more than the original events. To achieve greater and greater level of integration, thus, it is necessary to create the conditions for more and more complex sets of events to act as joint causes. At its roots, this hypothesis may look hopelessly over simplistic since it may look to amount to nothing more than a process implementation of hierarchies of AND gates. Yet, its strength may lay in such apparent simplicity.

Consider two events, $x_1$ and $x_2$. They take place respectively at times $t_1$ and $t_2$. Suppose that they happen independently of each other and that their individual occurrence does not exert any mutual influence. Now, suppose that they have the possibility to interact with a certain causal structure[3] C. Because of C, $x_1$ and $x_2$ trigger the occurrence of a further event $e$. Thus, in themselves, the two events are isolate and autonomous. However, because of C, they trigger the occurrence of a further event $e$. Finally, it is fair to assume that the physical structure that realizes C is akin to an AND gate.

As a clarificatory example, consider the "distributed key" of an Atomic rocket launch system in which the launch is triggered by two different "keys" that have to be turned at the same time. In fact, they are not two keys, but one key in two pieces. The two pieces are not keys since, if alone, they are not able to unlock anything. The hypothesis is that, because the $x_i$ have a joint effect $e$, the two originally independent events are integrated into a whole $w$. $w$ does not exist until $e$ occurs. In this model, integration is the upshot of a feedforward causal relation that produces a joint effect. The point here is to suggest an elementary casual structure to underpin integration.

In causal terms, it makes sense to distinguish "the cause and that without which the cause would not be a cause" (Yablo, 2004, p. 119). The former may be taken to be an event actually occurring while the latter may be just a state of affairs. The latter may be formalized in terms of conditions C such that $W \wedge C \;\Box\!\!\rightarrow e$ (which may unfolded in three conditionals $W \wedge \neg C \rightarrow \neg e$, $\neg W \wedge C \rightarrow \neg e$ and $\neg W \wedge \neg C \rightarrow \neg e$).

Getting back to the example of the Rocket launcher, before being put into the "lock," both "key pieces" are as unrelated as $\{x_i\}$. They are physically not integrated. Their relation is only in the eye of the beholder, so to speak. However, after being inserted into the security lock, they action integrate to form a whole — physically and causally, they are a whole.

A final consideration regards the temporal order. It may seem strange that $e$ takes place at a time $t_e$, while the $\{x_i\}$ occur at times $t_i$, with $\forall i, t_i < t_e$. Does such an order entail that the occurrence of $e$ changes the past? In some sense, yes. This should not be surprising and indeed, it is a sign of integration. It is to be expected that integration is capable of putting together events that occur in different moments in time.

It should be acceptable to admit that, at minimum, the occurrence of $e$ changes the causal role of the past. However, here a stronger claim is here considered — namely that, because of $e$, the past got physically

[3]Something that philosophers may call either a causal circumstance or a combination of causal *enabler* and causal *enobler* (Yablo, 2004).

integrated into something more than a simple sum of scattered events. Consider that no causation moving backward in time is required.

In physics, due to the limits of causal propagation, causal processes (and thus physical phenomena) are extended in time. This is the difference between real processes and pseudo-processes outlined by Reichenbach (1971). The point is that integration must occur across a discrete time interval since any causal process is temporally extended. Thus, any physical process must complete in time. Therefore, when something begins to unfold, its nature is not wholly defined until it reaches some natural causal ending. Nothing goes backward in time — the past is, of course, past. However, what the past is may well be defined by what happens in the present. If one considers a physical phenomenon as something extended in time, then it can be the case that what happens along such a temporal line redefines the causal role of the phenomenon. A less puzzling way to describe this temporal series is to refer to the physical process waiting to get to an end as an extended present. Is it a chance that one of the most difficult aspects of consciousness has always been related with the ontology of the moment present?

However, is joint causation as a physical criterion for integration? In this regard, different scholars have raised objections based on the so-called overdetermination argument (sometimes dubbed the exclusion argument). According to this argument, given a physical phenomenon all its causal powers are drained by its smallest part. Thus any wholes would be overdetermined by its parts. This means that, if the existence of a whole is defined in causal terms, wholes do not exist since they have no real causal powers. Wholes are overdetermined by their parts (Hudson, 2003; Merricks, 2001; Walter, 2008; Kim, 1993, 2005). Yet, do micro-constituents drain all causal powers? Here, the hypothesis that joint causation may be an exception to the overdetermination argument is put forward and one of the reasons to consider this possibility is offered by introspective analysis of phenomenal experience. It is a fact that conscious experience seems to show a different picture of the world — no matter what arguments may lead us to claim the contrary. Experience of the world suggests that objects and other macro-physical stuff have indeed a causal role. Conscious agents seem to have causal influence. This chapter puts forward this hypothesis — if joint causation is the condition that brings a whole into existence, macro-physical objects have causal influence. Joint causation, which is crucial for the existence of wholes (and in many phenomenal and cognitive cases), might be a special yet crucial case. Intuitively, joint causation is causally tantamount to logical conjunction — it glues together separate elements.

Is joint causation the building block on which more complex integration dwells? Consider Fig. 1, the AND gate outcome takes place if and only if both inputs take place. Let $x_1$ and $x_2$ be the input events and $e$ be the output event. $e$ takes place if and only if both $x_1$ and $x_2$ take place. Trivially, the occurrence of either $x_1$ or $x_2$ is not enough to trigger the outcome $e$. Only the joint (and contemporary) occurrence of both $x_1$ and $x_2$ causes $w$ to take place. Given the physical structure of the AND gate, (C) the joint occurrence of both $x_1$ and $x_2$ is the only case in which $x_1$ and $x_2$ produce an effect *as a whole*. No circularity is involved. To recap, $(x_1 \wedge x_2)$ is the cause of $e$. If the AND gate is properly built, there is no way to separate causally $x_1$ from $x_2$, and vice versa. Their causal role cannot be disentangled. As to the occurrence of $e$, both events are necessary and neither one is sufficient.

A reply to the overdetermination argument is the following. As separate atomic events $x_1$ and $x_2$ do not overdeterminate the gate output. Only their union $(x_1 \wedge x_2)$ causes $e$. If this were the case, joint causation might constitute a whole. Taken by themselves, $x_1$ and $x_2$ are unable to be the

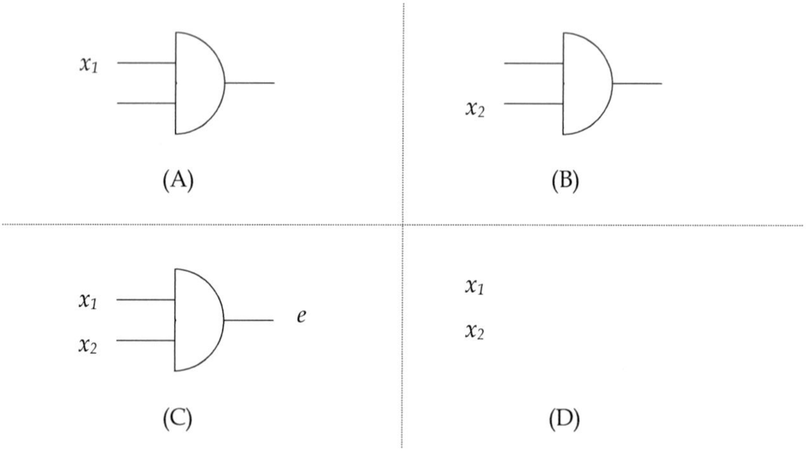


**Fig. 1.** An AND gate and the integration of events by means of a joint effect $e$. Only in (C) the two events constitute a causal whole. In (A) and (B) their individual occurrence does not trigger any effect. In (D) the lack of a proper causal structure prevents their joint occurrence from producing any effect.

cause of a joint effect. $(x_1 \wedge x_2)$ would not exist without the occurrence of $e$. Thus, $x_1$ and $x_2$ get integrated into $w = (x_1 \wedge x_2)$.

The core of the hypothesis is that something as simple as joint causation may be the building block of dynamic integration. Why is the whole $w = (x_1 \wedge x_2)$ different from the occurrence of $x_1$ and $x_2$? First, $x_1$ and $x_2$ can occur without $e$ taking place. Thus, in many circumstances, $x_1$ and $x_2$ take place and $w$ (that is $x_1 \wedge x_2$) does not (since $w$ does not exist causally if $e$ does not occur). Thus, $x_1$ and $x_2$ may occur without $e$, and thus without $x_1 \wedge x_2$, occurring. For instance, if the gate were damaged, the occurrence of the two inputs would ne be enough to trigger the output. The joint effect e integrates its causes into a whole.

Hence, the core hypothesis is that

> *a whole $w$ occurs whenever there is joint causation, namely when there is a set of events $\{x_1, \ldots, x_N\}$ such that they cause jointly an effect $e$ by means of a state of affairs C. The occurrence $e$ depends counterfactually on the occurrence of $(x_1 \wedge \ldots \wedge x_N)$. If these conditions obtain, a whole $w = (x_1 \wedge \ldots \wedge x_N)$ occurs.*

Here, the integration of $\{x_1,\ldots, x_N\}$ is assumed to be tantamount to the fact that $\{x_1,\ldots, x_N\}$ have the causal role of a whole. According to this hypothesis, a whole results from a causal process between a set of events and a context that allows them to produce a joint effect.

This account of integration requires adopting a temporally oriented view. Integration and whole-constitution are two sides of the same dynamic causal process. Therefore, integration is a temporal phenomenon — something entirely meaningless in a static domain.

### 6.7. Integration and Time

The causal account places the ontology of wholes, and objects by and large, in a temporal dimension. Everything exists in time as a result of causal processes spread in time and space. The fact that the ontology of the world is intrinsically temporal is consistent with the identity between world and experience. In fact, one's experience of the world is always embedded in time and it is always causally entangled with some preceding cause.

In Fig. 2(a), three blobs are represented in $t = 0$. A blob represents any simple one may like — be it an event, a particle, a composite object. In the figure, they have been depicted as the three pacman-like figures

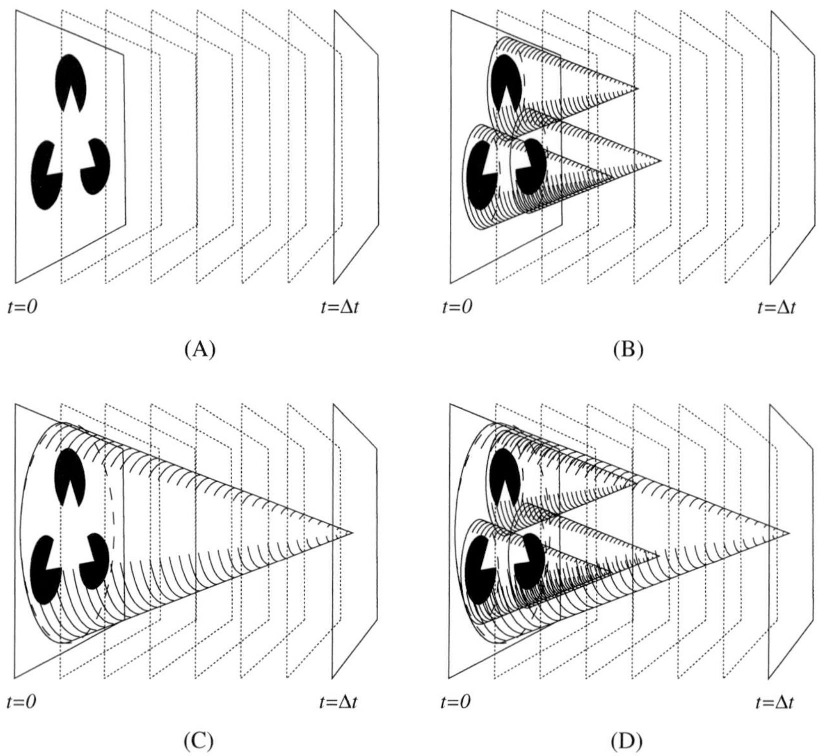


**Fig. 2.** Wholes requires causation and time.

composing the Gestaltic Kanisza's triangle so to emphasize their propensity to be perceived as a whole. At $t = 0$, they have not yet produced any effect. At that instant, there is no whole. They are isolated and separate processes. In Fig. 2(b), the three blobs have produced an effect but there is no joint effect yet. No whole has taken place yet. The case in Fig. 2(b) is akin to the baseball case. In figure c, finally, a joint effect occurs. As a result, the whole is brought into existence thanks to some additional structure — akin to an AND gate — not shown in the picture. The whole, though, is temporally located at $t = 0$ although it is brought into existence by the joint effect occurring at $t = \Delta t$. In Fig. 2(d), all causal processes are shown. The causal processes are separate and unique for each existing entity — be it a part or a whole.

This approach defines an unambiguous version of Restricted Composition steering between No Composition and Unrestricted Composition.

There is only a limited number of wholes, each corresponding to those situations whereas an actual effect takes place. Although such a number is likely much greater than that resulting from applying common-sense, it is smaller than that admitted by Unrestricted Composition. Surely certain combinations do not enjoin to any causal effect. Of course, if it turns out that, because of nomological and contingent factors, there is always a joint effect for whatsoever combination of events, the presented causal account still hold. In fact, it provides a causal account for Unrestricted Composition. The advantage is that actual wholes have a causal role — such as those considered here — in shaping both nature and experience.

The suggested causal account is applicable to all cases, from perception — faces, words, tunes, chairs, stars, planets, books, living beings, persons and patterns — to neural and cognitive processes — symbols, thoughts, neural activities, logical operations, and decisional processes. Once the physical circumstances that bring into existence a whole are fixed, it is fair to suppose that they underpin both the macro wholes like objects and the functional/processual wholes such as neural activities. Parts do not have to take place in the same instant — they are co-present thanks to the occurrence of causal processes.

A causal account of wholes — and thus of objects and experiences — has several advantages. A conspicuous one is that it allows existence into a temporal perceptive merging seamlessly experience and reality. Objects are no longer static entities contrasted with the flow of experience in time. Both objects and experience are part of the causal flow.[4] Speaking of which, the flow of experience is not a sequence of isolated events and frames, rather it is a continuous causal unfolding of unities like the aforementioned William James' "drops or buds." Hence, if a whole is something that exists because a causal process occurred between a set of events so as to produce a joint effect, a whole is something that has an intrinsic temporal character because all causal processes have a temporal nature due, probably, to the spatio-temporal fabric of the universe.

The problem of wholes shifts from a timeless analytical perspective to a temporally oriented view. A whole corresponds to the joint cause of a causal process in time and space. Hence, wholes are not entities definable in any *a priori* way. They are contingent causal moments. Consequently, if there were no time, there would be no wholes — no time, no wholes. Let me sketch a few interesting consequences.

[4]Existence and becoming are much closer in this account.

First, imagine that it were possible to freeze time. According to a widely spread opinion, if time were halted, everything would remain frozen like in a snapshot. Raindrops and snowflakes would remain still in the air. Cars and bystanders would stay motionless. Patterns and shapes would remain were they are. This view is false. In reality, if one could freeze time, everything would disappear. For one, in such a timeless instant, there would be no sounds since sounds require time. There would be no words since words are composed by sound waves. There would be no neural activity since neural activity is implemented by means of chemical sequences spanning time. There would be no light since light rays travel in time. Finally there would be no wholes since every whole requires time to take place. There would be no patterns, no shapes, and no objects.

Second, different wholes have different temporal durations. For instance, a face needs the time required for light to go from one person to another plus the time required by the neural machinery of the beholder to allow the face to eventually produce a joint effect. Other wholes can be faster, the two events at the inputs of an electronic AND gate produce an effect in a very short span of time. Each whole has its own specific time corresponding to the time requested by the corresponding joint causal process to occur. Thus, the narrower the span of such a temporal window, the fewer wholes are going to be included.

Third, reality seems made of elements and wholes taking place at different temporal lengths. At the same time the causal processes of different wholes are spread on different time spans. An interesting example is offered by the human perceptual system where many different wholes, corresponding to many different collections of events in the environment, produce effects in different instants (Zeki, 1978; Zeki & Moutoussis, 1997; Bartels & Zeki, 1998). For instance, movement takes more time to produce an effect than color in the brain. A bright spot would produce a very fast response while a face would take a longer time. So there are fast wholes and slow ones.

Fourth, another issue to consider is the fact that, since wholes do not exist but rather take place, a persisting whole needs to be continuously rehearsed. For instance, the two events at the input of the AND gate do not correspond to an everlasting whole. They need to take place again and again if an enduring whole is expected. They must be repeated in order to produce again and again the gate output. Whenever a whole endures, it is because it takes place repeatedly. Consider a frame on a film. In order to show a still object, a series of identical frames has to be projected. If, so to speak, one needs again the same whole, the same causal process has to occur. In order to do so, the same context is required. This is what happens when one looks at a pattern. Whenever one looks at it, a new causal process takes place.

Fifth, wholes are locatable both in space and time although their location is neither like a four-dimensional worm nor point-like. There could be either wholes whose elements are already scattered in time (like a piece of music made of sounds that are scattered in time) or wholes whose elements take place at the same time (like the black spots on a piece of paper constituting Kanisza's triangle). In both cases, the causal relation and thus the event that "completes" the whole must take place, at least, a little after the last event of the collection.

Sixth, the suggested approach provides a solution for spatiotemporally spread gerrymandered wholes. A whole can be constituted by elements scattered in time and space as long as there is a causal process that allows to such scattered elements to cause a joint effect as in the case of a constellation or a memory or a dream or a hallucination.

Does the presented account raise the issue of thorny issue of backward causation? It is not necessarily the case. In short, the problem is the following (Fig. 3). Consider two initial events $x$ and $y$. They occur at a time $t_1$. For instance, they may be the two inputs of an AND gate or the parts of a Gestalt shape. Anyway, at the time in which they occur, there is no certainty that they will be able to produce a joint effect $e$. An unfathomable event might always pre-empt the occurrence of $e$. Suppose that everything goes well and that $e$ occurs at $t_2 > t_1$. At that moment, the ontology of the joint cause is sealed. The occurrence of $e$ pulls into existence the whole $W = x \wedge y$ only after $t_2$ but the temporal location of the whole $W$ cannot be anything else but $t_1$.

Such a strange causal and temporal loop is not necessarily a critical issue. Rather it may highlight the unsaturated and potential nature of reality until the occurrence of something seals a portion of the past — in this case, the whole. It is as though there were a causal entanglement that gets to a completion when the effect actually states the causal role of the whole $W$.

In the model, $W$ does not happen at $t_1$ until $e$ takes place. It is as if $W$'s present were still open and unsaturated. No matter how long does it take for the rest of the universe to allow $e$ to take place, *the ontology of $W$ is not*

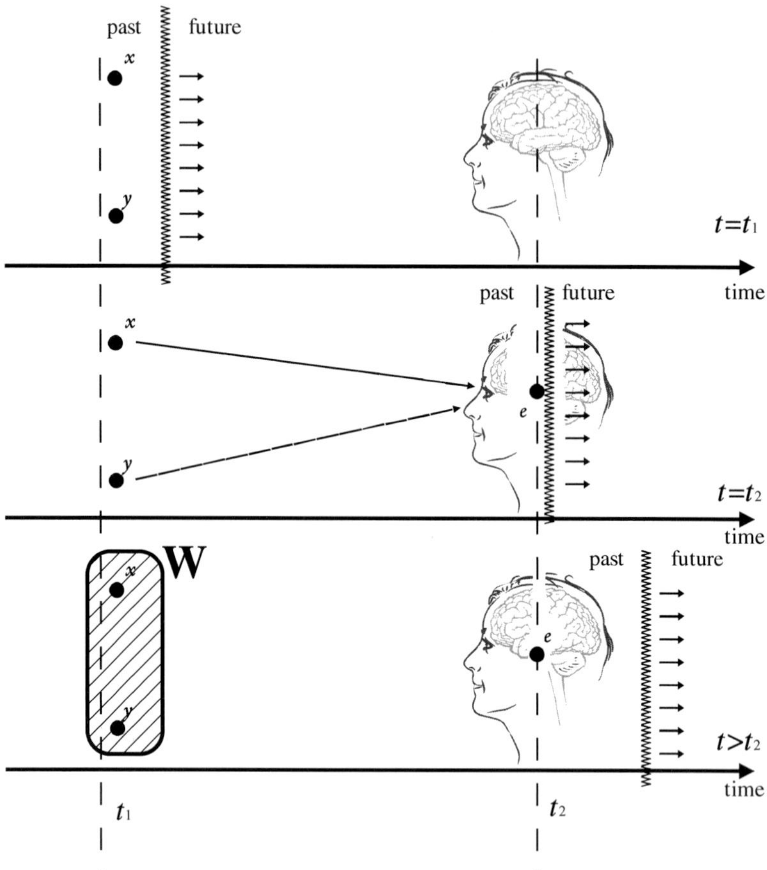


**Fig. 3.** There is no backward causation. The existence of an object (a whole) is not fixed until its elements produce an effect in a subsequent instant of time.

*fixed at time* $t_1$ *until e occurs at time* $t_2 > t_1$. This statement is not as bizarre as it seems. In fact, if the causal criterion for existence is taken seriously and if the causal processes are contingent, any entity *x* is defined only when its effects occur — their occurrence can take place at a later time.

Until *e* does not occur, *W* is still un-happened and thus *W*'s ontology is open since *W* has not existed yet. Thus, if *W* had not existed yet, how is it possible for *W*'s time to have gone by? It has not. *W*'s present and *W*'s ontology are like opening brackets. They are not even brackets. An opening bracket is not a bracket until the closing bracket seals the pair. *W* is just a possibility that, if it does not get to an end, remains only a possibility.

However, when *e* takes place, the bracket closes. *W* has happened as the joint cause and its ontology, like a pair of brackets, gets fixed. *W* becomes a moment of reality. But *W* is still located in the instant $t_1$, no matter how much later $t_2$ is.

Thus, for *W*, no time has elapsed until *e* occurred since no *W* existed. No time has gone by for *W*, because there was yet no *W*. Thus, this model does not endorse backward causation. There is no need for anything to go from *e* at $t_2$ back to *W* at $t_1$. *W* remains indefinite until *e* occurs. There is no backward causation since from *W*'s perspective, *x*, *y* and *e* are co-present. A dramatic example is offered by light. Consider the photon that starts on Alpha Centauri and ends in my eye. For the photon, the two moments are taking place in the same present. For an external observer, such moments are separate, but for the internal time of photon at the same moment, which is an elementary causal process in itself, nothing divides the start from the end of the journey. There is no internal event that distinguishes temporally and causally its emission on Alpha Centauri from its absorption in my eyes. Moreover, while the example of the photon is particularly convincing, a similar case can be made for all causal processes. Until the final effect has not occurred, the cause has not taken place yet and thus its present is open. When the effect occurs, the cause is defined and the brackets close collapsing the present. However, once again, there is no backward causation because, from the perspective of the causal process, there is no internal time. Therefore, there is no need to pull time backward. It appears to backward causation because the process is seen from an external perspective, which is uninfluential for the causal process in itself. Inside the causal process going from *W* to *e*, there is no time flow, thus there is no backward causation.

Thus, when does a whole take place, really? Although the whole is temporally located where its parts are, the whole is undefined until the final joint effect occurs. Metaphorically speaking, it is like getting a degree. Assume that you passed all exams and prepared and printed you thesis. Unpropitiously, because of some unexpected event, you miss the final defence, on the very last day. Or, even more tragically, you defend it but your university collapses just before the faculty staff signs your degree. You have yet not obtained a degree, although all the parts of your degree occurred. Although the underlying relevant phenomena are practically the same (studying, passing exams, writing your final dissertation), what you do becomes a degree only after the dissertation, which is the final effect. The effect changed the past, by positing a joint cause (your degree).

In a sense, the occurrence of the whole is potential until it produces an effect. So, the whole takes place earlier than the effect, but only after the effect actually occurred. There is a time when the whole has not happened yet, but after the occurrence of the effect, the whole had happened some time earlier. Strange but apparently true.

A whole does not exist *per se*. It requires the interaction with a proper situation where it can produce an effect as a whole.

## 6.8. Conscious Aspects of Physical Integration

Does this causal account satisfy the proposed criteria for integration in the case of consciousness? Here, the case of conscious perception is a test bed. More complex cases — such as free will, self-consciousness, and altered states of consciousness — will not be considered.

First, consider the issue of scaling up. It is fair to maintain that the presented account is capable of gracefully scaling up at different levels of complexity, different temporal scales, and different cognitive levels. Different wholes may have radically different temporal durations. Some examples from perception may be helpful here. Consider face perception. To perceive a face, a complex series of causal processes are needed to allow the features of the face to act as a whole in one's cortex (Liu *et al.*, 2002; Reddy & Kanwisher, 2006). During development and learning, a conscious subject becomes coupled with increasingly larger and more extended wholes — letters becomes words, single words form sentences, sentences constitute speeches, sounds make music, single events coalesce into complex stories, and so forth. So, the key to implement the kind of integration that a conscious agent manifests might consist in a system capable of embody larger and larger "AND gates" (more properly, joint causal structure) between events in the environment and internal events. Of course, these internal events ought not to remain isolate but, in turn, ought to coalesce into internal joint causal processes that embody the agent history. However, the account also suggests a way to quantify the degree of integration in a mathematical way akin to Tononi's approach. The account may thus be applied both to almost atomic cases of integration, such as the example of the AND gate, and to very complex multi-layered hierarchical causal processes like those involved by face-recognition and consciousness.

Second, consider the issue of avoiding circularity. The proposed account does not dwell on any suspicious notions that may hide a reference to the mind. Causation is a fundamental and neutral feature of the physical world. The proposed account does not refer to any mentalistic or covert mentalistic notion.

Thus, the presented hypothesis cuts across the distinction between consciousness and physical reality — the perceived world made of macrophysical entities is the physical environment as integrative processes single it out because of embodied nervous systems. Interestingly, this view suggests a physical world made of elements and wholes taking place at different temporal scales — there may be wholes very short and others much longer, according to the length of the integration required to bring them into existence. An interesting example is offered by the chrono-architecture of the human perceptual system where different percepts, corresponding to different integration processes, produce a final cortical effect in different instants (Zeki, 1978; Zeki & Moutoussis, 1997; Zeki & Bartels, 1998). So far, great care has been taken to avoid any characterization that may disguise descriptions for explanations. It would be all too easy to refer to neural systems as paradigmatic examples of integrated physical systems. If an account of integration has ever to be provided, it has to be provided in neutral terms. This strategy, whether successful, should allow avoiding any circularity. Causation is a tempting option.

Moreover, causation may be used to model neural activities. It is easy to see that joint causation is a common feature of neural networks. The neuron itself is glorified AND gate and all the subsequent hierarchically higher levels can be modeled as AND gates (Tanaka, 1997; Mountcastle, 1997; Horton & Adams, 2005; Zeki, 2001; Kanwisher, 2001). Obviously, integration across scales is the result of the dynamics of neural systems that continuously arises from the causal processes realized by neurons and carried on by electrical field across synapses. This dynamic continuity results in correspondingly large fields of influence for augmenting cognitive processes in assemblies of neural networks (Poznanski, 2002).

Third, consider the two remaining criteria — physical grounding and causal efficacy — they are a natural consequence of the proposal since it is based on physical processes with a causal role.

In addition, the proposed account may help to solve the debate between internalist and externalist explanation of consciousness. According to the former, consciousness is a property of the neural activity alone. While the structure of the brain is the result of the interaction with the environment, once the brain is up and running, according to internalism, the neural inside the brain is sufficient to produce consciousness. Many neuroscientists adopt this stance. Learning, development, and external world are only contingently necessary in order to produce the proper neural structures (Revonsuo, 2006; Crick, 1994; Koch, 2004; Block, 2005; Tononi, 2008; Tononi & Koch, 2008; Dehaene *et al.*, 2003). This position has been recently labeled as brainbound (Block, 2005). In contrast, other authors have stressed the fact that the meaning of brain states cannot be reduced to the activity inside the brain but it depends on the external world. Thus, mental states must somehow be constituted by something external to the neural system (Rockwell, 2005; Noë, 2004; Hurley, 1998; O'Regan & Noë, 2001; Honderich, 2006; Velmans, 2007; Putnam, 1975). The issue of integration might provide evidence in favor of the brainbound stance insofar consciousness seems to be organized into integrated wholes that have a neural grounding. The presented approach, concerning perception, suggests a different interpretation. Causal processes do not begin at the sensory level. Causal processes originate in the environment and then integrate in the cortex; however, they remain spread since their external origin. Consider face recognition. Subjects do not recognize certain configurations of chemical activity in their retina. Rather they recognize faces. The causal process that gets to an end in one's fusiform gyrus originates on the face of another individual in the environment.

This is where the causal account just presented is of advantage because it is not committed to neural processes alone. If what matters is the causal geometry of joint causes and their effects, there is no reason to limit integration processes to neural activity alone. After all, neural events are physically discontinuous and mediated — as Cajal rightly surmised — by synapses. If what matters is the causal configuration and not the physiological implementation, one may set aside the biological boundaries and consider whether the integration does take place on a larger scale than the body. In this way, the physical underpinnings of consciousness may include events that do not fit exclusively inside one's body thereby providing an explanation as to how consciousness perceives the external world (Manzotti, 2006, 2011). An objection to this form of phenomenal externalism based on the integration of processes larger than the body is that it contradicts the local supervenience of the mind on neurons (Kim, 1998). However, this objection is not totally convincing since the principle of local supervenience is based on causal considerations. If it were possible to show that causal processes integrate at different scales, there would be no reason to stick to the skin boundary. To a certain extent, the body is transparent to causation.

At the end, a proposal for further work is put forward. Joint causation could be used to estimate the degree of integration of a neural network in terms of a two dimensional index. On the one hand, the total number of joint causal processes could be measured. On the other hand, the complexity and the hierarchical order of such joint causal processes would provide a second dimension. In this way, it would be possible to compare the level of integration of two systems according to two related but complementary parameters — the total amount of wholes and their hierarchical level of complexity. For example, one agent may have a huge number of percepts but a limited hierarchical complexity. All percepts may be at the same level. Another agent may have the same total number of percepts but distributed on a more developed hierarchical structure — that is many wholes may be the results of higher-level integration.

## 6.9. Conclusions

The presented account suggests to start from elementary casual process and to see whether they have the resources to combine wholes that match the empirical evidence about consciousness. The key hypothesis is that integration is rooted at a fundamental level of elementary causal structure. It is then exploited by biological and neural structures.

This proposal challenges popular computational — or functional, or neural, or cognitive — notions of integration. Since the ontology of these levels remains uncertain (Landauer, 1996; Floridi, 1999; Searle, 1980), a physical foundation is conceivable. Causation is arguably the hallmark of the physical. In principle, therefore, all neural and cognitive levels are reducible to the causal geometry of the underlying physical processes. Likewise, if consciousness is a physical phenomenon, it must correspond to some natural carving of the physical world (borrowing Block's formulation). The issue of integration is germane both to phenomenal integration and to macro-physical entities. Causation allows avoiding any commitment to specific explanatory levels like the neural, the functional, the cognitive, or the computational one.

## References


Aleksander, I. & Gamez, D. (2010) Informational theories of consciousness: A review and extension. In *BICS 2010*. Madrid.

Alkire, M. T., Hudetz, A. G. & Tononi, G. (2008) Consciousness and anesthesia. *Science*, **322**, 876–881.

Amari, S.-I., Beltrame, F., Bjaalie, J. G., Dalkara, T., De Schutter, E., Egan, G. F., Goddard, N. H., Gonzalez, C., Grillner, S., Herz, A., Hoffmann, K.-P., Jaaskelainen, I., Koslow, S. H., Lee, S.-Y., Matthiessen, L., Miller, P. L., DaSilva, F. M., Novak, M., Ravindranath, Y., Ritz, R., Ruotsalainen, U., Sebestra, V., Subramaniam, S., Tang, Y., Toga, A. W., Usui, S., Van Pelt, J., Verschure, P., Willshaw, D. & Wrobel, A. (2002) Neuroinformatics: The integration of shared databases and tools towards integrative neuroscience. *J. Integr. Neurosci.*, **1**, 117–128.

Baars, B. J. (1997) *In the Theater of Consciousness. The Workspace of the Mind*. Oxford: Oxford University Press.

Baars, B. J. & Ave, D. (1997) In the theatre of consciousness. global workspace theory, a rigorous scientific theory of consciousness. *J. Conscious. Stud.*, **4**, 292–309.

Baars, B. J. & Franklin, S. (2003) How conscious experience and working memory interact. *Trends Cogn. Sci.*, **7**, 166–172.

Baars, B. J., Gally, J.A. & Edelman, G. M. (2011) Biology of consciousness. *Front. Psychol.*, **2**, 4.

Barrett, L. F. & Satpute, A. B. (2013) Large-scale brain networks in affective and social neuroscience: Towards an integrative functional architecture of the brain. *Curr. Opin. Neurobiol.*, **23**, 361–372.

Bartels, A. & Zeki, S. (2005) The Chronoarchitecture of the cerebral cortex. *Philos. Trans. R. Soc. Lond. B*, **360**, 733–750.

Bartels, A. & Zeki, S. (1998) The theory of multistage integration in the visual brain. *Philos. Trans. R. Soc. Lond. B*, **265**, 2327–2332.

Bayne, T. (2004) Self-consciousness and the unity of consciousness. *The Monist*, **87**, 219–236.

Bayne, T. (2008) The unity of consciousness and the split-brain syndrome. *J. Philos.*, **105**, 277–300.

Bayne, T. & Chalmers, D. J. (2003) *What is the Unity of Consciousness?* A. Cleeremans, ed., Oxford: Oxford University Press.

Bigelow, J. & Pargetter, R., (2006) Real work for aggregates. *Dialectica*, **60**, 485–503.

Bishop, R. C. (2006) Downward causation in fluid convection. *Synthese,* **160,** 229–248.

Block, N. (2003) Do causal powers drain away? *Philos. Phenomenol. Res.,* **67,** 133–150.

Block, N. (2005) Review of Alva Noë's "Action in Perception." *J. Philos.,* **102,** 259–272.

Chauvet, G. A. (2006) On the integration of physiological mechanisms in the nervous tissue using the MTIP. Synaptic plasticity depenending on neurons-astrocytes-capillaries interactions. *J. Integr. Neurosci.,* **5,** 443–482.

Chauvet, G. A. (2004) *The Mathematical Nature of the Living World. The Power of Integration.* Singapore: World Scientific.

Cleeremans, A. (2003) *The Unity of Consciousness: Binding, Integration, and Dissociation.* Oxford: Oxford University Press.

Couzin, I. D. (2009) Collective cognition in animal groups. *Trends Cogn. Sci.,* **13,** 36–43.

Couzin, I. D. (2007) Collective minds. *Nature,* **445,** 715.

Crick, F. (1994). *The Astonishing Hypothesis: The Scientific Search for the Soul.* New York: Touchstone.

Davidson, D. (1969) The individuation of events. In: N. Rescher, ed. *Essays in Honor of Carl G. Hempel: A Tribute in the Occasion of His Sixty-Fifth Birthday.* Dordrecht: D. Reidel Pub. Co., 295–309.

Dehaene, S., Sergent, C. & Changeux, J.-P. (2003) A neural network model linking subjective reports and objective physiological data during conscious perception. *Proc. Nat. Acad. Sci. USA,* **100,** 8520–8525.

Dennett, D. C. (1987) *The Intentional Stance.* Cambridge, MA: MIT Press.

Diacu, F., (1996) The solution of the n-body problem. *Math. Intellig.,* **18,** 66–70.

Di Lollo, V. (2012) The feature-binding problem is an ill-posed problem. *Trends Cogn. Sci.,* **16,** 317–321.

Dorigo, M. & Blum, C., (2005) Ant colony optimization theory: A survey. *Theor. Comput. Sci.,* **344,** 243–278.

Dorigo, M., Bonabeau, E. & Theraulaz, G. (2000) Ant algorithms and stigmergy. *Fut. Gener. Comput. Syst.,* **16,** 851–871.

Dowe, P. (2000) *Physical Causation.* New York: Cambridge University Press.

Engel, A. K. & Singer, W. (2001) Temporal binding and the neural correlates of sensory awareness. *Trends Cogn. Sci.,* **5,** 16–25.

Fass, D. (2006) Rationale for a model of human systems integration: the need of a theoretical framework. *J. Integr. Neurosci.,* **5,** 355–372.

Feinberg, T. E. (2009) *From Axons to Identity: Neurological Explorations of the Nature of the Self.* New York: Norton.

Feinberg, T. E. (2012) Neuroontology, neurobiological naturalism, and consciousness: a challenge to scientific reduction and a solution. *Phys. Life Rev.,* **9,** 13–34.

Feinberg, T. E. (2011) The nested neural hierarchy and the self. *Conscious. Cogn.,* **20,** 4–15.

Feldman, J. (2013) The neural binding problem(s). *Cogn. Neurodynam.*, **7**, 1–11.

Floridi, L. (1999) *Philosophy and Computing*. London: Routledge.

Fodor, J. A. (1974) Special sciences: The disunity of science as a working hypothesis. *Synthese*, **28**, 97–115.

Gamez, D. (2010) Information integration based predictions about the conscious states of a spiking neural network. *Conscious. Cogn.*, **19**, 294–310.

Gamez, D. (2014) The measurement of consciousness: a framework for the scientific study of consciousness. *Front. Psychol.*, **5**, 714.

Gangopadhyay, N. & Kiverstein, J. (2008) Enactivism and the unity of perception and action. *Topoi*, **28**, 63–73.

Goldstone, R. L. & Gureckis, T. M. (2009) Collective behavior. *Topics Cogn. Sci.*, **1**, 412–438.

Haggard, P., Clark, S. & Kalogeras, J. (2002) Voluntary action and conscious awareness. *Nature Neurosci.*, **5**, 382–385.

Honderich, T. (2006) Radical externalism. *J. Conscious. Stud.*, **13**, 3–13.

Horton, J. C. & Adams, D. L. (2005) The cortical column: a structure without a function. *Philos. Trans. R. Soc. Lond. B*, **360**, 837–862.

Hudson, H. (2003) Alexander's Dicta and Merricks' Dictum. *Topoi*, **22**, 173–182.

Hurley, S. L. (2003) *Action, the Unity of Consciousness, and Vehicle Externalism*. A. Cleeremans, ed., Oxford: Oxford University Press.

Hurley, S. L. (1998) *Consciousness in Action*. Cambridge, MA: Harvard University Press.

James, W. (1904a) A world of pure experience. *J. Philos. Psychol. Sci. Meth.*, **1**, 533–543.

James, W. (1904b) Does "consciousness" exist? *J. Philos.*, **1**, 477–491.

James, W. (1911). *Some Problems of Philosophy: A Beginning of an Introduction to Philosophy*. Nebraska: University of Nebraska.

James, W. (1890) *The Principles of Psychology*. New York: Henry Holt and Company.

Kanwisher, N., (2001). Neural events and perceptual awareness. *Cognition*, **79**, 89–113.

Kelso, S. J. A. (1995) *Dynamic Patterns: The Self-Organization of Brain and Behaviour*. Cambridge, MA: MIT Press.

Kim, J. (2003) Blocking causal drainage and other maintenance chores with mental causation. *Philos. Phenomenol. Res.*, **67**, 151–176.

Kim, J. (2006) Emergence: Core ideas and issues. *Synthese*, **151**, 547–559.

Kim, J. (1999) Making sense of emergence. *Philos. Stud.*, **95**, 3–36.

Kim, J. (1998) *Mind in a Physical World*. Cambridge, MA: MIT Press.

Kim, J. (1992) Multiple realization and the metaphysics of reduction. *Philos. Phenomenol. Res.*, **52**, 1–26.

Kim, J. (2005) *Physicalism, or Something Near Enough*. Princeton: Princeton University Press.

Kim, J. (1993) *Supervenience and Mind*. Cambridge: Cambridge University Press.

Klarreich, E. (2006) The mind of the swarm. *Science News*, **170**, 347–349.

Koch, C. (2004) *The Quest for Consciousness: A Neurobiological Approach*. Englewood: Roberts & Company Publishers.

Koffka, K. (1935) *Principles of Gestalt Psychology*. New York: Harcourt Brace and Howe.

Köhler, W. (1947) *Gestalt Psychology An Introduction*. New York: Liveright.

Landauer, R. (1996) The physical nature of information. *Phys. Lett. A*, **217**, 188–193.

Lewis, D. K. (1986) *On the Plurality of Worlds*. London: Blackwell.

Lewis, D. K. (1991) *Parts of Classes*. Oxford: Blackwell.

Libet, B. (1985) Unconscious cerebral initiative and the role of conscious will in voluntary action. *Behav. Brain Sci.*, **8**, 529–566.

Liu, J., Harris, A. & Kanwisher, N. (2002) Stages of processing in face perception: An MEG study. *Nat. Neurosci.*, **5**, 910–917.

Manzotti, R. (2006) A process oriented view of conscious perception. *J. Conscious. Stud.*, **13**, 7–41.

Manzotti, R. (2008) A process-oriented view of qualia. In: E. Wright, ed. *The Case for Qualia*. Cambridge, MA: MIT Press, pp. 175–190.

Manzotti, R. *et al.*, (2012) Cognitive integration through goal-generation in a robotic setup. In: A. Chella *et al.*, eds. *Biologically Inspired Cognitive Architecture 2012*. Dordrecht: Springer, pp. 225–231.

Manzotti, R. (2012) The computational stance is unfit for consciousness. *Int. J. Mach. Conscious.*, **4**, 401–420.

Manzotti, R. (2011) The spread mind. Seven steps to situated consciousness. *J. Cosmol.*, **14**, 4526–4541.

Marras, A. & Yli-Vakkuri, J. (2008). *The "Supervenience Argument": Kim's Challenge to Nonreductive Physicalism*. S. Gozzano & F. Orilia, eds., Frankfurt: Ontos-Verlag.

Maturana, H. R. & Varela, F. J. (1980) *Autopoiesis and Cognition: The Realization of the Living*. Dordrecht (Holland): D. Reidel Pub. Co.

Megill, J. (2012) A defense of emergence. *Axiomathes*, **23**, 597–615.

Merricks, T. (2001) *Objects and Persons*. Oxford: Oxford Clarendon Press.

Metzinger, T. (2003) *Being No One: The Self-model Theory of Subjectivity*. Cambridge, MA: MIT Press.

Mirolli, M. & Baldassarre, G. (2013). *Intrinsically Motivated Learning in Natural and Artificial Systems*. G. Baldassarre and M. Mirolli, eds. Berlin, Heidelberg: Springer.

Mountcastle, V. B. (1997) The columnar organization of the neocortex. *Brain*, **120**, 701–722.

Noë, A. (2004) *Action in Perception*. Cambridge, MA: The MIT Press.

O'Regan, K. J. & Noë, A. (2001) A sensorimotor account of vision and visual consciousness. *Behav. Brain Sci.*, **24**, 939–973.

Passino, K. M. *et al.*, (2014) Swarm cognition in honey bees. *Behav. Ecol. Sociobiol.*, **62**, 401–414.

Poznanski, R. R. (2002) Towards an integrative theory of cognition. *J. Integr. Neurosci.*, **1**, 145–156.

Putnam, H. (1975) *Mind, Language and Reality*. H. Putnam, ed. Cambridge: Cambridge University Press.

Reddy, L. & Kanwisher, N. (2006) Coding of visual objects in the ventral stream. *Curr. Opin. Neurobiol.*, **16**, 408–414.

Reichenbach, H. (1971) *The Direction of Time*. Berkeley: University of California Press.

Revonsuo, A. (1999) Binding and the phenomenal unity of consciousness. *Conscious. Cogn.*, **8**, 173–185.

Revonsuo, A. (2010) *Consciousness. The Science of Subjectivity*. Hove: Psychology Press.

Revonsuo, A. (2006) *Inner Presence. Consciousness as a Biological Phenomenon*. Cambridge, MA: MIT Press.

Rockwell, T. (2005) *Neither Ghost nor Brain*. Cambridge, MA: MIT Press.

Rosanova, M. *et al.*, (2012) Recovery of cortical effective connectivity and recovery of consciousness in vegetative patients. *Brain*, **135**, 1308–1320.

Roskies, A. L. (1999) The binding problem. *Neuron*, **24**, 7–9.

Santana, P. & Correia, L. (2010) A swarm cognition realization of attention, action selection, and spatial memory. *Adap. Behav.*, **18**, 428–447.

Searle, J. R. (1980) Minds, brains, and programs. *Behav. Brain Sci.*, **1**, 417–424.

Seth, A. K. *et al.*, (2008) Measuring consciousness: relating behavioural out neurophysiological approaches. *Trends Cogn. Sci.*, **12**, 314–321.

Shanahan, M. P. (2005) Global access, embodiment, and the conscious subject. *J. Conscious. Stud.*, **12**, 46–66.

Shanahan, M. P. (2010) *Embodiment and the Inner Life. Cognition and Consciousness in the Space of Possible Minds*. Oxford: Oxford University Press.

Shanahan, M. P. (2012) Knotty-centrality: finding the connective core of a complex network. *PLos One*, **7**, e35679.

Shanahan, M. P. & Connor, D. (2008) *Modeling the Neural Basis of Cognitive Integration and Consciousness*. Proceedings ALife XI.

Simons, P. M. (1987) *Integral Wholes*. P. M. Simons, ed., Oxford: Clarendon Press.

Singer, W., (2001). Consciousness and the binding problem. *Ann. NY. Acad. Sci.*, **929**, 123–46.

Sundman, K. F. (1912) Mémorie sur Le Problème Des Trois Corps. *Acta Mathem.*, **36**, 106–180.

Tanaka, K. (1997) Columnar organization in the inferotemporal cortex. *Cerebral Cortex*, **12**, 469–498.

Tegmark, M. (2015) Consciousness as a state of matter. *Chaos, Solitons & Fractals*, **76**, 238–270.

Theraulaz, G. & Bonabeau, E. (1999) A brief history of stigmergy. *Artific. Life*, **5**, 97–116.

Thompson, E. (2004) Life and mind: From autopoiesis to neurophenomenology. A tribute to Francisco Varela. *Phenomenol. Cogn. Sci.*, **3**, 381–398.

Tononi, G. (2004) An information integration theory of consciousness. *BMC Neurosci.*, **5**, 42.

Tononi, G. (2008) Consciousness as integrated information: A provisional manifesto. *Biol. Bull.*, **215**, 216–242.

Tononi, G. & Koch, C. (2008) The neural correlates of consciousness: an update. *Ann. NY Acad. Sci.*, **1124**, 239–261.

Tononi, G. & Massimini, M. (2008) Why does consciousness fade in early sleep? *Ann. NY Acad. Sci.*, **1129**, 330–334.

Treisman, A. M. (1998) The binding problem. *Curr. Opin. Neurobiol.*, **6**, 171–178.

van Inwagen, P. (1987) When are objects parts? *Philos. Perspec.*, **1**, 21–47.

van Inwagen, P. (1990a) Four-dimensional objects. *Noûs*, **24**, 245–255.

van Inwagen, P. (1990b) *Material Beings*. New York: Cornell University Press.

Velmans, M. (2007) Reflexive Monism. *J. Conscious. Stud.*, **15**, 5–50.

Walter, S. (2008) The supervenience argument, overdetermination, and causal drainage: assessing Kim's master argument. *Philos. Psychol.*, **21**, 671–694.

Wang, Q.-D. (1991). The global solution of the n-body problem. *Celest. Mech. Dynam. Astron.*, **26**, 73–88.

Williams, L. M. (2006) An integrative neuroscience model of "significance" processing. *J. Integr. Neurosci.*, **5**, 1–47.

Williams, L. M. *et al.*, (2008) The integrate model of emotion, thinking and self regulation: an application to the "paradox of aging." *J. Integr. Neurosci.*, **7**, 367–404.

Wolfe, J. M. (2012) The binding problem lives on. *Trends Cogn. Sci.*, **16**, 307–308.

Yablo, S. (2004) Advertisement for a sketch of an outline of a prototheory of causation. In: N. Hall, L. A. Paul and J. Collins, eds. *Causation and Counterfactuals*. Cambridge, MA: MIT Press, 119–137.

Zeki, S. (1978) Functional specialization in the visual cortex of the rhesus monkey. *Nature*, **274**, 423–428.

Zeki, S. (2001) Localization and globalization in conscious vision. *Ann. Rev. Neurosci.*, **24**, 57–86.

Zeki, S. (2003) The disunity of consciousness. *Trends Cogn. Sci.*, **7**, 214–218.

Zeki, S. *et al.*, (1999). The neurological basis of conscious color perception in a blind patient. *Proc. Nat. Acad. Sci. USA*, **96**, 14124–14129.

Zeki, S. & Moutoussis, K. (1997) Temporal hierarchy of the visual perceptive systems in the Mondrian world. *Philos. Trans. R. Soc. Lond. B*, **264**, 1415–1419.

Zschorlich, V. R. & Köhling, R. (2013) How thoughts give rise to action — conscious motor intention increases the excitability of target-specific motor circuits. *PloS one*, **8**, e83845.

Biophysics of Consciousness: A Foundational Approach
R. R. Poznanski, J. A. Tuszynski and T. E. Feinberg
Copyright © 2017 World Scientific, Singapore.

# 7

# THE HOLOINFORMATIONAL HOLOINFORMATION FOUNDATIONS OF CONSCIOUSNESS

*Francisco Di Biase*

*Department of Neurosurgery, Clínica Di Biase and Santa Casa Hospital, Barra do Piraí, Rio, Brazil, 27120120*

*Department of Post-Graduation UGB Universidade Geraldo Di Biase — Volta Redonda, Rio, Brazil*

## Abstract

This chapter addresses a quantum-informational holographic model of consciousness based on the holonomic neural networks of Karl Pribram, in the holographic quantum theory developed by David Bohm, and in the non-locality property of the quantum field described by Umezawa. We consider this model an interconnection between brain and mind by means of quantum microsites named *dendrons and psychons*, as proposed by John Eccles. A dynamic concept of consciousness is assumed as a holoinformational flux interconnecting the holonomic informational quantum brain dynamics, with the quantum informational holographic nature of the universe. This self-organizing flux is generated by the holographic mode of treatment of neuronal information and can be optimized through altered states of consciousness that underlie the coherence of cerebral waves. In brain mapping studies performed during the occurrence of these harmonic states the spectral array of brain waves is highly synchronized and perfectly ordered like a unique harmonic wave, as if all frequencies of all neurons from all cerebral centers play the same symphony. This highly coherent brain state generates the non-local holographic distributed informational cortical field of consciousness that interconnect the entire human brain with the quantum-holographic universe. The comprehension of this holonomic quantum informational nature of brain–mind-interconnectedness allows us to solve the old mind-matter Cartesian hard problem in a more transdisciplinary and integrated paradigm.

*Keywords*: Holoinformation; consciousness; quantum information; non-locality; holonomic neural networks; quantum holographic theory; coherent brain state.

## 7.1. Introduction

We are living a special moment in human history with the emergence of a fantastic integral quantum-holographic brain-cosmos extended vision (Di Biase & Amoroso, 2008). The knowledge about the physics of reality that we have today shows that the universe is a vast and immense informational quantum-holographic field harmoniously organized in all levels of complexity. This universal interconnectedness not limited by space-time is a field of non-local information that interpenetrates everything in the cosmos instantaneously. Bohm (1983), Bohm & Peat (1987) and Bohm & Hiley (1993) in their Quantum-Holographic Theory of the universe have demonstrated that a cosmos organized this way has non-local properties of indivisibility. During the 1950's until the 1980's Bohm developed a quantum-holographic theory of the universe, that explains everything the Copenhagen's Bohr model explains and goes further eliminating indeterminism and wave-particle paradoxes. Nevertheless, unfortunately for the history of Physics, during the 1950's Bohm was implicated in a political persecution by senator MacCarthy and he had to leave the USA, being his theory not recognized in USA as the best development of quantum physics until then. As Bell (1987) wrote:

> *"the De Broglie–Bohm's idea seems ... so natural and simple, to resolve the wave-particle, dilemma in such a clear and ordinary way, that it is a great mystery ... that it was so generally ignored."*

According to Bohm & Hiley (1993), De Broglie in his model of quantum mechanics developed a new kind of field in which the activity is dependent of the informational content that is conducted to all the experimental field.

Bohm developed his quantum theory adding to these field equations a Quantum Potential that satisfies Schrödinger's equation, that depends on the form but not on the amplitude of the wave function, creating a quantum model of the universe in which the quantum potential, carries *"active information"* that *"guides"* the particle along its way. The quantum potential has inedited characteristics unknown up to then, because being different from the other nature's forces, it is subtle in its form, and does not decay with distance. Such interpretation allows communication between this "pilot wave" and the particle, to be processed in a higher speed than light, unveiling the quantum paradox of non-locality, i.e., of instantaneous causality. For Bohm, differently from Bohr, the elementary particles do not have dual nature wave/particle but are particles all the time, and not only when observed. Actually, the particle originates from a global quantum field fluctuation, being its behavior determined by the quantum potential that carries information about the environment of the particle in-forming its motion. According to Bohm & Hiley (1993) since the information in the potential is very detailed, the resulting trajectory is so extremely complex that it appears chaotic or indeterminist. Attempts to measure the particles' properties, changes the quantum potential, destroying its information.

In Bohm's quantum-holographic theory there is an occult *implicate order* (a spectral dimension of frequencies) that is continuously unfolding in a *explicate order* (our manifested space-time universe) and enfolding again in the implicate order by means of a non-local information that he named *holomovement*. Later, he also proposed a *superimplicate order,* as he explains to Weber (1982):

> *"the implicate order is a wave function, and the superimplicate order or superior informational field, is a function of the wave function, i.e., a super wave function that makes the implicate order non-linear by organizing it in complex and relatively stable structures."*

For Bohm (1983) this superimplicate order allows us to:

> *"understanding consciousness, energy and matter as expression varieties of a same informational order."*

In Bohm's theory everything, fields, particles and even our holographic space-time itself, originates from an implicate spectral dimension, through a non-local information flux which he calls the *holomoviment*.

As we will see later, Pribram (1991, 1997) developed a holographic theory of consciousness showing experimentally that the fields of distributed electromagnetic activity in the brain cortex are quantum holographic non-local informational distributed patterns. Pribram's neural network equation is similar to Schrödinger's wave equation of quantum physics;

$$i\hbar\frac{\partial\Psi(q;t)}{\partial t}=\left(-\frac{\hbar^2}{2M}\frac{\partial^2\Psi(q;t)}{\partial q^2}+V(q)\right)\Psi(q;t)$$

where $(q, t)$ space-time variables; $\hbar$ is Plank's constant; M is the mass of the particle; $i=\sqrt{-1}$ is a complex number; $V$ is the potential; and $\Psi$ is the wave information.

In the above Schrödinger equation of the ontological formalism, one makes the replacements $\hbar\rightarrow\nu$, $M\rightarrow 1$, and $V\rightarrow U_{ex}$ there results comparable holonomic formalism's Neural wave equation (from Pribram (1991)):

$$i\nu\frac{\partial\Psi}{\partial t}=\left(-\frac{\gamma^2}{2}\Delta+U_{ex}\right)\Psi$$

where $U_{ex}$ is the extrinsic static energy of the Debye layer formed immediately beneath the dendritic membrane, $\nu$ is a proportionality constant (>0) and $\Delta=\nabla^2$ is the Laplacian.

The holographic system of information is mathematically and technologically distributed in every part of the system as we can see in the non-local broadcast diffusion of radio, TV and internet information. Bohm's universe and Pribram's brain are connected by this distributed non-local quantum holographic informational mode, so consciousness has been present in our universe since the beginning of creation on the various levels of nature's unfolding and enfolding informational order. This foundation of consciousness in the very profound non-local informational organization of our quantum-holographic universe and at the same time in the quantum-holographic organization of our brain-mind, is the basis for the understanding of our holoinformational quantum model of consciousness (Di Biase & Rocha, 2000; Di Biase & Amoroso, 2008; Di Biase, 2009).

The discovery of holographic information surfaces in quantum physics and biological systems are in tune with the concept of Smolin's (2000) holographic principle that proposes the universe is a network of holograms, and each hologram contains coded in it information about all the others. According to this principle, the area of any surface in space can be seen as a screen.

> *"that is really nothing but the capacity of that surface as a channel of information."*

The geometry, as measured in terms of area and volume, is a measure of the capacity of these screens to transmit information. For Smolin (2000) space is only a way of talking about all the different channels of information passing from observer to observer. In other words, the universe is a non-local informational holoarchy.

## 7.2. Non-Local Information Paradigm

The concept of non-local information is implicit in the Quantum Field Theory developed by Umezawa (1993) that unifies the electromagnetic, nuclear and gravitational fields, in one subjacent indivisible totality that until then were analyzed as separated. The quantum field does not exists as a physical reality in space-time as the gravitational and electromagnetic fields of the classical Newtonian physics giving it a peculiar non-local property, i.e., it is not localized in any region of space-time. When a non-local phenomenon occurs it instantaneously influences others regions of the space-time without the need of any transfer of energy or classical information. This non-local information connects everything in the universe from quantum physics-chemistry, quantum biology and quantum mind, through quantum consciousness and quantum cosmology. The cosmic evolution seems to be fine-tuned for the emergence of life, mind and consciousness, as an inexorable consequence of an intelligent non-local informational universe. This self-organizing holoinformational field is continuously emerging from the quantum field, a *plenum* (not a *vacuum*) that permeates all the cosmos, full of quantum information, energy and virtual particles popping out of nothing every billion of trillionth of a second. This quantum field *plenum* is a kind of intelligent cosmic DNA scattering *"in-formation"* (active information with meaning that *in-forms* the reality) through all the universe, creating galaxies stars and supernovas with its thermonuclear furnaces and generating the atoms of carbon, nitrogen, and oxygen, that are the foundation of all life. This fine-tuned biosignature of the non-local informational quantum holographic field is so fundamental for the cosmic evolution and the emergence of life and consciousness that information must be seen as a cosmic organizational principle with a *"status"* like matter, energy, space and time.

### 7.3. Information Self-Organization Consciousness and Negentropy

Chalmers (1995) states that information is an essential property of reality, as matter and energy, and that conscious experience must also be considered a fundamental feature, irreducible to anything more basic. He argues that each informational state has two different aspects, one as conscious experience, and the other as a physical process in the brain, that is, one internal/intentional and the other external/physical. This view finds support in the present developments of the so-called 'Information Physics', developed by Zurek (1990). This Information Physics developed in the 1990's demonstrates that beyond the Law of Conservation of Energy there is a more fundamental Law of Conservation of Information. In the process of developing a new Quantum Information Theory, Zurek (1990) proposed that the physical entropy would be a combination of two magnitudes that compensate each other: the observer's ignorance, measured by Shannon's statistical entropy, and the disorder degree of the observed system, measured by the algorithmic entropy which is the smallest number of bits needed to register it in the memory. During the measurement, the observer's ignorance is reduced, as a result of the increase in bit numbers in its memory, remaining, however, constant the sum of these two magnitudes, that is, the physical entropy.

In this context, the equivalence/identity between order, negentropy and information as stated by Brillouin's celèbre theorem, is the way that allows us to build upon and understand the whole irreducible and natural flow of order transmission in the universe, organized in a meaningful and intelligent informational mode. In the classical thermodynamic theory of Boltzmann, the definition of order is probabilistic and dependent on the entropy concept, which measures the degree of disorder of a system, reducing to uncertainty the immense dimension of natural meanings. Brillouin's equation that describes the energetic entropy of Boltzmann and at the same time the quantity of information (negentropy) of Shannon is logarithmic, i.e., holographic, with potency laws invariants in relation to the scale.

For Atlan (1972, 1979), Di Biase (1981), Di Biase & Rocha (2000), Di Biase & Amoroso (2008) and Di Biase (2009b), entropy should not be understood as a disorder measure, but much more as a measure of complexity and information. To do this, it is necessary to consider that information implies a certain ambiguity, meaning the bit capacity of a physical system as Shannon and Weaver (1949) put it, or the semantic content (meaning) conducted by the bits during a communication. In the classical information theory, the organization, the order expressed by the amount of information in the system (Shannon's *H* function) is the information measure that is missing, the uncertainty about the system. Relating this uncertainty, this ambiguity to the variety and the non-homogeneity of the system, Atlan (1979) could solve certain logical paradoxes of self-organization and complexity, widening Shannon's theory and defining organization in a quantitatively formal mode. Atlan showed that the system's order corresponds to a commitment between the maximum informational content (i.e., the maximum variety) and the maximum redundancy, and showed also that the ambiguity can be described as a noise function, or even a time one, if we consider the time effects as related to the random factors accumulated by the environment's action. Such ambiguity, peculiar to biological self-organizing systems as brain and mind can be manifested in a negative way (*"destructive ambiguity"*) with the classical meaning of disorganizing effect, or in a positive way (*"autonomy producer ambiguity"*) that acts by increasing the relative autonomy of a part of the system in relation to the others, reducing the system's natural redundancy and increasing its informational content.

Atlan (1979) developed this self-organizing theory of complexity for biological systems. Jantsch (1980) has shown that cosmological evolution is also a self-organizing process, with the microevolution of the individual systems (holons) co-evolving toward macrosystemic collective structures better organized, and with a big reduction in the amount of these collective systems. This whole self-organizing process represents, actually, a universal expression of a bigger acquisition of variety or informational content that is a consequence of a reduction of redundancy in the totality of the system-universe.

Seager (1995) states that consciousness, self-organization and information are connected at the level of semantic significance, not at the level of bit capacity, and that as the classical theory of information is situated at the level of bit capacity it would seem unable to provide the proper connection to consciousness. To set this connection, we have to reach a more radical view of the nature of consciousness and of information. Seager (1995) still reminds us that in the famous two-slit experiment, and in the quantum eraser experiment, what is at stake is not the bit capacity, but the semantically significant correlation of *information laden* distinct physical systems, in a non-causal mode.

Recently, Vedral (2010) developed a quantum informational theory of the universe in which everything that exists including us, can be understood in informational terms. He retakes the profound correlation between the entropy concept of disorder and the information concept of order of Léon Brillouin that relates information and negentropy (order) applying this correlation reformulated to the quantum universe. In the Quantum Information Theory the *bit* defined by the binaries choices yes/no is changed in *qbit* (quantum bit) that permits possibilities of much bigger choices as both/and, that originates from the phenomenon of quantum entanglement. Quantum entanglement makes particles that interacted in any epoch of its life to stay correlated forever by means of *quantum non-local information*.

## 7.4 Linking Quantum Information to Consciousness and Physics

Wheeler (1990) describes an elegant information-participatory universe with his *"the it from bit"* concept that united quantum information theory to consciousness and physics:

> *"...every it — every particle, every field of force, even the space-time continuum itself — derives its function, its very existence, entirely — even if in some contexts, indirectly — from the apparatus-elicited answers to yes-or-no questions, binary choices, bits."*

Wheeler's *it from bit* is an idea that every item of the physical world has an immaterial source and explanation. Reality arises for him in the last analysis from the yes–no question and the registering of equipment-evoked responses. Wheeler (1990) states that:

> *"all things physical are information-theoretic in its origin and that this is a participatory universe."*

In the same paper, Wheeler gives the example of a photon being detected by a photodetector under watch, when we ask the yes–or–no question when the counter registers a click and we often say *'a photon did it'*. But as Wheeler (1990) states:

> *"we know perfectly well that the photon existed neither before the emission nor after the detection... The yes or no that is recorded constitutes an unsplitable bit of information."*

Wheeler (1990) developed *"it from bit"* model where quantum information is more fundamental than energy, matter and space-time. This has total relevance for the consciousness holoinformational model of Di Biase (Di Biase & Rocha, 2000; Di Biase & Amoroso, 2008; Di Biase, 2009b, 2011) with consciousness as primarily a non-local informational system.

Matzke (1994) makes an analogy with the particle/wave duality seeing the paradoxical aspects of consciousness more understandable if we adopt energy/information duality just as early in the 20th century the particle/wave duality was insightful for physics. He states that consciousness requires the adoption of an energy/information duality, such as quantum states and consciousness. Although I see no possibility to do the paradigmatic jump to a quantum-holographic model of universe full of information and consciousness using Bohr's Copenhagen model with its paradoxes, the wave-particle dilemma, its non-deterministic mathematical structure and its incompatibility with Einstein's Relativity Theory. Again, as Bell (1987) states:

> *"Bohm's idea is so natural and simple, to resolve the wave-particle dilemma in such a clear and ordinary way, that it is a great mystery... that it was so generally ignored."*

If we adopt Bohm's quantum-holographic informational non-local model of the universe (Bohm, 1983; Bohm & Hiley, 1993) and Pribram's quantum-holographic informational non-local model of consciousness together in a Holoinformational quantum model of consciousness we will see the consciousness paradox (Chalmers (1995) hard problem) disappearing naturally.

## 7.5. Organisms and Brains are Macroscopic Non-Local Quantum Systems

In the living world, non-local coherence is just present as in the quantum and in the cosmos scale. In living organisms the coordination of functions inside the organisms is ensured by quantum coherence as we can see in the instantaneous correlation between parts and molecules and also between the organism and the external milieu. This instantaneous quantum information transfer are observed in organic molecules in entangled quantum states, in quantum tunneling, in Bose–Einstein condensates, in the Frölich effect and in superradiance states occurring in brain structures as microtubules, synapses and the cerebrospinal fluid.

According to Schrödinger (1944), in living organisms we must replace the concept of *mechanical order* that make order from disorder, by the notion of *dynamic order*, that produces order from order, from complex organization and information. This difference between mechanical and dynamical order, according to Schrödinger was first proposed by Max Planck who already made this distinction (Di Biase, 1981). This complex informational organization order is the non-local informational order that explains the living matter and is not based on mechanical molecular chance collisions and interactions, but in a system-wide correlations involving even distant parts that could not have time for mix in a mechanical process, as during emergence states of a fight/fleet behavior in which trillions and trillions of molecular reactions are occurring inside the trillions of the organism cells almost instantaneously. This organic coherence is only possible through the mobilization of non-local information and energy far from thermodynamic equilibrium.

Ho (2008) suggests the organism maintain itself in a negentropic state through the superposition of a non-dissipative cyclic process with entropy balance out of zero, and a dissipative irreversible process with entropy production greater than zero. The cyclic non-dissipative loop coupling with the irreversible energy loop frees the living organism from immediate thermodynamic constraints. Ho (2008) has been also demonstrating that highly polarized multiple layers of liquid crystalline water molecules form dynamically coherent units with the macromolecules, enabling them to function as quantum molecular energy machines that transform and transfer energy with close to 100% efficiency. According to her the liquid crystalline continuum of intimately associated polarized water and macromolecules extends throughout the extracellular matrix into the interior of every single cell, enabling each cell, ultimately each molecule, to intercommunicate with each other.

Rakovic (2009), points out that *Quantum-Holographic and Classically-Reduced Neural Networks can model psychosomatic functions*. For him the prevailing scientific paradigm considers information processing within the central nervous system as occurring through hierarchically organized and interconnected neural networks and biological neural networks that goes down sub-cellular cytoskeleton level, making an interface between neural and quantum levels. He states that with the Feynman propagator version of the Schrödinger equation, it is possible to describe the quantum level by analogous mathematical formalism as Hopfield-like quantum-holographic associative neural network. Also he points that this opens the fundamental question of how the quantum parallel processing level gives rise to classical parallel processing level, that is related to the quantum decoherence.

## 7.6. Quantum-Holographic Neuronal Dynamics and Classical Neuronal Computer Network Interconnection

How a self-organizing quantum mind can overlap **quantum decoherence** and maintain a persistent coherent state for a long time, at room temperature? It is very well established experimentally today that the molecules of chlorophyll responsible for the photosynthesis process that transforms light photons in chemical energy of plants, can do this with extraordinary efficiency, for about 750 sampling frequency, compared with 1 to 1.5 sampling frequency of chemical-bond vibrations. This is due to the action of a protein called antenna protein that holds the chlorophyll molecule sustaining the quantum coherence state and suppressing decoherence, by reinducing coherence in decohering parts of the chlorophyll molecule, as shown by Lee *et al.* (2007). This show us that the capability to suppress decoherence at environment temperature is a common process in nature. So the capability to suppress quantum decoherence must be seen as a natural process in the wet brain, and we must work with the possibility that neurons and glia can sustain a quantum coherent state for milliseconds in the organized cellular complex molecular system full of proteins macromolecules, small molecules, ions and water. It is well known that in the vicinity of these macromolecules there is ordered water, and that proteins with a cavity in its 3D structure can hold one or a few water molecules by means of hydrogen bonds. Beratan *et al.* (2005) had already demonstrated that quantum chemical computation shows that these ordered water molecules within and between two proteins separated by 12 to 16 ångstrons permits the occurrence of quantum coherent electron transfer. This quantum coherence can propagate through non-local information transfer in the nervous system and in the body by quantum entanglement and superradiance. Also, as biological self-organized systems these molecular systems have a huge structural and functional redundance that facilitates the non-local interconnection of all its parts.

## 7.7. In-formation in Self-Organizing Dissipative Structures

Prigogine and Stengers developed an extension of thermodynamics that shows how the second law can also allow the emergence of novel structures, and indicates the ways in which order can emerge from chaos (Prigogine & Stengers 1979, 1988). This type of self-organization generates dissipative structures that are created and maintained through the energy's exchanges with the environment in non-equilibrium conditions. These dissipative structures are dependent upon a new order, called by Prigogine *"order from fluctuations"*, which corresponds to a *"giant fluctuation"* stabilized by the exchanges with the environment. In these self-organizing processes the structure is maintained through an energy-information dissipation that displaces itself, simultaneously generating (*in-formating*) the structure through a continuous process. The more complex the dissipative structure, the more information is needed to keep its interconnections, making it consequently more vulnerable to the internal fluctuations, which means a higher instability potential and higher reorganization possibilities. If fluctuations are small, the system accommodates them and does not change its organizational structure. If the fluctuations reach a critical size, however, they cause disequilibrium in the system, generating new intra-systemic interactions and reorganization. As Prigogine & Stengers (1988) says:

> *"The old patterns interact between themselves in new ways, and establish new connections. The parts reorganize themselves in a new whole. The system reaches a higher order."*

Our brain is a dissipative self-organizing conscious quantum computer that creates new synaptic connections and patterns, every 90–120 minutes, by means of neuroplasticity, that is a result of a continuous quantum-molecular fluctuations dialogue between the mind, the brain and the DNA, always reaching new intra-systemic interactions and reorganization, and a higher order.

## 7.8. Consciousness Quantum Holographic Self-Organization and In-formation

Pribram (1991, 1997) has been demonstrating an analogy between the fields of distributed neural activity in the brain and the wave patterns in holograms. His neural network equation is similar to Schrödinger's wave equation of quantum physics with the addition of Bohm's quantum potential that guides, by means of active information, the particle alongside its course. As any elementary particle is united to the whole cosmos by means of a quantum active non-local information potential, capable of changing the structure of the universe, information then can be understood as nature's fundamental process, as Stonier (1990, 1997) put it. This "*active*" (Bohm) non-local information that organizes the particle's world reveal that the whole nature is informational, organized in a meaningful way. In the brain, this informational process is dissipative non-local quantum holistic, based on quantum-holographic neural fields, and at the same time local classical Newtonian and mechanistic, based on classical neural networks. So, as Di Biase (Di Biase, 2000; Di Biase & Amoroso, 2008; Di Biase, 2009) has been demonstrating in the last years it is a holoinformational field (non-local and local).

This view is crucial to understand the holoinformational nature of consciousness and intelligence in the universe. Matter, life and consciousness are meaningful activities, intelligent quantum-informational processes, order transmitted through the cosmic evolution, originated from a generating non-local informational field beyond our perception limits.

A universe *plenum* of non-local quantum potential *in*-formation with meaning (active information) is an intelligent universe *functioning like a mind*, as Sir James Jeans already had observed. So, as consciousness has always been present in all nature's levels of organization, matter, life and consciousness cannot be considered as separated entities, capable of being analyzed under a fragmentary Cartesian–Newtonian framework. Actually, consciousness must be considered as a fundamental property of the universe like information, energy, matter, and space-time and must be seen as an irreducible quantum non-local information distributed in a holistic cosmic way, generating self-organization, complexity, intelligence and evolution. Such view of a holoinformational intelligent "continuum," a fundamental generating order with a quantum-holographic informational creative flow permeating the whole cosmos, permits to understand the basic nature of the universe as an intelligent self-organizing unbroken wholeness. A kind of universal extended consciousness unfolding in an infinite holoarchy. As a quantum-holographic system this universal consciousness is distributed in every part of the cosmosphere, and in our quantum-holographic brain-mind. As part of this universal holographically distributed system our mind-body contains the active *in*-formation of the whole indivisible cosmos (Di Biase, 2009).

## 7.9. Eccles Interactive Dualism and Pribram's Monism

Eccles (1986, 1989, 1993) and Eccles & Beck (1992) described that the brain fine fibers structures, which he called *dendrons*, composed of pre-synaptic teledendrons, synapses and post-synaptic dendrites connections, that postulated to interact with the mind side of the interaction by way of units called *psychons*. He proposed that these psychons could operate on synapses through quantum processes, and with Beck (Eccles & Beck, 1992) developed a beautiful and logical quantum interpretation of the synaptic function. Pribram (1991, 1997, 2011b) demonstrated that Eccles' dendrons make up receptive fields in cortical sensory units, and that as sensory receptive fields they can be mapped in terms of wavelets, or wavelet-like patterns such as Gabor Elementary Functions.

Gabor (Pribram, 2011b) called these units Quanta of Information, because he used the same mathematics as Heisenberg in describing the quantum microphysics. Pribram (2011b) states that these quanta of information define the processes occurring in the material brain, but he points that Gabor invented his function, not to describe brain processes, but to find the maximum compressibility of a telephone message that could be sent over the Atlantic Cable without destroying its intelligibility. So, "*Gabor's function thus describes both a unit of brain processing and a unit of communication*," affirms Pribram, demonstrating that *brain is material, communication is mental and that the same mathematical formulation describes both.* Pribram (2011b) writes:

> *"The elementary structure of processing in Eccles' material dendron is identical to the elementary structure of processing of a mental (communication) psychon. There is a structural identity to the dual interactive process."*

Pribram (2011a) proposes a monistic basis for Eccles dualism, stating that:

> *"there is a interactive mind/matter duality that is a 'ground' from which both matter and mind are "formed" and the "dual" emerges."*

That ground functions as a potential reality similar to Heisenberg potential world and this informational flux provides the ontological roots from which our experience regarding matter as well as mind (psychological processing) itself become actualized in space-time.

Actually if we take the reductive path, we end up with psychology, not particles. In fact, it is a psychological process, mathematics, that describes the relationships that organize matter, he says. Pribram (2011b) states:

> *"that this root, through constrained by measures in spacetime, needs a more fundamental order, a potential order that underlies and transcends spacetime. The spectral basis of both matter and communication portrayed by the Fourier relationship delineate this claim."*

The holonomic brain theory of Pribram and the holographic quantum theory of Bohm, added with Laszlo's *akashic field* shows that we are part of something much more wider than our individual mind. Our mind is a subsystem of a universal hologram, accessing and interpreting this holographic universe. We are fractal-like holographic harmonic systems interacting continuously with this unbroken self-organizing wholeness. We are this holoinformational field of consciousness, and not observers external to it. The external observer's perspective made us lose the sense and the feeling of unity with the cosmos, generating the immense difficulties we have in understanding that we are one with the whole and not part of it.

In this holoinformational model of consciousness the non-local quantum-informational flux in a continuous holomovement of expansion and enfoldment between the brain and the implicate-superimplicate order, is a universal extended consciousness self-organizing as human mind. This non-local quantum-holographic universal extended consciousness is our own brain-mind consciousness seeing itself through our eyes and our brains, interconnecting us in a participatory holistic and indivisible way to all levels of the self-organizing multiverse.

## 7.10. Quantum Brain Dynamics

Experimental research developed by Pribram other consciousness researchers like Hameroff (1994), Hameroff & Penrose (1996), Jibu & Yassue (1993, 1995) and Ho (2008) confirm the existence of a Quantum Brain Dynamics in neural microtubules, in synapses, in the molecular organization of the cerebrospinal fluid, and in the intracellular medium matrix. This Quantum Brain Dynamics can generate Bose–Einstein condensates and the Fröhlich effect. Bose–Einstein condensates consist of atomic particles, and the Fröhlich effect of biological molecules, that can assume a high level of coherent alignment, functioning as a highly ordered and unified informational state, as seen in lasers and superconductivity. Sir John Eccles's psychons operate on synapses by way of quantum coherence processes. These quantum dynamics show us that the interaction process between what Eccles calls *dendrons* (the brain side) and *psychons* (the mind side) are not limited to the synaptic cleft, as stated by him, but have a much wider embodiment throughout the whole brain.

Pribram (1991, 1993) demonstrates good evidence that Eccles's dendrons make up receptive fields in cortical sensory units. Dendrons are composed of pre-synaptic teledendrons, synapses and post-synaptic dendrites, and they compose the fine fiber structure wherein brain processing occurs. I see the quantum holographic interactions between brain and cosmos as a natural extension of Eccles's ideas of a interactionism between dendrons and psychons (Di Biase & Amoroso, 2008).

The study by Jibu & Yasue (1993, 1995) along with Umezawa on quantum brain dynamics have demonstrated that brain dynamics consists of quantum brain dynamics (i.e., quantum mode) and classical brain dynamics (i.e., classical mode), and Jibu & Yasue (1993) claim:

> *"quantum brain dynamics is the fundamental process of the brain given by quantum field dynamics of the molecular vibrational fields of water molecules and biomolecules."*

Stuart *et al.* (1978) introduced, in quantum brain dynamics, the notion that the quanta of the molecular vibrational field of biomolecules are *corticons*, and the quanta of the molecular vibrational field of water molecules are exchange *bosons*. Quantum coherence can propagate through these vibrational fields of biomolecules and water molecules by non-local information transfer, quantum entanglement and superradiance. As we already said, biomolecular systems are self-organized systems with a huge structural and functional redundance, facilitating the creation of the quasi-crystalline medium that provides the interconnection of the molecular quantum computer networks dynamics with the neuronal classical computer network.

The dissipative quantum brain model is the extension to the dissipative dynamics of the many-body model proposed by Ricciardi & Umezawa (1967) who wrote:

> *"the extended patterns of neuronal excitations may be described by the spontaneous breakdown of symmetry formalism of Quantum Field Theory."*

Umezawa in the same paper states yet that:

> "in condensed *matter physics any particular information is carried by certain* any *material ordered pattern maintained by certain long range correlation mediated by massless quanta."*

For him, this is the only way to memorize some information, as memory is a printed pattern of order supported by long range correlations. I see this information and memory mediated by *massless quanta* as our brain-mind consciousness, understood here as going on information in the very moment of its manifestation.

### 7.11. A Universe of Consciousness

In my conjecture the interconnectedness between brain and cosmos is an instantaneous holistic nonlocal quantum holographic extended connection and I proposed the concept of a holoinformational flux, from which both mind and matter are informed. The quantum holographic brain dynamic pattern is conceived as an active part of the universal quantum-holographic informational field, and capable of generating an informational interconnection that, as I have already stated, is simultaneously non-local quantum-holistic (mind-universe extended holographic connection), and local Newtonian-mechanistic (brain-mind neural networks connections), i.e., holoinformational. This self-organizing flux is generated by the holographic mode of treatment of neuronal information and can be optimized through altered states of consciousness and the relaxation response, that underlie the coherence of cerebral waves. In brain mapping studies, which I regularly do in meditating patients, I find that during the occurrence of these harmonic states, the spectral array of brain waves are highly synchronized and perfectly ordered like a unique harmonic wave, as if all frequencies of all neurons from all cerebral centers play the same symphony. It is this highly coherent brain state that optimizes the non-local holographic distributed informational neuronal cortical field that generates the state of consciousness.

## References


Atlan, H. (1972) *L'Organization Biologique et la Théorie de L'Information.* Paris: Hermann.

Atlan, H. (1979) *Entre le Cristal et la Fumée, Essai sur LEOrganization du Vivant.* Paris: Seuil.

Bell, J. (1987) *Speakable and Unspeakable in Quantum Mechanics.* Cambridge: Cambridge University Press.

Beratan, D. J., Balabin Lin, & Beratan, D. N. (2005) The nature of aqueous tunneling pathways between electron-transfer proteins. *Science,* **310**, 1311–1313.

Bohm, D. (1983) *Wholeness and the Implicate Order.* New York: Routledge.

Bohm, D. & Hiley B. J. (1993) *The Undivided Universe.* London: Routledge.

Bohm, D. & Peat, F. D. (1987) *Science Order, and Creativity. A dramatic New Look at the Creative Roots of Science and Life.* New York: Bantam Books.

Chalmers, D. J (1995) Facing up to the problem of consciousness. *J. Conscious. Stud.,* **2**, 200–219.

Di Biase, F. (1981) Auto-organização nos sistemas biológicos. *Ciência e Cult.,* **33**, 1155–1159.

Di Biase, F. (2009a) Quantum-holographic informational consciousness. *NeuroQuantology,* **7**, 657–664.

Di Biase, F. (2009b) A holoinformational model of consciousness. *Quant. Biosyst.,* **3**, 207–220.

Di Biase, F. (2013) Quantum information self-organization and consciousness — a holoinformational model of consciousness. *Journal of Nonlocality Vol II, Nr 2.*

Di Biase, F. & Amoroso R. L. (2008) Holoinformational consciousness: An extension of interactive dualism with anticipatory parameters. In D. M. Dubois (ed) *Int. J. Comput. Anticipatory Syst.,* **22**. CHAOS, Liége, Belgium.

Di Biase, F. & Rocha M. S. (2000) Information, self-organization and *consciousness: Toward* a holoinformational theory of consciousness. In: Amoroso R. L., ed. *Science and the Primacy of Consciousness: Intimation of a 21st Century Revolution.* Oakland: Noetic Press.

Eccles, J. C. (1989) A unitary hypothesis of mind–brain interaction in the cerebral cortex, *Proc. R. Soc. Lond. B,* **240**, 433–451.

Eccles, J. C. (1993) Evolution of complexity of the brain with the emergence of consciousness. In: K. Pribram, ed. *Rethinking Neural Networks: Quantum Fields and Biological Data.* Manwah: Lawrence Erlbaum.

Eccles, J. C. (1986) Do mental events cause neural events analogously to the probability fields of quantum mechanics? *Proc. R. Soc. Lond. Biol.*, **227**, 411–428.

Eccles, J. C. & Beck, F. (1992) Quantum aspects of brain activity and the role of consciousness. *Proc. Natl. Acad. Sci. USA*, **89**, 11357–11361.

Hameroff, S. R. (1994) Quantum coherence in microtubules: A neural basis for emergent consciousness? *J. Conscious. Stud.*, **1**, 81–91.

Hameroff, S. R. & Penrose R. (1996) Orchestrated reduction of quantum coherence in brain microtubules: A model for consciousness. In: Hameroff, S. R. Kaszniak, A. W. & Scott, A. C. eds. *Toward a Science of Consciousness: The First Tucson Discussions and Debates*. Cambridge, MA: MIT Press Cambridge.

Ho, M-W., (2008) *The Rainbow and the Worm, The Physics of Organisms*, 3rd ed. Singapore: World Scientific.

Jantsch, E. (1980) *The Self-Organizing Universe*. New York: Pergamon Press.

Jibu, M. & Yasue, K., (1993) The basics of quantum brain dynamics. In: K. Pribram, ed. *Rethinking Neural Networks: Quantum Fields and Biological Data*. Manwah: Lawrence Erlbaum.

Jibu, M. & Yasue, K. (1995) Quantum brain dynamics and consciousness. In: *Advances in Consciousness Reasearch*. Amsterdam/Philadelphia: John Benjamins Publishing Company.

Lee, H., Cheng, Y. C. & Fleming, G. R. (2007) Coherence dynamics in photosynthesis: protein protection of excitonic coherence. *Science*, **316**, 1462–1465.

Matzke, D. (1994) Consciousness: a new computational paradigm, In: *Toward a Science of Consciousness*. Cambr., MA: MIT Press.

Pribram, K. (1997) *Languages of the Brain*. Monterey, CA: Wadsworth Publishing.

Pribram, K. (1991) *Brain and Perception: Holonomy and Structure in Figural Processing*. Hilsdale, N. J., Erlbaum.

Pribram, K. (2011a) *Além do Dualismo Cérebro-Mente*. In: Di Biase, F., Pribram, K. & Amoroso, R. eds. Fronteiras da Consciência, Homenagem ao Centenário de Sir John Eccles. Parana, Brazil: Editora CRV.

Pribram, K. (2011b) O Substrato Neural da Consciência. In: Di Biase, F., Pribram, K. & Amoroso, R. eds. Fronteiras *da Consciência, Homenagem ao Centenário de* Sir John Eccles. Curitiba, Paraná, Brazil: Editora CRV.

Prigogine, I. & Stengers, I. (1979) *La Nouvelle Alliance*. Paris, France: Editions Gallimard.

Prigogine, I. & Stengers I. (1988) Entre le Temps et L'Eternité. Paris, France: Fayard.

Rakovic, D. (2009) *Integrative Biophysics, Quantum Medicine, and Quantum-Holographic Informatics: Psychosomatic-Cognitive Implications*. Belgrade: IASC & IEPSP.

Ricciardi, L. M. & Umezawa, H. (1967) Brain and physics of many-body problems *Kybernetik*, **4**, 44–48.

Schrödinger, E. (1944) *What is Life?* Cambridge University Press.

Seager, W. (1995) Consciousness, information and panpsychism. *J. Conscious. Stud.*, **2**, 272–288.

Shannon, C. E. & Weaver, W. (1949) *The Mathematical Theory of Communication.* Urbana, University of Illinois Press.

Smolin, L. (2000) *Three Roads to Quantum Gravity.* London: Weidenfeld and Nicolson.

Stonier, T. (1990) *Information and the Internal Structure of the Universe.* Reading, MA: Springer Verlag, New Addison-Wesley.

Stonier, T. (1997) *Information and Meaning. An Evolutionary Perspective.* UK: Springer.

Stuart, C. I. J., Takahashi Y. & Umezawa H. (1978) On the stablity and non-local properities of memory. *J. Theor. Biol.*, **71**, 605–618.

Umezawa, H. (1993) *Advanced Field Theory.* New York: AIP Press.

Vedral, V. (2010) *Decoding Reality — the Universe as Quantum Information.* Oxford University Press.

Weber, R. (1982) The enfolding unfolding universe: A conversation with David Bohm. In: K. Wilber, ed. *The Holographic Paradigm.* New Science Library, Boulder CO.

Wheeler J. (1990) Information, physics, quantum: The search for links. In: Zurek, W. H. ed. Complexity, *Entropy and the Physics of Information.* Reading, MA: Addison-Wesley.

Zurek, W. H. (ed.) (1990) *Complexity, Entropy and the Physics of Information,* Santa Fé Institute, Studies in the Science of Complexity, Vol. 8. Redwood City, CA: Addison-Wesley.

Biophysics of Consciousness: A Foundational Approach
R. R. Poznanski, J. A. Tuszynski and T. E. Feinberg
Copyright © 2017 World Scientific, Singapore.

# 8

# NEURAL TRANSITION DYNAMICS AND CONSCIOUS PERCEPTIVE STATES

*Gustav Bernroider*

*Department of Ecology and Evolution*
*Neurosignaling & Neuroecology Unit*
*University of Salzburg, Hellbrunnerstr 34, A-5020 Salzburg, Austria*

## Abstract

I present a testable relationship between brain dynamics and conscious experience. The focus is on how transmembrane ion currents can become associated with distinct electron de-localizations and excitations within ion channel proteins. Conditional on a micro-psychistic premise, stating that if experience emerges from a particular physical process then this process must build on physical properties that are experiential themselves, specific electronic ion-filter states, amplified into the macroscopic domain of neural masses, can become causally related to conscious perceptive states (cPSs).

*Keywords*: Conscious perceptive states; brain dynamics; neural signals; ion channels; channel filter states; ion channel states; electron de-localizations; n-pi* electron transitions; dynamic dissipative self-assembly (DDSA); neural mass models; micro-psychism.

## 8.1. Introduction

The ability to experience is the essential part of our nature and at the center of all life forms. However, we know that experience comes along in two shades. With *conscious* experience, a fundamental and purely subjective ingredient, the way it feels to have this experience, is "added" to the process of experiencing and perceiving. It is this purely subjective ingredient that remains remarkably resistant to any causal relation between the physical architecture of the experiencing agent on one hand and the subjective phenomenon of this experience on the other. Much has been said about the principal limitation of any objectivistic science to get closer, or even to "explain" the essence and intimacy behind subjective experience. At this point it seems to be particularly timely to revisit the core of these arguments. Instead, stripped of the usual philosophical categorization (see for e.g., Chalmers (1995, 2003)), I intend to present a testable conceptual relationship between brain dynamics and conscious experience. An attempt such as this can be expected to provoke several objections in the face of an extremely rich debate on the relation of physics and subjectivity. The question about the inferential and causal properties of such a relation between these "two worlds" becomes particularly critical and challenging for the plausibility of the present approach. In addition, reflections of this kind ultimately lead us to the puzzling ontology of reality and nature.

I will initially set out with a discussion of the main motivation and general context of this chapter. What follows can be seen as a guide through the rest of the chapter, provided by a relational model that combines cause and effects and offers, what I consider a plausible route to conscious perceptive brain states. This route will gradually cross many physical action orders or dimensional scales, down to the quantum physical level. Finally it will end at a very particular and demanding type of dynamics, a *Prigogine–Nicolis type* of dissipative, dynamic self-assembly (DDSA). I will raise plausible arguments which identify ionic configurations within channel proteins as candidate structures that realize DDSA type phenomena. This is on the purely physical side. In order to make this concept "viable" and "experiential" I set a basically "micro-psychistic"premise, which suggests that there are physical phenomena from which experience emerges that are themselves experiential. I argue that DDSA type processes in ion channels may just provide this property.

## 8.2. Neural Context and Dualities

There are several obstacles that need to be addressed if the intention is to find a relation between the brain's physical properties and the phenomenon of experience. I want to provide some formal background initially and pay special attention to the relational structure between dynamic variables and observables in physics on one hand and their "meaning," i.e., those properties that are usually shifted into a phenomenological background of physics, characterizing "subjectivity." This discussion involves aspects from a general theory of causation as defined for example by the Structural Causal Modeling (SCM) as suggested by Pearl (2010) and builds on the formal notations introduced by Scheines (1997). One of the conclusions drawn from this section and important for the following parts is that we should clearly draw a demarcation line in neuro-physics between purely *associational* and *causal relations* among physical magnitudes. Whereas associational relations are statistical in nature, characterized by distribution functions, causalities necessitate premises and are solely within the jurisdiction of a cognitive agent and cannot be deduced from observational data alone.

The dualism behind the subject/object dichotomy at the level of the brain is reflected by this difference between purely physical properties behind the concept of *information* on one hand and the "*meaning*" of information on the other. Information processing is embodied in the physical organization of the brain at different levels and is basically a statistical conception. The relations it builds on are associational. A cognitive phenomenon is however quite different. An excellent demonstration of these dual aspects behind information was highlighted by the historical Chomsky–Shannon debate on the structure of language (for the present context, see for example the discussion provided by Bialek (2002). The argument of Chomsky (1956) being that language is "more than just statistics" as suggested by Shannon (1949). The "meaning" behind a message is unrelated to the frequency of occurrence of words, as could be expected from the physical nature behind information. However, there is one physical concept that can shed light on this difference: *Energy*!

Energy is central in physics and perhaps the most critical conception in the structure-function relation in brain science. Whenever energy becomes embedded into statistical physics, it is interwoven with the concepts behind *information* and *entropy*. Applied to the brain, a recent reflection of this situation is the attempt of Friston "to explain" the brain by a "free-energy principle," an information theory quantity that measures the discrepancy between the statistical nature behind observations and the "expectations" provided by the organization of a perceptive brain (Friston, 2009). While I can agree with the author's claim that this principle can help to shed some light on a substantial number of adaptive and possibly self-assembling properties in the brain of a perceiving agent, I cannot agree with his claim to "explain perception." Friston's approach misses the "meaning," the essence behind subjective experience and is completely located within the "associational structure" of the brain. Within the picture of the Chomsky–Shannon debate on language, the free energy principle could be put upon Shannon's side. What is missing is the essence behind brain function: a plausible relation to consciousness and conscious perceptive states (cPSs).

In the subsequent chapter on brain dynamics I will further argue that the "free energy principle" is at variance with the kind of thermodynamics which most plausibly underlies perception. The metastable states that the free energy principle suggests do play a role but can be seen to characterize the statistical (i.e., associational) structure of the effective (functional synaptic) network associated with perception, rather than a decisive and causal relation to the experience behind perception itself. In short, the free energy principle describes a process characteristic that is necessary but not sufficient to explain a cPS. Instead the viable level of experience and life are more likely to reside in an even more demanding type of dynamics, a *Prigogine–Nicolis type* of dynamic self- assembly which is characterized by a continuous energy supply and dissipation at very small scales (Nicolis & Prigogine, 1989).

This will lead us to the core of the present discussion in Sec. 8.5 on dynamic self-assembly at the quantum scale. I suggest that the organization of DDSA and non-equilibrium processes across excitable membranes may just provide a key element to explain a main paradox in brain science: The striking and puzzling many to one relation between structure and function in the brain, as Park & Friston (2013) have outlined very convincingly. DDSA features the creation of a manifold of assembled state configurations emerging from single process dynamics. A remarkable complexity can result from one and the same organization. A set of physical and biological arguments address ionic configurations within channel proteins as candidate structures which meet these criteria as previously suggested by Summhammer *et al.* (2012) and Bernroider & Summhammer (2012). As the signaling membranes atomic configuration is forcefully kept away from energy minima by an energy supply-dissipation pair, ordered nanoscaled structures emerge reproducibly, which can account for the richness of neural function based on a single structure. I conclude that at least a part of the dialectic between the brain's physical state and the brain's phenomenal property behind conscious experience, may reside in the elaboration of DDSA type quantum scale dynamics in ion channels.

Finally, building on the previous Secs. 8.3–8.5, I argue that the relation between both domains, the physical and the phenomenal, can in fact be very tight. Building on some premises, the associational relation between the physical descriptions of the brain with the perceptive phenomenology behind brain functions can become causal. The building blocks involve two aspects: Premises, and a rational interpretation of the nature of involved transitions, the relations among the constituent variables that should be "explained." The premises proposed must in some way oppose "strong and orthodox physicalism," because they grant some basic experiential properties to physical phenomena. This view could be put into the context of *micro-psychism*, a moderate and less radical version of panpsychism, as proposed by Strawson (2006) and very competently analyzed by Simons (2006). Although, quite often labeled as "counterintuitive," I advocate the view that micropsychisms are within the realm of physics, particularly quantum physics.

Physics hosts much more than the reductive orthodox view usually offers. Properties in physics are emergent properties and this is particularly striking in quantum physics. There properties can be gained or lost and they can be stripped off their carriers as, for a recent example, is impressively demonstrated by a special set of interventions called *"fractionalization,"* where new states of matter emerge, leaving behind those properties (charge, spin, statistics) that characterize their constituents as demonstrated by Barkeshli *et al.* (2014). The question of micro-psychistic emergence will therefore become a central issue in the final Sec. 8.6. In fact, "property emergence" is a fundamental part of physics with a decisive role for the present essay. This final part adds the spices to an otherwise expensive but taste-less soup.

## 8.3. Brain Relations

How can the physical description of the brain be related to the phenomena of consciousness? Can we at least capture a relational structure that eventually may provide a template for cause-effect analysis in the study of consciousness? The enormous fragmentation of brain science into different disciplines and technical specializations has become a frequent stone of contention in the study of higher level brain function and, particularly, the phenomenon of consciousness.

As I mentioned in Sec. 8.1, we need to combine two rather opposing inferential strategies in order to proceed with our understanding along this line: Causal relations building on some assumptions, conjectures or premises and associational relations building on joint probability distributions of the observed variables. The strategy I want to use here allows a combination of both types of inferences and is based on the SCM theory as developed in Pearl (2010) and Scheines (1997).

In this notation, the set $\{X_1 \rightarrow X_2, X_2 \rightarrow X_3\}$ with two pairs of vertices (variables) causally related by edges (arrows, transients) can be written as a Directed Acyclic Graph (DAG): $X_1 \rightarrow X_2 \rightarrow X_3$. In this case the vertex $X_2$ is said to block or *d*-separate the causal connection from X1 to $X_3$. The way to relate a causal *d*-separated DAG to a probability distribution is to say that the *d*-separated variables $X_1$ and $X_3$ are independent conditional on $X_2$. Using the symbol $\perp$ for independence, one may express this as $\{X_1 \perp X_3 \mid X_2\}$. In Fig. 1, two examples for possible causal relations among the brain's physical state, a "cPS" and a "report state" (Rs) are shown.

The two examples of causal relations in Fig. 1 demonstrate the indispensability of premises and assumptions to set up relations combining cause and effect in the brain. Whereas in the first line of Fig. 1 it is assumed that a kind of cPS reflecting awareness, results from a particular brain state and must causally precede a report about this state (e.g., a report through language), the situation is the other way around in the second line. There, some kind of motory Rs results from a brain state and causes a cPS.

| (A) | (B) |
|---|---|
| (1) Bs → cPS → Rs | { Bs ⊥ Rs \| cPS } |
| (2) Bs → Rs → cPS | { Bs ⊥ cPS \| Rs } |

**Fig. 1.** (A) "D-separated" causal graphs (DAGs) relating 3 vertices (variables) by two edges (connectors, "→") and (b) the resulting set of independencies. On the right side (B) we use the symbol "⊥" to indicate "independence" in the probabilistic sense, e.g., the notion { Bs ⊥ Rs | cPS} means: Bs and Rs are independent conditional on cPS. In the example above Bs could encode a "Brain state, a cPS and Rs." The independence terms result from the "blocking" action of interjacent variables, cPS in line (1) and Rs in line (2), in the corresponding DAGs (based on the notation used by Scheines (1997) and Pearl (2010)). Note, the two lines (1, 2) offer different causal relationships between the brain's physical state (Bs), a cPS and a Rs (see main text).

However, both causal assumptions are principally tenable in the view of an enormous literature dealing with the neuro-functional and neuropsychological basis of cognition and emotion. If, for the moment, we insert "phenomenal knowledge" or "conscious representation" e.g., in the sense of Prinz (2007) for a cPS and interpret the umbrella term "Rs" as a state of some bodily changes or motor execution pattern which are associated with a cPS, then the d-separated DAGs in Fig. 1 dissociates a widely agreed difference between cognitive (line 1) and emotional (line 2) processes. Both processing domains, cognition and emotions, are considered as types of perceptive phenomena (see Prinz (2005, 2007); Panksepp (2005)). However, the relational structure behind cognition suggests that the brain state is conditionally independent from a Rs, the condition being a "cPS," whereas with emotions, the brain states are conditionally independent from a "cPS," the condition being a Rs. This reflects a "Jamesian interpretation" of emotions, as bodily changes preceding the experience of emotions (James, 1884) or in the view of LeDoux, as "body routed survival mechanisms" that are conditional for the "experience of emotions," i.e., "feelings" (LeDoux, 1996). Phrased more simply, the DAGs in Fig. 1 imply, that we can 'think without reporting' and we can have "emotions without feelings."

Arguably, I am oversimplifying a highly complex issue here. The intention is however, to demonstrate that SCM can substantially help to discern core relations among the argumentative structure of mind–brain studies. In particular, we can dissociate causal from associative relations and identify premises and conclusions together with their ontological claims and we can put this all into a precise formal language. In Fig. 2, I presents a core structure for what I believe is essential to connect the brain's physical properties to the phenomena of subjectivity and consciousness, a question central to this edition. The structural model demonstrated in this figure relates brain properties to each other. Put into kinship relations neural development for example is considered as a parent process that gives rise to the daughter "geometrical connectivity" in the brain. This detailed path is diagrammatically outlined as a structural equation model (SEM) in Fig. 3 and relations (1, 2).

In Fig. 3, the external factors, U(Cg) and U(Dev), which influence the specific property in question, are symbolized by dashed lines. The SEM associated with this diagram can be written as:

$$\text{Dev} = \text{U (Dev)} \tag{1}$$

$$\text{Cg} = \beta.\text{Dev} + \text{U (Cg)} \tag{2}$$

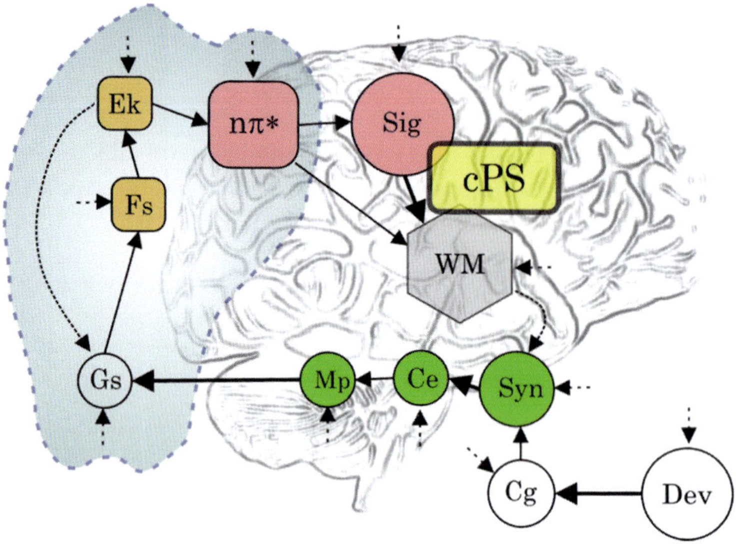


**Fig. 2.** SCM of brain states and their directed relations to conscious perception (cPS) as proposed in this chapter. The nodes encode: Cg (geometrical connectoms), Ce (effective connectoms), Dev (development), Ek (ion kinetics), Fs (channel filter states), Gs (channel gating states), np (electronic delocalization), Mp (membrane potential, Sig (Signal spread), Syn (Synaptomics), Wm ("working memory"). The blue shaded area on the left side symbolizes a single ion channel protein, hosting the indicated variables and transitions. All other variables map throughout the macroscopical brain. The brown colored nodes represent processes at the quantum chemical scale within the filter region of an ion channel protein. The bold arrow symbolizes the transition from (Sig) to (Wm) and is proposed to be accompanied by conscious experience (cPS). The size of the nodes reflects the "relative contribution" of causes to effects, the length of directed edges (arrows) symbolize the amount of intervenient variables separating causes and effects not contained in the model, the arrow strength reflects the "path coefficient" that quantifies the causal effect between successive nodes.

and the path coefficient $\beta$ relating the two variables, as the covariance

$$\text{Cov (Dev, Cg)} = \beta. \tag{3}$$

Linearity is simply assumed for formalizing the path relation and does not necessarily reflect the type of relation between the two vertices. The claim behind the "sub-DAG" in Fig. 3 is a causal relation between neurogenetic processes (Dev) and a structural connectom (Cg). In other words,

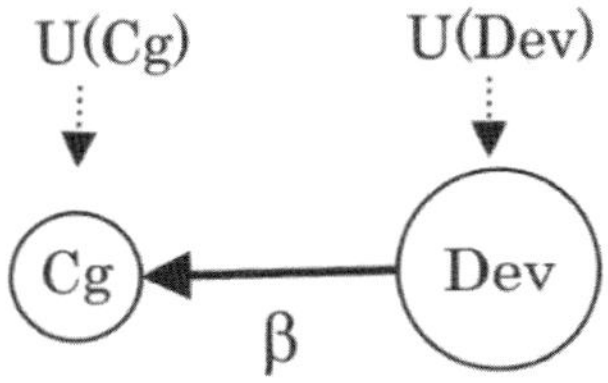


**Fig. 3.** A section through the path depicted in Fig. 2 showing a causal DAG, relating "Development" (Dev) to geometrical connectoms (Cg) in the brain, with (unobserved) influences or errors denoted as U(Cg) and U(Dev). The symbol "$\beta$" is used for a path-coefficient that quantifies the effect of "Dev" on "Cg," employing the covariance measure Cov (Dev, Cg) = $\beta$. The resulting SEM is discussed in the main text.

development causes the alignment of neural morphologies into a skeleton of structural or geometric connections, forming for example the geometry behind cortical microcircuitries. Again, this simple causal hypothesis involves a large number of highly debated aspects about the role of chemospecific mechanisms during development (Zipursky & Sanes, 2010). The important question, for example, to what extent connectivities among cells are random is implicitly contained in the question to what extent the external variables Ui differ from residual terms, i.e. are uncorrelated with the regressors of the model. For example, Song *et al.* (2005) and Stepanyants & Chklovskii (2005) suggest that "highly non-random features of synaptic connectivity" are embedded within a "sea of weaker ones."

A further look along the causal path depicted in Fig. 2 unravels the problem more clearly. There we make the useful distinction between geometric connectivities (Cg), i.e., morphological appositions among cells, and effective or functional connectivities (Ce), based on putative synaptic contacts and measurable transmissions. Coined into the language of causal structural modeling, both types of connectoms are d-separated (or blocked) by the formation of synaptic transmission (Syn), i.e., forming the d-separated DAG:

$$\text{Cg} \rightarrow \text{Syn} \rightarrow \text{Ce and the independencies } \{\, \text{Cg} \perp \text{Ce} \mid \text{Syn}\}. \qquad (4)$$

The model (4) states that geometrical connectivity, resulting from development (Dev) is an indirect cause of efficient connectivity. Both variables, Cg and Ce will correlate and the amount of correlation is at the center of interest for attempts to employ arbor morphologies to study neural functionality, for example as intended within the Blue Brain Project at Lausanne (see for example News & Analysis, Science, 334, pp. 748–749, (2011)). This group reports that (Cg) can predict an average of 74% of (Ce) properties, as measured by synaptic locations within nine types of cortical connections (Hill *et al.*, 2012). Yet, phrased more commonly, the reason why neural morphologies reflect about 3/4 rds of the brain's specific signaling structure is due to the developmental condition, that not all close approaches of neural arbors actually form functional synapses among these cells.

I have discussed the examples of property relations above mainly for demonstrative purposes. Naturally, the route to consciousness is much more complex and will be further outlined in the following chapters. The intention here is to introduce a way to clearly define the problem landscape in functional brain studies and to show how causal and probabilistic approaches can be connected, at least in principle. This of course goes much further and the reader is advised to visit Scheines (1997), Spirtes *et al.* (2000) and Pearl (2010) for a deeper involvement into causal inferences.

There are several restrictive issues regarding the above model, which need to be mentioned at this stage. First, the way that d-separated DAGs relate to probabilities is given by a (causal) *Markov condition*, i.e., only a systems present state is relevant for its future state or as Scheines (1997) puts it "a variable is independent of every other variable conditional on all of its direct causes." However, the DAG relations discussed above are unidirectional sections through much more extended paths as shown in Fig. 2. In a broader context, the 'parents' in relations (2) and (4) are true descendants and we largely ignore a most common motive in biological relations, *reciprocal causation*. Opposed to the more traditional unidirectional causation, biological reasoning now tends to involve reciprocal arguments. For example in evolutionary models it can be seen that a selected trait can evolve together with the source of selection, giving rise to reciprocal dependencies (Laland *et al.*, 2011). However, empirical studies must usually "cut out" a single directed sequence of cause and effects, so that based on the causal Markov condition, estimated probabilities can be related to causal influences. The DAG in relation (4), combining geometric connectoms to efficient connectoms under the condition of synaptic contacts gives a nice demonstration of this strategy.

Generally, the study of cause and effect within experimental approaches means to study the response to manipulations and this in turn means to dissect "Markovian relations" from a history of reciprocal causal influences. Yet, from another view, it is mainly due to this empirical restriction why many neuroscientists and neuro-philosophers are calling for new grounds to proceed with our understanding of a basically "holistic brain." Modeling cause and effect, together with theory and simulation might be able to reconcile these different perspectives gradually.

### 8.4. Neurodynamics at Different Scales

Let us now proceed with a discussion of the neurophysical organization behind the cause and effect relations suggested in Fig. 2. Such a discussion is extremely challenging in the view of several, very unique properties behind the brains organization and its function. Karl Friston came straight to the point in addressing the enormous functional divergence from structure as the "many to one" paradox in brain science (Park & Friston, 2013). Other highly demanding features seem to be related to this unique aspect behind structure function relations in the brain. For example, due to its extensive connectivity at the cellular scale, the functional volume of even a single cell can occupy a substantial part of the total organ volume, making the functional volume of a single cell about the same as the mass–volume of the entire brain (Lichtman & Denk, 2011). An additional and highly susceptive point is also reminiscent of this unique situation: Many functional disorders do not really precipitate in the histology of the brain!

How can we explain the many-to-one relation of function and structure? To cut it short: In my view this relation cannot be resolved at a single physical scale, i.e., the scale of cells and their connectoms. The term "scale" within the present context is understood as a dimensional attribute of true physical action order. It goes far beyond the cellular organization, which itself "scales" into hierarchical mosaics of macro- and microcircuits. The range of physical scales, setting out from the entire brain down to the molecular and sub-molecular level, involves about 10 orders of magnitude of spatial scales and 30 orders of physical, i.e., *Lagrange type* actions (Bernroider & Roy, 2004, 2005), see Fig. 4. The claim that I want to make here is, that the physical representatives established at these different scales, i.e., atoms, molecules, cells and the entire organ, have their own and indispensable functional significance. This significance goes far beyond the traditionally accepted bottom-up (e.g., from ion channels) and top–down (from neuronal ensembles) contribution of properties to establish the cellular signaling level as is widely held within

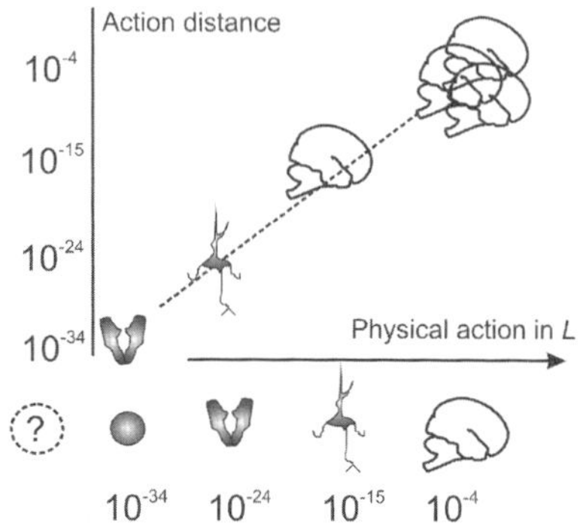


**Fig. 4.** Physical action orders and their biological realizations as a predictor of quantum-physical action distance in the brain (modified after Bernroider & Roy (2004)). The diagram sets out at the single atomic level and extends to the entire brain. Both variables, predictor and response, are explained in the main text (Sec. 8.5).

neuronal doctrine based approaches and the "brain as a computer" metaphor. Instead the brain seems to preserve and integrate the loss and gain of mutually incompatible attributes over many physical dimensions. This conjecture of course requires further explanations.

In the many- to-one relation, the fixed architecture, i.e., the efficient connectom of the brain at a given time (excluding the effect of (Dev) in Fig. 2) represents the "one." The "many" refers to the richness of neural function, the huge functional repertoire, operationally available 'in real time', at any given moment of perception and cognition. This manifold must reside in the dynamical degrees of freedom hosted by this structure. Therefore process dynamics is more likely to offer the key to the phenotypic qualities of the brain as opposed to the static geometry of the brain. This is not to say that we cannot learn important aspects from the brain's architecture or that this architecture is not tightly bound to its function. By comparison, we can also learn much about an organism's life history, its somatic appearance, body-posture and movement, foraging strategies, social and reproductive behavior, by a careful study of its skeleton. However, when it comes to individualizing the skeleton up to a unique and living organism, the correlations turn out to be unresolvable in the view of a causal structure.

Whereas the importance of integrating the characteristics of mass-velocities of physical entities (the bottom up models) over different scales has its tradition ("from chemistry to biology"), it is less obvious that together with masses also their characteristic time and energy changes. It is the change of these two, mutually incompatible observables time and energy which characterizes dynamics. The link between dynamics and the functional pattern that is connected with it is the process of *self-assembly* (Nicolis & Prigogine, 1989; Hermans *et al.*, 2013), where parts of the system arrange into ordered structures over different time scales. Whereas the structural composites which "carry" or organize an organism such as tissues and cells may equilibrate over a certain time scale (ranging from days to life-times) "life" is enabled by assemblies that are forcefully kept away from equilibration by the cooperation of energy "supply" and "dissipation." Now it can be found that the thermodynamical landscape, provided by free-energy gradients and their characteristic time intervals, changes conspicuously as we cross the organizational levels from cells to molecules and atoms. In Fig. 5, these changes are sketched with respect to the brain's functional organization and are demonstrated along 30 action orders expressed as energy-time products.

Along the path of relations discussed in Sec. 3 above, we can discern basically three types of dynamics as we proceed from (Ce) into the domain of a single molecule (e.g., an ion channel protein), i.e., along the path $\{(\text{Ce}) \rightarrow (\text{Mp}) \rightarrow (\text{Gs}) \rightarrow (\text{Fs}) \rightarrow (\text{Ek}) \rightarrow (\text{np}^*)\}$.

First, we find "partial equilibration" reflecting the more or less static connectom at the level of cells and their connections (E, Fig. 5) and transient intermediates (E, D), forms of static self-assembly which are used in modeling neural networks. Next, as we enter the molecular range (D, C,) we find transiently trapped states with even more shallow energy wells characteristic for "static self-assembly" and finally, at the atomic and electronic scale (B, A), we end up at "dynamic self-assembly" characterized by no particular energy minima. Also available from Fig. 5 is the observation, that the different thermodynamic patterns can be seen to be basically self-contained and hierarchically organized into a top-down principle. As we apply ever increasing higher resolutions to spatial scales, we finally arrive at DDSA, where a characteristic structure forms when the system is kept away from equilibration by a continuous energy supply and dissipation pair (A, Fig. 5). Besides biological realizations, such dynamic behavior has also been reported more recently with synthetics, such as ferro-fluids on super-hydrophobic surfaces (Timonen *et al.*, 2013).

Energy dependent electron delocalization, which characterizes the interplay between electrons and atoms, is among the candidate processes with DDSA type dynamics in proteins. This phenomenon is at the center

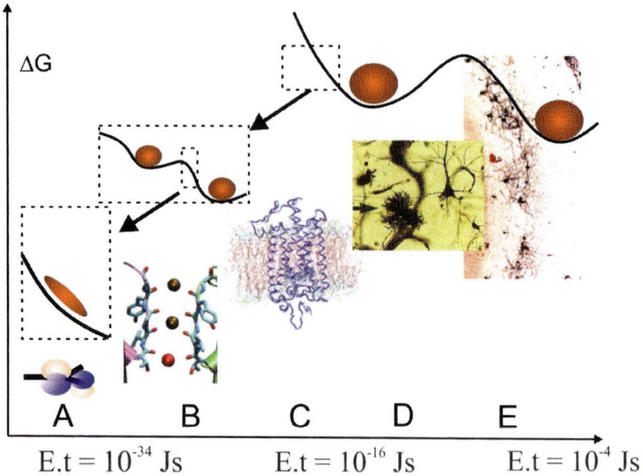


**Fig. 5.** The characteristic change in dynamics across 30 physical action orders in the brain (symbolized by their Gibbs free energy curves). Process characteristics are shown above and the associated physical arrangements are shown below the curves, starting from cell ensembles (E), single cell level (D), ion channel molecules (C), molecular subdomains (e.g., channel filter, B) down to electronic delocalizations in the atomic environment of the filter region (A). Note: The change in dynamics can be seen to depend on an increasing resolution of free energies starting from "near equilibration" (E, D) to transiently trapped processes (C, B), down to dynamic dissipative dynamics (A) (stippled boxes).

of my proposal for a physical connection to consciousness. Electrostatic interactions have long been recognized to be central in the living organism, in particular with respect to the function of proteins (Krishtalik *et al.*, 1997). These highly polar but usually low-dielectric molecules gain their response to external electric fields mainly by their integration as ion-conducting and "voltage-gated" structures within the cells signaling plasma membrane. In the following section, we will describe some electronic transitions that are generally essential for protein structure and particularly important for the association of electrical membrane signals with emergence at the quantum scale.

### 8.5. Dynamics at the Molecular and Atomic Scale

A decisive step during evolution has led to the endo-symbiotic internalization of energy transducing membranes characterizing prokaryotic cells into the mitochondrial membranes of eukaryotic cells. This step has rendered the cell's plasma membrane available for the organization of membrane signaling in excitable cells. Nerve cells have organized the propagation of voltage changes as cell signals by the cooperative organization of charge conducting proteins, i.e., ion channels, across their membranes. With atomic resolution studies of the engaged proteins (e.g., see mini review by MacKinnon's (2003)), it has now become apparent, that the overall organization of signaling events along nerve cells can be expected to comprise physical events established across at least 30 action orders (Fig. 4) (Bernroider & Roy, 2004, 2005). Naturally these events are difficult to reconcile into a coherent model. However, it has become clear that ion conducting, voltage gated membrane proteins occupy a key role and the structural dynamics of these molecules can now be characterized at the atomic level.

**Ion channels**: The building blocks for neural signaling are ion channels. They generate and regulate the concerted translocation of charges bound to ions across the cells plasma membrane and their physical properties are tightly and causally related to the established electrical signaling pattern of the brain, i.e., to the sequence of action potentials along axons. Thus ion channels are at the core of a causal relation between the proteins transition dynamics and the established pattern of signaling at the cellular level. They deserve particular attention in an attempt to explain the relation of the brain's physical structure with its functional phenomenology.

A channel protein presents a remarkable machinery (Fig. 6). It connects ion selectivity (for $K^+$ channels by a factor of $10^4$ over $Na^+$), with diffusion limited ion permeation, of up to $10^8$ ions.$s^{-1}$. After the determination of the molecular structure at atomic resolutions of the bacterial KcsA model channel by MacKinnon's group, an enormous number of reports involving diffusion models, Brownian dynamics, classical and ab initio molecular dynamics followed (for an early review see Kuyucak *et al.* (2001)).

In sum it is found that the proteins functional repertoire resides in basically two sets of properties: The first aspect is given by conformational changes, regulating the access of ions to the protein's pore region by "primary" or "cytoplasmatic" gating ((Gs) in Fig. 2) and regulating the states of conduction in the so-called selectivity filter of the channel ((Fs) in Fig. 2). The second aspect involves electrostatic interactions among the channel atoms and conduction ions. A tight cooperation among both aspects is generally assumed to account for ion selectivity within the charged carbonyl lining of the selectivity filter (C in Fig. 6).

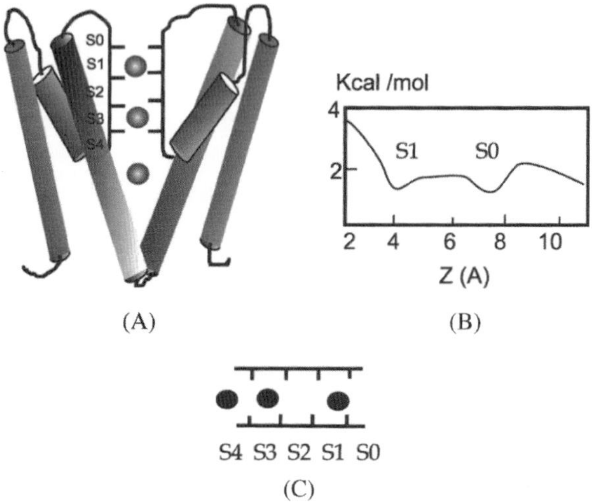


**Fig. 6.** A schematic section through the channel and selectivity filter of a model KcsA channel (A). In (B) a free energy profile along the z-coordinates of the filter region and in (C) ion locations along five binding sites (S0–S4) within the filter are shown (after Summhammer *et al.* (2012)).

A more detailed discussion of this highly complex situation is beyond the present intention. I will restrict the intention to one, but largely overlooked aspect behind selective ion conduction in nerve cells, which is essential for the present intention: Because channel proteins are responsible for the generation of electrical membrane signals and the maintenance of the membranes resting potentials by their charge-conductive properties, the focus has been traditionally rested on their role for ion translocation. What has been overseen is the effect of ion occupancy on the "hosting" protein itself, on the permanent ingredient of the signaling membrane. If we assume a characteristic open time of about 1 ms during an action potentials de- or repolarization for a "space-clamped" channel and an average firing rate between 10 and 100 spikes/s in cortical neurons, a single channel will reside in the closed conformation most of the time. The question becomes pressing whether there is some functional role for the protein besides regulating ion currents. Here, I propose that changes in the conductive states of the channel will also change the electronic configuration of the linings of the filter region in proteins and this leaves behind a distinct electronic signature with functional significance.

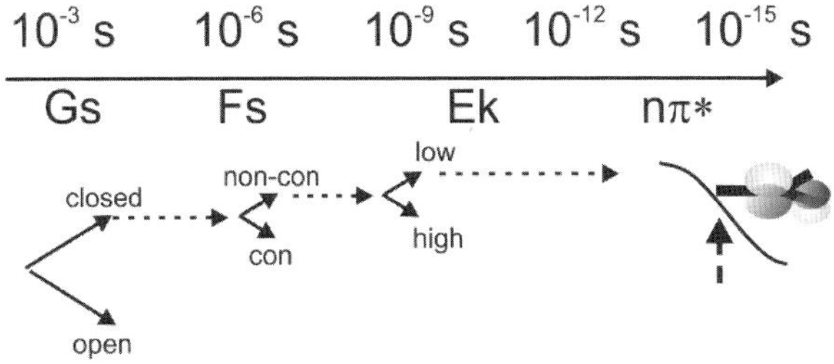


**Fig. 7.** Time scales (top) in seconds, events (below) and their associated state alternations within one channel molecule. Gs, pore or primary gates (open or closed), Fs, filter states (conductive or non-conductive) and Ek, filter occupancies. It is proposed that only the path from (closed → non-con → low) leads to the transition nπ* and DDSA type dynamics.

**Channel dynamics**: Across functional domains the channel operates on very different time scales with different alternating states of molecular sub-domains (Fig. 7).

Primary or cytoplasmatic gating occurs in the range of ms. Filter states can alter within ms to $\mu$s (e.g., about 140 $\mu$s in the Berneche & Roux model (2005)). Thus filter states may change many times during an open or closed main gate state. The non-conducting filter state in turn is associated with a low concentration of ions (Morais-Cabral *et al.*, 2001), which change position between the S0 to S4 location many times during this interval. Transitions e.g., from S1 → S2 occur in the range of 60–100 ps (along $10^{-10}$ s) (Guidoni & Carloni, 2002). Finally, the path depicted in Fig. 7 ends with state transitions at the ps and fs scale and electronic transitions that will be discussed below.

Within the atomic scale charge polarization and charge transfer in the KcsA channel model have been studied in great details (Guidoni & Carloni, 2002; Bucher *et al.*, 2006; Kariev *et al.*, 2007). These studies show that the electronic structure in the filter region of channel proteins changes considerably, depending on the ion-occupancy and their translocation between the five binding sites in the filter. These binding sites are essentially organized by a dynamic potential landscape (Fig. 6B) derived from the negative charges of carbonyl bound oxygen and positive charged ions together with effects from the overlaid gradient of the transmembrane electric potential. The basic electrostatic interaction between ions and the surrounding C=O dipoles is the Coulomb potential energy. The potential landscape is of the type *"static self-assembly"* with transiently trapped states (type B in Fig. 2) and trap sizes of up to 50 kT, i.e., in the range of 1–2 eV. It should be noted here that the forces at this scale are Coulombic in nature but are due to quantum mechanical effects (Kuyucak, 2001; Summhammer *et al.*, 2012). This brings me to the main point of my proposal, to the changes induced in the electronic structure of ion hosting channel proteins during the regulation of charge conduction in ion channels.

**Electronic delocalization in channel proteins**: We focus on the causal path {Fs → Ek → nπ*} as shown by the relational structure in Fig. 2. The DAG is d-separated by (Ek), by the cation interaction with the negative potential provided by carbonyl-oxygen in the filter region of ion channels along the conserved TVGYG amino acid sequence which lines the filter. The path suggests that filter states (Fs) will change ion-interactions (Ek) and the latter will cause electronic transitions in coordinating oxygen surroundings. Along this conjecture, filter states are indirect causes of electronic de-localizations, conditionally independent from electronic states, but can be expected to correlate. What is the evidence and what is the consequence?

First, electronic de-localizations as a consequence of non-covalent binding forces are well established. The cation-π interaction for example, plays an important *functional role* in ligand recognition during neurotransmission in the brain (Zacharias & Dougherty, 2002). If a particular electronic transition in channel proteins should be (causally) related to the brain's signaling structure, we would expect a tight relation of these electronic signatures with the electric conductivity of voltage-gated ion channels. Strong supportive empirical evidence stems from observations about the correlation of the (n → π*) electronic transitions in carbonyl groups with the "loading state," the low or high concentration of ions in the selectivity filter of the $K^+$ channel (Bartlett *et al.*, 2010). The (n → π*) transition in carbonyls delocalizes a non-bonding lone pair of electrons of the oxygen from the p-rich lone pair ($n_p$) to the higher energy level of an antibonding (π*) orbital (Fig. 8) (Atkins & de Paula, 2006). This type of electronic charge-transfer transition is abundant in proteins and thought to contribute to the proteins stability in general.

The observation that carbonyl (n → π*) transitions dissociate $K^+$ channel filter states stems from systematic observations of bond orbital properties reflected by the Ramachandran plot (where every measurement presents a value of a pair of bond angles in polypeptide chains, e.g., in α-helices). The authors found a run of four (n → π*) interactions in the

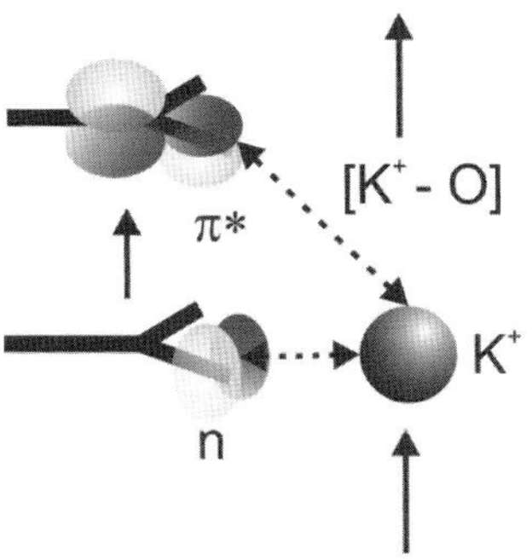


**Fig. 8.** (n → π*) transition in a carbonyl group induced by moving ions (arrows) and the associated changes in carbonyl coordination distance ($K^+$–O).

filters low concentration state vs. almost no such transitions in the high concentration state (i.e., the ion conductive state) (Bartlett *et al.*, 2010). The present claim builds on the causal relation {Fs → Ek → nπ*}, where the filter states are indirect causes of np*transitions and are expected to correlate. The conditional independence among these two variables is implied by the ion's motional degrees of freedom within the filter (Ek). Evidence for a causal effect of ionic motional energies on the above transitions comes from our previous quantum-mechanical description of ion motion within the confining potential of the filter region (Summhammer *et al.*, 2012).

Two aspects are important at this point. First, the interaction forces between protein bound charges and the charges of moving ions within the protein will drop along ($1/r^2$) with "$r$" denoting the charge separation distance. Charge induced polarizations naturally depend on interaction distances among the charges (Fig. 8). The second aspect is the motional coupling between C=O groups and ions including thermal oscillations. The latter phenomenon can be expected to extract kinetic energy from the ion, i.e., exert a 'cooling effect' in the proteins filter region. The sink of this energy and its reservoir is likely to provide the energy for coupling the filter states with the pore gate of the channel through the so-called 'p-loop' of the filter region. In our previous work we show that oscillation frequencies of carbonyls can discriminate kinetic energy minima of ions, in particular dissociate homo-specific ions from hetero-specific ion replacements (e.g., $Na^+$ in a $K^+$ filter). The temporal evolution of kinetic energies within 9 ps is demonstrated in Fig. 9 next page (from Summhammer *et al.* (2012)). There we find that only at "efficient cooling

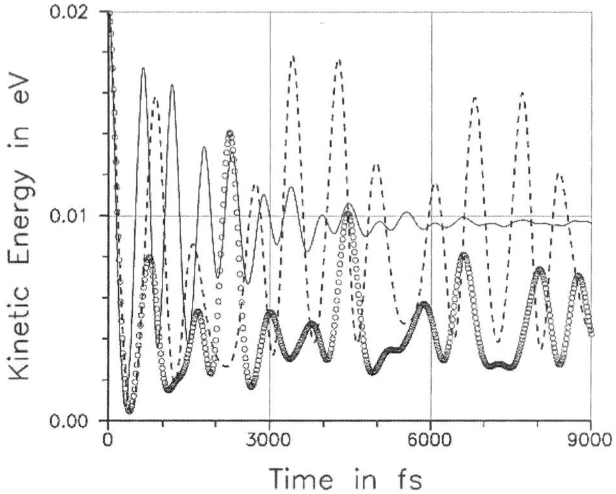


**Fig. 9.** Kinetic energy evolution of $K^+$ ions for three different C=O oscillations.
*Note*: Only for the lower curve (circles), which shows oscillations at "cooling-frequencies" of carbonyl oxygens (446 GHz for $K^+$ ions), the potential energies are not returned to the kinetic part during the next oscillation period (from Summhammer *et al.* (2012)).

frequencies" (446 GHz for $K^+$ ions) the potential energy is not returned to the kinetic part during subsequent oscillation periods of the ion. One can expect that these energies become invested in the electronic transitions along the C=O bonds as discussed above.

The dynamics that characterizes energy supply and dissipation during electronic transitions, e.g., from lone pairs to conjugated systems as suggested here, is most likely from the DDSA type. This type of dynamics allows self-assembly of structures when the system is kept off an energy minimum by a continuous or oscillatory energy supply. The kinetic energy profile of ions in the low concentration state of the channel can just provide this supply as the ions bounce between long-distance and short-distance interactions with the C=O lone pairs (Fig. 8). During the "close approach" the Coulomb force will attract lone pairs, whereas during the transition into a long-distance coordination the forces can relax and the gained energy induce the conjugated state on the C=O bond.

Taken together the above section discloses a highly plausible and evidence-based relation between the signaling characteristic of the brain, provided by channel-protein controlled ion conduction on one hand, and electronic signatures within these proteins on the other. Ion conduction leaves behind physical traces on the permanent ingredient of the signaling membrane, the ion-channel protein. This turns channel proteins into a unique zone of functional convergence, combining for example neural mass and neural field properties, a distinction which was recently suggested by Pinotsis *et al.* (2014). Information based signaling along a fixed structural connectivity (the "one" in Friston's sense) becomes connected with a variety of electronic assemblies at the quantum scale (the "many" in Friston's sense) by the temporal changes of membrane polarizations! *But why and how can this all become (causally) related to an experiential state of the brain?*

## 8.6. Closing the Loop

An answer to the above question will naturally present the more speculative part of this chapter. First, I want to close the chain of arguments as outlined in Fig. 2 and come to the essence of the final motif that connects physical states with the phenomenon of a "cP" in Fig. 10). The graph in Fig. 10 connects electronic transitions as introduced above ($n\pi^*$ for example) to a process property which I call SIG and "working memory" (wm), an umbrella term for an early stage of maintaining and manipulating information (Prinz, 2007). In the theory I endorse the transition from SIG to (wm) is accompanied by conscious experience.

The proposed connection builds on knowledge rather than associative information. It cannot be deduced from data alone. This situation is also reflected by the graphs topology: A d-separation or causal Markov condition applied to the relation in Fig. 10 does not provide the set of conditional independencies that we would expect from a causal DAG (see Chapter 3). The reason for this lays in the circularity of its connections.

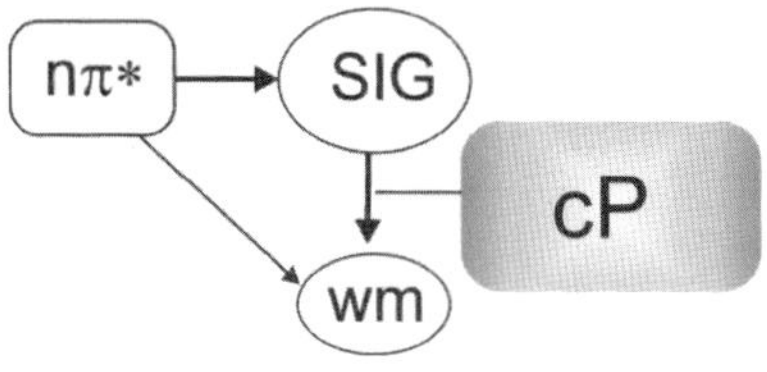


**Fig. 10.** The "final motif" demonstrating the relation of physics in ion channels ($n\pi^*$) to working memory (wm) through a process characteristic called SIG (discussed in the text). It is proposed that the transition from SIG to (wm) is associated with conscious experience.

Without independencies it is difficult to relate probability distributions to causal claims. Scheines would call this relation "unfaithful" to the causal graph that is generated (Scheines, 1997). However, there are ways to make this relation more "faithful." If for example, we can exclude coincidence of a balance between negative and positive effects along the directed route in this graph and the decisive property (cP) is attached to only one path, to the conditional path separated by SIG to wm, the relation becomes more faithful and testable.

**SIG**: The vertex "SIG" in Fig. 10 symbolizes a particular process characteristic that we have implemented about 10 years ago (Bernroider *et al.*, 1996). However, to explain its details goes beyond the scope of the present article. In sum SIG amplifies and focuses neural signals along basically unknown routes to single time events. It generates a dynamical state space and follows the path of wave-like signals changing the phase of signals during propagation. The dynamical structure behind SIG focuses signals into one point and the propagation of signals is modeled similar to the solution of path integrals in quantum mechanics. One can apply a "lens metaphor" for light propagation in a type of quantum-electrodynamical realization to capture the effect of SIG. The brain as a lens metaphor was a frequently expressed picture in Karl Pribram's work (Pribram, 1991) and Karl has inspired us to implement this model numerically by his holonomic interpretation of perception during the early times of neurophysical modeling. Today support for this model comes again from Karl Friston's group as they suggest a highly useful distinction between neural mass and neural field models (Pinotsis *et al.*, 2014). In this view, opposed to space dependent field models, *mass models* characterize activity over time, assuming that the result of an assemble activity is mapped into a single point. This reflects the intention of SIG. SIG is a mass amplification that finally produces a complex state vector characterizing the state of the entire system, e.g., the whole brain during a perceptive event. From a psychobiological point of view the action behind SIG comes close to the functional role of "attention."

## 8.7. Conclusion

Let me conclude with a summary of the proposal that I have made in this chapter:

**Final conjecture:** If electronic de-localizations within the oxygen cage of the filter domain in voltage gated ion channels, induced during the non-conductive, low-occupancy filter states, become "mass-amplified" (e.g., by a SIG type of process), the transition to working memory is associated with *phenomenal conscious experience*. If the route to working memory does not involve the accumulation of identical electronic excitations, the perceptive signals remain "subliminal." The changes in electronic configurations as suggested here are realized along carbonyl linings in the selectivity filter of ion channels.

Avoiding "miraculous emergence," the above conjecture builds on a micro-psychistic argument stating that if experience emerges from a particular physical process then this process must build on physical properties that are experiential themselves. However, the conjecture does not suggest that electronic signatures in molecules become conscious as such, but I do suggest that they carry "the seeds" of experience as Simons (2006) has put it. The transition to conscious experience requires a strong proliferation and accumulation of identical electronic states by transient or oscillatory correlations among widely distributed structures, i.e., along the channel proteins of a segregated ensemble of cells sharing the same sensory input. The process dynamics that could handle this task, is suggested by the SIG model. However, this is a modeling approach and the physics, or better the physiology behind it, still needs to be established. The computational structure behind SIG however makes several predictions on how a large number of molecular states could synchronize into a massive pattern established across many synaptic distances of neurons. If neurons engaged in identical sensory representations (e.g., for a particular angle within the segregation of orientation sensitive visual cortex) can arrange their effective connectivity in a way, so that they can maximize their recurrent input by their own firing pattern, ion-conduction will eventually synchronize the oscillatory states within the channels filter region. The filter states are *static*, transiently trapped states in the closed channel configuration (e.g., S1S3 and S2S4 alternating $K^+$ states in the filter). Ion-oxygen interaction can then change into a type of DDSA of electronic excitations. The resulting excitations are organized along the lining C=O bonds and are transiently stable and reproducible by a constant energy consumption from the kinetic energy term of ion oscillations in the filter. Put more simply, "cooling" of ions in the filter region nourishes particular electronic excitations at the transition from static to dynamic self-assembly. Over the short time-scales that characterize energy transfer times suggested here, quantum coherent states seem to be feasible even within the warm environment of the filter region (Vaziri & Plenio, 2010; Salari *et al.*, 2011).

Generally I have addressed the signal transducing cell membrane to host the phenotypic characteristic of experience. A conception that is pursued by an increasing number of scientists in an attempt to reconcile physical, biological, psychological and modeling aspects into a coherent view on this challenging problem (as reviewed by Cocchi *et al.* (2014)). In addition it should be noted that the mass-amplification of electronic states as required in my conjecture about conscious experience is reminiscent to the idea of mass accumulations in the original Orch OR model of Hameroff & Penrose (1996).

Furthermore, the role of the neural connectom within the present outline is radically different from the traditional engineering view and neural network conception. Rather than interpreting action potential patterns solely as carriers of a "neural (computer) code," they are found to play a decisive role for associating information based properties, the neural field properties, with the ingredients of experience, the "meaning" behind information. This latter role for neural activities at the cellular scale can be found in the way how membrane potentials can repetitively "reset" channel states, which in turn can lead to oscillatory coherences of these states across large scaled neural ensembles. In this view the role of neurons within functional ensembles is better described by the property of *"embeddedness,"* a term introduced recently by Vlachos *et al.* (2012). Taken together, I suggest that the brain can combine information based properties with experiential phenomenology and this combination can in principle be explained by physical properties established at the quantum scale propagating into the classical scale of neural networks.

## Acknowledgment

I gratefully acknowledge the continuing inspirations in course of the many highly encouraging discussions with my colleague Hans Summhammer from the University of Vienna.

## References

Atkins, P. & de Paula, J. (2006) *Physical Chemistry of the Life Sciences,* 1st edn. New York: W. H. Freeman and Company.

Barkeshli, M., Berg, E. & Kivelson, S. (2014) Coherent transmutation of electrons into fractionalized anions. *Science*, **346**, 722–725.

Berneche, S. & Roux, B. (2001) Energetics of ion conduction through the $K^+$ channel. *Nature*, **414**, 73–77.

Berneche, S. & Roux, B. (2005) A gate in the selectivity filter of potassium channels. *Structure*, **13**, 591–600.

Bernroider, G., Ritt, F. & Bernroider, E. W. N. (1996) Neural phase-time coding from the view of an electrodynamic perturbation problem. *Forma*, **11**, 141–159.

Bernroider, G. (1999) On form, mind and matter (with special reference to Crystal Souls by Ernst Haeckel, translated by A. Mackay). *Forma*, **14**, 185–198.

Bernroider, G. & Roy, S. (2004) Quantum-classical correspondence in the brain: scaling, action distances and predictability behind neural signals. *Forma*, **19**, 55–68.

Bernroider G. & Roy S, (2005) Quantum entanglement of $K^+$ ions, multiple channel states, and the role of noise in the brain. *Proceedings SPIE*, **5841**, 29.

Bernroider, G. & Summhammer J. (2012) Can quantum entanglement between ion transition states effect action potential initiation? *Cogn. Comput.* **4**, 29–37.

Bernroider, G. (2012a) Is there space for the quantum domain in consciousness research? Comment on consciousness, biology, and quantum hypotheses by Baars and Edelman. *Phys. Life Rev.*, **9**, 301–302.

Bernroider, G. (2012b) Image in action. In: B. Zavidovique and G. Lo Bosco, eds. *Common Grounds: The Role of Perception in Science*. Singapore: World Scientific, pp. 103–110.

Bialek, W. (2002) Thinking about the brain, In: H. Flyvberg, F. Jülicher, P. Ormos and F. David, eds. *Physics of Biomolecules and Cells*. Berlin: EDP Sciences, Springer-Verlag, pp. 70–78.

Bucher, D., Raugei, S., Guidoni, L., Dal Peraro, M., Rothlisberger, U., Carloni, L. & Klein, M.L. (2006) Polarization effects and charge transfer in the KcsA potassium channel. *Biophys. Chem.*, **124**, 292–301.

Chalmers, D. (1995) *The Conscious Mind, in Search of a Theory of Conscious Experience*. New York: Oxford University Press.

Chalmers, D. (2003) Consciousness and its place in nature, In: S. Stich and T. Warfield, eds. *The Blackwell Guide to Philosophy of Mind*. Blackwell Publishers, pp. 102–142.

Chomsky, N. (1956) Three models for the description of language. *I.R.E. Trans. Inf. Theory*, **2**, 113–124.

Cocchi, M., Tonello, L. & Gabrielli, F. (2014) Cell membrane and consciousness: A journey through biology, mathematics and philosophy. *Ann. Depression Anxiety*, **1**, 1–3.

Friston, K. (2009) The free-energy principle: A rough guide to the brain? *Trends Cogn. Sci.*, **13**, 293–301.

Guidoni, P. & Carloni, P. (2002) Potassium permeation through the KcsA channel: A density functional study. *Biochim. et Biophys. Acta*, **1563**, 1–6.

Hameroff, S. & Penrose, R. (1996) Conscious events as orechestrated space-time selections. *J. Conscious. Stud.*, **3**, 36–53.

Hermans, T. M., Frauenrath, H. & Stellacci, F. (2013) Droplets out of equilibrium. *Science*, **341**, 243–244.

Hill, S. L., Wang, Y., Riachi, I., Schürmann, F. & Markram, H. (2012) Statistical connectivity provides a sufficient foundation for specific functional connectivity in neocortical neural microcircuits. *Proc. Natl. Acad. Sci. USA*, **109**, 2885–2894.

James, W. (1884) What is an emotion? *Mind*, **9**, 188–205.

Kariev, A. M., Znamenskiy, V. S. & Green, M. E. (2007) Quantum mechanical calculations of charge effects on gating the KcsA channel. *Biochim. Biophys. Acta*, **1768**, 1218–1229.

Krishtalik, L. I., Kuznetsov, A. M. & Mertz, E. L. (1997) Electrostatics of proteins: Description in terms of two dieletric constants simultaneously. *Proteins*, **28**, 174–182.

Laland, K. N., Sterelny, K., Odling-Smee, J., Hoppitt, W. & Uller, T. (2011) Cause and effect in biology revisited: Is Mayr's proximate-ultimate dichotomy still useful? *Science*, **334**, 1512–1516.

LeDoux, J. E. (1996) *The Emotional Brain*. New York, USA: Simon and Schuster.

Lichtman, J. W. & Denk, W. (2011) The big and small: Challenges of imaging the brains's circuits. *Science*, **334**, 618–622.

MacKinnon, R. (2003) Potassium channels. *FEBS Lett.*, **555**, 62–65.

Morais-Cabral, J. H., Zhou, Y. & MacKinnon, R. (2001) Energetic optimization of ion conduction by the $K^+$ selectivity filter. *Nature*, **414**, 37–42.

Nicolis, G. & Prigogine, I. (1989) *Exploring Complexity*. New York, USA: Freeman and Company.

Panksepp, J. (2005) On the embodied neural nature of core emotional affects. *J. Conscious. Stud.*, **12**, 158–184.

Park, H. J. & Friston, K. (2013) Structural and functional brain networks: From connections to cognition. *Science*, **342**, 579.

Pearl, J. (2010) An introduction to causal inference. *Int. J. Biostat.*, **6**, 7.

Pinotsis, D., Robinson, P., Beim Graben, P. & Friston, K. (2014) *Front. Comput. Neurosci.*, **8**, 149.

Pribram, K. H. (1991) *Brain and Perception*. Hillsdale, New Jersey: Lawrence Erlbaum Association.

Prinz, J. (2005) Are emotions feelings? *J. Conscious. Stud.* **12**, 9–25.

Prinz, J. (2007) Mental pointing, phenomenal knowledge without concepts. *J. Conscious. Stud.*, **14**, 184–211.

Salari, V., Tuszynski, J., Rahnama, M. & Bernroider, G. (2011) Plausibility of quantum coherent states in biological systems. *J. Phys. Conf. Ser.*, **306**, 012075.

Scheines (1997) An introduction to causal inference. In: V. McKim and S. Turner, eds. *Causality in Crisis?* University of Notre Dame Press, pp. 185–200.

Shannon, C. E. (1949) *The Mathematical Theory of Communication*. Urbana, USA: University of Illinois Press.

Simons, P. (2006) The seeds of experience. *J. Conscious. Stud.*, **13**, 146–150.

Song, S., Sjöström, P. J., Reigl, M., Nelson, S. & Chklovskii, D. B. (2005) Highly non-random features of synaptic connectivity in local cortical circuits. *Plos Biol.*, **3**, 507–519.

Strawson, G. (2006) Realistic monism, why physicalism entails panpsychism. *J. Conscious. Stud.*, **13**, 3–31.

Summhammer, J., Salari, V. & Bernroider, G. (2012) A quantum-mechanical description of ion motion within the confining potentials of voltage-gated ion channels. *J. Integr. Neurosci.*, **11**, 123–135.

Stepanyants, A & Chklovskii, D. B. (2005) Neurogeometry and potential synaptic connectivity. *Trends Neurosci.*, **28**, 387–394.

Timonen, J. V. I., Latikka, M., Leibler, L., Ras, R. H. A. & Ikkala, O. (2013) Swichtable static and dynamic self-assembly of magnetic droplets on superhydrophobic surfaces. *Science*, **341**, 253–257.

Vaziri, A. A., Plenio, M. B. (2010). Quantum coherence in ion channels: Resonances, transport and verification. *New J. Phys.*, **12**, 085001.

Vlachos, I., Aertsen A. & Kumar, A. (2012) Beyond statistical significance: Implications of network structure on neural activity. *Plos Comput. Biol.*, **8**, e1002311.

Zacharias, N. & Dougherty, D. A. (2002) Cation-π interactions in ligand recognition and catalysis. *Trends Pharmacol. Sci.*, **23**, 281–287.

Zipursky, S. L. & Sanes, J. R. (2010) Chemoaffinity revisited: Dscams, protocadherins, and neural circuit assembly. *Cell*, **143**, 343–353.

Biophysics of Consciousness: A Foundational Approach
R. R. Poznanski, J. A. Tuszynski and T. E. Feinberg
Copyright © 2017 World Scientific, Singapore.

# 9

# THE TWO-BRAINS HYPOTHESIS: IMPLICATIONS FOR CONSCIOUSNESS

*D. Bercovich,*[†] *G. Goodman,*[‡] *L. A. Cacha** *and R. R. Poznanski*[§,||]

[†] *Department of Molecular Genetics, Faculty of Life Sciences, Tel-Hai Academic College, Galilee, 12110 Israel*

[‡] *Galil Genetic Analysis, Katzrin, 12900 Israel*

**Laser Centre, IBNU SINA ISIR, Faculty of Science*

[§] *Department of Clinical Sciences, Faculty of Biosciences and Medical Engineering, University Teknologi Malaysia 81310 Johor Bahru, Malaysia*

[||] *Laboratory of Biological Modeling, The Rockefeller University, 10065 New York, USA*

## Abstract

The human brain is a product of gravitation, quantum biophysics and evolution by survival; a "universe," with 85 billion neurons and a hundred trillion synapses; more than stars in our galaxy. In the context of the unresolved, unpredictability of modern physical theory and as a stimulus and guide for research, we postulate the novel Two-Brains[1] Hypothesis (TBH). The animal "brain" evolved in parallel as two fundamentally different, interdependent organs, complementary in the Bohr sense: one, electroionic, tangible, known and accessible, the other electromagnetic, intangible, difficult to access, but a relatively stable, structurally and

[1] The non-grammatical term "Two-Brains," emphasizes the Copenhagen principle of the special, complementary nature of the two organs: totally different, but inseparable and interdependent.

functionally integrated 3D compendium of differently induced fields arising from and interacting closely with specific counterpart regions of the electro ionic brain. Natural selection ensured normally efficient, anatomic, metabolic and functional integration of the two brains. Understanding aspects of them such as mind, consciousness, aging and pathologies is a major human challenge and it is suggested that research be undertaken on the possible nature, workings and potential of the two interdependent entities by which *Homo sapiens* may have risen and may yet fall.

*Keywords*: Two-Brains Hypothesis; electroionic organ; electromagnetic organ; natural selection of mutual field induction; Bohr complementarity; consciousness; CNS spintronics.

## 9.1. Introduction

### 9.1.1 *Unpredictable and problematic: The quantum in biology*

Alongside conjectures on the Universe, fundamental problems in physics of immediate practical importance to humans remain unresolved. Quantum and gravitational theory are still not united, or particle-wave duality explained. Following Einstein's confirmation of the particulate nature of energy and Rutherford's of the existence of atoms, Bohr proposed that electrons occupy closed, fixed orbits around the nuclei of atoms while gaining or losing energy by instantaneous "jumps" to a higher or lower orbit. These findings and those such as Schrödinger's representation of those electrons as a wave-like, 3D "cloud" of charge density and other phenomena of quantum mechanics, less or more weird and based on probabilities, mystify and made difficult accessibility of quantum physics to non-physicists.

Fortunately, wave mechanics enabled intuitive understanding of molecular bonding as overlap of smeared-out electron clouds and soon became a standard tool in biochemistry. However, though life science research also came to enjoy technological fruits of the new physics e.g., in electron microscopy, X-ray crystallography, chemical spectroscopy, fMRI and electro-encephalography, wariness of quantum denial of the objectivity of natural physical laws has persisted. Thus, though quantum theory has advanced understanding, for example of biological light gathering (in photosynthesis and bacteria) (Engel *et al.*, 2007; Prokhorenko *et al.*, 2006) and of bird navigation (photon induced reading of earth's magnetic field) (Gauger *et al.*, 2011), understanding of complex biological systems such as conduction of action potentials along axons, intertwined nucleic acid chains, gene function, whole cells and whole organs such as the brain, remains mainly in the realms of classical physics and chemistry. The problems for quantum analysis in biology were already recognized in the 1930's and especially defined by Bohr (1933).

### 9.1.2. *Complementarity and biology*

Bohr proposed that though particles and waves are mutually exclusive in the means of their measurement, both are essential and complementary for understanding their interdependence and suggested that the principle may also assist in describing complex biological phenomena. However, traditionally, biological study has been reductionist: breaking down systems such as chromosomes, cells and even the brain, to the simplest molecular level. So as biological complexity was not experimentally accessible to quantum analysis, Bohr (1933) considered life "underivable from physics and chemistry." Denying this was mysticism, when considering higher levels of biological organization and behavior he and his supporters nevertheless regarded "purpose" as necessary and convenient in the thought process (Stent, 1989; McKaughan, 2005). A century later, the philosophy and other problems associated with quantum complementarity as being possibly applicable to biology (Mazzocchi, 2010) have done nothing for understanding of the brain, particularly consciousness, a major key in human behavior. Here, we will propose a new direction of thought.

## 9.2. Consciousness

### 9.2.1 *The "hard problem"*

In recent decades, progress has been made in understanding the mind, a controversial term but convenient here to describe the functional totality of memory storage and recall, cognition and perception. Consciousness remains a conundrum. Attempts to explain it have been highly controversial. Thus, according to Chalmers (1995) awareness, often equated with consciousness, is only a functional aspect of it, not consciousness itself. Termed by him the "hard problem," he asks why and how widely distributed detectors in the awake brain, constantly sense different features such as size, color, tone, shape, sound and motion and bind them together with associated thoughts and memories into innumerable private conscious experiences "qualia" (Chalmers, 1995). These merge and emerge in a stream of thought and awareness that can be shared in conversation with others, though they remain uniquely personal to the individual.

Other explanations of the nature of qualia differ widely. Crick & Koch (1990) suggested 35–75 Hz neural oscillations in the cerebral cortex are the basis of consciousness, Edelman (2007) linked it to the self-concept and Jackendoff (1987) described underlying computational processes as a basis for consciousness. Chalmers himself, roots conscious experience in information, notes consciousness has quantum characteristics for some, while also implying that it is beyond derivation from physical theory.

### 9.2.2. *Quantum-gravity to the rescue?*

Avoiding Kantian "purpose," the Penrose & Hameroff (1996) model of consciousness posits closely spaced, quantum mechanical, coherent molecular condensations in the long microtubule structures based on tubulin and well known to extend within neuronal dendrites and axons. The condensations are considered stable for a microperiod long enough for physiological information transfer over distance to synapses which are crossed by quantum tunnelling (for which there is still no evidence), thus supposedly influencing brain activity regionally or as a whole. Slight deviation in superposition of the tubulin moieties is said to be caused by a quantum-gravity-induced, minute change in the geometry of space-time which, on becoming critical causes collapse (reduction) of the wave function by objective reduction, i.e., not by the orthodox "subjective reduction by observation." The collapse is said to be an event irreversible in time and constantly occurring in a sequence of such events flowing in time. Thus, a unified quantum-gravity principle novel to biology was proposed to underlie consciousness (Hameroff & Penrose, 1996). Was this the fundamental, non-reductionist principle that Bohr sought over 60 years before?

The possible decoherence of condensation caused by random, room-level thermal influences, is indeed claimed to be prevented by organisation of the aqueous environment around and within the microtubule framework (Hagan *et al.*, 2002). However, microtubules naked of that function are widespread, and intra-axon sheathing of microtubules by the water molecules said to be preventing thermal decoherence, would restrict other functions, such as in intra-axonal transport. The model remains much argued and very controversial (Tegmark, 2000). Controversial too, have been proposals concerning electromagnetic (EM) fields in the brain.

### 9.2.3. *Electromagnetism to the rescue?*

Among theories of consciousness developed in the last two decades involving EM fields in the brain are for example, findings that synchronized firing in large groups of neurons create fields strong enough to feed back to the neuron groups which are their source (Smith, 2009). There is evidence that such synchronization is strongly correlated with attention, awareness and conscious perception. This has suggested to McFadden that oscillations at certain frequencies in the firing neurons, perturb a global EM field external to them, creating in it a pattern representing neuronal information. This is combined and fed back to the neurons as consciousness (McFadden, 2013).

Different EM field theories disagree as to how a conscious EM field actually effects brain function. For example, whereas in McFadden's theory the EM field in a feed-back loop deterministically modifies electric charges across neural membranes and thus the probability that specific neurons will fire, for Pockett (2000) a causal link between the EM field and willed actions is not necessary. Moreover, it is not actually clear how a "patterned," overall EM field encompassing a concatenation of data, is sensed in a bound, intimate conscious experience. That drawback is not really answered by a claim that evolution eliminated the problem (McFadden, 2013). Nevertheless, evidence suggests that endogenous, extracellular electric fields in the brain feed back onto the electric potential across the neuronal membrane thus influencing action potentials, independently of synapses (Frohlich & McCormick, 2010). Moreover recently, experimentally applied external fields of physiological strength and frequency, changed firing characteristics of specific neurons in rat brain cortical slices (Anastassiou *et al.*, 2011). As Smith (2009) concludes, the neural correlates of consciousness are unresolved.

### 9.2.4. *Natural selection of induction*

Nature has from archaic times relied on ion flow to create a charge differential across membranes which run in sequences, can initiate action at a distance; and electrostatics for example, has provided animals with electric darts for defence. However, when a varying flow of charge exists in whichever form, it is accompanied by a magnetic field, yet apart from what seem to be relatively special situations, such as the magnetism used in navigation by some animals, magnetic fields seem to have been mainly regarded as side products of chemical activity; perhaps because biological magnetic fields seem so small, inaccessible and masked by the earth's field. Yet humans have increasingly succeeded in their exploitation, for clinical imaging, whether, for example for "reading" metabolic activity via local presence and flow of heme, or nervous activity via charge-derived magnetic fields (Goodman & Gershwin, 2011). However, it seems unlikely that natural selection did not precede man in a full exploitation of the physical laws of induction.

Recently, it has been proposed that fatty myelin-protein membranes wrapped by satellite Schwann cells in numerous turns almost totally ensheathing the axons of peripheral motor neurons in vertebrates and some invertebrates, are not, as has been thought for many decades, primarily or just insulation against charge leaking from the axons. Nor are they though, now just means for 'jumping' action potentials more rapidly along their membranes. Instead, it was argued that the axon's electric field threads the Schwann wrapping and induces in it a secondary, magnetic field influencing the axon membrane, its flow and complex content and its extracellular environment (Goodman & Bercovich, 2013). Though more obviously effective at the node of Ranvier and paranode, it was suggested that the secondary field is also induced in the very long, sub internode section of the axon and in the remarkably stable extra-cellular gap between axon and its internodal sheath. These peripheral secondary fields could be the first known example of such animal adaptation associated with induction. At the same time, it was also suggested that it was unlikely that such selection for induction was only peripheral and not central. This has now been formulated and is considered further here (Goodman *et al.*, 2014).

## 9.3. The Two-Brains Hypothesis (TBH)

The animal "brain" evolved as two fundamentally different, but complementary, interdependent organs: one, electroionic, tangible, known and accessible, the other electromagnetic, intangible, currently unidentified and difficult to access. Natural selection ensures normally efficient, anatomic, metabolic and functional integration of the two brains (Fig. 1).

### 9.3.1. *Axon-myelin EM induction in the CNS*

An obvious source of induction in the brain would be similar to that found peripherally. However, whereas wrapping in the periphery by a given Schwann cell is limited to a single axon, in the brain, axons are very much

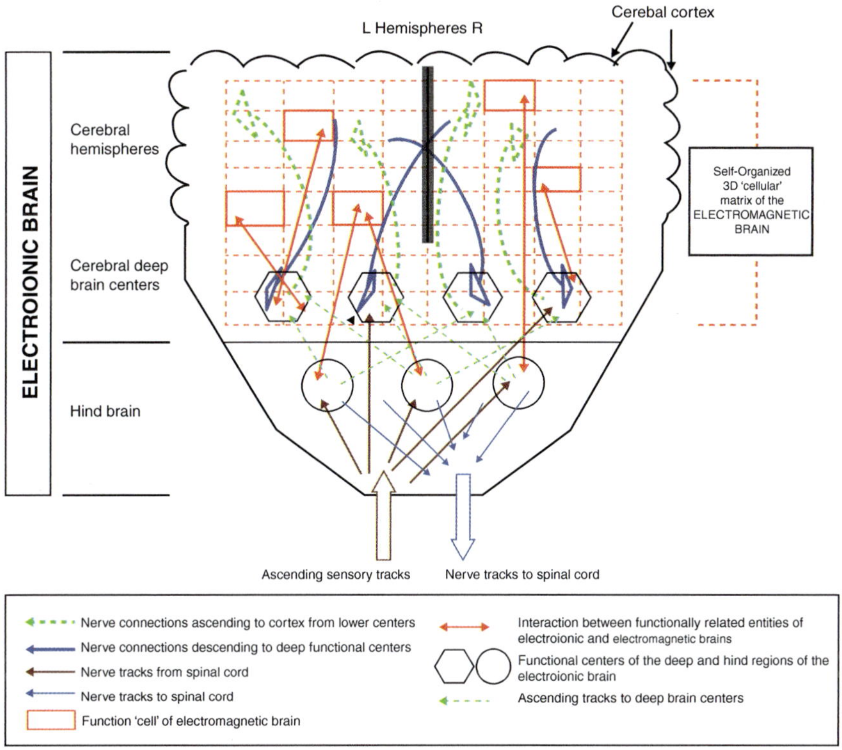


**Fig. 1.** A schematic expression of the Two-Brain Hypothesis according to which an electroionic and an EM brain interact together as independent but complementary organs, thereby enhancing higher processes of recall, perception, thought, judgment and decision in a process engendering a uniquely individual consciousness. Reproduced with permission from the Israel Medical Association and Goodman *et al.*, 2014.

shorter and wrapping by an oligodendrocytic cell (CNS equivalent of the peripheral Schwann cells) can extend around a number of axons. This partly reflects the exceptional crowding in the brain resulting in a premium on space and energy and a need for dissipation of heat due to the intense metabolism. It is not clear that increased axonal signaling speed afforded by satellite-supported, "jumping" of action potentials as in the periphery, is justified in the conditions of the brain. Indeed, it was reported recently that energy for sheath production, maintenance and operation in CNS white matter, costs more than the energy saved by "jumping" of potential (Harris & Attwell, 2012). However, those costs may be warranted in the brain for a critical function quite different to any in the peripheral nervous system, viz. EM induction.

### 9.3.2. *EM brain: structure*

Unlike a generalized, patterned EM field in the brain (McFadden 2013), proposed here is that the EM brain is a relatively stable, structured but flexible, integrated 3D grid of specialized regions of EM field arising from and interacting closely with specific counterpart regions of the electroionic brain. We suggest that the EM brain evolved as a self-organized, EM small-world network of the type first modelled in 1998 by Watts and Strogatz as dynamic constructs, artificial and neural (in *C. elegans*, a nematode). Interesting here, is that the model exhibited enhanced signal-propagation speed, computational power and synchronizability. Even networks with no material topological structure can be of this type and their development is usually bottom up (Watts & Strogatz, 1998). Accordingly, it is proposed that the structure of the induced EM brain is based on evolution over aeons of an independent functional integration of magnetic fields separately and secondarily arising from dendrite and oligodendrocyte electroionic activity, according to the intensity and frequency characteristics of their electric fields. How the fields are integrated is unknown, though frequency and dynamic continuity rather than "all-or-nothing" is an attractive conjecture.

Confirmation or denial will come as newer research provides methodology and anatomical detail different from that which over a century has described the more accessible electroionic brain. The magnetic fields of the neuronal dendrites massively present in the brain were once thought to be inconsequentially minute. Recent model simulation suggests otherwise. Development of direct, spatially and temporally accurate MRI measurement of brain magnetic neuronal activity now seems possible, independent of heme-indicated metabolism (Blagoev *et al.*, 2007). This was demonstrated as 3D magnetic fields represented by vortex like clouds of vectors (Fig. 2) corresponding to the simulated electrical activity of the dendrites of a monkey brain pyramidal cell.

### 9.3.3. *Magnetic data storage?*

Remarkably too, the white matter of the cerebral hemispheres now studied by diffusion MRI in humans and four primate species, was shown to be not just a chaotic mass of brain axons as has been usually thought. Rather, it is an orderly, dense, criss-crossing at 90° to each other, of parallel layers of axons, held side by side with some intertwined bands. Thus, in

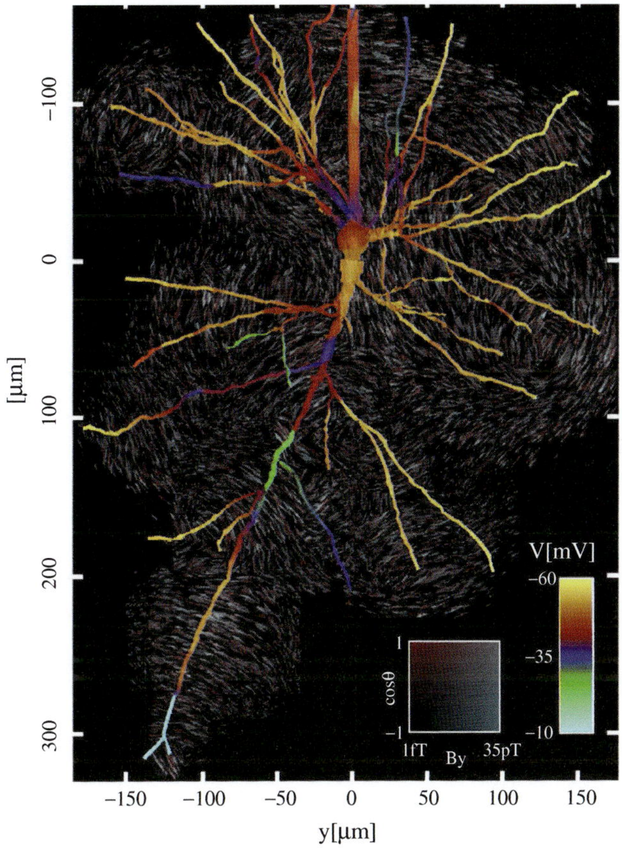


**Fig. 2.** Reconstructed pyramidal neuron from macaque monkey with the simulated electrical activity and corresponding magnetic field at 30 ms after start of the simulation. The neuron is viewed from a point in the +*z* direction looking down on an *x*–*y* plane. In the coordinate system, the soma is at (0, 0, 0). Dimensions of the box that contains the neuron are 480 microns in *x* by 350 microns in *y*, by 180 microns in *z*. The potential in the dendrites is color coded. The two axes are labelled in microns (μm), whereas the color scale represents millivolts (mV). The 3-D magnetic field is represented by vortexlike clouds of vectors whose colors represent direction and whose lengths represent relative magnitude. Reproduced with permission from Elsevier and Blagoev *et al.*, 2007.

the deep white matter of the occipital lobe of the rhesus monkey, the parallel grid structure of pathways was not limited to particular 2D surfaces but extended throughout entire 3D volumes, no diagonal paths were observed and no brain pathways were observed without sheet structure.

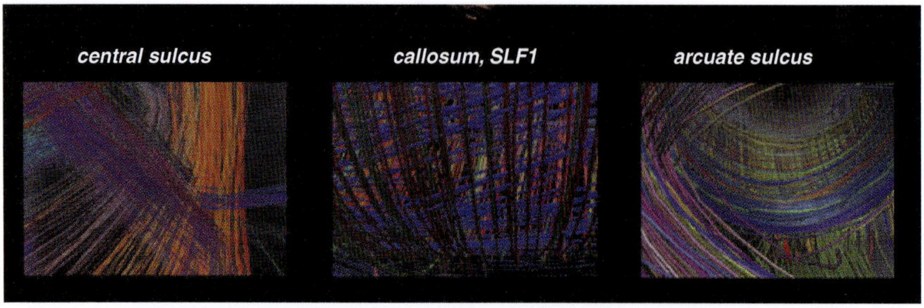


**Fig. 3.** Continuous grid structure of rhesus frontal lobes. All cortico-cortical pathways show highly curved elements in a continuous sheet of interwoven paths in two nearly perpendicular orientations aligned with gyral topography in the acuate sulcus and callosum, but oblique in central sulcus. Reproduced with permission from the American Association for the Advancement of Science and Weeden *et al.*, 2012.

The grid structure was found in all orientations and anatomic curvatures and coherency and continuity with the three principal axes of development of the grid, was maintained at all scales from single small voxel to hemisphere: a telling ubiquity (Wedeen *et al.*, 2012; Fig. 3).

It is proposed that, depending on presence or absence and direction of flow in the axons at the points at which cerebral myelin-protein sheathed axons cross, EM vortices may be engendered. Two different characteristics present in EM vortices, spin direction and core polarity are well known (Lloyd *et al.*, 2012), both switchable by the very low levels of field and frequency changes that have been associated naturally with, and by external application to the nervous system (Kammerer *et al.*, 2011). Appropriate to the CNS scale, the vortices involved in the latter work (experimental and simulated) ranged from 10 nm to 300 nm. A 3D EM matrix of such vortices in the brain may enable natural digital mechanisms analogous to magneto-electronic random access memories and processors. Evolution of such an induced field in the layered, wired architecture of the electroionic brain would have increased overall efficiency of brain exploitation and space use, without increasing demand on the local cardio-vascular system for gaseous exchange and heat removal.

Similarly, the brain dendritic masses mentioned above as having active magnetic fields were thought to be involved in data storage, because of their sub-unitary nature (Blagoev *et al.*, 2007). As part of a widely integrated induced process, this would be much more energy efficient than membrane-based nervous activity, and much faster. Indeed, speed would have been of the essence in an evolution of the EM brain, parallel to that of long term memory — a major advantage to the first creatures surviving long enough to benefit.

### 9.3.4. *Data recall*

Access to memory storage must be increasingly rapid as data quantity increases in line with evolution of capacity for perception, analysis, judgement, action and communication.

Moreover, accuracy of recall would have been increasingly vital for performance and preventing energy waste by mistakes. For this and structural stability, the relatively simple EM brain may have an advantage over one based directly on complex genetic, molecular pathway processes. Though originating in and interacting with the latter, the physical nature of the EM organ, once formed, may in general have been less prone to attrition by random problems, than the more archaic, labile, chemical-based processes on which short-term memory depends.

The speed adequate for motor reactions, is surely much less than in the conscious processes in which memories, mentality in its widest sense, individual psyche and reactions to changing dynamic internal and external environments, merge and emerge, seemingly in a stream. It is suggested that the electroionic nervous system in the brain, rapid electric synapses through CNS gap junctions or tuned up speed as between neurons in the *C. elegans* hermaphrodite, could not provide the signal transmission speeds required for the advanced vertebrate CNS. EM brain recall, abstraction and permutation in complementarity with the slower electroionic brain, has resulted in an evolutionary peak, human consciousness: a candidate for satisfaction of Bohr's prediction for biology — without his concern for "purpose."

Differences in function between the two brains, does not necessarily exclude overlap or similar activity when advantageous: a short term memory mainly associated with the electroionic brain and long term memory with the EM brain, may not be absolute. However, the electromagnetic brain may be generally more sensitive to the extracorporeal, the electroionic brain to the internal environment, while the electroionic brain evolved earlier, for action, the EM brain perhaps later for the recall, analysis and formulation now at the heart of human consciousness and creativity.

### 9.3.5. *"Consciousness" as a shared "conversation"*

What and how animals other than man, experience in consciousness is likely very limited. Rooted in archaic selection, animal interaction usually entails relatively simple, conscious, vocal signalling and body language. In the case of man alone, thought processes of individuals and in communication between them, involves complex language and symbolism. In search of the reality that is consciousness, is it necessary to go beyond physical theory? It may be considered that interaction between the relatively slow sensory system which informs and is responded to by the electroionic brain for action, and the incomparable speed of the EM brain in recall, abstraction and permutation, is a process analogous to conversational intimacy between two individuals each of whom perceives the outlook of the other, allowing a synthesis objective to both. This "conversation" is between the centers in the electroionic brain known to be involved in "language" and emotion and the EM brain about which we know so little. When we know much more about both and their "conversation" the immediacy of consciousness may be more understandable.

Speculatively, though evolution of language may have lessened development of direct inter-personal communication between human EM brains, this presumably would not be the case in other vertebrates; for example, in marine communication at a distance, or for advantage in swarming, particularly in situations where chemicals could be ineffective. Thus although Pocketts' view that EM-based animal consciousness extends across the universe is tenuous it may yet be proved realistic on a modest scale. For example, extracorporeal unification of EM brain fields due to synchronous firing in huge numbers of different avian brains, may extend across close-formation swarming of tens or hundreds of thousands of starlings and their EM brains. This may lead, for example to central tranquilizer release and calming during gathering and night roosting. Signs of such extracorporeal EM fields may be seen in the effect of power pylons and radio masts as seemingly suggested by Fig. 4. Though magnetism is involved in navigation of many animals, the mechanism for this in many is unknown; a barely developed EM brain may be involved. The question arises: Has linguistic ability and modern means of communication at a distance actually deprived mankind of a potential for direct interpersonal contact via the EM Brain? We suggest that there is no reason why such a lack should be permanent. If not, its physical nature and possible relatively rapid openness to epigenesis, suggests its development as not inconceivable.

**Fig. 4.** Bird swarm close to high voltage electric pylon (source: Neels Castillon).

### 9.3.6. *Cosmos and consciousness*

The enigma of consciousness has encouraged flights into realms appropriate to mysticism and philosophy rather than science. Some of the theories go far into the universe. Seeking to find laws unique to biology, when realizing the problems associated with applying quantum physics to the complexity of living systems, Bohr may have been closer to rarefied regions than he cared to accept. There are more recent examples of such venturing. Chalmers was not prepared to equate awareness with consciousness and seems to have finished up with something not entirely distanced from mysticism. More adventurous are those who consider it a short step from consciousness to the universe.

Penrose and Hameroff as mentioned, regard consciousness as the result of coherent quantum processes in microtubules of brain neurons. However even they also consider that, as a form of their theory of the collapse ("objective reduction") of coherent biological quantum processes is related to the fundamentals of quantum mechanics and space-time geometry, so also consciousness is part of the basic structure of the universe (Hameroff & Penrose, 2014). This seems like one giant step for mankind, derived from exceedingly small evidence from man — especially as they grant that we do not know what consciousness is. It could be added we know increasingly less about space, theories about which expand as do(es) the universe(s) — so fast that we cannot know what is "beyond": bubbling universes, with an arrow of time perhaps opposite to ours; or nothing, just in our minds — "stringed" and many dimensioned, an infinitely studied *ad infinitum*. Naturally, questions of free will and choice must also find a place in their mix, though more as a side show. Those can be central, as for Pockett: an issue for philosophy and psychology rather than anatomy.

Defining EM theory in terms of 3D patterns of extracellular, local field potentials which necessarily feedback from whence they came, Pockett has concluded that behavior precedes consciousness of it and that our predisposition to see events as caused by what went before is why we tend to think our consciousness causes our actions; a reasonable thought. However, she too strays into a giant step for mankind, by declaring that there is something to be said for regarding the whole of the universe as pervaded by consciousness. What is meant by this is not clear. As we are part of the universe, then it must pervade all our brains. Though not rejecting this brave leap totally as a possibility, a smaller, more modest step seems appropriate. We suggest that the EM brain of the TBH provides a suitable medium for what may best be considered as potential ultra—conscious, interpersonal human contact. However, Earthly phenomena of other kinds may be eventually clarified rather more quickly by the TBH, such as sudden recovery from long term, severe coma — currently controversial, some out-of-body experiences and not least, various pathologies.

## 9.4. Conclusion

Here we considered the TBH which predicates two interdependent, complementary organs, one electroionic, tangible, well known and accessible, the other electromagnetic, intangible and difficult to access. The latter organ conceives integration in unknown fashion of EM fields combining glial secondary induction with vorticular electronics. Thus, fundamental rethinking and research on consciousness will be advanced with reconsideration of conflicting and debated concepts and data under a novel light. Fortunately, newer means for measuring and analysing endogenous, induced EM fields, such as for example the possible use of the 1-MV Field-Emission Electron Microscope, together with simulation, and other, newer, experimental techniques (Blagoev *et al.*, 2007; Frohlich & McCormick, 2010; Ozen *et al.*, 2010; Anastassiou *et al.*,

2011; Kammerer *et al.*, 2011; Sudre *et al.*, 2011; Lloyd *et al.*, 2012; Gross *et al.*, 2013) are now available to the life sciences. Bohr may not have worried in vain.

## References

Anastassiou, C. A., Perin, R., Markram, H. & Koch, C. (2011) Ephaptic coupling of cortical neurons. *Nat. Neurosci.*, **14**, 217–223.

Bohr N. (1933) Light and life. *Nature*, **133**, 421–423, 457–459.

Blagoev KB, Mihaila B, Travis BJ, Alexandrov LB, Bishop AR, Ranken D, Posse S, Gasparovic C, Mayer A, Aine CJ, Ulbert I, Morita M, Müller W, Connor J & Halgren E. (2007) Modelling the magnetic signature of neuronal tissue. *Neuroimage*, **37**, 137–148.

Chalmers DJ. (1995) Facing up to the problem of consciousness. *J. Conscious. Stud.*, **2**, 200–219.

Crick, F. & Koch, C. (1990) Toward a neurobiological theory of consciousness. *Semin. Neurosci.*, **2**, 263–275.

Edelman GM. (2007) Naturalizing consciousness: A theoretical framework. *Proc. Natl. Acad. Sci. USA*, **100**, 5520–5552.

Engel GS, Calhoun TR, Read EL, Ahn T, Manal T, Cheng Y, Blankenship RE & Fleming G.R. (2007) Evidence for wavelike energy transfer through quantum coherence in photosynthetic systems. *Nature*, **446**, 782–786.

Frohlich F & McCormick DA. (2010) Endogenous electric fields may guide neocortical network activity. *Neuron*, **67**, 129–143.

Gauger EM, Rieper E, Morton JJL, Benjamin SC & Vedral V. (2011) Sustained quantum coherence and entanglement in the avian compass. *Phys. Rev. Lett.*, **106**, 040503.

Goodman G & Gershwin ME. (2011) Physics, biology and the origin of life: the Physicians' view. *Isr. Med. Assoc. J.*, **13**, 719–724.

Goodman G & Bercovich D. (2013) Electromagnetic induction between axons and their Schwann cell myelin-protein sheaths. *J. Integr. Neurosci.*, **12**, 475–489.

Goodman G, Gershwin ME & Bercovich D. (2014) Mars can wait: Facing the challenges of our civilization. *Isr. Med. Assoc. J.*, **16**, 744–747.

Gross J, Baillet S, Barnes GR, Henson RN, Hillebrand A, Jensen O, Jerbi K, Litvak V, Maess B, Oostenveld R, Parkkonen L, Taylor JR, van Wassenhove V, Wibral M & Schoffelen JM. (2013) Good practice for conducting and reporting MEG research. *Neuroimage*, **65**, 349–363.

Hagan S, Hameroff SR & Tuszynski JA. (2002) Quantum computation in brain microtubules: decoherence and biological feasibility. *Phys. Rev. E*, **65**, 061901-1-11.

Hammeroff S & Penrose R. (2014) Consciousness in the universe: a review of the 'Orch OR' theory. *Phys. Life Rev.*, **11**, 39–72.

Hameroff S & Penrose R. (1996) Orchestrated reduction of quantum coherence in brain microtubules: a model for consciousness. *J. Conscious. Stud.*, **3**, 36–53.

Harris, JJ & Attwell D. (2012) The energetics of CNS white matter. *J. Neurosci.*, **32**, 356–371.

Jackendoff, R. (1987) *Consciousness and the Computational Mind.* Cambridge, MA: MIT Press.

Kammerer, M., Weigand, M., Curcic, M., Noske, M., Sproll, M., Vansteenkiste, A., Van Waeyenberge, B., Stoll H., Woltersdorf, G., Back, C. H. & Schuetz, G. (2011) Magnetic vortex core reversal by excitation of spin waves. *Nature Comm.*, **2**, 279.

Lloyd, S. M., Babiker, M., Yuan, J. & Kerr-Edwards, C. (2012) Electromagnetic vortex fields, spin, and spin-orbit interactions in electron vortices. *Phys. Rev. Lett.*, **109**, 254–801.

Mazzocchi, F. (2010) Complementarity in biology. A reassessment in relation to molecular-reductionist and systemic approaches. *EMBO Reports*, **11**, 339–343.

McFadden, J. (2013) The CEMI field theory: closing the loop. *J. Conscious. Stud.* **20**, 153–168.

McKaughan, D. J. (2005) The influence of Niels Bohr on Max Delbruck: Revisiting the hopes inspired by "Light and Life". *Isis*, **96**, 507–529.

Ozen, S., Sirota, A., Belluscio, M. A., Anastassiou, C. A., Stark, E., Koch, C. & Buzsaki, G. E. (2010). Transcranial electric stimulation entrains cortical neuronal populations in rats. *Neurosci.*, **30**, 11476–11485.

Pockett, S. (2000) *The Nature of Consciousness: A Hypothesis. Lincoln*. NE Writers Club Press.

Prokhorenko, V. I., Nagy, A. M., Waschuk, S. A., Brown, L. S., Birge, R. R. & Miller R. J. D. (2006) Coherent control of retinal isomerization in Bacteriorhodopsin. *Science*, **313**, 1257–1261.

Smith, C. U. M. (2009) The 'hard problem' and the quantum physicists. Part 2: Modern times. *Brain Cognition*, **71**, 54–63.

Stent, G. S. (1989) Niels Bohr's legacy to contemporary biology. *Genome*, **31**, 11–15.

Sudre, G., Parkkonen, L., Bock, E., Baillet, S., Wang, W. & Weber, D. J. (2011) rtMEG: A real time software interface for magnetoencephalography. *Comput. Intell. Neurosci.*, **2011**, 1–7.

Tegmark, M. (2000) The importance of quantum decoherence in brain processes. *Phys. Rev. E*, **61**, 4194–4206.

Watts, D. J. & Strogatz, S. H. (1998) Collective dynamics of 'small-world' networks. *Nature*, **393**, 440–442.

Wedeen Van, J., Rosene, D. L., Wang, R., Dai1, G., Mortazavi, F., Hagmann, P., Kaas, J. H. & Tseng, W. I. (2012). The geometric structure of the brain fiber pathways. *Science*, **335**, 1628–1634.

Biophysics of Consciousness: A Foundational Approach
R. R. Poznanski, J. A. Tuszynski and T. E. Feinberg
Copyright © 2017 World Scientific, Singapore.

# 10

# THE ORIGINS OF THE BRAIN'S ENDOGENOUS ELECTROMAGNETIC FIELD AND ITS RELATIONSHIP TO PROVISION OF CONSCIOUSNESS

*C. G. Hales*

*Neuroengineering Laboratory,*
*Department of Electrical and Electronic Engineering*
*University of Melbourne, Australia*

## Abstract

As a potential source of consciousness, the brain's endogenous electromagnetic (EM) field has much to commend it. Difficulties connecting EM phenomena and consciousness have been exacerbated by the lack of a specific conclusive biophysically realistic mechanism originating the EM field, its form and dynamics. This work explores a potential mechanism: the spatial and temporal coherent action of transmembrane ion channel currents which simultaneously produce electric and magnetic fields that dominate all other field sources. Ion channels, as tiny current filaments, express, at a distance, the electric and magnetic fields akin to those of a short (transmembrane) copper wire. Following assembly of appropriate formalisms from EM field theory, the chapter computationally explores the scalar electric potential produced by the current filaments responsible for an action potential (AP) in a realistic hippocampus CA1 pyramidal neuron. It reveals that AP signaling can impress a highly structured, focused and directed "sweeping-lighthouse beam" that "illuminates" neighbors at mm scales. Ion channel currents thereby provide a possible explanation for both EEG/MEG origins and recently confirmed functional EM coupling effects. Finally, a physically plausible EM field decomposition is posited. It reveals objective and subjective perspectives intrinsic to the membrane-centric field dynamics. Perceptual "fields" can be seen to operate as the collective action of virtual EM–boson composites (called qualeons) visible only by "being" the fields, yet objectively appear as the familiar EM field activity. This explains the problematic evidence presentation and offers a physically plausible route to a solution to the "hard problem."

*Keywords*: Maxwell's equations; electromagnetic field; compartmental model; consciousness; pyramidal neuron; hard problem.

## 10.1. Introduction

The involvement of science in the study of consciousness was recently restated as follows (Barrett, 2014):

> *The key question in consciousness science is: "Given that consciousness (i.e., subjective experience) exists, what are the physical and biological mechanisms underlying the generation of consciousness?"*

In the same paper, physicist Adam Barrett clarified a fact of brain tissue: if you were to write down a list of all the fundamental physics involved in an account of intelligence, cognition, behavior and consciousness, that list would have a length of one: the fundamental EM (bosonic) class of the standard model of particle physics. With the Barrett perspective, it is not a matter of whether it is EM that delivers consciousness, it is a matter of which portion of the brain's EM is doing it. But standard-model fundamental EM, in its specification from the quantum level upwards, is completely mute on consciousness. Hopefully, this chapter will take us a little further along the road toward dealing with that paradox.

The endogenous EM field is a natural expression of myriad field sources. This multitude of nanoscale sources superpose by vector field summation. The field system has an intricate fine structure, mirroring the fine micro-structure of the tissue as a system of field sources placed in space. Relatively weak but spatially large, the totality of the field system permeates the tissue, including all intracellular and extracellular spaces (ECSs). Thus expressed, it has a seamless unity and behaviors of its own, over and above any one originating source. In a very real, physical way, the brain's EM field is a single, huge, EM entity of some kind. A captive of the neural substrate and the cranium that hosts it, the endogenous EM field is a "whole," with an operational life of its own, not obviously reflected in its contributing parts. This is the endogenous EM field of the brain.

Ultimately the field system (electric and magnetic) exits the brain, permeating the intervening cranial tissue, bone and scalp. There it becomes visible in EEG (electric field) and MEG (magnetic field) measurements. If you measure ECoG on the brain surface you are measuring spatially/temporally averaged potentials on the brain's surface produced by the endogenous electric field. Deep in tissue, if you measure the "local field potential" (LFP) then you are measuring a much finer-scale average potential produced by the endogenous electric field.

In connecting the endogenous EM field with consciousness, what we are saying is that some selected portions of the endogenous EM field produce consciousness: the experienced first-person perspective. The endogenous EM field is the loud "signal," arising from specific coherent atomic/molecular-level sources, that dominates the burbling incoherent field "noise" of the rest of the atomic/molecular-level EM phenomena. We cannot say which parts of the EM field are delivering consciousness until we figure out what actually delivers the EM field in the first place. What are these dominant atomic/molecular-level EM field sources? The answer to this question forms the bulk of the rest of this work.

While the earliest specific EM field theory of consciousness dates to 1942 (Köhler, 1960), discourse was essentially mute until the late 1980s arrival of the "scientific study of consciousness." A generic term for the science of consciousness is "The ABC Correlate of Consciousness." The most prominent in the science so far is ABC = "Neural Correlates of Consciousness" (NCC) (Chalmers, 2000; de Graaf *et al.*, 2012; Hohwy, 2012; Metzinger, 2000; Molyneux, 2010; Mormann & Koch, 2007; Neisser, 2012; Velmans & Schneider, 2007). A more recent popular contender is the ABC = "Information Integration Theory" (IIT) correlate of consciousness (Balduzzi & Tononi, 2008, 2009; Barrett, 2014; Tononi, 2008, 2004). Another famous contender is the ABC = "Global Workspace Theory of Consciousness" (GWS) (Baars & Franklin, 2009; Baars, 1997; Baars *et al.*, 2003; Shanahan & Baars, 2005; Velmans & Schneider, 2007).

In positing endogenous EM fields as the origins of consciousness we have ABC = "Electromagnetic Correlates of Consciousness" (EMCC). That is the sense in which EMCC are posed as an answer to the (Barrett, 2014) question that puts it in a critically superior position by virtue of all other "ABC theories of consciousness" actually being EM field theories in disguise. An early field theory contender is Benjamin Libet, who proposed a "Conscious Mental Field" that is (a) not in any category of known physical fields, such as EM, gravitational, etc., (b) may be viewed as somewhat analogous to known physical fields and yet, (c) cannot be observed directly by known physical means (Libet, 1994). Karl Popper also lent support to a non-specific field theory of consciousness (Libet, 1996; Lindahl & Arhem, 1994; Popper *et al.*, 1993). The EM-field proposition presented here eliminates all need to posit exotic field systems. As such, the exotic field propositions become irrelevant. Exotic quantum mechanical effects have a long history of being posed as a solution to consciousness. The most recent and loudly touted form is "quantum computation in microtubules," and recent empirical work suggests some evidence of quantum effects in microtubules. That evidence's claim to be related to consciousness is, so far, lacking (Hagan *et al.*, 2002; Hameroff, 1998, 2012; Hameroff & Penrose, 2014). The later EM-field proposition presented here eliminates all need to posit exotic quantum effects. The quantized (bosonic) EM field impressed on space implicitly supplies all the necessary quantum effects. Exotic quantum effects thereby become irrelevant.

The main EM field theories arose in late 1990s and early 2000s with the work of Susan Pockett (*"The Conscious EM field theory — CEMF"*) (Pockett, 1999, 2000, 2002, 2007, 2012), John-Joe McFadden (*"CEMI — Conscious Electromagnetic Information theory"*) (McFadden, 2002a, 2002b, 2006, 2013) and E. Roy John (*"The coherent EM field theory"*) (John, 2001, 2002, 2003, 2005, 2006).

This chapter adds critical physical details to the body of support for all EM-field theories. For example, it provides a direct basis for the "wave mechanics" of the (McFadden, 2002) treatment. It does this by (a) providing specific plausible, physical origins for both the electric *and* magnetic fields that are (b) consistent with all the empirical evidence used to support all field theory treatments to date (including a causal role) and that (c) also have a plausible mechanism underlying subjectivity.

In dealing with the "hard problem" of an experienced first-person perspective, and in particular with EM fields as its deliverer, we must recognize *"being the EM fields"* and *"scientifically observing the EM fields"* as fundamentally different perspectives. The former is a first-person perspective/consciousness (a faculty enabling scientific observation in the first place). The latter is what the fields look like when you use consciousness to scientifically observe them. Because of the fundamental categorical distinction between these two perspectives, heightened skeptical attention to assumptions must be maintained. We cannot assume, for example, that causal relations between two points/events, intensity (field strength, or perhaps signal/noise ratio), temporal persistence (or otherwise) or spatial separation remain subject to the usual historical/cultural assumptions.

What is motivating this vigorous attention to the EM-field system? One good reason is the fact that some aspect of the dynamic EM-field system has finally been held accountable for causal effects in tissue. EM coupling therefore joins action potential (AP) signaling to become the second signaling system in the brain.[1] The specific mechanism responsible for EM coupling is currently unknown. In any event, EM signaling must join AP signaling with its own formalism in neural modeling. Apart from its consciousness-related contribution, it is one of the goals here to lay down the groundwork needed to eventually add the appropriate EM field effects to mainstream neural models.

The dominant view throughout the 20th century was that the EM field is epiphenomenal. That is, it had no functional/causal role in tissue, and was assumed to merely reflect activity like the sound of a heart does for heart function. This view is shown diagrammatically in Fig. 1(a). The epiphenomenal view was not universal, as EM-field effects have been routinely reported all along, leading some researchers to expect the "closed loop" behavior of Fig. 1(b) (Anastassiou *et al.*, 2010; Aronsson & Liljenstrom, 2001; Bikson *et al.*, 2004; Bishop & Oleary, 1950; Faber & Korn, 1989; Francis *et al.*, 2003; Jefferys, 1981, 1995; Jefferys & Haas, 1982; McIntyre & Grill, 1999; Noebels & Prince, 1978; Parra & Bikson, 2004; Rall & Shepherd, 1968; Reato *et al.*, 2010; Taylor & Dudek, 1982; Terzuolo & Bullock, 1956). The cultural partitioning of views, along open/closed loop EM field lines, was sufficiently unsettling that it once lead to a call for resolution (Bullock, 1997). EM field feedback, despite evidence to the contrary, can still be found declared hypothetical (Weiss & Faber, 2010).

Now, sufficiently inspired, technologically feasible and compeling empirical work supports the Fig. 1(b) closed-loop view (Anastassiou *et al.*, 2011; Frohlich & McCormick, 2010; Reimann *et al.*, 2013). The feedback

---

[1]Sometimes called ephapsis or ephaptic coupling, although this name is likely to be ultimately regarded as a misappropriation of a term used for a medical condition.

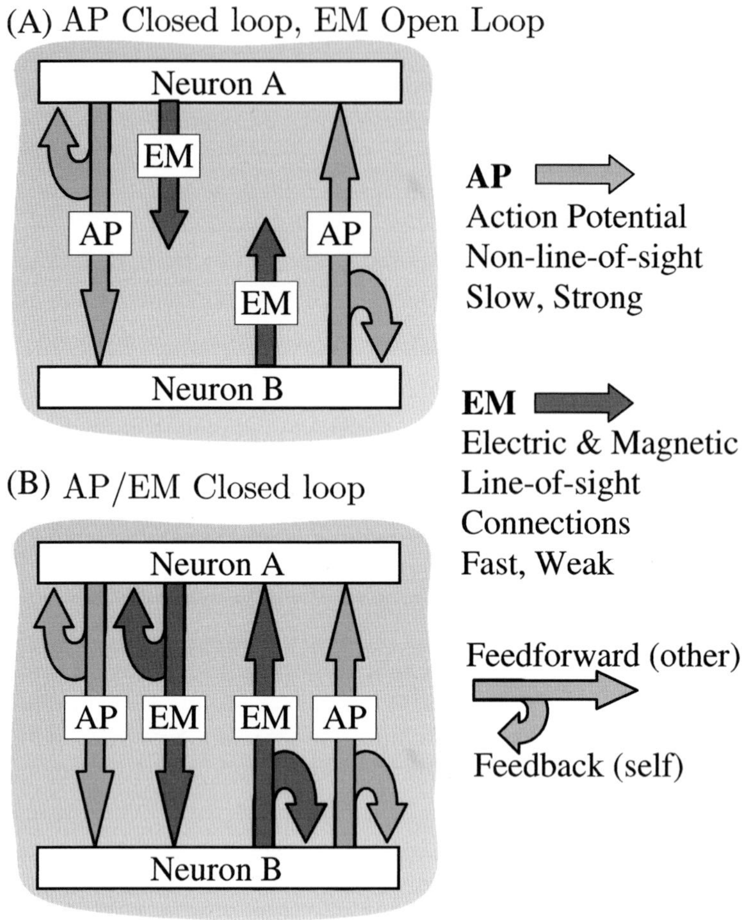


**Fig. 1.** (A) The traditional open-loop view holds that the fields are epiphenomenal. (B) The closed-loop view is that which accounts for recent empirical evidence in certain tissues. Based on figures in Frohlich & McCormick (2010).

effect can no longer be claimed epiphenomenal. It is not a matter of whether it happens. It is a matter of how the effect arises. The originating mechanism provided in this chapter is consistent with the closed-loop form of Fig. 1(b), including, eventually, its quantification. The EM feedback is a result of the action of later Eqs. (11) and (14) for the dynamic electric and magnetic fields, respectively. Based on Eq. (14) it may also be possible that, at very small distances, even the tiny magnetic field can have a self-feedback causal effect. Just like wires carrying current can attract each other, ion channels firing together may be attracted to each other, moving laterally in the plane of the membrane to collocate into clumps of channels.

Not all tissue needs be expected to operate as per Fig. 1(b). If function can be satisfied without EM feedback, then the anatomy and physiology of the cells may self-organize to simply not make use of it (lower cell density, for example, reduces EM coupling effects). In that case the functional equivalent of Fig. 1(a) would result. It becomes a matter of the requirements of the tissue. It is therefore an obvious expectation that tissue functional EM field effects will be found to operate on a spectrum of roles, with Figs. 1(a) and 1(b) forming the extrema of possibilities. It is hoped that the potential for a sophisticated and subtle feedback mechanism is at least inprinciple demonstrated here.

### 10.2. Nervous Tissue and EM: Fundamentals

Consider Figs. 2(a) and 2(d) which depict the physical reality of cortical gray matter formed by the intimate interweaving of thousands of neurons and astrocytes, each of which has an intricate branched structure spanning hundreds/thousands of $\mu$m. Setting the intracellular molecular machinery aside, the tissue is essentially water electrolyte radically fragmented/partitioned by neuron and astrocyte cell membrane (nonconducting lipid bilayer) at the sub-micron level. The so-called "extracellular space" (ECS, shown in red in Fig. 2(a), and detailed in 2(c), is actually a nanoscale sheet/tunnel matrix" largely 40–110 nm across (maximum) and approaching even closer when involved in a synapse (Kinney *et al.*, 2013; Mishchenko *et al.*, 2010).

While the ECS gap is very narrow, the ECS has a high tortuosity, leading to its volume fraction being typically cited as between 10% and 20% of cerebral volume (Bach-Y-Rita, 2001; Bignami, 1991; Geddes & Baker, 1967, p. 273; Nicholson & Phillips, 1981; Sykova, 1997, 2004; Sykova & Nicholson, 2008; Sykova *et al.*, 1999; van Harrenveld, 1968). This formidable complex is that which generates static and dynamic endogenous EM fields, and upon which the EM field impinges and, eventually, exits the scalp. Understanding the origins of the various components of the endogenous EM fields, as an expression of Figs. 2(a) and 2(d) material, is the central goal of this work. Prior to that, however, we have some fundamental physics and chemistry to assemble.

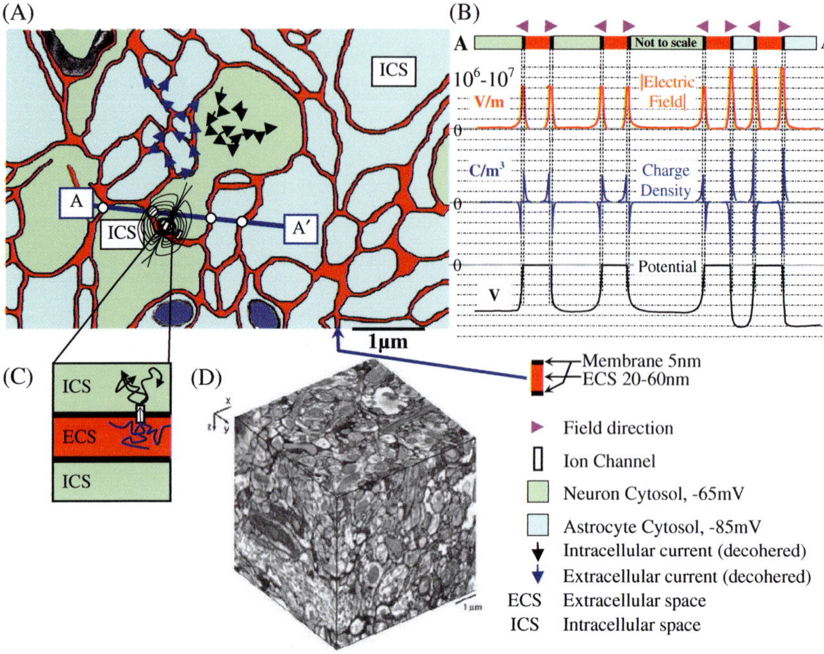


**Fig. 2. EM-field coherence origins in nervous tissue**. (A) is a notional portion of a slice through (D) that reveals EM-field system origins. Transporters create tortuous folded sheet-charge transmembrane dipoles shown in (B) astride each membrane boundary (black) of the ECS fluid (red). Line A–A′ shows a typical line-of-sight profile of (equilibrium/resting conditions) static charge, electric field strength and potential at each intersection with membrane (Aidley, 1998; Dayan & Abbott, 2001; Hille, 2001; Johnston & Wu, 1995). Ion channel activity in (A), highlighted in (C), produces electric and magnetic fields as described elsewhere here. The ▶ arrow heads indicate the direction of the resting electric field. Potentials are referenced to the ECS as zero volts. Blue/black arrows and (C) show the same transmembrane current decohered in the ECS and ICS. The foam structure serves to decohere overall current flow at the scale of µm and above. Not to scale. (A) based on (Nicholson & Sykova, 1998). (D) based on (Briggman & Denk, 2006).

### 10.2.1. *Tissue and macroscopic vs. microscopic EM field theory*

There are two fundamental sets of Maxwell's equations: the microscopic form and the macroscopic form (Craig & Thirunamachandran, 1984; Jackson, 1999; Nunez & Srinivasan, 2006). These are tabulated in

Appendix 1. The traditional material/constitutive/macroscopic form of Maxwell's equations (that have such things as conductivity, permittivity, permeability) has been derived from the microscopic form (in which there are no materials, just atoms/charges behaving in regular ways) (Jackson, 1999). Brain tissue is a non-homogenous, anisotropic, nonstationary, non-linear, far-from-equilibrium complex system detailed at the nanoscale. It does not take much analysis to see that the conditions under which Jackson validates the macroscopic form are violated by brain tissue. However, the macroscopic form has been used beneath these "Jackson" limits for so long in so many contexts, that we must suppose that by careful description of components of the tissue, and by "acting-as-if "macroscopic material EM is applicable, we can glean some understanding. That understanding, however, does not alter the fact that, ultimately, it is likely that atomic level molecular dynamics incorporating a full set of Maxwell's microscopic equations (convective/diffusion form, electric and magnetic fields and the Lorentz force) will ultimately be the only form fully revealing of tissue endogenous EM-field origin and role.

### 10.2.2. *Electrolyte is not a "conductor"*

Assuming that Maxwell's macroscopic EM equations apply, the brain is not a formal conductor. The bulk of the tissue is water (insulator) with ions flowing through it. Therefore, it is actually a convector (note: convection does not mean diffusion) (Sadiku, 2001, p. 163). The appropriate EM equations suited to the brain are like those of gas plasma. For decades, neuroscience practitioners have used voltage/current relations to characterize tissue as having certain conductance in certain circumstances. This usage of Ohm's law for voltage/current relations, while very useful in modeling, does not permit Ohm's Law to be used to characterize EM field expression. The tissue is fundamentally based on expression of dynamic non-zero charge density established at microscopic scales and then micromanipulated on a microsecond timescale over huge areas/volumes, with macroscopic effects. Ohm's law is fundamentally based on zero charge density at an ultra-fine scale and cannot be used to accurately express EM fields in brain tissue (Jackson, 1999, p. 706). What this means is that conductivity is fundamentally the wrong material property to use in assessing the relevant EM equations for excitable cell tissue. More fundamental forms of Maxwell's equations must be used. These are Eqs. (2) and (3) and the terms in the dotted ellipse.

When cited, however, the conductivity is quite instructive and still remains predictive of voltages and currents in typical laboratory measurements. A classic example is in Katz (1966), who cites ICS electrolyte conductivity at 3.3 $Sm^{-1}$, which contrasts with the ECS conductivity of 5 $Sm^{-1}$, which is very similar to the bulk conductivity of sea water. This is to be expected. To satisfy the needs of the adjacent ICS on both sides of the ECS (tens of nm apart as per Fig. 2(c)), the ultra-fine ECS must have approximately twice the average charge density associated with a single membrane. With twice as much charge to flow for a given impressed electric field, the conductivity of the ECS is expected to be roughly double the ICS. Note that both these conductivities are more than "bulk brain tissue" conductivity, which tends toward that cited for cerebral spinal fluid: 1 $Sm^{-1}$ (da Silva & Van Rotterdam, 2005). ECS and ICS electrolytes are very poor conductors, but considerably better than pure water ($5{:}5 \times 10^{-6}$ $Sm^{-1}$), which is a very good insulator.

### 10.2.3. *Transmembrane ion speed and conduction speed in copper conductor*

In Fig. 2(c), when electrolyte ions rapidly transit the membrane through an ion channel pore, patch clamp recording of a single open channel reveals a current of the order of $10^{-12}$ A (Coulombs per second) (Hille, 2001, p. 90). Given the transmembrane pore length of 5–7.5 nm, and that a Coulomb consists of roughly $6.241 \times 10^{18}$ univalent charges, a patch clamp current measurement corresponds to a passage of the order of $10^6$ univalent charges per second (a transit time of roughly $1\mu s$). Assuming simple ion channel kinetics with an average "open" duration of 1 ms (a single spike), of the order of 1000 single-file charges will transit a single ion pore with an average speed of the order of cm/s.

Copper uses electrons as the charge carrier in a crystalline matrix. It is a formal conductor in the proper sense of the laws of EM (Ohm's law predicts the fields in the material). The comparison with convective motions of large ionized atoms must be carefully made. The measured conductivity of copper at 20°C is ~ $5.9 \times 10^7$ $Sm^{-1}$. When you work out typical electron drift velocities in copper at physiologically realistic temperatures you discover it to be of the order $\leqslant$mm/s to cm/s with typical voltages and material geometries. This means that although the ion channel pore is a charge transport mechanism totally unlike copper, that the charge transit speeds of an ion channel pore are comparable to that of copper. This can only be attributed to the details of the quantum effects encountered between the pore walls and the ion during its highly constrained passage through the pore. The fact that the mass of the ion is millions of millions of times more than that of an electron attests to the highly energetic drive forces at work in the pore that accelerate it to a terminal drift velocity akin to copper.

### 10.2.4. *ECS/ICS ion transport through water*

Here we look at Fig. 2(a) ECS/ICS ion behavior (blue/black arrows, respectively). Basic physical chemistry (Hille, 2001) tells us that ion mobility relates to the local electric field intensity as follows:

$$v_{\text{av}} = u_s \left|\mathbf{E}(\mathbf{r},t)\right|, \tag{1}$$

where $v_{\text{av}}$ ($ms^{-1}$) is the average speed for an ion with a mobility $u_s$ $ms^{-1}/vm^{-1}$. Mobility captures the classic drift motion of basic EM texts on conduction (Sadiku, 2001). The ions, on average, move in the direction of **E**. In reality the ions have a much greater total speed (many orders of magnitude), due to thermal (kinetic) energy, upon which this average motion is superimposed. This ECS/ICS random ion motion is shown (greatly exaggerated) in Fig. 2(c). For weak electrolyte solutions such as the ECS/ICS, using the mobility of $K^+$ of $7.62 \times 10^{-8}$ $ms^{-1}/vm^{-1}$ (Hille, 2001, p. 317), assuming they are released from the membrane entrapment shown in Fig. 2(b), and for a typical local electric field magnitude, $|E|$, of 10 $Vm^{-1}$ applied parallel to the membranes, the speed of the $K^+$ ion is, from Eq. (1), $7.62 \times 10^{-8}$ $ms^{-1}$. This means that in the period of one AP (1 ms), when the fields are active, an ECS or ICS ion will move a distance of a fraction of a nm. Field motivated charge motion in the ECS/ICS, over timescales of importance — APs — is vanishingly small. The bulk of the motion is thermal noise. All else being equal, it would take many APs for an ion to traverse a distance comparable to the ECS width of tens of nanometer. Notice the huge difference in average speed between the ECS/ICS and the transmembrane pore speed (Sec. 10.2.3). This vast disparity is the basis for the dominance, in field expression, of transmembrane currents. Note that ion transport could, however, be greatly facilitated by pure thermal diffusion and charge concentration gradient flows (convection). These effects are expected, by this analysis, as a major mechanism behind localized ion motions around ion channels. What is important here is the contribution to *EM fields* when such motion occurs.

### 10.2.5. *ECS/ICS and membrane as a dielectric*

The lipid bilayer membrane dielectric constant is highly anisotropic, inhomogeneous and dependent on chemical composition and frequency. However, for typical purposes in neural modeling of EM fields, it can generally be accepted as around $\varepsilon = 2.1$ (Dilger *et al.*, 1979; Huang & Levitt, 1977). The relative permittivity of water is well documented and is generally accepted as roughly $\varepsilon = 80$ at physiological temperatures and signal frequencies. The lipid bilayer, as a non-conductor and dielectric, has a long history of being modeled as a capacitor. High-frequency nonlinearity effects aside (Bédard & Destexhe, 2008; Poznanski & Cacha, 2012), the lipid bilayer has an effective capacitance of roughly 1 $\mu$F/cm$^2$ or $10^{-2}$F/m$^2$ (Hille, 2001).

### 10.2.6. *Fundamental EM equations*

Setting aside brain tissue for the moment, the most general differential form of Maxwell's microscopic equations can be treated as a system of sources and converted to Eqs. (2) and (3) form that explicitly shows the origins of the field system in any configuration of atoms in space or even free space (see Appendix A for symbol definitions and units).

Static | Dynamic

$$\mathbf{e}(\mathbf{r},t) = \frac{1}{4\pi\varepsilon_0}\nabla \iiint_u \frac{\eta(\mathbf{r}',t)}{|\mathbf{r}-\mathbf{r}'|} d^3r' + \frac{1}{4\pi}\nabla \times \iiint_u \frac{\partial \mathbf{b}(\mathbf{r}',t)}{\partial t}\frac{1}{|\mathbf{r}-\mathbf{r}'|} d^3r' \tag{2}$$

Electrostatic | Conduction | Convection

$$\mathbf{b}(\mathbf{r},t) = \frac{\mu_0}{4\pi}\nabla \times \iiint_u \frac{\mathbf{j}(\mathbf{r}',t)}{|\mathbf{r}-\mathbf{r}'|} d^3r' + \frac{1}{4\pi c^2}\nabla \times \iiint_u \frac{\partial \mathbf{e}(\mathbf{r}',t)}{\partial t}\frac{1}{|\mathbf{r}-\mathbf{r}'|} d^3r' \tag{3}$$

Magnetostatic | Radiation

Constitutive | Free Space

There is only one set of originating charges. In being positioned in space they have a charge density $\eta$ that creates an *electric* field via Eq. (2) left. If these charges also move then they also create a current density $\mathbf{j}$ that results in a magnetic field via Eq. (3) left. If the magnetic field changes rapidly then it creates an electric field via Eq. (2) right. If the electric field changes rapidly it creates a magnetic field via Eq. (3) right. In Eqs. (2) and (3), there are no materials. Adaptation of the equations to a material configuration, involves segregating charge carriers into collections of "bound" and "free" charges. These become the material. The kind of EM phenomena that apply in any given material circumstance depends on which Eqs. (2) and (3) terms apply. In addition to these is the Lorentz force, where a charge q moving with velocity **v** within an electric field **e** and a magnetic field **b** will experience a force **f** as follows:

$$\mathbf{f}(\mathbf{r},t) = q\left(\mathbf{e}(\mathbf{r},t) + \mathbf{v}(\mathbf{r},t) \times \mathbf{b}(\mathbf{r},t)\right) \tag{4}$$

Now, focusing in on the particular form used for brain tissue EM fields, the important aspects of Eqs. (2)–(4) are:

(1) The tissue is a quasi-static convector. The applicable terms are inside the dashed oval. There may be tiny effects due to the Eq. (2) right term. However, this term is assumed to be zero at this stage because the rate of change of the magnetic field is too small to contribute to Eq. (2). Equation (3) right-hand term is similarly too small, and is additionally divided by a huge number, $c^2$, making it a vanishing contribution to the field system. Brains do not radiate anything that is currently detectable by our instruments or that might have a functional relevance within tissue.

(2) If you have a system of sources, say one entire neuron, N1, and its ECS fluid, somehow maintained in otherwise empty space, and if the neuron fired, it would express quasi-static fields **e** and **b** on otherwise empty space well away from the neuron. This is the mechanism by which EEG and MEG are evidenced in space outside the tissue.

(3) The "EM coupling" effect of brain tissue is literally action at a distance provided by Eq. (4). If there happened to be another isolated neuron, N2, sitting in the field of neuron N1, then its fields would be modulated by the field system of N1. This is the mechanism of EM coupling. It does not need a tissue medium.

(4) Magnetic field activity is driven by current *density*, not current (Eq. (3)). This means that magnetic fields from ICS/ECS currents (Fig. 2(a)) blue and black arrows), carrying the identical overall currents originating in transmembrane activity, yet dispersed and randomized over relatively huge volumes, will generate negligible magnetic fields.

## 10.3. The Origin of Endogenous EM Fields in Nervous Tissue

### 10.3.1. *Coherence: The key to understanding tissue fields*

The key to understanding the dominant mechanism responsible for the dynamic component of both the electric and magnetic fields is spatiotemporal *coherence*. Ion motion synchronized in position, direction and time produces coherent currents. The resultant field system will dominate all other charge positions/motions. It is obvious how a plaque of hundreds of physically adjacent ion channels, all conducting ions at the same time implements exactly this very coherence.

### 10.3.2. *The three origins of tissue fields*

Using the idea of coherence, and considering this over short and long terms, and then adding the background transmembrane static electric field system, we immediately isolate three mechanisms that can account for the entire EM-field system (electric and magnetic fields) over all times and over all spatial scales. These are depicted in Fig. 3 (in order of causality) and Fig. 4 (in order of spatial scale) and are detailed later.

In Fig. 3(a), we see the background static electric field (black outlines) impressed on space by the elaborate foam structure of Fig. 2(a)/2(d). Next, in Fig. 3(b), are myriad coherent AP-related (including synaptic activity) ion channel currents that modulate Fig. 3(a) transmembrane field at specific locations, producing a dynamic magnetic field and creating a dynamic transmembrane electric dipole field. Next, in Fig. 3(c), we have the result of many ion channels persisting in particular places in the membrane. These result in polarized regions of adjacent ECS and ICS that have an electric field effect of their own caused by non-homogeneous ECS/ICS space-charge equilibration effects. Systems of weak diffuse evanescent monopoles are to be expected.

A different view of Fig. 3 is shown in Fig. 4 by progression over micro–meso–macro spatial scales instead of temporal scales. In Fig. 4(a) (microscopic), a small patch of neural membrane shows (yellow) the highly confined membrane-transverse *static* electric field system of Fig. 3(a). Across this is the transmembrane (ion channel) dynamic current that produces a local magnetic field and an electric dipole. Figure 4(b) shows many of these coordinated microscopic fields dotted around the neuron. The electric and magnetic fields superpose in space. Finally, if synchronous circumstances prevail, after persistent transmembrane

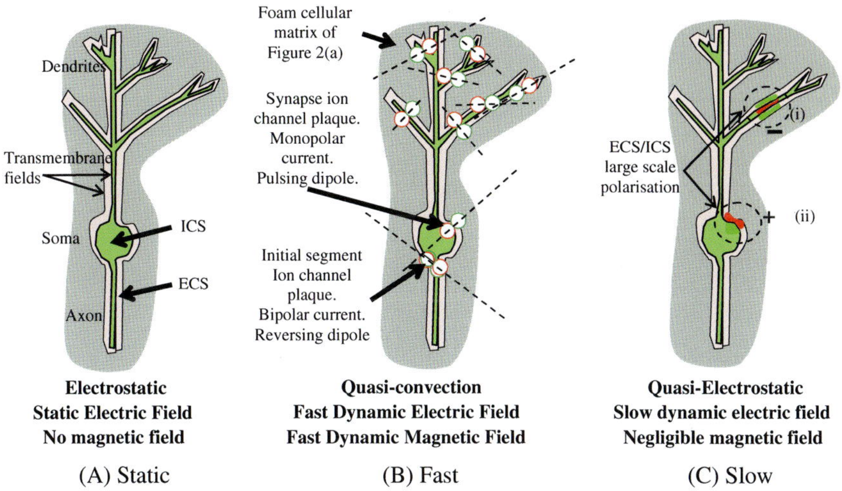


**Fig. 3.** (A) The huge static electric field localized to membrane. (b) Ion channels cause pulsing/reversing dipoles that express electric and magnetic fields over large distances. (C) Persistent (B) activity causes spatially large localized polarization effects that are reset over a longer time frame. ECS (pink) is shown greatly exaggerated.

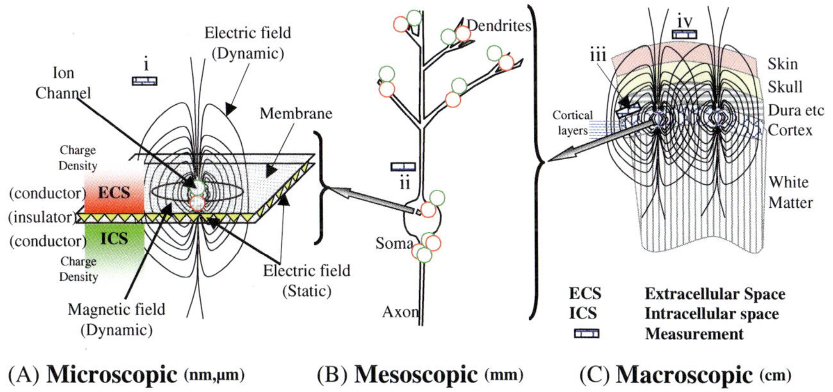


**Fig. 4.** Unifying the EM-field system from its origins to organ level. (A) The actual origins of electric and magnetic field expression at the nanometer scale. Static background electric field (no ion channel activity) is modulated by dynamic electric and magnetic fields that result from ion channel activity. (B) shows the typical depiction of the same field system (electric only) for a layer-spanning cortical neuron. (C) is the macroscopic scale which shows large-scale charge polarization that can express electric fields. Not to scale.

current at the same locations, a regional ECS/ICS polarization effect may result in an *electric field* expressed at the mesoscopic scale. These also superpose on all the other electric fields, and become visible at the macroscopic whole organ level of Fig. 4(c).

To complete the picture, imagine active brain tissue operating normally, but in which all synapse and AP-related ion channels suddenly disappeared or stopped operating. All the electric and magnetic fields associated with Fig. 3(b) would vanish, taking most of the EEG and all the MEG with them. The only remnant EEG would be from Fig. 3(c) large-scale charge imbalance. As this slowly neutralizes, all residual EEG and other tissue potentials would vanish into low-level tissue chemistry field noise. The message is that unless there is (a), there will be no (b). Unless there is (a) and persistent (b) there will be no (c). Note that (b) may arise in both sub-threshold and thresholded AP-related dynamics. (c) results from the sustained activity of (b).

### 10.3.3. *Mechanism 1: Background static field*

The underlying *static* electric field system, depicted as solid black lines in Figs. 2(a) and 3 result from charge profiles shown roughly in Fig. 2(b), which in turn leads to the electric field and scalar potential profiles shown. This is the most energetically dense field system in any tissue: the neuron and astrocyte transmembrane electrical field, which is of the order $10^6$–$10^7$ V/m in the direction of ECS→ICS (Freeman, 1975; Le *et al.*, 1994; Maggio *et al.*, 2008; Peterka *et al.*, 2011; Pethig, 1986; Romijn, 2002). The static background electric field of astrocytes, caused by a membrane potential at roughly –85 mV, is about 20 mV more negative than neurons (–65 mV) (Chen & Nicholson, 2000; Kuffler & Nicholls, 1976; Magistretti *et al.*, 2002). Different ion species have different charge densities on each side of the membrane. The concentration difference for each, and in combination, is a determinant of the resting cell potential (Aidley, 1998; Dayan & Abbott, 2001; Hille, 2001; Johnston & Wu, 1995). This idea can also be extended to localized intracellular stores for specific ion species such as calcium.

The necessary charge partitioning is the result of charge transporters (pumps), shown as Fig. 5(iii), at locations that are not documented in any useful way in the literature. They act asynchronously (no ion current coherence) to restore equilibrium conditions via the conceptual reuptake cycle shown in Fig. 5. The ion transport shown as Figs. 5(ii)/5(iii) produces no functional field system (electric or magnetic) because the

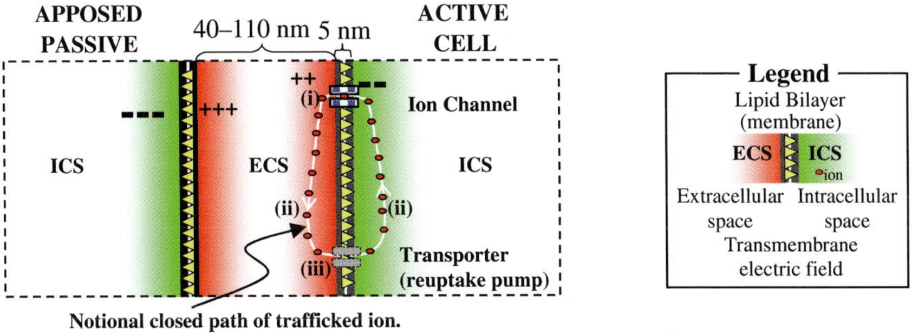


**Fig. 5.** Background transporters (iii) charge up the membrane, resulting in the para-membrane charge density profiles ■ and ■ that give the cells their (negative) resting potential. These act against the loss of ions due to active ion transport (i) through ion channels. Pumping (reuptake) (iii) could occur a long way from (i) the site of active trafficking. (i) FAST, COHERENT. (ii)/(iii) SLOW, INCOHERENT. (i) generates a field system that will stand out over (ii) and (iii), which are background noise.

current density/magnitude is small, diffuse and incoherent, producing only field noise. The only field-relevant current is the dramatically cohered transmembrane current at the specific locations of ion channels shown as Fig. 5(i).

### 10.3.4. *Mechanism 2: Fast, one AP*

Figure 3(b) depicts Figs. 4(a) and 5 transmembrane ion-channel activity (singly or as a colocalized group — a plaque), powered by the static electric field system, creating a spatially large, relatively weak dynamic electric field $\mathbf{e}(\mathbf{r}, t)$ and magnetic field $\mathbf{b}(\mathbf{r}, t)$ system as outlined in Fig. 6. The magnetic field, circulating in the plane of the membrane, lasts only as long as the ion channel current. We have already seen the dramatic difference between the transmembrane ion speeds and the ECS/ICS ion speeds. The speed of ion motion in the ECS/ICS is at least three or four orders of magnitude slower than the transmembrane ion trafficking speed (not the least of which is due to waters of hydration being dragged along with each ion (Hille, 2001)). ECS/ICS ion collision-free paths are basically limited to the size of water molecules (~0.3 nm). This is roughly 20 times smaller than the length of the (nm wide) transmembrane pore, which dramatically limits the direction of the ion trajectory.

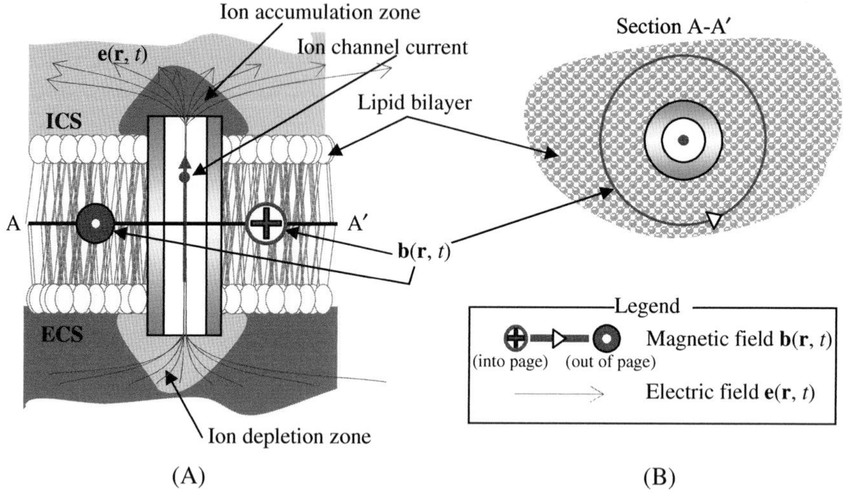


**Fig. 6.** The dynamic electric field, **e**(**r**, *t*), and magnetic field, **b**(**r**, *t*), characteristic of a transmembrane current filament. (A) Shows the membrane-transverse electric field arrangement. (B) Shows the magnetic field expected from the dynamics of the membrane-transverse current, shown in the plane of the membrane along section A–A′. Relatively rapid transmembrane ion traffic creates localized charge disparities in the regions proximal to the channel ends that disperse slowly because of the extremely poor charge mobility in the bulk electrolyte (ECS/ICS). The electric field lines project out into the ECS/ICS. The magnetic field strength directly follows the time course of the transmembrane current.

Ions are therefore transported across the membrane at a rate far higher than the adjacent ECS/ICS can replenish. The net result is a rapid localized depletion/accretion of charge: a local electric field dipole expressed against/astride the background transmembrane electrostatic field. This effect has been suggested in the literature previously (Destexhe & Bedard, 2012). The transmembrane charge transport dynamics are similar to that of say, a copper wire. Indeed the entire field system is simply what would be expected around a small length of wire, or within a diode, and should be unsurprising to any electrical engineer. Such dynamic electric dipoles can result from APs and/or synaptic current flows. AP currents reverse, so the dipole can reverse, reflecting the bipolar nature of the AP current. In contrast (chemical and electrical) synapses have a monopolar current, which means their dipole merely pulses.

#### 10.3.4.1. *Sub-threshold behavior*

Note that subthreshold oscillations involve transmembrane ion traffic and thereby potentially express a functionally relevant microscopic dynamic EM field contribution. Such oscillations and their fields can be expected to operate in the future mature neural modeling regimes and are outside the scope of the later computational examination of field origins. At this stage we seek merely to recognize that the proposed Fig. 3(b) mechanism also allows subthreshold oscillation as part of the "fast" field contributions, even though by definition the "AP" has not occurred yet.

### 10.3.5. *Mechanism 3: Slow, many AP*

Figure 3(c) depicts the result of an extended barrage of AP-related activity, say a whole group of collocated inhibitory or excitatory synapses. This can result in a large, more diffuse, but confined ECS/ICS charge dipole. It is basically a greatly magnified depletion/accretion zone of the kind in Fig. 6(a). Because of differing ECS/ICS charge mobilities (Sec. 10.2.4) and geometries, the ECS and the ICS can equilibrate back to normal charge concentrations at different rates in differing physical locations depending on the reuptake pump locations and numbers. For example, Fig. 3(c)(i) ICS is shown equalizing faster than the ECS portion, converting it from a dipole to a negative electric monopole. For the same reasons, and at some distance, monopole Fig. 3(c)(i) forms. The two pseudomonopoles may then form a weak and fading dipole field system linking the positive and negative monopoles. This is the only way the traditional large-scale electric polarization, e.g., (da Silva & Van Rotterdam, 2005; Kandel & Schwartz, 2000; Nunez & Srinivasan, 2006; Speckmann & Elgar, 2005) can work in practice. While the localized monopoles exist, weak quasi-static electric fields form a bulk tissue regional atmosphere in which Figs. 3(a) and 3(b) field systems operate until overall neutrality is restored or reversed by ion pumps and electrolyte charge flows. Note that both Figs. 3(b) and 3(c) may have an EM-coupling functional effect on signaling. The relative contribution of each is not something examined in this chapter.

## 10.4. Compartmental Models and the Three Mechanisms

Here the three Fig. 3 fundamental mechanisms are related to the currents in traditional compartmental neural models. The microscopic current mechanisms and the traditional compartmental model are shown side by side in Fig. 7. In Fig. 7(b), the membrane is a capacitor **C** in parallel with a nonlinear circuit element **g**, and its reversal potential "battery" **E**. One parallel branch per ion channel type. All ECS ⇔ ICS ion species (sodium, potassium, etc.) and capacitive currents are covered.

A compartmental model is constructed to replicate transmembrane voltage **V**(**t**) and overall currents, not the electric/magnetic fields. In using the lumped circuit element model, EM fields are abstracted away. In reality only a subset of the currents labeled 1 . . . 5 in Fig. 7(b) actually generate the coherent EM field. Components **g** and **E** are actually inside the dielectric of **C**. Also hidden within **E** current is a further division into a large number of small currents associated with ion transporters involved in the charge reuptake (notionally elsewhere on **C**) cycle of Fig. 5. As already discussed these currents restore the background static field across the membrane. This is how Fig. 3(a) mechanism 1 is hidden in the model within the current labeled 4 in Fig. 7.

Figure 3(b) electric and magnetic field system is created by the activity inside **g** and **E** (current labeled 3) that is actually inside **C** in Fig. 7(b). In contrast, the ECS current (Fig. 7 current 2) and the ICS current (Fig. 7 current 1) both produce only field noise as discussed earlier. The previously discussed dipole/monopole effects behind the Fig. 3(c) electric fields can be thought of as a result of persistent charge

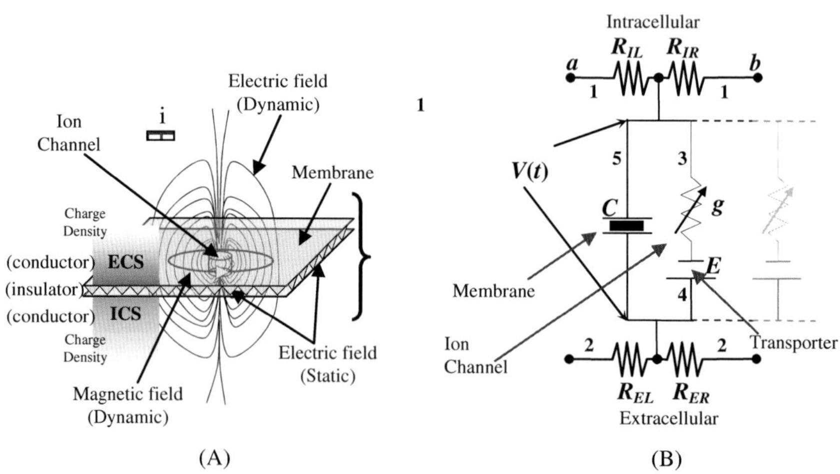


**Fig. 7.** The generally accepted fundamental unit of the AP signaling system in compartmental "lumped-circuit" element equivalent-circuit models.

stored in $\boldsymbol{C}$, captured by nonlinear effects in $\boldsymbol{g}$ that asymmetrically spread, possibly to other compartments through the resistors, thereby creating the possible monopole activity described above.

By connecting compartment to compartment (by components $R_{IL}$, $R_{IR}$, $R_{EL}$ and $R_{ER}$ in Fig. 7(b)) end to end, branching and terminating, complex generalized cell topologies can be modeled (Aidley, 1998; Dayan & Abbott, 2001; Hille, 2001; Hodgkin & Huxley, 1952a, 1952b, 1952c, 1952d, 1952e; Hodgkin *et al.*, 1952; Izhikevich, 2007; Johnston & Wu, 1995). The compartment model is quite general. In addition to AP ion channels, it allows for both electric and chemical synapses and subthreshold oscillations, each of which produces its own distinctive field system component in space. Space is abstracted away in Fig. 7(b). In the new EM field-inclusive models that account for both EM and AP coupling, space will have to be restored because the EM field generated by one compartment impacts (line-of-sight) other compartments and alters its $\boldsymbol{V}(\boldsymbol{t})$ by means not shown in Fig. 7(b).

Note that ion channels can be sparsely located in the membrane, part of a plaque in a chemical synapse (synaptic cleft plaque), part of a plaque in an electrical synapse (gap junction) or part of a plaque involved in APs such as the axon hillock/initial segment densities. They could be on dendrites, axons or soma or nodes of Ranvier (salutatory axon propagation, ringed densities separating myelinated axon segments). In each case, rapid coherent transmembrane ion transit can be held accountable for the EM field production in the manner described here.

## 10.5. A Computational Exploration of the Coherent Dipole System Created by Transmembrane Current Filaments

Exploration of the abstract mathematics of current filaments in a uniform conducting medium revealed filaments express tiny dipoles of exactly the kind in Fig. 6. To get a basic understanding of the EM-field system due to the Fig. 6 transmembrane currents (Fig. 3(b) electric and magnetic fast fields system) all we need to do is drive a spatially located set of current filaments with compartmental neuron model currents. The strategy is, therefore, to:

(a) Delete all tissue. Replace it with an infinite uniform conductance.
(b) Use a compartmental model to generate all the necessary currents of the kind underlying Fig. 7(b).

(c) Drive (a) with the filamentary source current(s) of (b) placed manually at known positions in space. This mimics ion channel activity. How the filamentary source is energized is irrelevant.

This simple configuration of materials and abstract current sources, appropriately driven by tissue-mimetic transmembrane currents positioned in space, must in principle produce brain-mimetic transmembrane dipole sources of the kind in Fig. 6. Of course, in reality much more complex electrodiffusion transport phenomena are at play. This gross simplification is a starting place only. Nevertheless, it will be shown to produce field systems arguably of a form similar to that we can expect in tissue. It can do this relatively economically. To see how to proceed consider Fig. 8 where we formalize the concept of a filamentary current in an infinite uniform conductance, yet configured in a transmembrane context.

Figure 8 shows a single current filament embedded in an infinite uniform conductivity $\sigma$. Charge $q_{m,k}$ is forced along a very short known path (mid-position $\boldsymbol{r}_{m,k}$, direction $\boldsymbol{n}_{m,k}$) by means unspecified. In doing so it constitutes a known source current $i_{m,k}(t)$ (a known current density) that may vary in time. That is, the amount of charge transiting the very short length $L_{m,k}$ of the filament $k$ is variable over time $t$. To connect the individual filament to a system of filaments applicable to typical brain models, consider that filament $k$ is merely one of a given cohort of a group of $K_m$ filaments associated with a known neuron compartment $m$, which in turn is merely

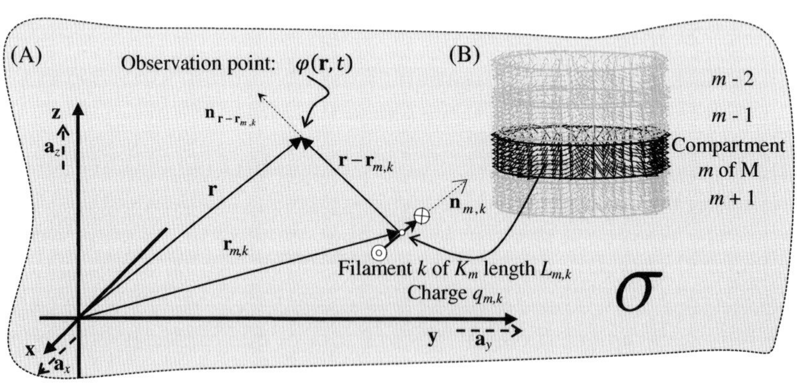


**Fig. 8.** A notionally infinite conductive medium of uniform conductivity σ. (A) Current $i_{m,k}(t)$ filament $k$ within a cohort of filaments (B) $K_m$ of compartment $m$ of $M$, located at position $\tau_{m,k}$, and of length $L_{m,}$ $k$, contributes to the field system (in this case, scalar potential $\varphi$) at position **r** at time $t$.

one of a large number, $M$, of such compartments located in space. One can immediately see the relationship between a physical neural membrane, the ion channels penetrating it and the typical scenario of a classic compartmental model of a neuron. Remember, however, that in Fig. 8 there is no membrane. There are only filamentary currents embedded in the positions determined as if there was a neuron membrane.

We are ultimately interested in the electric field $\mathbf{E}(\mathbf{r},t)$ ($Vm^{-1}$) at a chosen observation point $\mathbf{r}$ (meters, m) at a known time $t$ (seconds, s). That field is the vector sum of the contributions from each of the multitude of filamentary sources in all compartments throughout a model neuron. To compute this field all we need, initially, is to get the scalar potential $\varphi(\mathbf{r}, t)$ (in Volts, V). The electric field is directly related to the potential by

$$\mathbf{E}(\boldsymbol{r},\ t) = -\nabla\varphi(\mathbf{r}, t). \tag{5}$$

Fortunately there is a relatively well-developed form of Maxwell's equations specialized to the circumstances of current sources in an infinite conductance. It is called the "volume conduction" model. It was first developed by Plonsey and colleagues in the 1960's for biological material (Clark & Plonsey, 1966, 1968; Malmivuo & Plonsey, 1995; Plonsey, 1964; Plonsey & Collin, 1961; Plonsey & Heppner, 1967; Plonsey & Fleming, 1969). A quasi-magnetostatic specialization of the fundamental macroscopic EM equations says that the scalar potential (Volts) at location $\mathbf{r}$ and time $t$ is

$$\varphi(\mathbf{r}, \mathrm{t}) = \frac{1}{4\pi\sigma} \oiiint_V \frac{I_V(\boldsymbol{r},t)}{|(\boldsymbol{r}-\boldsymbol{r}')|} d^3r', \tag{6}$$

where

$$I_V(\mathbf{r},t) = \nabla \cdot \mathbf{J}_S(\mathbf{r},t) \tag{7}$$

and the volume current source density, $I_V$ ($Am^{-3}$), contained within volume $V$, has an interpretation as a "current moment per unit volume". Mathematically the volume current density results from the behavior of a non-ohmic vector source current density $\boldsymbol{J}_S$ ($Am^{-2}$) (Plonsey & Collin, 1961; Plonsey & Fleming, 1969). Conductivity $\sigma$ is in $(\Omega m)^{-1}$ and is a linear, time invariant material property that effectively renders brain material a passive, infinite, uniform continuous conducting media. The method of derivation of Eq. (6) involves an assumption that Ohm's law applies, followed by the imposition of a known source current density that carefully violates Ohm's law. In pure conduction, the divergence of the current density is specified as zero. The Ohm's law violation allows the divergence to be non-zero in a localized region. At distance from the disturbance, the volume conduction formalism still applies. A further assumption is that all disturbances in the current density settle out to zero much more quickly than the spatiotemporal scales of interest. In this way, the material is essentially free of all dynamics related to the permittivity.[2] This results in Eq. (6).

### 10.5.1. *Customizing volume conduction to filament sources*

To reach the final equation used in the computations, the Fig. 8 collection of infinitesimal current density filaments is constructed to act as sources in Eq. (7). An individual instantaneous current density filament, corresponding to notional ion channel $k$ in compartment $m$, can be expressed as (Jackson, 1999; Plonsey & Fleming, 1969)

$$\mathbf{J}_{m,k}(\mathbf{r},t) = q_{m,k}v_{m,k}(t)\delta(\boldsymbol{r}-\boldsymbol{r}_{m,k}(t)), \tag{8}$$

where charge $q_{m,k}$ (Coulomb) is that of the charge species suited to a particular ion channel. The punctate nature of the ionized electrolyte atom is captured by the Dirac delta function. The instantaneous velocity $V_{m,k}(t)$ ($ms^{-1}$) relates to the charge position $\mathbf{r}_{m,k}(t)$ as per

$$v_{m,k}(t) = \frac{d\boldsymbol{r}_{m,k}(t)}{dt}. \tag{9}$$

Charges travel single file (Hille, 2001) through the pore with an average velocity determined by the pore current. The direction of the average velocity is identical to the pore direction. The pore location defines the beginning of the journey through the pore, which, for a known pore current $i_{m,k}(t)$ (A), involves a known transit time for a particular pore length $L_{m,k}$.

Using an assumption that filaments have length $L_{m,k}$ that is relatively small compared to the distances of interest in the field expression, the total potential at an observed point of interest $\mathbf{r}$ at time t is given by:

$$\varphi(\boldsymbol{r},t) = \frac{-1}{4\pi\sigma}\sum_{m=1}^{M}\sum_{k=1}^{K_m}\left[\frac{L_{m,k}i_{m,k}(t)(n_{m,k}\cdot \boldsymbol{n}_{m,r-rm,k})}{\left|(\boldsymbol{r}-\boldsymbol{r}_{m,k})\right|^2}\right]. \tag{10}$$

---

[2]This does not mean that permittivity is unimportant in tissue. It means that for purposes here permittivity is regarded as unimportant. Future work can examine the impact of permittivity.

The total number of compartments is $M$. The total (variable) number of filaments for a particular compartment $m$ is $K_m$. Vector $\mathbf{r}_{m,k}$ locates a filament in space, with a normalized direction vector $\mathbf{n}_{m,k}$. The vector $(\mathbf{r} - \mathbf{r}_{m,k})$ locates our observation point $r$ with respect to each filament in the cohort, has its own normalized direction vector as shown, and must be large compared with $L_{m,k}$. This distance limitation is consistent with the original distance limit in the volume conduction formalism of Eq. (6). Based on well-traveled compartmental neuron modeling, we know the transmembrane currents. By our choice of compartmental geometry and the positioning of current filaments in the surface of a compartment (ion channels), we can compute Eq. (10) for a collection of observation points in space. Alternatively we can use Eq. (5) on Eq. (10) and compute the electric field instead.

$$\mathbf{E}(\boldsymbol{r},t) = \frac{1}{2\pi\sigma}\sum_{m=1}^{M}\sum_{k=1}^{K_m}\left[\frac{L_{m,k}i_{m,k}(t)(\boldsymbol{n}_{m,k}\cdot\boldsymbol{n}_{m,r-rm,k})}{\left|(\boldsymbol{r}-\boldsymbol{r}_{m,k})\right|^3}\right], \tag{11}$$

where

$$\boldsymbol{n}_{m,k} = \frac{(\boldsymbol{r}-\boldsymbol{r}_{m,k})}{\left|(\boldsymbol{r}-\boldsymbol{r}_{m,k})\right|}. \tag{12}$$

Inserting the filamentary current Eq. (8) into fundamental Eq. (2) (left):

$$\mathbf{B}(\boldsymbol{r},t) = \frac{\mu_0}{4\pi}\nabla\times\iiint_U \frac{\mathbf{J}(\boldsymbol{r}',t)}{\left|(\boldsymbol{r}-\boldsymbol{r}')\right|}\mathrm{d}^3\, r' \tag{13}$$

Yields, using the same approximations used for the electric field,

$$\mathbf{B}(r,t) = \frac{\mu_0}{4\pi}\sum_{m=1}^{M}\sum_{k=1}^{K_m}\left[\frac{L_{m,k}i_{m,k}(t)(\boldsymbol{n}_{m,k}\times(\boldsymbol{r}-\boldsymbol{r}_{m,k}))}{\left|(\boldsymbol{r}-\boldsymbol{r}_{m,k})\right|^3}\right]. \tag{14}$$

We now have established a basic set of Eqs. (10), (11) and (14) describing the scalar potential, electric field and magnetic field that is expressed by the *same* cohort of current filaments, at a distance relatively large compared to the length of the current filaments. This allows us to get a basic feel for the way we can expect the field system to operate at-distance in tissue.

### 10.5.2. *Neuron model and experiment preparation*

To explore the scalar potential field system, the AP-related currents in a large neuron are examined with Eq. (10). We therefore need to establish transmembrane currents and a detailed geometry. To achieve this we use a third-party compartmental model of a rat hippocampus CA1 pyramidal neuron, with a known provenance and a history of prior use. The single neuron is identified as D151 and is available via the NEURON (Hines & Carnevale, 1997) site http://www.neuron.yale.edu. It is called the "EAPS" package and is found under ModelDB database number 84589. EAPS stands for "Extracellular Action Potential Simulation." It uses a different field computation method called LSA. Here, the LSA portions of the EAPS package were left unused. Neuron D151 is highly documented and has been microscopically engineered by the EAPS originators with the best physiological information available at the time.

The important qualification here is that there are no chemical or electrical synapses in the model. What is provided with D151 is the contribution to the field system expressed on space by the currents in the ion channels involved in a single AP. Based on the equations, significant field contributions can be expected from synaptic plaques with coherent currents. These have not been included here. Therefore, the full field system expression will ultimately need all synapses and their currents, located and oriented accurately in space. Their field contribution will vectorially add to the field system produced by this work. Meanwhile, we can get a basic understanding of expectations by simply allowing the AP currents to tell their part of the field-system story.

There are complications. The D151 model includes 12 different ion channel types delivering five different kinds of currents. The NEURON compartments are modeled as 1128 conical frusta (Fig. 10(b)). The underlying ion channel densities account for the best known (at the time) physiologically realistic variations throughout the D151 structure (Gold *et al.*, 2007, 2006). Figure 9 shows five views of neuron D151 along with its geometric (spatial) mean computed without the axon. The axon is an artificial construct with standardized dimensions and properties (Gold *et al.*, 2007, 2006).

To provide input data for the field computation, a single D151 AP was triggered. The EAPS package, by default, triggers an AP by slowly modulating a few $Na^+$ synapse conductances, thereby depolarizing the cell without introducing erroneous synapse or soma current artifacts into the

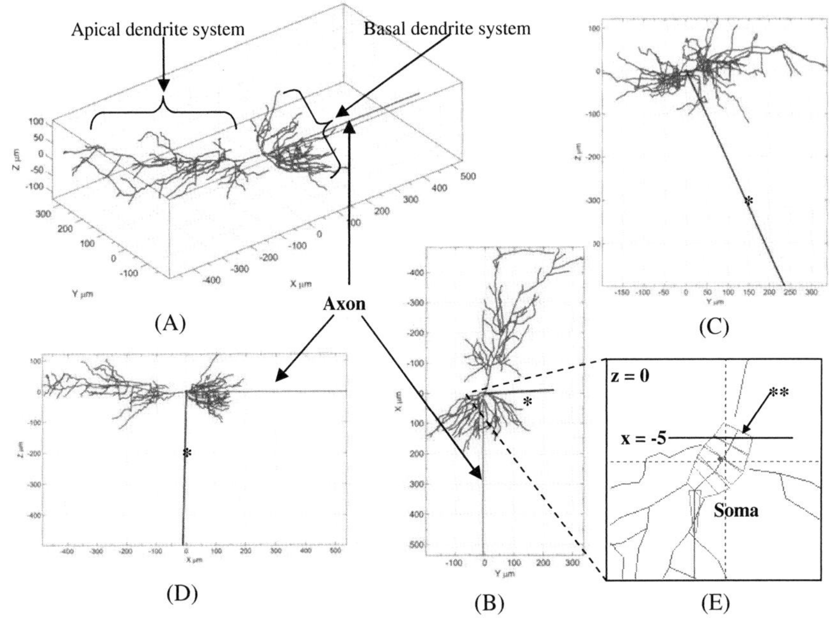


**Fig. 9.** Rat hippocampus CA1 neuron D151 shown in (A) isometric view, (B) *XY* plane viewed from +Z, (C) *ZY* plane viewed from +X and (D) *ZX* plane viewed from the – *Y*-direction. The * indicates the single vector representing the net geometry of the apical and dendrite compartments when the compartment's frustum centerline is treated as a vector. (E) The soma shown as fully expressed compartments in an exploded view of the origin and environs. ** frustum 1, a soma compartment which is the subject of Test 1. The soma compartment diameters average roughly 8 $\mu$m and their height is roughly 4 $\mu$m.

EM-field system. The EAPS package current report was modified to decompose the total compartment current, $I_{Tot}$, into $I_{Na}$, $I_K$, $I_{cap}$, $I_{Ca}$ and $I_{pas}$ (sodium, potassium, capacitive, calcium, passive). Additional preprocessing of current data involved their reverse-projection back onto the 1128 compartment three-dimensional (3D) structure. Note that the D151 model current dynamics are realistic, exhibiting the usual orthodromic and antidromic current propagation. This results in activity in all five classes of current. The dominant sodium, potassium and capacitative components overshadow the calcium and passive currents.

For test preparation purposes, the EAPS package was enabled, with its AP trigger delay at its default $t = 10$ ms. The delay allows the system to numerically stabilize to resting conditions. The EAPS cumulative trigger mechanism means that the AP occurs around 12 ms and continues until roughly 15 ms, during which all compartment currents go through the stereotypical variations associated with the various ion current types (see Fig. 10(a)). The balance of the 25 ms period allowed the system to settle. The recorded currents cover the period 0 ms to 25 ms (with variable time intervals of roughly 50–100 $\mu$s). This resulted in a 159-sample time-series for all six currents for every compartment in the model. That done, the EAPS program plays no further role. The EAPS "out of the box" functionality is minimally disturbed and is left that way so that other researchers can easily replicate and expand on the results. The 3D decomposed and projected compartment transmembrane currents are displayed for compartment 1 in Fig. 10(a). The procedure that projects the D151 NEURON currents back into the original 3D geometry does not retain the ion channel density information except in sofar as the underlying NEURON simulation parameters, and hence the transmembrane currents, are affected. Overall current density variations are expected to track the ion channel density variations throughout the D151 structure.

Experiments involved the subsequent feeding of the recorded currents and geometry information into a collection of customized MATLAB programs. The most important of these custom MATLAB programs uses Eq. (10) to compute the scalar potential produced by the recorded currents when appropriately apportioned into filamentary sources positioned in space according to the geometry of the studied notional membrane compartments. In its implementation, the total transmembrane compartment current, $I_{Tot}$, was divided equally throughout a single cohort of current filaments allocated to the compartment. Although it was considered, it is impractical, at this stage, to individually allocate filaments to each current type and individually locate all the different filament types in different ways. There is simply no available information or obvious strategy by which this process can be administered. In view of the overall intent to merely demonstrate the basic principles of field expression by filaments, it was simply decided to use a single set of filaments that collectively conduct $I_{Tot}$. This choice has no effect on the basic conclusions of this work.

The filaments have a length equal to the cross-sectional width of a typical lipid bilayer penetrated by a large ion-channel protein ($L_{m,k} = 7{:}5$ nm filament length was used). Having allocated filaments to each compartment, each filament is then located with one end embedded in the surface of its host compartment and the other oriented orthogonally to it. What

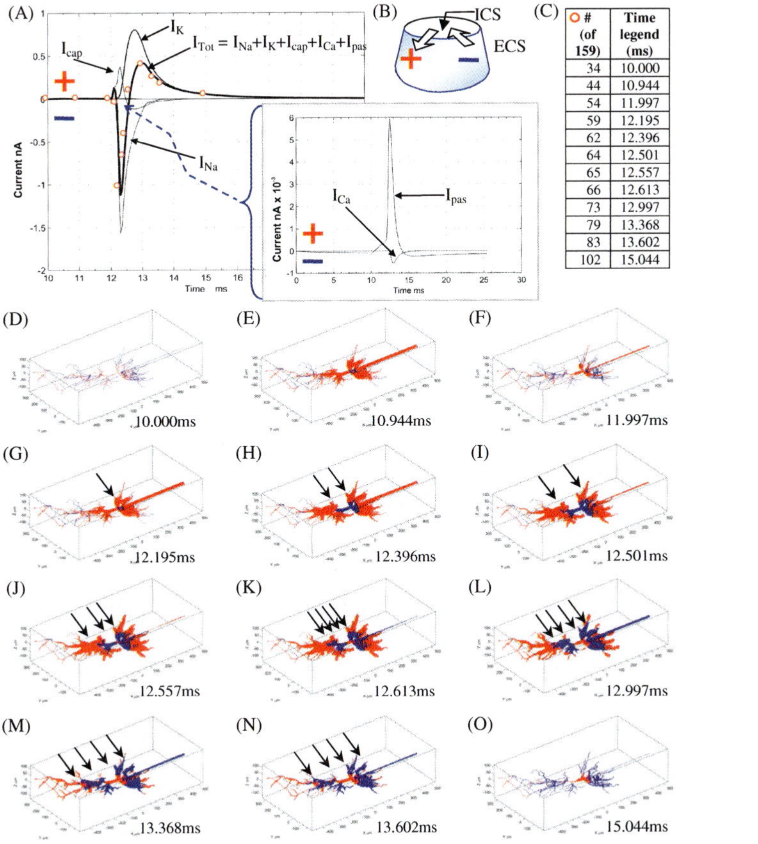


(C)

| ○ # (of 159) | Time legend (ms) |
|---|---|
| 34 | 10.000 |
| 44 | 10.944 |
| 54 | 11.997 |
| 59 | 12.195 |
| 62 | 12.396 |
| 64 | 12.501 |
| 65 | 12.557 |
| 66 | 12.613 |
| 73 | 12.997 |
| 79 | 13.368 |
| 83 | 13.602 |
| 102 | 15.044 |

**Fig. 10.** Test 1. Supplementary videos S2. (A) Transmembrane currents of D151 compartment 1 of 1128. The specific compartment is marked ** in Fig. 9(e). The dark line is the total transmembrane current $I_{Tot}$. Currents $I_{Na}$, $I_K$, $I_{cap}$, are the dominant sodium, potassium and capacitative components, respectively. Currents $I_{Ca}$ and $I_{pas}$, shown at an expanded scale, are the smaller calcium and passive currents. The time course proceeds according to the EAPS default setting discussed in the main text. (B) shows a typical conical frustum, indicating that the conventional positive current corresponds to outward flow of positive ions. (C) The table contains the time sample associated with the red circles in (A). The left column is the sample sequence number. The right column gives the actual time associated with that sequence number. For example, the 62nd sample occurred at time 12.396 ms. These times (and others) appear in later results and figures. (D)–(O) show the total compartmental transmembrane current propagation throughout cell D151 color-coded as per (B). The thickness of the line is indicative of current magnitude. Arrows indicate the emergent wavefronts discussed in the text.

remains the same throughout this variability is the total transmembrane current. Mathematically, the NEURON model 3D frusta are discontinuous notional constructs in a continuous conductive medium in space with Eq. (10) conductivity set at 1/3 $Sm^{-1}$ (da Silva & Van Rotterdam, 2005). This is order-of-magnitude accurate and is a determinant of field magnitude rather than the more interesting structure and dynamics. Despite this high level of material abstraction, useful realism results from the physiological accuracy of the underlying current dynamics and cell morphology.

It is not the intention of the experiments to validate the NEURON EAPS D151 behavior or accuracy, but to accept it as given for the purposes of later field computation using Eq. (10). Likewise it is not the intent of the D151 scalar potential computation to accurately predict the potential in any particular empirical context. The electric field and magnetic field are not computed, although it is a straightforward exercise. A series of six experiments examine the spatiotemporal expression of the scalar potential field, in various contexts, during the passage of a single AP. All experiments accept, as input, the "spatialized" NEURON/EAPS current recordings described above. The overall test regime is as follows:

- Test 1: Visualize D151 transmembrane currents for interpretive purposes.
- Test 2: Examine basic scalar potential produced by regular arrays of current filaments. Uses a single D151 soma compartment.
- Test 3: Compute spatial scalar potential. Visualize slices through the 3D scalar potential field system produced by the entire D151 structure.
- Test 4: Decompose test (3) result into basal, apical and somatic contributions to see how they superpose to create the test (3) result.
- Test 5: Variability of scalar potential with filament position. Varies the filament positions to examine the scalar potential sensitivity to filament position.
- Test 6: Artificial neuron, neuron process orientation sensitivity. While keeping current filaments in the identical positions in the surface of the cell, move whole compartments in space to see the effect it has on the scalar field morphology and dynamics. Uses a particular D151 basal dendrite and the D151 soma (and the same currents each compartments normally conducts) to create a small artificial neuron for which the scalar potential is computed with the soma oriented three different ways.

### 10.5.3. *Test 1: Overview of compartment currents*

Figure 10 shows the results of the NEURON current computation and a spatialized, color-coded depiction of the compartment currents as expressed over time. Figure 10(A) shows the time progression of the individual and total transmembrane AP current for compartment 1 shown as ** in Fig. 9(E). The Fig. 10(A) behavior is typical of all 1128 compartments as delivered by the NEURON EAPS simulation. The dominance of the sodium and potassium currents is clear. Figure 10(B) shows the sign convention (and color code): positive current is conventionally taken to be an outward flow of positive ions. A flow of negative charge inward, *into* the ICS, is also a positive current. In Figs. 10(D)–10(O), the compartment is represented by a single line (along the axis of each frustum). Line color reflects the sign of the current. Red for positive current. Blue for negative. Line thickness indicates the magnitude of the current. Figure 10(c) shows a subset of the times at which Figs. 10(D)–10(O) is plotted. These are indicated by the small red circles in Fig. 10(A). Two supplementary videos (S1 and S2) are provided to illustrate the progression of currents throughout the neuron.

The AP current begins in the axon initial segment and immediately spreads dromically, quickly down the axon and more slowly along basal dendrites and antidromically, via the soma into the apical dendrites. The soma region cannot be seen in the stills because it is obscured. To witness the soma and initial segment, a zoomed video of the soma area is provided in supplementary S2.

Emergent behavior is evident in the progression of interfacial regions of differing total transmembrane current polarity. The sequence reveals two moving dipole fronts spreading apart in the basal and apical dendrites. These wave-fronts are arrowed in Figs. 10(D)–10(O). The triggering of the AP is signified by the early RED appearance in Figs. 10(D)–10(F). This reflects the build up of soma potential by non-specific synapse (EAPS package) manipulation and the initial positive current shown in Fig. 10(A). The change to blue in Figs. 10(G) and 10(H) then reflects the deep negative swing due to sodium current. The later red in Figs. 10(J)–10(O) reflects the return overshoot (potassium) region of the current. This activity is largest in and around the soma. Collectively expressed through the coherent action of physically adjacent branching compartments, it is easy to see the creation and progression of a pair of red/blue wave-fronts of spreading virtual dipoles. While one compartment is red, a nearby compartment is blue because it is in a different part of Fig. 10(A) waveform. An evanescent virtual dipole is created in space, existing only for the duration of a localized spatial current polarity disparity. It is these currents that, via Eq. (10), produce the detail of the scalar potential field of the following tests.

Finally, note what is not shown: ECS and ICS currents. These are the exact same currents shown in Fig. 10, and are represented in the *horizontal* portions of the isolated lumped-element model shown in Fig. 7(B). This current has the same current total as the *vertical* portions of Fig. 7(B). Conservation of charge and Norton's equation/Kirchoff's law means that the currents in the horizontal and vertical portions must add up to the same numerical value. The important difference is in the comparative transport dynamics and spatiotemporal coherence between the vertical and horizontal currents. It is the current density of each branch in Fig. 7(B) that distinguishes the vertical and horizontal branches. Despite the ECS/ICS currents being numerically identical, and reflective of the same compartmental voltage, because the current density differs, their contribution to the *field system* is completely different. Only Fig. 7(B) transmembrane (vertical) current contributes any structured field contribution. The highly confined, rapid (think gunshot) transmembrane current density represented by ion-channel-mediated charge transport constitutes a current density that dominates the field system.

As discussed above, the ECS and ICS currents are additionally decorrelated on greater spatiotemporal scales to ensure that they continue to contribute only field noise at the greater scale. It is only when the physical instantiations of the charge transport are reinstated that one can appreciate the degree to which the same current, flowing in a different way, can result in an entirely different field system in tissue. This is how the compartmental lumped-element circuit model loses contact with the field system aspect of tissue function.

### 10.5.4. *Test 2: The scalar potential field of individual filaments*

Prior to full scale scalar field computation, here we examine the microscopic scalar potential expressed by a regular array of individual filaments located on a single compartment. This will establish the nature of the basic building block of the scalar field system for the whole of neuron D151. To do this, neuron D151 frustum number 1 of 1128 (Fig. 9(E)**) is isolated and populated with eight rows of three filaments equally distributed radially around the frustum surface. This is depicted in Fig. 11. The dotted line is the centerline of the frustum. The dashed lines form a rough outline of frustum 1. The filament lengths are 7.5 nm and are shown greatly exaggerated in Fig. 11.

The total compartment current of Fig. 10(A) is evenly distributed throughout the 24 filaments. The potential for the lone frustum is computed in three perspectives (*XY*, *XZ* and *YZ*), using Eq. (10) over the interval 11.839 ms to 15.044 ms to a spatial resolution of 0.1 $\mu$m. The results are expressed in still-frame (Fig. 12) and three supplementary videos S3(*XY*), S4(*XZ*) and S5(*YZ*). Selected snapshot frames (from the times indicated in Fig. 10(C)) show that the computations covered a volume 25 × 15 × 20 $\mu$m centered on $[x = 0, y = 2.5, z = 0]$ $\mu$m. Slices through this volume were taken as follows: *XZ* plane at $y = 0$ $\mu$m, *YZ* plane at

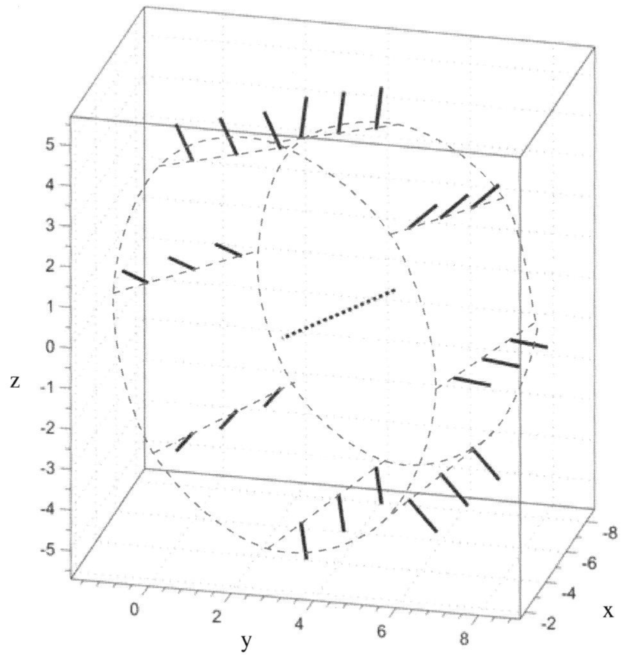


**Fig. 11.** Regular filament structure for Test 2. Filaments (0.0075 $\mu$m long) are shown exaggerated, not to scale. The dotted line is the frustum centerline. The dashed line is a rough (not to scale) outline of frustum 1 marked ** in Fig. 9(e). The scale is in $\mu$m.

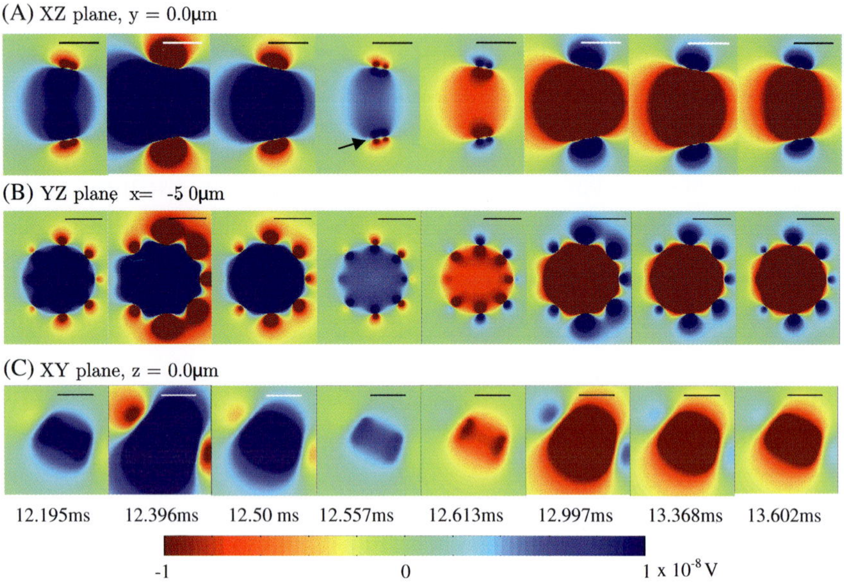


**Fig. 12.** Test 2. Supplementary videos S3–S5. LFP plots of slices (A) $y = 0$, (B) $x = -5$ and (C) $z = 0$ for the single frustum shown in Fig. 11(E). The arrow points to the filament-width planar region that would be the location of the lipid bilayer in the real material. Calibration bar = 5 $\mu$m.

$x = -5$ $\mu$m, $XY$ plane at $z = 0$ $\mu$m. The electric field lines are orthogonal to isopotential contours (of equal color). Note that Fig. 12(B) slice intersects obliquely with compartment 1, explaining why the scalar field pattern is non-uniform across the slice.

It is easy to see the contribution of individual filaments in Fig. 12(B). A single filament would produce a pulsing/reversing dipolar plume of potential. Figure 10(A) tells us that the current at 12.195 ms is slightly negative. In the progression of Fig. 12(B), the plume increases in magnitude until it goes through the current reversal, is increasingly positive and then settles down. This is the simple reversing dipole behavior of a uniform conductor impressed by a current filament carrying Fig. 10(A) waveform. More interesting is the way adjacent coherent current filamentary dipoles superpose their scalar potential fields. Importantly, notice that the field superposition naturally preserves a zero-potential boundary located where the membrane would be in the real tissue (remember, there is no actual membrane present, only uniform conductor). Despite there being the same conductivity throughout space, the current naturally avoids the plane that would be occupied by the membrane in real tissue. The location of the membrane is shown by the arrow in Fig. 12(A) 12.557 ms still-frame.

Test 2 therefore reveals that compartment-shaped current filament disposition produces a compartment-shaped potential. Figures 12(A) and 12(C) also show how the ICS-shaped potential would bleed longitudinally into the adjacent compartments if they were present. Overall, test 2 provides insight into how a large collection of similar compartments, all with spatially stationary filaments driven by coherent currents, may superpose individual fields in space to produce a complex unified and coordinated field system. A metaphorical way to view the images and videos is that of a pressure vessel first relieving and then re-establishing the pressure.

### 10.5.5. *Test 3: Full D151 scalar field*

Here we explore the collective scalar potential field expressed by the entire D151 neuron with a randomly positioned set of filaments. Each D151 compartment is allocated filaments based on a notional average "filament density" of 1 per $\mu m^2$ over the compartment outer (known) surface area, but with an enforced minimum of 10 and a maximum of 25 filaments on any individual compartment. This ensures some level of spatial distribution of the current for small (low external surface area) compartments and minimizes computation time, respectively. These constraints resulted in 19,066 filaments. Each filament is embedded orthogonally to the frustum surface and is located by randomly determining a circumferential position and height within its host compartment. Figure 13 shows a close up view of the D151 soma region that reveals some of the filaments used in test 3.

Following the generation of the cohort of filaments positioned throughout D151 on the compartmental surfaces, test 3 computed Eq. (10) potential within planar slices through the complete field system for the entire neuron, with the currents used in the previous tests divided, on a per-compartment basis, equally amongst the filaments on each compartment. The potentials were computed on a 1.0 $\mu m$ grid centered on the origin in the following specific planes and specific times: $XY$ plane, $[(-550 \leq x \leq +550), (-550 \leq y \leq +550), (z = 0)]$ $\mu m$, times: 11.112 ms $\leq t \leq$ 15:044 ms and $XZ$ plane, $[(-550 \leq x \leq +550), (-550 \leq z \leq +550), (y = 0)]$ $\mu m$, times: 11.782 ms $\leq t \leq$ 14:451 ms. Video presentation is the best way to appreciate

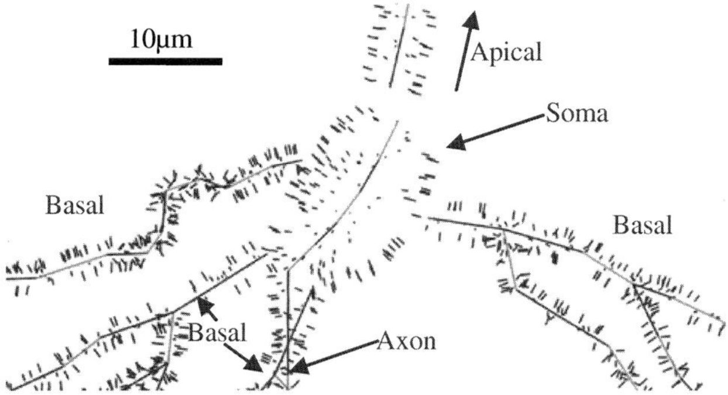


**Fig. 13.** Test 3. An isometric view of a subset of the random filament positions. The "soma" arrow points at the location of the same frustum #1 shown in Fig. 9(E) marked **. Filaments (0.0075 $\mu$m long) are shown exaggerated, not to scale. The line is the frustum centerline.

the result. Supplementary videos are provided; one each for the *XY* (S6) and *XZ* (S7) planes. The *YZ* plane was not computed due to limitations of computer time. A series of snapshots from the *XY* plane video are provided in Fig. 14 for the times indicated in Fig. 10(C). The XY slice at $z = 0$ $\mu$m passes obliquely through the soma, which is located near the origin. Straight arrows are superimposed to indicate non-rotating fields that are shrinking/growing at a rate roughly proportional to arrow length. Curved arrows indicate rotating fields that may have an excursion in the *Z*-direction.

Test 3 *XY* and *XZ* videos reveal that AP-derived potentials can create a mm-scale unified, rotating lobe pattern in space. Figure 14 shows the field system spatial extent to be much larger than the basal dendrite structure. The overall scalar potential field is a rough dipole centered on and rotating around the soma/hillock/initial segment locale.

Careful observation of the video reveals that the net interaction of all the filaments produces a lobed field structure that (i) establishes itself laterally, (ii) rotates, (iii) shrinks slightly when it aligns with the axon, (iv) grows again when laterally aligned and finally (v) contracts back to the resting state. Note that the behavior is only a slice through a more detailed 3D lobe-trajectory.

Every time an AP is triggered, D151 bathes itself in a large rotating electric field. Therefore, a real cell will have its own transmembrane potential modified to some extent. The field will also impact all the local

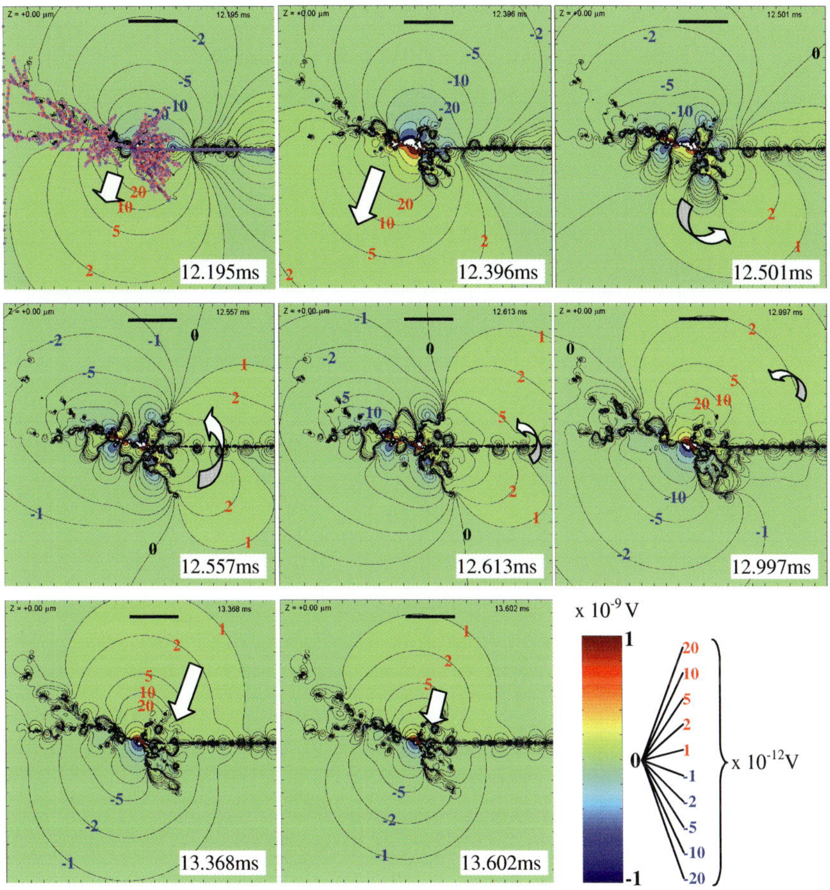


**Fig. 14.** Test 3. Supplementary videos S6 ($XY$), S7($XZ$). (A) Left–right, top–bottom time progression of LFP in the plane $XY$ at $Z = 0$ computed for the 19,066 filaments randomly positioned on neuron D151 during the single AP. The top left panel has the D151 structure superimposed. The straight arrows depict growth/shrinkage rates of a relatively stationary field pattern. The curved arrows indicate a rotating field pattern that has an excursion in the Z-direction (into/out of the page). The scale bar is 200 $\mu$m.

neighboring neurons in the same way. The local field/charge environment around all membranes will be modified in ways that are not in the neural model that generates the original currents. The influence (already characterized above as mediated by the Lorentz force) can be viewed as a feedback mechanism. In constructing the scalar potential field system in the manner of the NEURON EAPS model, the result can be regarded as open loop. If the field effects were built into the model then it would better be classed as closed loop, and could produce a field system differing from that shown in Fig. 14 if the closed loop modified the transmembrane currents.

Note also that in the volume depicted in the test 3 videos and images, there are 40,000–50,000 neurons all mutually interacting with each other in a line-of-sight manner. This result is consistent with an expectation that the filamentary formalisms might be responsible for EM coupling effects. The results show that the scalar potential drops off rapidly over a half mm or so, to be lost in the noise of the ECS and ICS field systems. However, proximal to the neuron out to about 200 $\mu$m, that potential, with sufficient synchrony between nearby neurons, could add up to become a functionally active potential capable of modifying firing thresholds.

Supporting this conclusion is Fig. 15, which shows that the dipole-like field behavior of Fig. 14 is consistent with the Eq. (10) expectation that the scalar potential has a time-course, at any particular point in space, which resembles the transmembrane current of the dominant nearby contributing current filaments. For points near soma, the potential should resemble the AP transmembrane current time course. When dominated by synapses (in a future simulation), the potential should resemble the post-synaptic current time course. In between, the time-course should resemble a mixture of the two waveforms. The results are therefore consistent with existing empirical measurements. The six plotted potentials shown in Fig. 15 are typical of "LFP" plots seen regularly in the literature (Holt & Koch, 1999; Einevoll *et al.*, 2013), and with empirical observation (Gold *et al.*, 2009).

### 10.5.6. *Test 4: Decomposition of experiment 3 results*

Test 3 filaments were decomposed into three subsets: apical, basal and soma. The same slices through the scalar potential are recomputed for the three subsets of the D151 filaments. The computation is otherwise identical to test 3. The soma/axon subset (compartments 1–17) includes all filaments attached to the soma, hillock and other axon segments. Subset 18–507 includes all basal dendrites. Subset 508–1128 includes all apical dendrites. It was verified that the sum of the decomposed potentials sums to the potentials in test 3. In this test, only the *XY* plane slices were computed for the three filament subsets. Three supplementary

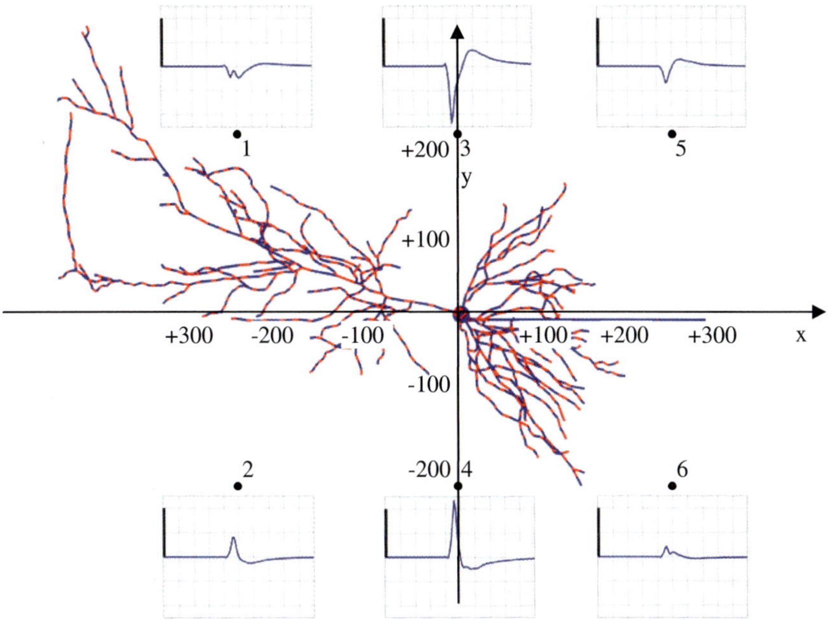


Position (μm)

| | x | y | z |
|---|---|---|---|
| 1 | –250 | 200 | 0 |
| 2 | –250 | –200 | 0 |
| 3 | 0 | 200 | 0 |
| 4 | 0 | –200 | 0 |
| 5 | 250 | 200 | 0 |
| 6 | 250 | –200 | 0 |

**Fig. 15.** Test 3. Scalar potential time course at specific locations 1, . . . , 6 as per the table. Distances in $\mu$m. Vertical scale bar is 0–40 pV. Horizontal extent 10 ms $\leq t \leq$ 15 ms. D151 compartments are colored to reveal their collocation.

videos (S8-soma, S9-basal with S9-basal) are provided. Stills from these videos are included in Fig. 16.

Figure 16 reveals that the filament subsets each provide a rotating dipole. The basal and apical dipoles contra-rotate roughly in the *XY* plane. The third rotates orthogonally (roughly in the *YZ* plane) and is associated with the soma/axon complex. The decomposition indicates that temporally coherent dipole activity sums into temporally coherent dipole activity, providing a form of scale invariance. If one continued to decompose the filaments into ever smaller subsets, the result is more

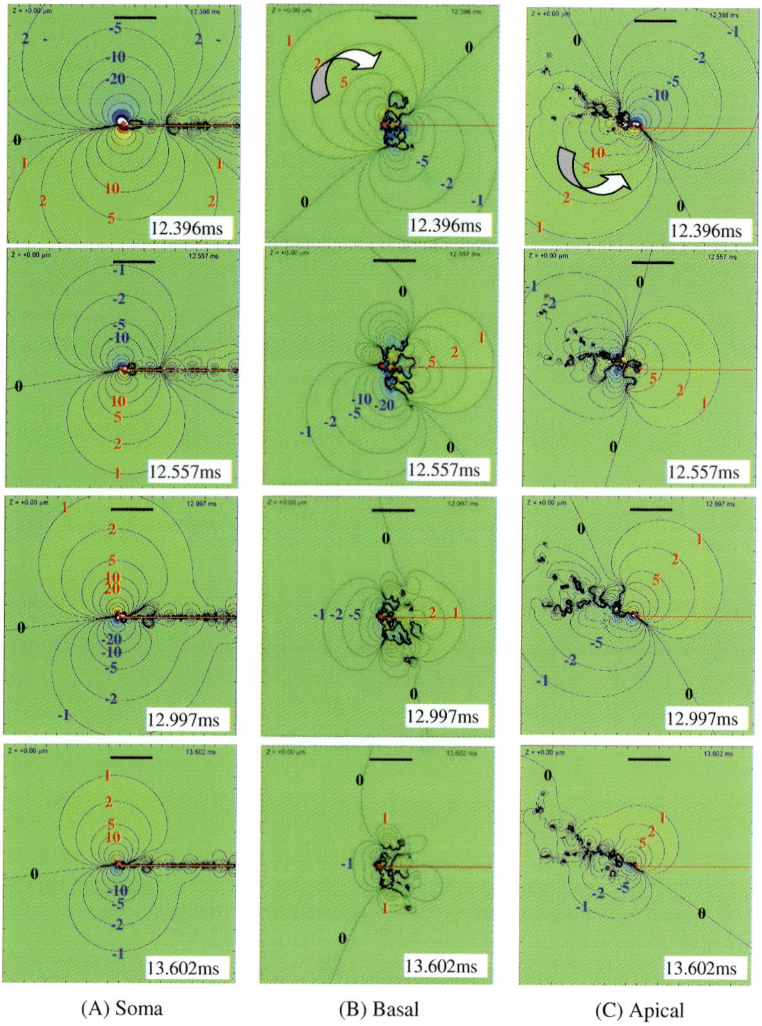


**Fig. 16.** Test 4. Supplementary videos S8, S9, S10. The test 3 $XY$ field system is decomposed by region. (A) The soma/axon field system. (B) Basal dendrite field system. (C) Apical dendrite field system. The soma/axon compartment structure is superimposed in red (it lies in the $XY$ plane). The soma/axon field system rotates in the $YZ$ plane. The basal and apical field systems contra-rotate (directions given by the arrows) largely in the $XY$ plane, but with trajectory including $Z$ excursions (into and out of the page). The contour densities reflect penetrations of this $XY$ plane at $z = 0$ by the respective processes. The scale bar is 200 $\mu$m. Color scale as per Fig. 14.

dipoles until finally the individual filaments are reached, all of which express fields of the kind revealed in test 2. This is an expected property of the original vector field equations, where the overall field system inherits the fundamental dipole-like character of the original active element: a current filament that originates a small scalar field dipole.

### 10.5.7. *Test 5: Variability with filament position*

This experiment explores how filament position affects the overall field expression by generating multiple new sets of the same filament quantities (per compartment) as test 3, but positioned differently on the D151 compartments. 10 new sets of 19,066 randomly positioned filaments were generated. Everything else is the same as test 3. The same total transmembrane current acts identically throughout the cell, again divided equally between current filaments. Because computation of the complete time series required prohibitive amounts of computer time, for each set of filaments, only the potential in the *XY* plane at $z = 0$ $\mu$m is computed, and only for $t = 12.195$ ms and $t = 12.501$ ms. This enables appraisal of (a) potential differences at specific times to be compared across filament sets and (b) the field system progression over time for each filament set. The results for 4 of the 10 filament sets are shown in Fig. 17.

In test 5, we discover that the macroscopic collective electric field structure and time-course is hypersensitive to the specifics of current filament distribution patterns. The result indicates that if a different filament set was used in test 3, a different set of videos would have been generated. The significance of this result is that many different field patterns are consistent with one set of coordinated transmembrane compartmental currents.

Therefore, what has been acquired in test 3 cannot be claimed to be the final/fixed field behavior for the given cell morphology. A particular field pattern will result from a particular filament set, but only knowledge of filament details at a specific time can determine the final field pattern at that time. This supports an expectation that, in real cells, ion channel locations and densities can be expected to result in a specific field pattern. However, without detailed ion channel disposition data, the specific pattern cannot be predicted. This flexibility in the field expression can be interpreted as a novel degree of configuration freedom in neurons. The result also means that full characterization of any particular neuron cannot only include current measurements. This result suggests that to describe a neuron in a way that includes its field generation behavior, all

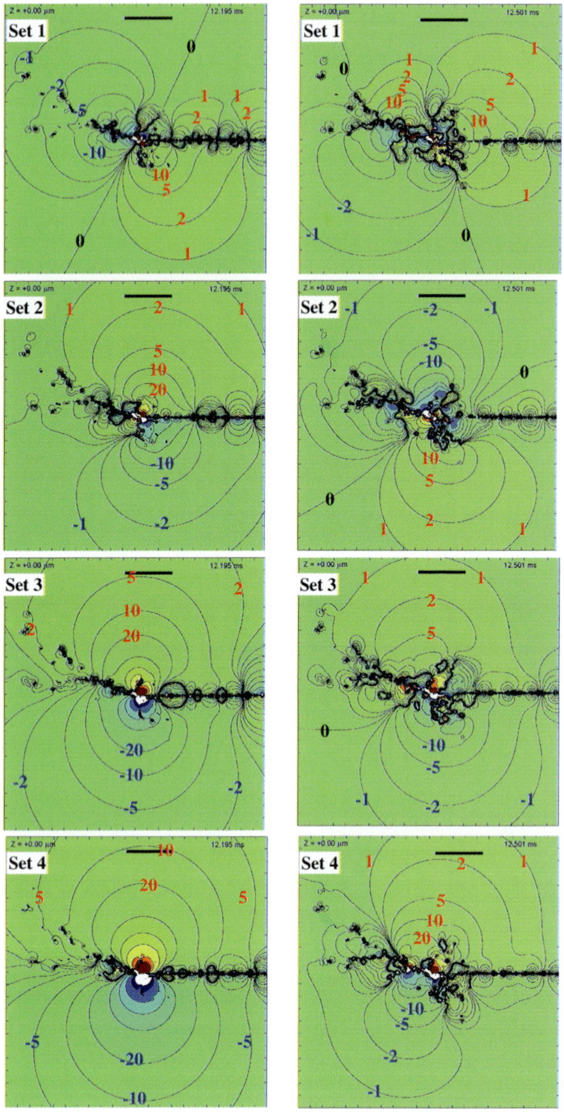

(A)12.195 ms (B)12.501 ms

**Fig. 17.** Test 5. Scalar potential variability with filament position. The potential is computed at times (A) 12.195 ms and (B) 12.501 ms for 4 of 10 different sets of randomly positioned filaments. Similar variability exists for the remaining six sets, and were not included for space reasons. Test 3 planes are shown. The scale bar is 200 $\mu$m. Color scale as per Fig. 14.

active ion channel types and densities must be measured throughout the entire structure of the neuron.

### 10.5.8. *Test 6: Variability with cell morphology*

To test the effect of cell morphology on field expression, a simplified structure was constructed using only one D151 basal process and its soma, hillock and initial segment. An arbitrarily chosen single dendrite was used to create a small artificial neuron. To do this, we aligned the chosen dendrite compartments (pulled the dendrite straight), located it in the *XY* plane (starting at its original start-point) and realigned it into the $z = 0$ plane at an angle of 45° to the $x$-axis. The dendrite structure was then mirrored in the YZ plane, producing a second dendrite. This resulted in a V-shaped neuron with the soma located at the apex of the V (Fig. 18(A)). Following population with a uniform filament mesh over the surface of each frustum, the structure is driven by the appropriate original D151 compartment currents. Holding the filament population size and spatial distribution constant, the resultant *XY* plane field system is computed for three different soma orientations covering a rotation of 180° around the V apex. A video for each soma position is provided. In this way, we have controlled out all variability unrelated to pure cell morphology.

The result is that identical filament sets moved by cell morphological shift creates large changes in the field expression. Figures 18(E)–18(G) show how the field system of the two-dendrite artificial neuron varies with soma orientation as per Figs. 18(B)–18(D). Soma rotation results in the reorientation of the plane of rotation of the field pattern from the *ZX* plane to the *XY* plane and can result in a reversal of rotation direction in the *XY* plane. These behaviors can only result from the alteration of the relationship of the soma and its dendrites. Overall this means that fixed filament locations on a compartment can express radically different field systems dependent on the relative positioning of whole compartments. This tells us that even if ion channel types and densities were rigidly elaborated on the membrane, the field structure is still radically dependent on the cell morphology.

### 10.5.9. *Discussion*

Section 10.5 has delivered a large computational exploration of how a neuron-shaped system of coordinated microscopic membrane-transverse

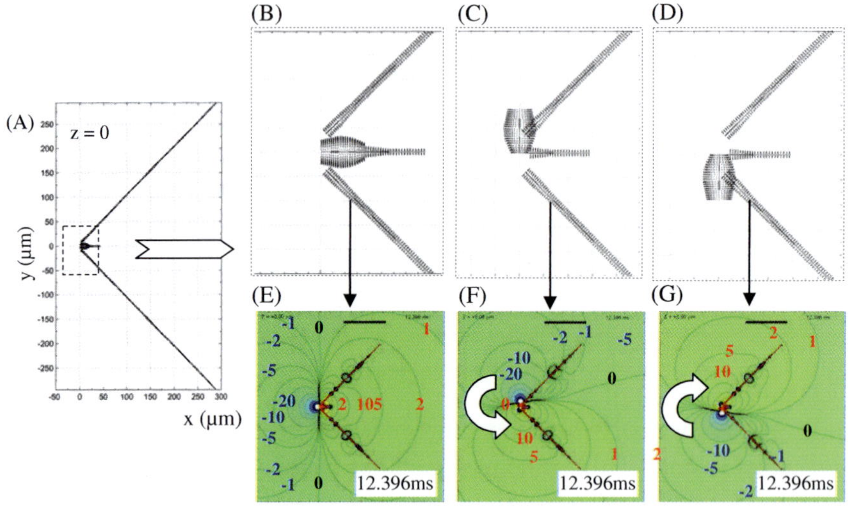


**Fig. 18.** Test 6. Supplementary videos S11, S12, S13 provided. Scalar potential variability with cell morphology. (A) The overall artificial neuron morphology configured as per (B), which is a close-up view that reveals the regular arrays of filaments. Cases (C) and (D) are identical in all respects to (B) except that the soma is (C) rotated 90° counterclockwise around the left end base centerline and (D) rotated 90° clockwise around the left end base centerline. (E)–(G) The scalar potential at spatial slice $z = 0$ for time $t = 12.396$ ms for (B)–(D), respectively. The arrows indicate that (F) and (G) field patterns rotate in the $XY$ plane. No arrow is included in (E) because it rotates in the $YZ$ plane. Scale bar is 200 $\mu$m. Color scale as per Fig. 14.

current filaments expresses a scalar potential field in an infinite uniform conductance. The next task is to examine how this might inform knowledge of field origins and expression in real tissue. An earlier section established that transmembrane currents are reasonably expected to be the dominant source of dynamic field behavior. This was for reasons of basic physics: in the ECS and the ICS, the same currents are diffuse and decorrelated, lacking the current density and spatiotemporal coherence needed to produce anything more than field noise. The proposition is based on the simple idea that a transmembrane ion channel is, in effect, an intermittently conducting short length of "wire" that expresses the field system of a short length of wire. By virtue of a far higher current density, spatial coherence and temporal coherence, filamentary transmembrane current is expected to dominate the dynamic field system expressed by neurons. This idea runs counter to the view held for many decades, where ECS or ICS currents were thought to originate the endogenous field system. We now have the basic physics knowledge needed to doubt that position, especially when it so easily accounts for both the dynamic electric and dynamic magnetic fields with the same mechanism.

Section 10.5 exploration only considered the electric field system produced by AP related ion channels. Also absent in Sec. 10.5 computations was the membrane itself, its para-membrane charge density profiles and its huge electric field system. Despite these limitations and a rather extreme material abstraction, the analysis supports an expectation that, in real neurons, individual ion channels, by virtue of their large numbers spread spatially over the cell surface, through their collective action and production of small, highly localized and time-sequenced pulsing/reversing dipoles, can deliver a spatially large and unified dynamic electric field and magnetic field system. Such a field system extends spatially over distances large compared to their hosting neuron and many orders of magnitude larger than the filaments themselves. The analysis revealed that the field system produced by Eq. (10) is naturally suited to the presence of a membrane, even though there was no actual membrane involved in the material abstraction. This is highly suggestive of membrane-spanning filamentary dipoles as the originator of the dynamic part of the endogenous EM-field system of neurons.

The analysis revealed the field system shape and behavior to be hypersensitive to current filament spatial distribution in the membrane. In addition, even if filaments were always located in the same place on a neuron compartment, the actual relative compartment positioning (cell morphology/geometry) can radically alter the field expression. Once filament position and cell morphology is determined, the field system will be determined. Only knowledge of filament type, position in a compartment and compartment relative positions enable the determination of the final field pattern. This flexibility in the field expression can be interpreted as a novel degree of configuration freedom in neurons. In principle the same field expression can result from many different cell morphologies and filament distributions. The same cell morphology can express many different field systems. The role of this level of additional freedom in tissue development, normal function and in learning is not something that can be commented on at this stage.

Of particular interest is the possibility of highly directional and mobile sweeping "lighthouse illumination," by one cell, of thousands of nearby cells, with the lobes of the sweeping field system exhibiting focus, sweep and dwell that differ laterally and longitudinally. This is suggestive of the possibility that specific host cell morphologies and ion channel locations could be self-configured to provide some level of EM coupling directionally specialized for avoidance/preference for both host cell and/or (perhaps specially selected subsets of) thousands of its columnar and/or layer neighbors. Synchronized cell firings in a confined region may have an intricate complex interplay of EM field coupling that is unexplored at this stage.

## 10.6. The Subjective Frame of Reference and the Endogenous EM Fields

It is easy to demonstrate a reference-frame change that decomposes an objectively measurable EM-field system into two systems of EM fields that account for a subjective view of the field system. If you mentally shift your perspective from that of "looking at" an EM-field system to "being" the field system, then the decomposition facilitates the shift in perspective. When that is done, it becomes obvious what neural activity is responsible for the subjective view. Once formalized, in principle the reference frame shift could be computationally applied to the EM-field systems delivered in the previous section. This proposal is posited merely as a way of broaching a new concept and inspiring further exploratory work.

### 10.6.1. *The basis for the subjective view of EM fields*

Consider Fig. 19(A) where a static field system $X(\cdot)$ interacts with an impinging field system $A(\cdot)$. During the interaction, field system $X(\cdot)$ is affected in some way as $A(\cdot)$ arrives. This is indicated in the impact zone on $X(\cdot)$. The details of the interaction are moot. Next consider Fig. 19(B) where an enhanced field system $X'(\cdot)$ is acting as if it were $X(\cdot)$ interacting with field system $A(\cdot)$. *$A(\cdot)$ does not actually exist*. $X'(\cdot)$ literally replicates all the residual phenomena that would have existed had a real $A(\cdot)$ arrived. This is denoted as a virtual field system $[A(\cdot)]$. The crucial conceptual leap is when you consider *being* $X'(\cdot)$. That is, from the perspective of being

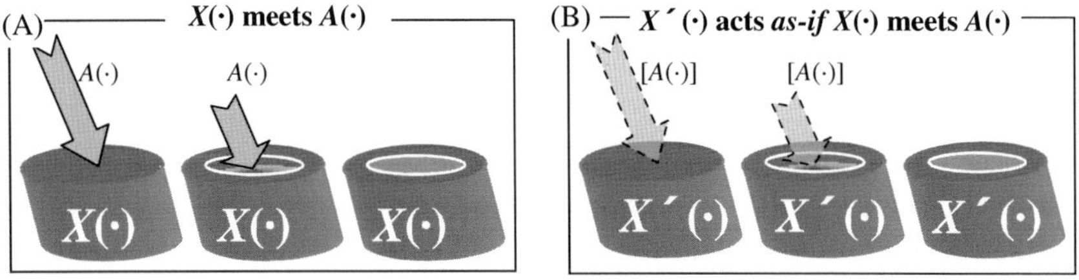


**Fig. 19.** Virtual EM-field decomposition.

field system $X'(\cdot)$, it is "as-if" you were $X(\cdot)$ being hit by an $A(\cdot)$. If, for whatever reason, subjectively, it was "like something" to be Fig. 19(A) (for reasons unknown and unspecified), then we are forced to accept that it must, to some extent be "like that" in Fig. 19(B), without $A(\cdot)$ actually arriving. $X'(\cdot)$ delivers all the field-energetics involved in an $X(\cdot)$ colliding with an $A(\cdot)$.

This virtual interaction is the key concept in understanding the first-person perspective of an EM-field system. This is meant in a literal physical sense. The physical residual EM-field phenomena resulting from the collision by $A(\cdot)$ actually exists in and around $X'(\cdot)$. The collision is, to that extent, literally reified. Indeed we have a right to actually name a particle. If $X'(\cdot)$ is a unified EM-field system maintained in space then it is, in effect, some kind of large complex composite boson or at least a bosonic entity of some kind (say a C-boson). So is $A(\cdot)$. In Fig. 19(B) we therefore have the arrival of a virtual C-boson. Indeed we can characterize the interaction as a collision between a C-boson $X'(\cdot)$ and a virtual $A(\cdot)$ C-boson, $[A(\cdot)]$. In attributing the delivery of a 1st person perspective, the virtual C-boson is responsible (by means unspecified) for qualia. We can therefore call the virtual boson a *qualeon*.[3]

### 8.6.2. *The brain's EM field and its virtual C-boson interactions*

To see how virtual EM phenomena are physically implemented in brain material (that is material behaving in the manner of the previous sections) is relatively straightforward. There is a gigantic 3D foam matrix of huge static electric fields impressed on space: *the transmembrane static electric*

[3]This name was originated by Richard Coleman.

*field*. This field system has the morphology of all the cell membranes in the tissue, including all neurons and astroctyes. We have depicted the reality of this "foam" previously in detail. This is a huge version of $X'(\cdot)$. Embedded throughout this foam matrix, in the cell membrane, at many locations and in the many configurations discussed previously, are ion channels delivering transmembrane currents that perform major localized transformations of the electric and magnetic field systems. We humans *are* literally this structure. To be human is literally to *be* this aggregation of EM field expressions: An elaborate static background field system impressed on space, punctured dynamically by myriad small field perturbations caused by ion channels. Such perturbations may exist in the form of legions of simultaneously firing chemical or electric synapses, sub-threshold oscillations and APs. Each of these circumstances is exactly the same as Fig. 19. One such instance is further detailed in Fig. 20.

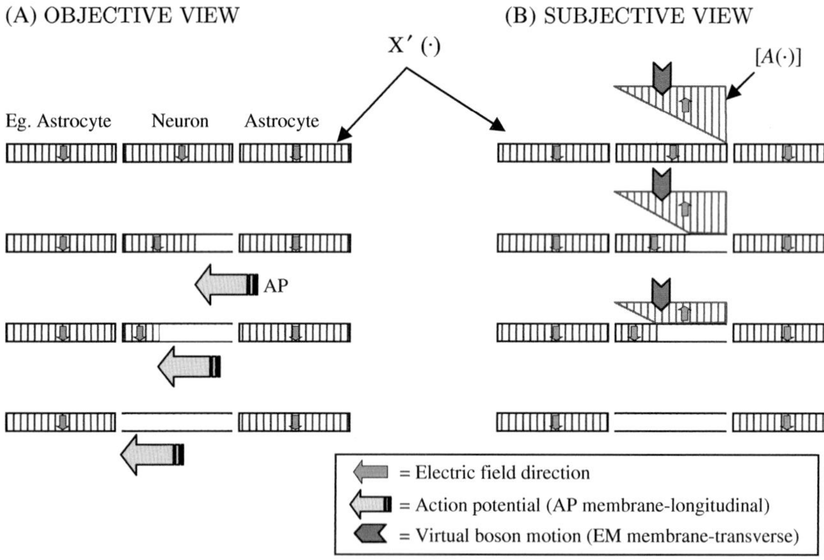


**Fig. 20.** (A) Objective view. The classical AP traveling wave is moving along the membrane axis. This is what the tissue looks like when measured. (B) Subjective view. From the perspective of "being" the membranes involved, (A) is identical to the membrane field remaining steady, but a transverse virtual boson arrive that literally modifies (collapses/reverses/restores) the electric field. It is "as-if" virtual boson $[A(\cdot)]$ collided with $X'(\cdot)$.

Figure 20(A) shows the objective (third person) situation. This looks like an abstraction of the simple progressively collapsing electric field on the central section of membrane (for example the "wave fronts" shown in Fig. 10). The left and right membrane sections do nothing. The situation can be seen in, say, two astrocytes astride a neuron. It could just as easily be a single neuron with different portions of its own membrane at different stages of field activity. Other neurons not currently firing would do as well as astrocytes. However, it is actually implemented, Fig. 20(A) captures the objective view of a progressive field collapse traveling membrane-longitudinally.

In contrast, in Fig. 20(B) is the exact same thing viewed from the first-person perspective. In all physical respects it is identical to Fig. 20(A). The arrival of the virtual C-boson [$A(\cdot)$] is posed as the fundamental mechanism upon which a private subjective percept is based. All that is claimed is that it is a plausible mechanism that is physically measurable and has enough flexibility to participate in a large complex of such interactions used, by the brain, to construct a "perceptual field" (like olfaction or vision). The main physical nuance is the recognition that the collapse of the longitudinal transmembrane field is identical to the transverse arrival, at a completely static membrane, of the electric field of a virtual C-boson of equal intensity and opposite direction. It is the virtual C-boson arrival that, from the perspective of $X'(\cdot)$, causes the field system collapse.

In the process of generating the field system Eq. (4) causal (Lorentz force) influences may be exerted on the surrounding tissue. If that particular activity also contributes to a first-person perspective, then we have a well-understood physics mechanism for consciousness potentially being functionally active in some sense yet to be detailed. The relation between the production of the first-person perspective and the EM coupling effect is of great interest in such issues as the role of consciousness in effecting behavior and free will. In the spirit of setting aside assumptions it cannot be assumed that production of a first-person perspective via the virtual C-boson interaction, and the EM-field coupling mechanism are necessarily related. The mere visibility of the virtual C-boson fields is all that is needed.

In real tissue undergoing normal realistic 3D real-time neural activity involving untold numbers of interacting virtual C-bosons, extraordinarily complex "appearances" could conceivably be constructed using the neuron/astrocyte foam as a substrate "stage" upon which their arrival is orchestrated. Their collective action is, from a first-person perspective, highly spatially structured and could be set up to replicate an enormous number of qualeons directed at any representational process that could be thought of in the brain. For example, in the case of audition, we know that the transmission of sound physically involves EM in the form of phonons (already known as a kind of boson) that propagate through a medium. Collectively and when organized appropriately in space, a set of qualeons could, in principle, make a brain "act-as-if" phonons arrived, say in the planar field perturbations of a multitude of appropriately physically organized synapses in the apical dendrites of neuron D151. The qualeons then become virtual phonons. In the case of vision, we know that sight involves the arrival of bosons called photons (light). Collectively and when organized appropriately in space, a set of qualeons could, in principle, make a brain "act-as-if" an interference pattern of photons arrived within the depths of the cortical layers involved in vision. The qualeons then become virtual photons of some kind or something appropriately similar to photons to be useful as a representation, say, some kind of hologram.

And so forth. Vision and audition are two examples that are relatively easy to imagine. What of more abstract perceptual "scenes" like emotion, touch, olfaction and so forth? It seems possible that abstract, completely naturally implausible qualeons could be created for use in such circumstances merely by positioning current filaments and compartments in space. The virtual C-bosons may be loosely related to the physical transduction of a particular sense organ, but they need not be. That being the case, it becomes feasible that we have a zoo of qualeons to identify and test for in brain tissue based on physical geometry, ion channel locations and their EM field activity. Such strange qualeons as the "touch-on," the "sad-on," the "orgasm-on," the "pain-on" and so forth, seem entirely plausible as something to actively search for in tissue as EM-field patterns in specific brain regions. By appropriately configuring neurons in space and then instigating their longitudinal membrane activity, in principle any spatiotemporally novel electrical/magnetic field structure could be devised that has a subjective view as a field (potentially an interference pattern) of qualeons. At the same time, all that any third-person would see when they looked at tissue is neurons arranged in space firing in the highly organized familiar ways seen for decades. This, it is posited, is the reason why the physics of consciousness has struggled for so long with its evidence-base for subjective experiences: a reference frame issue combined with naive disposal of EM effects in neural modeling activities. The mere identification of a reference frame change does not constitute an answer to why the subjective percept is conveyed by a virtual C-boson [$A(\cdot)$].

## 10.7. Concluding Remarks

What has been presented here is a confluence of observations that forms a natural and obvious route toward an account of consciousness that includes the following general features:

- (I) The standard model of particle physics tells us that fundamental physics of EM underpins all physics of the brain and its mind. Therefore, despite it being utterly mute on consciousness, the physics of EM, and the science of consciousness are somehow identities.
- (II) The endogenous EM-field of the brain, as a well explored and empirically cogent originator of consciousness, gains critical priority over all other accounts of consciousness.
- (III) The recognition of the three basic mechanisms that jointly originate the entire EM-field system. Transmembrane current filaments are the likely origins of both the dynamic electric *and* magnetic fields.
- (IV) The same field system automatically incorporates a causal (EM-field coupling) mechanism via the Lorentz force, thereby imbuing something delivering consciousness with a causal mechanism. This is the same causal influence recently confirmed in laboratory experiments.
- (V) The exact same field system can also be reinterpreted to be naturally decomposed into an interaction between a totally static field system and virtual EM "C-bosons," thereby revealing a mechanism by which consciousness may be described and understood in an empirically tractable way. It does not explain why there is consciousness.

It is posited that this confluence of multiple explanatory virtues, along with compatibility with all existing empirical results, makes such an EM-field theory a critically superior potential solution to the problem of a scientific account of consciousness. The real challenge is that the proposition is telling us that further progress involves some kind of formal, structured review of ourselves (scientists), by ourselves. But that is a matter for another day.

## 10.8. Supplementary Videos

To access the Videos, please follow the instructions below.

1. Go to: *http://www.worldscientific.com/r/9444-supp*
   (Note: This access URL should be activated ONLY ONCE.)
2. You will next be prompted to register an account/login.
3. Access will be activated upon your account registration/login.
4. For subsequent access, simply log in with the same login details and access: *http://www.worldscientific.com/worldscibooks/10.1142/9444#t=suppl*.

For enquiries, please email: sales@wspc.com.sg

| Test | Video | Description |
|---|---|---|
| 1 | S1 | D151 currents, isometric view, entire |
| 1 | S2 | D151 currents, isometric view, soma close-up |
| 2 | S3 | Compartment 1, regularly spaced filaments, *XZ* plane |
| 2 | S4 | Compartment 1, regularly spaced filaments, *YZ* plane |
| 2 | S5 | Compartment 1, regularly spaced filaments, *XY* plane |
| 3 | S6 | D151, randomly placed filaments, *XY* plane |
| 3 | S7 | D151, randomly placed filaments, *XZ* plane |
| 4 | S8 | D151 Test 3 *XY*: soma only |
| 4 | S9 | D151 Test 3 *XY*: basal only |
| 4 | S10 | D151 Test 3 *XY*: apical only |
| 6 | S11 | Artificial neuron, soma rotation 0 |
| 6 | S12 | Artificial neuron, soma rotation −90 |
| 6 | S13 | Artificial neuron, soma rotation +90 |

## Appendix A

The following lists show the differential form of Maxwell's macroscopic equations (MME) in SI units contrasted with the microscopic version (MmE) (Jackson, 1999, p. 248).

| Microscopic Differential MmE | Macroscopic Differential MME | Meaning |
|---|---|---|
| $\nabla \cdot \mathbf{e} = \frac{\eta}{\varepsilon_0}$ | $\nabla \cdot \mathrm{D} = \rho$ | Coulomb's law: The spatial (divergence) rate of change of field is proportional to spatial charge density. |
| $\nabla \times \mathbf{e} + \frac{\partial \mathbf{b}}{\partial t}$ | $\nabla \times \mathbf{E} + \frac{\partial \mathbf{B}}{\partial t} = 0$ | Faraday's law: The spatial (curl) rate of change of electric field results in a temporal rate of change of the magnetic field. |
| $\nabla \cdot \mathbf{b} = 0$ | $\nabla \cdot \mathbf{B} = 0$ | The non-existence of magnetic monopoles: The spatial (divergence) rate of change of the magnetic field is zero. |
| $\nabla \times \mathbf{b} - \frac{1}{\mathbf{c}^2} \frac{\partial \mathrm{e}}{\partial t} = \mu_0 \mathbf{j}$ | $\nabla \times \mathbf{H} - \frac{\partial \mathrm{D}}{\partial t} = \mathbf{J}$ | Ampere's law: The spatial (curl) rate of change of magnetic field combined with the temporal rate of change of the electric field determines the spatial current density. Maxwell added the displacement current term. |
| N/A | $\mathbf{D} = \varepsilon_0 \mathbf{E} + \mathbf{P}$ | **P** is the polarization of the macroscopic material medium. |
| N/A | $\mathbf{H} = \frac{1}{\mu_0} \mathbf{B} + \mathbf{M}$ | **M** is the magnetization of the macroscopic material medium. |

### Notes and SI units

| | |
|---|---|
| $\eta$, $\rho$ = volume charge density | (Coulomb/m$^3$) |
| $\varepsilon_0$ = free space permittivity | (Farad/m) or (Coulomb/(Vm)) |
| $\mu_0$ = free space permeability | (Henry/m) or (V/(amp m)) |
| $c$ = free space speed of light | (Meter/s) |
| **j**, **J** = vector current density | (Amp/m$^2$) |
| **e**, **E** = vector electric field | (Volts/m) |
| **b**, **B** = vector magnetic flux | (Weber/m$^2$) or (V/m$^2$) |
| **H** = vector magnetic field | (Amp/m) |
| **M** = vector magnetization | (Amp/m) |

**D** = vector displacement (Coulomb/m$^2$)
**P** = vector polarization (Coulomb/m$^2$)

and

$c^2 = 1/\varepsilon_0 \mu_0$.

## Acknowledgments

Thanks to A/Prof. David Grayden and A/Prof. Harry Quiney for their support during Sec. 5 exploration. Thanks also goes to the D151 neuron data assemblers and the originators of the EAPS NEURON model, without which Sec. 5 work could not have happened. This research was supported by a Victorian Life Sciences Computation Initiative (VLSCI) grant number VR0003 on its Peak Computing Facility at the University of Melbourne, an initiative of the Victorian Government.

## References

Aidley, D. J. (1998) *The Physiology of Excitable Cells*. New York: Cambridge University Press.

Anastassiou, C. A., Montgomery, S.M., Barahona, M., Buzsaki, G. & Koch, C. (2010) The effect of spatially inhomogeneous extracellular electric fields on neurons. *J. Neurosci.*, **30**, 1925–1936.

Anastassiou, C. A., Perin, R., Markram, H. & Koch, C. (2011) Ephaptic coupling of cortical neurons. *Nat. Neurosci.*, **14**, 217–223.

Aronsson, P. & Liljenstrom, H. (2001) Effects of non-synaptic neuronal interaction in cortex on synchronization and learning. *Biosystems*, **63**, 43–56.

Baars, B. J. (1997) In *The Theater of Consciousness*: The *Workspace of the Mind*. New York: Oxford University Press.

Baars, B. J. & Franklin, S. (2009) Consciousness is computational: The LIDA model of global workspace theory. *Int. J. Mach. Conscious.*, **1**, 23–32.

Baars, B. J., Newman, J .B. & Banks, W. P. eds. (2003) *Essential Sources in the Scientific Study of Consciousness*. Bradford.

Bach-Y-Rita, P. (2001) Nonsynaptic diffusion neurotransmission in the brain: Functional considerations. *Neurochem. Res.*, **26**, 871–873.

Balduzzi, D. & Tononi, G. (2008) Integrated information in discrete dynamical systems: Motivation and theoretical framework. *Plos Comput. Biol.*, **4**, e1000091.

Balduzzi, D. & Tononi, G. (2009) Qualia: The geometry of integrated information. *Plos Comput.* Biol., **5**, e1000462.

Barrett, A. B. (2014) An integration of integrated information theory with fundamental physics. *Front. Psychol.*, **5**, 63.

Bédard, C. & Destexhe, A. (2008) A modified cable formalism for modeling neuronal membranes at high frequencies. *Biophys. J.*, **94**, 1133–1143.

Bignami, A. (1991) Glial cells in the central nervous system'. *Discuss. Neurosci.*, **8**, 1–45.

Bikson, M., Inoue, M., Akiyama, H., Deans, J. K., Fox, J. E., Miyakawa, H. & Jefferys, J. G. R. (2004) Effects of uniform extracellular DC electric fields on excitability in rat hippocampal slices *in vitro*. *J. Physiol.-London*, **557**, 175–190.

Bishop, G. H. & Oleary, J. L. (1950) The effects of polarizing currents on cell potentials and their significance in the interpretation of central nervous system activity. *Electroencephalogr. Clin. Neurophysiol.*, **2**, 401–416.

Briggman, K. L. & Denk, W. (2006) Towards neural circuit reconstruction with volume electron microscopy techniques. *Curr. Opin. Neurobiol.*, **16**, 562–570.

Bullock, T. H. (1997) Signals and signs in the nervous system: The dynamic anatomy of electrical activity is probably information-rich. *Proc. Natl. Acad. Sci. USA*, **94**, 1–6.

Chalmers, D. J. (2000) What is a neural correlate of consciousness? In: T. Metzinger, ed. *Neural Correlates of Consciousness: Empirical and Conceptual Questions*. MIT Press.

Chen, K. C. & Nicholson, C. (2000) Spatial buffering of potassium ions in brain extracellular space. *Biophys. J.*, **78**, 2776–2797.

Clark, J. & Plonsey, R. (1966) A mathematical evaluation of the core conductor model. *Biophys. J.*, **6**, 95–112.

Clark, J. & Plonsey, R. (1968) Extracellular potential field of single active nerve fiber in a volume conductor. *Biophys. J.*, **8**, 842–864.

Craig, D.P. & Thirunamachandran, T. (1984) *Molecular Quantum Electrodynamics: An Introduction to Radiation-Molecule Interactions*. London, Orlando: Academic Press.

da Silva, F. H. L. & Van Rotterdam, AB (2005) Biophysical aspects of EEG and magnetoencephelogram generation. In: E. Niedermeyer and F.H.L. da Silva, eds. *Electroencephalography: Basic Principles, Clinical Applications, and Related Fields*. Philadelphia: Lippincott Williams & Wilkins, p. xiii.

Dayan, P. & Abbott, L. F. (2001) *Theoretical Neuroscience: Computational and Mathematical Modeling of Neural Systems*. Cambridge, Mass., London: MIT Press.

de Graaf, T. A., Hsieh, P.-J. & Sack, A. T. (2012) The `correlates' in neural correlates of consciousness. *Neurosci. Biobehav. Rev.*, **36**, 191–197.

Destexhe, A. & Bedard, C. (2012) Do neurons generate monopolar current sources? *J. Neurophysiol.*, **108**, 953–955.

Dilger, J. P., McLaughlin, S. G., McIntosh, T. J. & Simon, S. A. (1979) The dielectric constant of phospholipid bilayers and the permeability of membranes to ions. Science, **206**, 1196–1198.

Einevoll, G. T., Kayser, C., Logothetis, N. K. & Panzeri, S. (2013) Modelling and analysis of local field potentials for studying the function of cortical circuits. *Nat. Rev. Neurosci.*, **14**, 770–785.

Faber, D. S. & Korn, H. (1989) Electrical-field effects — their relevance in central neural networks. *Physiol. Rev.*, **69**, 821–863.

Francis, J. T., Gluckman, B. J. & Schiff, S. J. (2003) Sensitivity of neurons to weak electric fields. *J. Neurosci.*, **23**, 7255–7261.

Freeman, A. J. (1975) *Mass Action in the Nervous System: Examination of the Neurophysiological Basis of Adaptive Behavior through the EEG*. New York: Academic Press.

Frohlich, F. & McCormick, D. A. (2010) Endogenous electric fields may guide neocortical network activity. *Neuron*, **67**, 129–143.

Geddes, L. A. & Baker, L. E. (1967) Specific resistance of biological material-a compendum of data for biomedical engineer and physiologist. *Med. Biol. Eng.*, **5**, 271–293.

Gold, C., Henze, D. A. & Koch, C. (2007) Using extracellular action potential recordings to constrain compartmental models. *J. Comput. Neurosci.*, **23**, 39–58.

Gold, C., Henze, D. A., Koch, C. & Buzsaki, G. (2006) On the origin of the extracellular action potential waveform: A modeling study. *J. Neurophysiol.*, **95**, 3113–3128.

Gold, C., Girardin, C. C., Martin, K. A. C. & Koch, C. (2009) High-amplitude positive spikes recorded extracellularly in cat visual cortex. *J. Neurophysiol.*, **102**, 3340–3351.

Hagan, S., Hameroff, S.R. & Tuszynski, J.A. (2002) Quantum computation in brain microtubules: Decoherence and biological feasibility. *Phys. Rev. E*, **65**, 1–11.

Hameroff, S. (1998) Quantum computation in brain microtubules? The Penrose-Hameroff 'Orch OR' model of consciousness. *Philos. Trans. R. Soc. London S. A-Math. Phys. Eng. Sci.*, **356**, 1869–1896.

Hameroff, S. (2012) Quantum brain biology complements neuronal assembly approaches to consciousness: Comment on "Consciousness, biology and quantum hypotheses" by Baars and Edelman. *Phys. Life Rev.*, **9**, 303–305.

Hameroff, S. & Penrose, R. (2014) Consciousness in the universe: A review of the `Orch OR' theory. *Phys. Life Rev.*, **11**, 39–78

Hille, B. (2001) *Ion Channels of Excitable Membranes*. Sunderland, MA.: Sinauer Associates, Inc.

Hines, M. L. & Carnevale, N. T. (1997) The NEURON simulation environment. *Neural Comput.*, **9**, 1179–1209.

Hodgkin, A. L. & Huxley, A. F. (1952a) The components of membrane conductance in the giant axon of loligo. *J. Phys.-London*, **116**, 473–496.

Hodgkin, A. L. & Huxley, A. F. (1952b) Currents carried by sodium and potassium ions through the membrane of the giant axon of loligo. *J. Physiol.-London*, **116**, 449–472.

Hodgkin, A. L. & Huxley, A. F. (1952c) The dual effect of membrane potential on sodium conductance in the giant axon of loligo. *J. Physiol.-London*, **116**, 497–506.

Hodgkin, A. L. & Huxley, A. F. (1952d) Propagation of electrical signals along giant nerve fibres. *Proc. R. Soc. London. Ser. B, Biol. Sci.*, **140**, 177–183.

Hodgkin, A. L. & Huxley, A. F. (1952e) A quantitative description of membrane current and its application to conduction and excitation in nerve. *J. Physiol.-London*, **117**, 500–544.

Hodgkin, A. L., Huxley, A. F. & Katz, B. (1952) Measurement of current-voltage relations in the membrane of the giant axon of loligo. *J. Physiol.-London*, **116**, 424–448.

Hohwy, J. (2012) Neural correlates and causal mechanisms. *Conscious. Cogn.*, **21**, 691–692.

Holt, G. R. & Koch, C. (1999) Electrical interactions via the extracellular potential near cell bodies. *J. Comput. Neurosci.*, **6**, 169–184.

Huang, W. & Levitt, D. G. (1977) Theoretical calculation of dielectric-constant of a bilayer membrane. *Biophys. J.*, **17**, 111–128.

Izhikevich, E. M. (2007) *Dynamical Systems in Neuroscience: The Geometry of Excitability and Bursting*. Cambridge, Mass., London: MIT Press.

Jackson, J. D. (1999) *Classical Electrodynamics*. New York: Wiley.

Jefferys, J. G. R. (1981) Influence of electric-fields on the excitability of granule cells in guineapig hippocampal slices. *J. Phys.-London*, **319**, 143–152.

Jefferys, J. G. R. (1995) Nonsynaptic modulation of neuronal-activity in the brain — electric currents and extracellular ions. *Physiol. Rev.*, **75**, 689–723.

Jefferys, J. G. R. & Haas, H. L. (1982) Synchronized bursting of CA1 hippocampal pyramidal cells in the absence of synaptic transmission. *Nature*, **300**, 448–450.

John, E. R. (2001) A field theory of consciousness. Conscious. *Cognit.*, **10**, 184–213.

John, E. R. (2002) The neurophysics of consciousness. *Brain Res. Rev.*, **39**, 1–28.

John, E. R. (2003) A theory of consciousness. *Curr. Dir. Psychol. Sci.*, **12**, 244–250.

John, E. R. (2005) From synchronous neuronal discharges to subjective awareness? *Prog. Brain Res.*, **150**, 143–593.

John, E. R. (2006) The sometimes pernicious role of theory in science. *Int. J. Psychophysiol.*, **62**, 377–383.

Johnston, D. & Wu, S. M. -S. (1995) *Foundations of Cellular Neurophysiology*. Cambridge, Mass.: MIT Press.

Kandel, E. R. & Schwartz, J. H. (2000) *Principles of Neural Science*. New York: McGraw-Hill.

Katz, B. (1966) *Nerve, Muscle, and Synapse*. New York: McGraw-Hill.

Kinney, J. P., Spacek, J., Bartol, T. M., Bajaj, C. L., Harris, K. M. & Sejnowski, T. J. (2013) Extracellular sheets and tunnels modulate glutamate diffusion in hippocampal neuropil. *J. Comp. Neurol.*, **521**, 448–464.

Köhler, W. (1960) *Dynamics in Psychology*. WW Norton & Company.

Kuffler, S. W. & Nicholls, J. G. (1976) *From Neuron to Brain: A Cellular Approach to the Function of the Nervous System*. Sunderland, Mass.: Sinauer Associates.

Lee, K. Y. C., Klingler, J. F. & McConnell, H. M. (1994) Electric-field-induced concentration gradients in lipid monolayers. *Science*, **263**, 655–658.

Libet, B. (1994) A testable field theory of mind-brain interaction. *J. Conscious. Stud.*, **1**, 119–126.

Libet, B. (1996) Conscious mind as a field. *J. Theor. Biol.*, **178**, 223–224.

Lindahl, B. I. B. & Arhem, P. (1994) Mind as a force-field -- comments on a new interactionistic hypothesis. *J. Theor. Biol.*, **171**, 111–122.

Maggio, B., Borioli, G. A., Del Boca, M., De Tullio, L., Fanani, M. L., Oliveira, R. G., Rosetti, C. M. & Wilke, N. (2008) Composition-driven surface domain structuring mediated by sphingolipids and membrane-active proteins. *Cell Biochem. Biophys.*, **50**, 79–109.

Magistretti, P.J. & Ransom, B. R. (2002) Astrocytes. In: K.L. Davis, D. Charney, J.T. Coyle and C. Nemeroff eds. *Neuropsychopharmacology: The Fifth Generation of Progress*. Chapter 10. Lippincott Williams & Wilkins.

Malmivuo, J. & Plonsey, R. (1995) *Bioelectromagnetism: Principles and Applications of Bioelectric and Biomagnetic Fields*. New York: Oxford University Press.

McFadden, J. (2002a) The conscious electromagnetic information (CEMI) field theory. *J. Conscious. Stud.*, **9**, 45–60.

McFadden, J. (2002b) Synchronous firing and its influence on the brain's electromagnetic field: Evidence for an electromagnetic theory of consciousness. *J. Conscious. Stud.*, **9**, 23–50.

McFadden, J. (2006) The CEMI field theory: Seven clues to the nature of consciousness. In: J. A. Tuszynski, ed. *The Emerging Physics of Consciousness*. Berlin Heidelberg: Springer, pp. 385–404.

McFadden, J. (2013) The CEMI field theory closing the loop. *J. Conscious. Stud.*, **20**, 1–2.

McIntyre, C. C. & Grill, W. M. (1999) Excitation of central nervous system neurons by nonuniform electric fields. *Biophys. J.*, **76**, 878–888.

Metzinger, T. (2000) *Neural Correlates of Consciousness -- Empirical and Conceptual Questions*. Cambridge MA: MIT Press.

Mishchenko, Y., Hu, T., Spacek, J., Mendenhall, J., Harris, K.M. & Chklovskii, D.B. (2010) Ultrastructural analysis of hippocampal neuropil from the connectomics perspective. *Neuron*, **67**, 1009–1020.

Molyneux, B. (2010) Why the neural correlates of consciousness cannot be found. *J. Conscious. Stud.*, **17**, 168–188.

Mormann, F. & Koch, C. (2007) Neural correlates of consciousness. *Scholarpedia*, **2**, 1740.

Neisser, J. (2012) Neural correlates of consciousness reconsidered. *Conscious. Cogn.*, **21**, 681–690.

Nicholson, C. & Phillips, J. M. (1981) Ion diffusion modified by tortuosity and volume fraction in the extracellular micro-environment of the rat cerebellum. *J. Physiol.-London*, **321**, 225–257.

Nicholson, C. & Sykova, E. (1998) Extracellular space structure revealed by diffusion analysis. *Trend. Neurosci.*, **21**, 207–215.

Noebels, J. L. & Prince, D. A. (1978) Excitability changes in thalamocortical relay neurons during synchronous discharges in cat neocortex. *J. Neurophys.*, **41**, 1282–1296.

Nunez, P. L. & Srinivasan, R. (2006) *Electric Fields of the Brain: The Neurophysics of EEG*. Oxford, New York: Oxford University Press.

Parra, L. C. & Bikson, M. (2004) Model of the effect of extracellular fields on spike time coherence. *Proc. 26th Ann. Int. Conf. IEEE Eng. Med. Biol. Soc.*, 1–7. pp. 4584–4587.

Peterka, D. S., Takahashi, H. & Yuste, R. (2011) Imaging voltage in neurons. *Neuron*, **69**, 9–21.

Pethig, R. (1986) Ion, electron, and proton transport in membranes: A review of the physical processes involved. In: F. Gutmann and H. Keyzer eds. *Modern Bioelectrochemistry*. New York: Plenum Press, pp. 199–239.

Plonsey, R. (1964) Volume conductor fields of action currents. *Biophys. J.*, **4**, 317.

Plonsey, R. & Collin, R. (1961) *Principles and Applications of Electromagnetic Fields*. New York: McGraw Hill.

Plonsey, R. & Heppner, D. B. (1967) Considerations of quasi-stationarity in electrophysiological systems. *Bull. Math. Biophys.*, **29**, 657–664.

Plonsey, R. & Fleming, D. G. (1969) *Bioelectric Phenomena*. New York: McGraw-Hill.

Pockett, S. (1999) Anesthesia and the electrophysiology of auditory consciousness. *Conscious. Cognit.*, **8**, 45–61.

Pockett, S. (2000) *The Nature of Consciousness: A Hypothesis*. Writer's Club Press.

Pockett, S. (2002) Difficulties with the electromagnetic field theory of consciousness. *J. Conscious. Stud.*, **9**, 51–56.

Pockett, S. (2007) Difficulties with the electromagnetic field theory of consciousness: An update. *NeuroQuantology*, **3**, 271–275.

Pockett, S. (2012) The electromagnetic field theory of consciousness a testable hypothesis about the characteristics of conscious as opposed to non-conscious fields. *J. Conscious. Stud.*, **19**, 191–223.

Popper, K. R., Lindahl, B. I. B. & Arhem, P. (1993) A discussion of the mind-brain problem. *Theor. Med.*, **14**, 167–180.

Poznanski, R. R. & Cacha, L. A. (2012) Intracellular capacitive effects of polarized proteins in dendrites. *J. Integr. Neurosci.*, **11**, 417–437.

Rall, W. & Shepherd, G. M. (1968) Theoretical reconstruction of field potentials and dendrodendritic synaptic interactions in olfactory bulb. *J. Neurophys.*, **31**, 884–915.

Reato, D., Rahman, A., Bikson, M. & Parra, L. C. (2010) Low-intensity electrical stimulation affects network dynamics by modulating population rate and spike timing. *J. Neurosci.*, **30**, 15067–15079.

Reimann, M. W., Anastassiou, C. A., Perin, R., Hill, S. L., Markram, H. & Koch, C. (2013) A biophysically detailed model of neocortical local field potentials predicts the critical role of active membrane currents. *Neuron*, **79**, 375–390.

Romijn, H. (2002) Are virtual photons the elementary carriers of consciousness? *J. Conscious. Stud.*, **9**, 61–81.

Sadiku, M. N. O. (2001) *Elements of Electromagnetics*. New York: Oxford University Press.

Shanahan, M. & Baars, B. (2005) Applying global workspace theory to the frame problem. *Cognition*, **98**, 157–176.

Speckmann, E.-J. & Elgar, C. E. (2005) Introduction to the neurophysiological basis of the EEG and DC potentials. In E. Niedermeyer and F. H. Lopes da Silva, eds. *Electroencephalography: Basic Principles, Clinical Applications, and Related Fields*, Philadelphia: Lippincott Williams & Wilkins, pp. 17–29.

Sykova, E. (1997) The extracellular space in the CNS: Its regulation, volume and geometry in normal and pathological neuronal function. *Neuroscientist*, **3**, 28–41.

Sykova, E. (2004) Extrasynaptic volume transmission and diffusion parameters of the extracellular space. *Neuroscience*, **129**, 861–876.

Sykova, E. & Nicholson, C. (2008) Diffusion in brain extracellular space. *Physiol. Rev.*, **88**, 1277–1340.

Sykova, E., Roitbak, T., Mazel, T., Simonova, Z. & Harvey, A. R. (1999) Astrocytes, oligodendroglia, extracellular space volume and geometry in rat fetal brain grafts. *Neuroscience*, **91**, 783–798.

Taylor, C. P. & Dudek, F. E. (1982) Synchronous neural after discharges in rat hippocampal slices without active-chemical synapses. *Science*, **218**, 810–812.

Terzuolo, C. A. & Bullock, T. H. (1956) Measurement if imposed voltage gradient adequate to modulate neuronal firing. *Proc. Nate. Acad. Sci. USA*, **42**, 687–694.

Tononi, G. (2004) An information integration theory of consciousness. *BMC Neurosci.*, **5**, 42.

Tononi, G. (2008) Consciousness as integrated information: A provisional manifesto. *Biol. Bull.*, **215**, 216–242.

Van Harrenveld, A. (1968) The extracellular space in the vertebrate central nervous system. In: G. H. Bourne, ed. *The Structure and Function of Nervous Tissue*. New York: Academic Press, pp. 447–511.

Velmans, M. & Schneider, S. eds. (2007) *The Blackwell Companion to Consciousness*. Malden, MA; Oxford: Blackwell Publishing.

Weiss, S.A. & Faber, D.S. (2010) Field effects in the CNS play functional roles. *Front. Neural Circuits*, **4**, 15.

Biophysics of Consciousness: A Foundational Approach
R. R. Poznanski, J. A. Tuszynski and T. E. Feinberg
Copyright © 2017 World Scientific, Singapore.

# 11

# CONSCIOUSNESS AS A QUANTUM DYNAMIC EFFECT

*R. R. Poznanski,*,† L. A. Cacha,‡ J. Ali,‡,§ D. Aur,¶ J. L. Perez Velazquez,‖ M. Bahadoran,‡ P. P. Yupapin** and J. Yunus**

**Faculty of Bioscience and Medical Engineering, Universiti Teknologi Malaysia, 81310 UTM Johor Bahru, Malaysia*

*†Laboratory of Biological Modeling, The Rockefeller University, New York, NY 10065-6399, USA*

*‡Laser Centre, IBNU SINA ISIR, Universiti Teknologi Malaysia, 81310 UTM Johor Bahru, Malaysia*

*§Physics Department, Faculty of Science, Universiti Teknologi Malaysia, 81310 UTM Johor Bahru, Malaysia*

*¶Department of Comparative Medicine, Stanford University, Palo Alto, CA 94305, USA*

*‖Neurosciences and Mental Health Program, University of Toronto, Toronto, Ontario, M5G 1X8, Canada*

***Advanced Studies Center, Faculty of Science, King Mongkut's Institute of Technology, Bangkok 10520, Thailand*

## Abstract

Electromagnetic quantum field interactions provide a deeper understanding of the emergence of consciousness through an ontological interpretation of quantum mechanics. Quasiparticle interactions in brain matter can produce "virtual" particles, hitherto mathematical conveniences, which reflect the manifold images of brain activities in terms of effects on charged particles that have physical quantum effect similar to the Aharonov–Bohm effect in quantum field theory. Electromagnetic potentials in the quantum domain pervade the entire brain through causative interactions. The exchange of energy quanta during quantum field interactions emerge as distinct patterns of quasiparticles and their entanglement produces waves of "virtual" particles with *periodicities of discrete energies* tethered through a common quantum potential as they interact and collectively represent the emergence of self-awareness as a quantum dynamic effect carried on the magnetic vector potential throughout the brain. This is caused by the magnetic dipole moments sensitivity to the phase shifts entrained by the *zeitgeber* acting on dipolar molecules in interfacial water. The emergence of consciousness is inseparable (spatiotemporally) in the sense that the endogenous electromagnetic field (and therefore the forces acting on the particles) vanish yet neurophenomenologically separate from the physical through the magnetic dipole moments sensitivity to the phase shifts of the coherent system. The subjective aspect of consciousness or phenomenal consciousness facilitates *qualia* as quantum information through information-physical interactions that metastabilize into rhythms unfolding through resonance into access consciousness as cognizable associable representations of subjective experiences. The transition from subjectivity occurs non-reductively from the microlevel to the macrolevel as a mechanism of spontaneous breaking of symmetry resulting from "photon" behavior of interfacial water instantiated by macroergic effects in the nuclei of neurons through a continuous supply of metabolic energy. The emergence of consciousness in living organisms disappears once autopoietic processes in populations of neurons cease to function.

*Keywords*: Quantum dynamic effect, interfacial water, quantum causality, quasiparticles, DNA, photon behavior, periodicities of discrete energies, magnetic vector potential, zeitgeber, quantum potential, informational field, collective mode, dipolar waves, consciousness, qualia, subjectivity, neurophenomenology, resonance, non-reductive physicalism, virtual particles, quantum fields.

## 11.1. Introduction

In the 19th century, Helmholtz's grand dream of a natural science was one that unifies the objective and subjective views of the physical world. Heretofore, neurophysiologist Helmholtz derived the term of thermodynamic nerve energy. This intellectual scientific development was continued by Sigmund Freud who in 1895 wrote about neural networks driven by non-physiological energy forces that altered conscious processes through "periodicities of energies." However, during the turmoil of WWII, Kohler (1940) made a categorical error when he identified perceptual field with the epiphenomenal electric field of the electroencephalograph (EEG) in which the Coulomb forces are too weak to synchronize the observed oscillations, and which was voraciously opposed by Nobel laureate Sperry (1980) among others.

Kohler 's idea which was claimed as a mistake, found its way and gained support after WWII among mostly engineers and the notion of thermodynamic "nerve energy" was converted with "electric energy" transformed into life force as evidenced by EEG. This notion of converting led to a new paradigm of "information" namely "information processing" with the advent of discovery of DNA as an informational system in biological organisms and together with the Hodgkin & Huxley (1952) model of the nerve pulse led to the birth of Biological Cybernetics in the 1960s. From this field induced some application of artificial neural networks in 1970s that later were found to be incongruent with the workings of the brain and thus reformed into Computational Neuroscience in the 1980s. The evolution was inspired by a hybrid between the doctrine of biological computation in terms of the DNA as an informational system and brain computer metaphor that was uprooted from Shannon's information theory and posited to govern the different roles of information processing in cognition all because of Kohler 's mistake in 1940 during the Blitzkrieg.

The underlying foundation of Shannon's information theory as intended for communication is the stereotypical spike paradigm that assumes firing of spikes to form representation of an action, object, person or even subjective experience (Gelbard-Sagiv *et al.*, 2008), which according to Pribram (2013) is a neurophysiological dogma. Others adhere to a different information paradigm based on information-physical interactions (Bazyan, 2010; Aur *et al.*, 2011); while others recognize the importance of a field theory, insofar as a spike is a small fragment of information carried by population of neurons in a nerve packet (Freeman, 2003) or carried by the electromagnetic (EM) field as intrinsic "Gödelian" information (Cicurel & Nicolelis, 2015). The duality of information gives rise to the realization that the brain is not some sort of computer and that consciousness emerges from its coalescence.

Gerald Edelman whose main idea was that consciousness resides in a large assembly of neurons mostly within the thalamocortical system, dynamically orchestrating conscious awareness as postulated in the dynamic core hypothesis (Edelman & Tononi, 1995). Stemming from Edelman's vision of a natural consciousness spanning a decade or so before the dynamic core hypothesis, contemporary theories of consciousness (Baars, 1988, 1997; Tononi, 2008; Korzeniewski, 2010; Baars & Edelman, 2012; Dehaene, 2014) have posited that consciousness must result from the dynamic organization of neural systems or brain connectivity. Therefore, to answer questions about consciousness can only be done from inference on conscious and unconscious cognition (Sejnowski, 2015).

According to Greenfield (2001) unique signature of consciousness occurs through chemicals called "peptides" which are co-localized within conventional chemical messengers that can be trilingual in terms of these other control systems in the body and these circulate and feedback, so there is interaction of multi-lingual chemicals that iterate around the brain and eventually within a subsecond causing the neuronal size to shrink or expand and this determines the degree of consciousness. This explanation suggests that consciousness abodes in a neuronal assembly and functions as a "dynamic core" first envisaged by Edelman (see Tononi & Edelman, 1998). However, such theories have been unable to extract the essence of consciousness simply because existing methodologies of visualization in 3D and current brain mapping technologies are not at the scale of observation where consciousness is elemental.

What is meant by elemental consciousness? There is a difference between basic self-awareness requiring introspection through self-consciousness and phenomenal consciousness, in the sense as defined by Block (1998), in relating to raw experience of the color red, for example. This experience is called *qualia*. Phenomenal consciousness is the subjective aspect of consciousness or *qualia* that does not require language and cognition, yet is reportable to cognition. In this way, phenomenal consciousness acts as a facilitator of *qualia* and this can be said to occur at the quantum level through elemental consciousness involving dynamism, self-organization and life, which in this chapter we refer to as consciousness. Consciousness at the quantum level remains not explicitly accessible and it contrasts with "awaking" consciousness encapsulating both self-consciousness and access consciousness which are episodes of phenomenal consciousness available for cognitive processing.

Brains consist of massive, layered neural networks, which are far from thermodynamic equilibrium thus in such complex adaptive systems one needs to address the problem from an integrative neurobiophysics or neurophenomenological approach that begins at the quantum level in order to elucidate the biophysical basis of consciousness and not just the neural correlates of consciousness. We do not assume that quantum level phenomena is conscious, since phenomenal consciousness is possible only through specific brain regions that emerge between various fields inside neurons and across synapses that leads to the transfer of information in neural networks by way of access consciousness or core consciousness (Damasio, 1999). However, it is at the quantum level where electromagnetic quantum interactions need to be resolved rather than be hidden in chemical reactions as suggested by Greenfield (2001).

At the quantum level, consciousness is not a cognitive function, but a separate phenomenon of awareness. This view advocates that there is more to brain function that grounds cognition with the idea of consciousness. At the quantum level, information processing exists (Matsuno & Paton, 2000; Patel, 2001), but it is not the Shannon information processing like in communication theory. There is information exchange of energy between kinetic energy, classical potential energy, and the quantum potential energy within elementary constituents of matter (Hiley, 1995). This raises the stakes for an energy scale that defines the source of consciousness at the quantum mechanical level in terms of energy exchange in tandem with spatiotemporal scale for brain connectivity used in understanding cognition through the traditional electrochemical communication as modi operandi of brains. These beginnings underlay the roots of our deliberations for a quantum biophysical basis of consciousness that is based on interactionism within brain matter (Cacha & Poznanski, 2014). It is a monist version of interactionism (Popper & Eccles, 1977) used in the philosophy of mind which holds that, matter and mind are one yet they exert causal effects on one another. What are these causal effects? If consciousness is not a state of matter (Tegmark, 2015), but an action then what is the causative action-at-a-distance?

The ultimate understanding of consciousness must reflect upon the emergence of subjective experiences. A further important point is that subjective experiences have no significant bearing on the realization of raw perceptions, unlike conscious experiences that are being created by the actions and feelings. The self-referential qualities of consciousness that philosophers often call the experiential and subjective aspects occur in conscious life that appears to recognize "self" or self-awareness as opposed to any reflexive organism's behavior. Subjective experiences originate as quantum processing that due to indeterminism allows first person ontology. They arise unconsciously based on the exchanges of information and energy at every scale from quantum events to neural activity (Freeman, 2007) to the more complex conscious experiences (Damasio, 1999).

The totality of our consciousness is comprised of different levels in accordance with the Freudian interpretation: the conscious, the unconscious and the aconscious or subconscious (Orsucci, 2009). These levels of consciousness represent differing degrees of intensity of awareness. For instance, aconscious in reference to the emergence of this neurophenomenological entity as being a neurophenomenon that is not repressed (i.e., subconscious) and not recalled to conscious awareness as is associated with the commonly held view on unconscious (Freud, 1915). It should be emphasized that aconscious subjective experiences are not the conscious interpretation of the emerged conscious experiences because interpretation involves cognition, comprising cognitive semantics. Aconscious subjective experiences occur in living organisms, but in the majority of cases such experiences transpire without conscious interpretation before Shannon (extrinsic) information interacts with Gödel (intrinsic) information (Cicurel & Nicolelis, 2015).

The recursive look into the subjective nature of consciousness (or self-consciousness, cf. Korzeniewski, 2013) captures account of "selfhood" its own being, felt and perceived from within hence it is first-person ontology or subjectivity of experience that endures through time, but less so through space (Varela, 1999). The idea that self-consciousness is the projection of temporal, linguistic trajectories is meaningless (Varela, 1971). In contrast, there is an epistemological view which corresponds to third-person form of knowing. It focuses on physical concepts that have objective descriptions involving (biological and physiological knowledge). Further, physical conceptualization also includes psychological knowledge (such as findings from psychophysical experiments) whose attribution do not contain mental descriptions.

Accordingly, philosophers construed this knowledge argument that there is an explanatory gap in introspecting first- and third-person ontology (Levine, 1983). It is only a "gap" in the sense that there is only a rudimentary understanding between subjective experiences and the objective descriptions of those conceptual self-referential experiences. This can be confirmed as how a transcendental brain that is non-experiential can fathom subjective experiences that has first-person ontology, such as red experience when looking at a particular apple. However, when viewed from third-person perspective may require empirical validation. Such representation reflects the dynamics and structure, performing detection and discrimination of visual stimulus such as short or long wavelength light. To put it differently, the phenomenal concept to explain the experiential aspect of mind (Carruthers & Veillet, 2007; Levin, 2006, 2008) appears as a quick-fix solution to the mind–body problem defined to be the problem about how phenomenal concepts like subjective experiences and qualia relate to the brain. First-person ontology such as subjectivity, creativity, feelings, intelligence, self-awareness, and intentionality does not need an explicit mentioning of the mind, since it includes subjective experience as a fundamental property of the functioning of the brain that is performed through operations that have causal relations to physical processes that have an objective nature or third-person ontology.

Consciousness maybe quite subtle and complex, reducible to physical processes at either the quantum level or at the level of large-assemblies of neural populations. At the Planck scale, Hameroff and Penrose (2014) have hypothesized quantum gravity effects inside microtubules known as Orch-OR (orchestrated objective reduction) theory associated with quantum vibrations in microtubules — the primary correlate of consciousness. At the mesoscopic scale, Tononi (2012) hypothesized consciousness as resulting from large-scale integration of information in the brain resulting from its complex neuronal connections of large assemblies of specially selected groups or populations of neurons (Edelman & Tononi, 1995). However, both reductionist models and information theoretic models of consciousness (Balduzzi & Tononi, 2009; Tononi, 2008, 2010, 2012; Tononi & Sporns, 2003) are problematic.

The Orch-OR theory cannot explain the spontaneity of consciousness that requires the totality of *protoconciousness* to successfully bridge the gap between microscopic and macroscopic events in the brain; while the information theoretic models explain the qualitative character of conscious experience rather than consciousness itself (Peressini, 2013). Moreover, if consciousness is an emergent phenomenon that emerges from the functioning of the brain that supervenes upon the dynamical processes originating in the brain and realized in brain structures (Searle, 2000) then the information theoretic framework of consciousness has a conflict in epistemology. This argument of information theoretic integration is that physical signals that carry information, which are not conscious, is accomplished through interactions between physical signals in the third-person ontology, where all information is dynamic and the dynamical interactions lead to a dynamic core hypothesis (Tononi & Edelman, 1998), and that is a problem because information processing depends on a specified algorithm *a priori*.

Neural organization which is easier addressed experimentally at higher levels points to cognition arising at the level of massive neuronal interactions, as for example in the cortico-thalamic system (Baars & Edelman, 2012). Recordings from thalamocortical oscillations point towards brain waves recordings from EEG in certain frequency bands as neural correlates of consciousness since they arise in cognition. However, there is also a categorical error in identifying consciousness through DC electric fields of the EEG. Notably such methodologies pick up cognition (third person ontology), but not the occurrences of self-awareness (first person ontology) for the latter can only be observed at the quantum level. For example, Alkire *et al.* (2008) recorded the brain's electrical activity using EEG under anesthesia as consciousness faded there was a loss of synchrony between different brain regions. This supports a breakdown in global neuronal workspace (see Baars, 1988).

The "global workspace" theory (Baars, 1997), assumes that incoming Shannon information is first processed locally in separate brain regions unconsciously and conscious experience occurs when these signals globally spread throughout the brain, which then start firing spikes in synchrony. However, the major hurdle is that synaptic transmission is too slow for brain operations that are responsible for consciousness. Gap junctions allow neuron assemblies to work collectively at a much faster rate however, because this type of activity depends on coherent gap-junction signaling, the very extensive type of neuronal assemblies that mediate consciousness would be composed of synchronous neuron firing at high frequencies. This is not directly observable in experimental contexts, yet would more likely reflect neural correlates of consciousness. Therefore, the "global workspace" theory (Baars, 1997) provides an account of cognition, but does not address the deeper problem of what consciousness is. The "global workspace" theory and integrated information theories are similar. They base their theories on brain connectivity to answer questions about consciousness from inference on cognition that provide important insights into the relation between consciousness and cognition (Dehaene & Naccache, 2001; van Gaal & Lamme, 2012).

Searle (2007) refers to consciousness as a biological phenomenon which he labeled as "biological naturalism" to explain in naturalistic terms that there is nothing in the brain that supervenes upon physical processes (matter and energy). The basic ideas in biological naturalism are: (a) Consciousness cannot function causally in the production of physical behavior (i.e., consciousness has no causal powers); (b) Neuronal processes can function causal in the production of consciousness (i.e., consciousness is reducible to neuronal processes); (c) Consciousness is ontologically irreducible (i.e., consciousness has first person ontology).

Under (a) the firing of a neuron can always be explained in terms of the firing of other neurons that will be added to all of the other inputs of that neuron, to formulate behavioral responses to those firings, but not as a result of a subjective experience. To fit it together, processes such as neuronal firing lead to functions that are subjective experiences such as experience of painfulness — these experiences do not interact directly with physical processes such as amount of firing of pain related neurons. Under (b) consciousness is an ordinary physical system that is somewhat conscious or perceived as self-sensation as opposed to a separate property that the brain "gives rise to." The difference between "property dualism" and "biological naturalism" is that the former entails that consciousness is a non-physical emergent property of the brain that is "over and above" the neuronal phenomena, while under (c) consciousness is a brain state. This is comparable to considering consciousness as a state of matter (Tegmark, 2014) without trying to explain how this brain state arises. Biological naturalism has been extended to "neurobiological naturalism" by Feinberg (2012) and in Chapter 1.

Feinberg & Mallet (2013) argue that consciousness evolved over half a billion years across phyla of vertebrates and genes someway 'pattern' the proposed elements of consciousness through neural interactions. Unfortunately their work avoided the intricacies of cognition and quantum-based approaches to consciousness. We accept the evolutionary and genetic origins of consciousness and go further to claim that consciousness emerges from quantum processes that predispose on cognition and how it may arise in the brain through cognitive processes. Thus our purpose is to give a conceptual underpinning of how subjective experiences are facilitated by the complex interactions resulting from quantum processes that give rise to a "stream" of consciousness. Even though this does not imply that all living organisms have the potential for consciousness, it is obvious that rudimentary forms of self-awareness exist in lower forms of life (Feinberg & Mallatt, 2013; Chapter 2). Clearly from the perspective of evolution of organic matter there is an additional characteristic which is required for consciousness. The consensus is for biological organisms to possess autopoietic processes that can rely on its quantum state. This quantum state occurs at the submicrolevel where the transfer of quantum information through collective mode of creation and annihilation dynamics of energy quanta called "corticons" (Jibu & Yasue, 1995) of the cortical quantum field (Stuart *et al.*, 1979). The interaction between "corticons" as Goldstone bosons[1] creates and annihilates "virtual" photons in interfacial water throughout the brain.

Popper *et al.* (1993) suggested that the EM force field could represent the aconscious subjective experiences as a result of "interactionism" within brain matter. They proposed that in the aconscious, subjective experience capable of interacting through the EM force field acted upon by an unknown physical force could bring on conscious experiences. However, what they did not show was how the unknown physical force could achieve this in the brain. Rouleau and Dotta (2014) claim it is an EM *zeitgeber* that is any external or environmental cue which induces consciousness upon interacting with the brain.

A unified consciousness either exists at the entire cortical level (John, 2001) or primarily in the thalamocortical region (Llinas *et al.*, 1998). The fundamental conscious act from which arise subjective experiences (or *qualia*) cannot be spatiotemporally assigned because of discontinuities would lead to violation of consciousness as a unified phenomenon so theories of consciousness that account for subjective experience as spatiotemporal patterns of amplitude and phase coherent EEG activity are unlikely to be correct (see John, 2005). Lindahl & Arhem (1994) proposed a model with two levels of interaction: the first between a certain spatiotemporal pattern of action potentials and an EM field; second between the EM field and a conscious field representing a subjective experience without explaining the make-up of what constitutes a conscious field (Libet, 1996).

The EM theory of consciousness (McFadden, 2002b) proposed that information within neurons is pooled and integrated to form a conscious EM field. According to conscious EM information field theory (see McFadden, 2013 for an overview) the generation of consciousness requires synchronous firing of large numbers of synapses on cortical pyramidal cells (McFadden, 2002a). At present there is no *in vivo* experimental evidence to indicate a viable endogenous microscopic EM field

---

[1]The term generally used in high energy physics is Goldstone boson; while in solid state physics the term most usually encountered is "collective mode."

(although biomolecules are endowed with electric dipole moments and when their charges move an EM field is generated) since most if not all the experimental data are based on tissue/brain extracellular EM fields (Cifra *et al.*, 2011; Buzsaki *et al.*, 2012). Given that the strength of the extracellular EM field is very local, the encoding of any given conscious experience would be undetectable at a short distance away from where it was generated. This in turn means that neurons across the whole brain would simply not be "turned on" by these localized putatively conscious EM patterns. In short, the EM theory of consciousness fails to provide an explanation for a unified consciousness and does not explain how conscious awareness occurs.

Walker (1970) was first to suggest consciousness to be non-physical, but real quantity. According to McGinn (1995) the difficulty comes from consciousness being nowhere and everywhere in the brain. If one posits consciousness acquires "physical meaning" from quantum phenomena then as Romjin (2002), suggests "virtual" photons can be the carriers of consciousness. Others have criticized erroneously those "virtual" particles to be carriers of consciousness, based on synchronous firing of neurons would imply increased consciousness (see Lipkind, 2003). The "virtual" interaction is the key in understanding the first-person perspective of an endogenous EM field system (Hales, 2014). The basis for the subjective view according to Hales (2014) is when an interference pattern of photons arrived and thus become "virtual" photons (Romijn, 2002).

## 11.2. What are the Salient Features of Consciousness?

In order to begin elucidating the integrative "biophysics of being" (Varala, 1996) we need to first identify the salient features of consciousness. A triple-aspect monistic model of consciousness proposed by Cacha & Poznanski (2014) elaborates three aspects that are inseparable (spatiotemporally), yet neurophenomenologically separated: (i) the self-referential organization consisting of subjective concepts, (ii) the informational-hierarchical organization consisting of uncognizable/cognizable concepts and (iii) the relational organization consisting of objective concepts. The interconnectedness of the three organizations (informational-hierarchical, self-referential, and relational) constitutes a triple-aspect monistic model (see Fig. 1). The interrelationship between the three aspects is "sui generis." It cannot be causal in the ordinary sense of "physical causation" (as the action of forces between two bodies) because the aspects are not

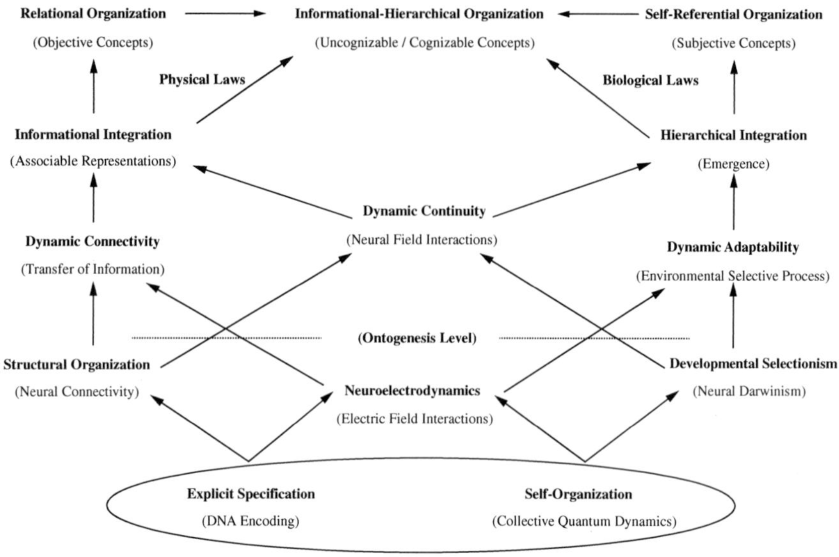


**Fig. 1.** A neurophenomenological segregation of consciousness in terms of non-reductive physicalism. Subjective experiences are expressed through a self-referential organization that manifests through a hierarchically based informational-hierarchical organization, which is causally augmented with a relational organization that supports semantics stored in the neural processes of the brain. Living organisms such as mammals and in particular primates have hierarchically driven informational-hierarchical organization that is sufficiently complex to sustain subjective experiences. The left side is associated with third-person ontology, while the right side reflects first-person ontology. Self-organization governs the complexity for the realization of conscious experiences. It is assumed to have originated at the quantum level and is therefore governed through discrete quantum processes. The dotted lines between structural organization, neuroelectrodynamics, and developmental selection indicate ontogenesis. Self-replication occurs through explicit specification in the offspring via deoxyribonucleic acid (DNA) encoding. Explicit specification remains the only brain operation that is hardwired through phylogenetic evolution.

spatiotemporally separated across any level. We are proposing that subjective experiences emerge from complex interactions in the brain (Solms & Turnbull, 2002) that are three aspects of one individual process.

It is important to realize that the sources of consciousness are intrinsically intertwined with functioning of the brain and understanding how qualia emerge from neural processes in the brain is called the 'hard problem' of consciousness. Chalmers (1996) claimed that subjective experiences are determined by, but not identical to, the organization of physical processes in the brain. As for example in "anomolous monism", subjective experiences supervene upon the physical processes in the brain (Davidson, 1980). It was Chalmers (1993, 1996) who first suggested that consciousness requires a fundamental theory that is likely to be one that combines both qualitative and quantitative elements (Seth *et al.*, 2006). However, in order to make progress it is by not reducing *qualia* to the objective entities of neural processes, but rather by including *qualia* as part of a hierarchical model where subjective experiences are assumed to emerge from collective quantum dynamics.

Quantum mechanics is a model to describe "the physical world" at the quantum level. The underlying perception is that quantum mechanics is not causative at other scales. The perception of quantized events as 'instabilities' occurring in the brain at higher levels of observation is not new yet it is these quantum 'instabilities' that play a crucial role in consciousness. In order to point to a causal physical link between consciousness and brain function, a causal closure in weaker form is necessary in which "virtual" interactions supervene on the physical interactions through Heisenberg's uncertainty principle. That is, that there can be no change at the microlevel without a corresponding change at the macrolevel and vice versa. This is how causation in the brain allows for objective concepts like cognitive semantics to emerge from quantum level neurophenomenology as an extension of Verela's framework of "biophysics of being" (Rudrauf *et al.*, 2003).

### 11.2.1. *Non-spatiotemporal confinement*

Hypothetical or "virtual" particles are emergent quantum level phenomenon that occurs in subatomic interactions (e.g. repulsion or attraction between charges can be due to exchange of "virtual" particles). Eccles (1992) first hypothesized on "virtual" particles in the evolution of consciousness or the unfolding of consciousness during quantum field interactions (see Appendix A). The exchange of "virtual" particles exchanging a wave of hypothetical or "virtual" particles, which are extremely ephemeral (Romijn, 2002) or evanescent (Jibu *et al.*, 1996; Jibu & Yasue, 1997) is not confined to the space-time domain, but to the frequency domain within the informational field. A "virtual" particle, generally, is an excitation in a quantum field caused by charged subatomic particles (e.g., quasifree electrons in interfacial water) interacting. As a result, brain produces an abundance of quasiparticles at room temperature which are charged subatomic particles of matter (e.g., quasifree electrons) surrounded in an ephemeral cloud of "virtual" particles. These quasiparticles constitute the fabric of consciousness (Beck & Eccles, 1992).

With reference to Fig. 2, the diagram elucidates the significance of the emergence of an upward/downward causation. It occurs at the submicrolevel as a corticial quantum field consisting of composite quasiparticle fields that through self-trapping of energy quanta give rise to microscopic phenomena at the quantum level as a collective mode of excitation resulting in a quantum coherent state. The dynamics of energy quanta may generate dipolar waves which are maintained by quasifree electrons moving along filaments of protein molecules in microlevel (**C**). Energy quanta as "virtual" biophotons[2] are created in submicrolevel system **A** by incoming energy from microlevel system **C** through microlevel system **B**. The incoming energy triggers the emission and absorption of energy quanta. If one "virtual" photon escapes from the interacting field because of Heisenberg's uncertainty principle, then the interacting field which is a quantum vacuum has gained a photon to the electric dipole field and this process would continue until many photons are trapped giving rise to a collective mode of excitations which are waves of "virtual" photons in "quantum water."

In theory, an electrolyte such as water is capable of supporting spontaneous self-organization of quantum phenomena from the result of quantum field interactions referred to as collective quantum dynamics or quantum coherence (Del Giudice *et al.*, 1988, 2005). Quantum properties of interfacial water can produce collective quantum dynamics by transforming any incoherent or disordered energy quanta into two distinct collective modes of excitations: (i) self-referential quantum information through waves of "virtual" particles at the submicrolevel and (ii) a microlevel event along the myriad of protein molecules within neurons including the extracellular matrices of the brain as a coherent wave of electric dipole oscillations.

The quantum spin of waves of "virtual" particles point in the same direction, so systems with different collective spin states will have *periodicities of discrete energies* tethered to a common frequency resulting in same energy. The microlevel events are realized in terms of electric dipole oscillations excited by an endogenous electric field that are the product of

[2]"Virtual" biophotons are "virtual" photons through sustained incoming energy transfer from submicrolevel (downward causation) in biological organisms.

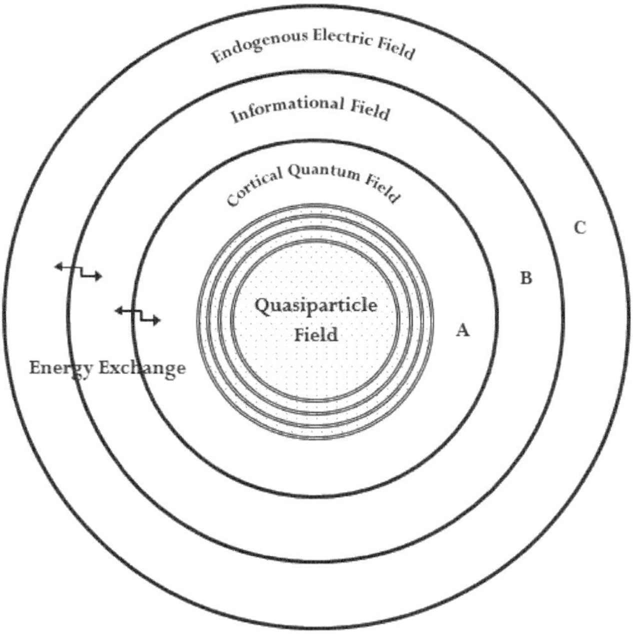


**Fig. 2.** A schematic illustration of an upward causation and downward causation intrinsic information through energy transfer between various quantum fields emerging from a cortical quantum field composed of interfacial water electric dipole field and many quasiparticle fields. The submicrolevel system (i.e., quantum level) **A** incorporates the dynamics of energy quanta in brain interacting with each other by absorbing and emitting "virtual" biophotons. The dynamics of energy quanta in system **A** results in a collective mode of excitations in terms of waves of "virtual particles." The microlevel system (i.e., at molecular level) **C** incorporates the collective modes of excitations along the filaments of protein molecules as dipolar waves which are maintained by quasifree electrons of interfacial water. The information field **B** involves dynamics that is interconnected by exchange of energy and intrinsic information transfer through protonic conduction in interfacial water.

collective quantum dynamics of the oscillating quantum fields described in quantum field theory as collective modes of excitations. As indicated in Fig. 2 the submicrolevel system **A** receives energy from the microlevel system **B** at the quantum level and microlevel system **C** at the molecular level. Systems **A** and **B** interact by exchanging energy. Exchange energy from **A** to **B** is relayed in the form of the collective mode of excitation

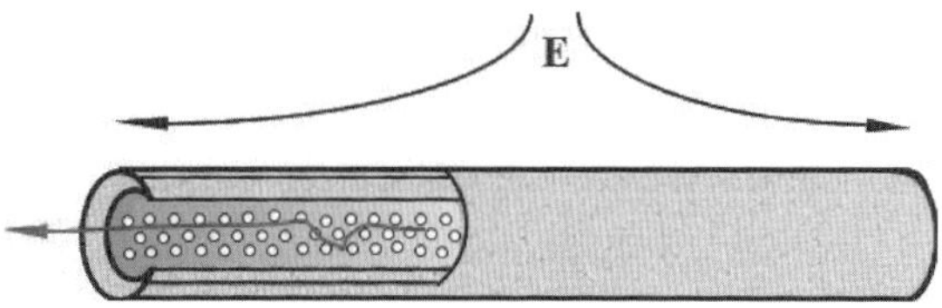


**Fig. 3.** A schematic illustration showing effect of an endogenous electric field (**E**) on polarized proteins within the ionized milieu of interfacial water depicting collective modes of excitations at the submicrolevel as local distortions of the electron density of quasiparticles (i.e. wave of "virtual particles") and at the microlevel where the dipole order propagates as one traveling condensate. One cannot directly measure an action, but only the quasiparticle patterns resulting from energy quanta interactions with charged subatomic particles (e.g., quasi-free electrons in interfacial water). The electric dipole oscillations instantiated through quantum disturbances are caused by interactions between quasiparticles constituting a quantum field and the endogenous electric field (**E**).

resulting in a coherent wave of excitations of energy quanta influenced by the endogenous electric field through energy exchange from **C** to **B** in terms of collective modes of oscillating quantum fields known as electric dipole oscillations.

Electric dipole oscillations occur when the dipole order propagates as one traveling condensate. How does "quantum water" feedback on waves of "virtual" particles? For example, a collective "virtual" biophoton emission within the ionized milieu of interfacial water proposes that the magnetic dipole moments in interfacial water molecules constitute a quantum field, composed of "virtual" biophotons as the energy quanta of the field. This composite quasiparticle field is postulated to interact by energy transfer as disturbances in the quantum field acted upon by endogenous electrical field in between molecular proteins in neurons, which are quasiparticle patterns that are depicted as dipolar waves (see Fig. 3).

### 11.2.2. *Self-referential quantum information*

How can consciousness be an emergent neurophenomenon and transform possibilities into actualities? According to standard quantum physics upward causation of material interactions cannot convert possibilities into actualities. However, in quantum brain dynamics based on quantum field theory, subjectivity becomes the unpredictability of the quantum field interactions according to an energy scale. So frequency or energy does in fact separate subjectivity from objectivity by interpreting subjectivity to originate from the indeterminism that occurs at the quantum level. This unpredictability differs from Copenhagen interpretation in terms of possibilities or probabilities of certain events. Under the de Broglie–Bohm causal interpretation, subjectivity seems to originate at the interface between quasiparticle interactions facilitating quantum information and quantized information or "qubits" resulting from sustaining such quantum interactions from the environment.

The meaning of a quantum field as interpreted by Bohm & Hiley (1993) is defined in terms of the quantum potential interpreted to carry quantized information from the environment that acts on the waves of "virtual" particles. The self-referential quantum information can therefore be interpreted to mean the particle and the field as aspects of the same entity, whereupon the quantum potential mediates quantum information that induces a change of form from within the system itself. In this case, the energy must come from within the particles, which are manifestations of quantum wave fields through the quantum potential. Eposito (1999) interpreted the quantum potential as the kinetic energy of the internal motion associated with the quantum spin of the particles.

Quantum systems are not mechanical that can be analyzed into interacting fundamental elements through reductionism, but holistic in nature (Bohm & Hiley, 1993) with interplay between the manifest aspect (3pp) and the subtle aspect (1pp) (Hiley, 1995). This is not dualism because the manifest aspect consists of some subtle features and in the subtle aspect there are more subtle features, so one can regard those features, which are not so subtle, as manifest aspect. Subtle features of physical phenomena are information based from active to passive to inactive and the continuous change of information is a process that appears to show the emergence of consciousness (Hiley, 1995).

If consciousness is elemental then it is hypothesized to be manifested by "virtual" waves through "quantum force" induced dynamic effect. Consciousness exists not in space–time, but as a "virtual" wave in frequency domain within the informational field. The bridge is made possible through quantum properties of interfacial water holding interacting entities together that can otherwise be de cohered with the environment (see Appendix B). The "virtual" waves that connect with the information from a vast number of neurons in the cerebral networks across the brain through the quantum potential is a 'quantum force' induced quantum dynamic effect that according to Bohm (1980) represents a wholeness that is private and intimate open only to the first person perspective (1pp).

All physical phenomena arising from brain matter consist of a manifest aspect, which is governed by electrochemical signaling and includes Shannon information that gets quantized as "qubits." Consciousness is sourced from the subtle aspect that can be actualized by self-referential quantum information mediated through the quantum potential. The information field between the subtle aspect and the manifest aspect is the wave of "virtual" particles facilitating the subtle aspect through quantum information colliding with "qubits" from the manifest aspect that are hypothesized to provide the essential linkage between 1pp and 3pp; between the explicate order of Newtonian realm and the implicate order of the quantum potential.

The resulting interferometry pattern emerging from the collision of self-referential quantum information and Shannon information as "qubits" gives rise to subjective experience or qualia that is instantiated by quasiparticle interactions as waves of "virtual" particles. This collision between quantum information and "qubits" as quantized information is believed to occur through proton nuclear spin ensembles in neural membranes and molecular proteins, which is a primary mechanism to mediate Shannon quantized information in the brain (Hu & Wu, 2004).

Spinning water molecules couple the intrinsic quantum information resulting in an *informational field* that occurs as collective modes of oscillating quantum fields across both intracellular and extracellular matrices of the brain. Interfacial water does not store intrinsic quantum information, but it acts as a conduit for endogenous electric field as a messenger from the informational field. For example, interfacial water contains a narrow zone of protons, as positively charged hydrogen ions, which can move very rapidly in water from one water molecule to the next. The principle of protonic conduction in interfacial water by way of the Grotthuss mechanism (Hassanali *et al.*, 2013) can transfer quantum and quantized information through proton nuclear spin ensembles in polarized proteins. The functional role of interfacial water has shown to act as an informational field where quantum information is transferred as energy like a dipolar laser (Del Giudice *et al.*, 1988).

At the molecular level information-physical interaction (Bazyan, 2010) is governed through dynamics of physical interactions that transform quantum information into a new format through the endogenous electric field. Physical interactions in proteins cause conformational changes to release energy. The energy profile of conformational rearrangements of proteins during their interaction occurs only in interfacial water, while "integration of information" (Tononi, 2008) can potentially occur elsewhere and therefore should not be used in the same context as information-physical interactions (Bazyan, 2010). In living organisms consisting of organic matter such physical interactions facilitate the self-referential quantum information.

### 11.2.3. *Biological organization and quantum causality*

Information-physical interactions in interfacial water are acted upon by a "quantum force" induced dynamic effect through an endogenous electric field attributed to the action of the magnetic vector potential as "action-at-a-distance." It is not like the EM field is accountable for causal effects in the brain, but rather the EM potentials (Szasz *et al.*, 2009). In particular, the magnetic vector potential enters into Schrödingers equation as a phase-shift affecting charged subatomic particles passing through a field free region (e.g., in the absence of interfacial water). The magnetic vector potential changes the phase of the charged subatomic particles oscillating at a common frequency creating a causative effect that can produce "action-at-a-distance" which is the basis of causative interactions in organisms referred to as quantum causality.

A system of correlated quasiparticles can exhibit non-locality when negative probabilities arise, but in our sense of the word "quantum" refers to brain environment that is describable through molecular quantum brain mechanics and the existence of a "quantum state" occurs in the occurs in the interfacial water that is isolated from the disruptive role of thermal collisions, mimicking the case of low temperature water ($T < 200\,\mathrm{K}$).

At the quantum level, quantum causality arises because the wave function[3] of two or more brain matter particles are entangled by the quantum potential (Bohm, 1952). The entanglement is not mechanical, but more subtle or undefinable because quantum phenomena produce an uncontrollable and unpredictable change without necessity of any change of energy (Hiley, 1995). Thus the wave function is taken to be an informational field (Bohm & Hiley, 1993) such that self-referential quantum information can arise linked to a common quantum potential, in biological systems where the causation is outside the realm of the system involving

[3]The wave function is a type of algorithm from which probabilities of particles can be determined.

a *zeitgeber*. In other words, brain matter particles that are coupled to a common quantum potential are non-separable, resulting in causative interactions or "action-at-a-distance." If there is quantum entanglement then it is broken when the entangled particles decohere beyond interfacial water spanning the spatial volume of the brain. And therefore there is no "cosmic" consciousness as one would propose if quantum entanglement is carried too far beyond the realm of brain-based consciousness.

Fundamental characteristic of living organisms also include the property of biological causality in its dynamic organization (Chauvet, 1993). The causative action in space, which arises from hierarchical structure, is reflected by an informational field. So, the hierarchy describes a specific physiological process at each level, coupled with processes evolving in adjacent levels. At the macrolevel biological causality is the representation of activity and how it results from the dynamic continuity in space and time, underlying information integration. The topology of the biological organism (see Chauvet, 1996, 2002) considers autopoietic process invoking both space-time and dynamic continuity, while geometry influences the spatiotemporal dynamics of the process. Thus geometry as a metaphor for qualia (Balduzzi & Tononi, 2009) does not entail the presence of an informational field and hence there can be no integration of information from the microlevel to the macrolevel and vice versa. This is a valid explanation why integration of information is unlikely to happen when only geometry is considered. The integration of information must allow for the propagation of causative interactions or "action-at-a-distance" according to the relationship between the topology of autopoietic processes and the geometry of the neural system. Furthermore, dynamic continuity removes the elusiveness of finding the mechanisms underlying consciousness-related binding (Revonsuo & Newman, 1999) as is required in the Orch-OR theory (see Chapter 14).

### 11.2.4. *Wave of 'virtual' particles as a "quantum force" induced dynamic effect*

In this chapter we consider replacing the wave function by a wave of "virtual" particles in order to describe quantum potential as a casual effect through hydrodynamic formulations of quantum mechanics where all quasiparticles in the quantum "fluid" follow the trajectories given by the de Broglie-Bohm pilot wave theory (Bohm, 1952). The quantum potential is argued to be causative in similar context to the Aharonov &

Bohm (1959) effect which demonstrates that the magnetic vector potential, rather than the electric and magnetic fields, are the fundamental quantities in quantum mechanics. When the classical EM Lorentz force vanishes the magnetic vector potential still affects the quantum potential via the remaining action of the "quantum force" caused by the local variations of the quantum probability density $\rho(x,t)$[4]:

$$f = -\nabla\Omega.$$

And we have redefined a quantum potential $\Omega$ for particles that have no spin, only mass and electric charge based on Madelung's hydrodynamic formulation of quantum mechanics and de Broglie-Bohm pilot wave theory (Bohm, 1952; Wallstrom, 1994):

$$\Omega = -(\hbar^2/2m)\ \nabla^2\sqrt{\rho}/\sqrt{\rho},$$

where $\rho$ is the probability density of the quantum "fluid" of quasiparticles, m is the mass, and $\hbar = h/2\pi$ is the reduced Planck's constant. In quantum field theory only wave of "virtual" particles are investigated.

The "quantum force" can be expressed (Maddox & Bittner, 2003):

$$f = \frac{\hbar^2}{4m}\left[\frac{\nabla^3\rho}{\rho} - \frac{\nabla(\nabla\rho\cdot\nabla\rho)}{2\rho^2} - \left(\frac{\nabla^2\rho}{\rho} - \frac{\nabla\rho\cdot\nabla\rho}{\rho^2}\right)\frac{\nabla\rho}{\rho}\right]$$

It will be discussed later that this causative "quantum force" tethers the quantum field oscillations to specific frequncy domains to collectively represent the collision of quantum information and quantized information in the informational field.[5] And represents itself through information-physical interaction at the molecular level. Bohm & Hiley (1993) treat $f$ on an equal footing to the Lorentz force $F = q[\mathbf{E} + \mathbf{v} \times \mathbf{B}]$ where $\mathbf{E} = -\partial\mathbf{A}/\partial t - \nabla\phi$ is the electric field, $\mathbf{B} = \nabla \times \mathbf{A}$ is the magnetic field, q is the charge $\mathbf{v}$ is velocity, $\mathbf{A}$ is the magnetic vector potential and $\phi$ is the scaler electric potential.

The momentum of a quantum system is $\nabla S$ where S is the phase of the system governing the rhythm of the quantum fluctuations of the system. Its dynamics is altered by the Schrödinger equation when it enters into a

[4] $\rho < 0$.

[5] Sensitivity is obtained from the phase shifts induced by the magnetic vector potential **A** are inscriptions into interfacing structures for information field.

coherence domain in the presence of the magnetic vector potential (**A**) since **A** only appears in the phase (S)

$$\nabla S = q\hbar \mathbf{A}$$

Magnetic vector potential is the mechanism for coherence through particular *periodicities of discrete energies*. The magnetic vector potential adds up to the phase of the system S so that the magnetic vector potential is able to drive the phase which in turn produces the influence through its relation to the momentum of the quantum system ∇S. As a consequence, the coherent system is affected by the magnetic vector potential. Magnetic vector potential is able to change the phase of a coherent system in the absence of a magnetic field. Classically $f = 0$, $\mathbf{B} = \nabla \times \mathbf{A} = 0$ but $\mathbf{A} \neq 0$.

## 11.3. Molecular Mechanisms of Quantum Causality

The brain as a complex adaptive system is composed of smaller parts interacting both within and across scale as a non-reductive physical system that is thermodynamically an open system with causative entropic forces (Wissner-Gross & Freer, 2013) emerging into spontaneous self-organizing phenomena. An entropic force acting in a system is a phenomenological force resulting from the entire system's statistical tendency to increase its entropy (i.e., energy dissipation by disordering or randomness in the system), but in far from equilibrium open systems like the brain the entropy decreases with increasing molecular order (i.e., high degree of coherence) leading to the generation of self-organization phenomena so that more information can be stored.

It is understood that self-organization is a process in which biological systems spontaneously develop order at higher level that provides a causal account for integration to be a physical phenomenon (Varela, 1979). An example which requires spontaneous self-organization of quantum events once there is critical amount of neurons leading to quantum phenomenon is called "objective reduction" (OR) (Hameroff & Penrose, 2003). Such a spontaneous self-organizing quantum phenomenon in all of the neurons of the brain instantiated by quantum gravity would result in the collapse of the wave function, resulting in a spontaneous downgrading of the collective modes of excitations in microtubules across a large number of neurons where subjectivity emerges spontaneously from the self-organization of quantum phenomena.

What differs between quantum phenomena in the human brain from quantum phenomena in other unconscious systems? Del Giudice and colleagues (see De Giudice, 1988) advocate quantum properties of interfacial water, but no reference to molecular mechanisms of quantum causality. It is not clear why the EM quantum field interactions should entail the defining characteristic of self-awareness, indeed a way it "feels" to, nor what feature of this physical process can actually convey a conscious feeling, nor how subjectivity emerges from quantum phenomena.

### 11.3.1. *Is there lasing behavior within the ionized milieu of interfacial water?*

In the brain the classical idea of electrodiffusion of ions appears to be supplemented with convective motions of large ionized atoms that lead to brain tissue behaving more like gas plasma (Hales, 2014). Therefore, it is the endogenous EM field and not a self-propagating EM field that governs the emission and absorption of energy quanta (photons) in gas plasma, which could in theory achieve coherency by means of superradiance (Dicke, 1954). In other words, superradiance is a specific coherent order of energy quanta where metabolic energy supplies incoherent photons and through superradiance transforms them into coherent photons by trapping of energy quanta in the system. A photon (i.e., energy quanta) that is emitted from living cells, including neurons, was first coined by Popp (1979) to be a "biophton." Biophotons are released from every cell in the body where chemical reactions take place as byproducts of metabolism (e.g., chemiluminescence and bioluminescence) are caused by chemical reactions (oxidation). These biophoton emissions refer to the phenomenon of constant and spontaneous emission due to metabolic activity, without excitation or enhancement due to incoherent thermal emission of disordered biophotons.

Excited states of macromolecular structures drive complex behavior in neurons can also be a catalytic process generally associated with the presence of an oxidative metabolism that accompanies the production of reactive oxygen species which participate in the regulation of a wide spectrum of biochemical reactions. The coherent order of energy quanta induced by biophotons interacting with electric dipoles of interfacial water molecules (see Mentre, 2012 for a review) in and around endogenous structures of neurons and the extracellular matrices is referred to as "lasing" behavior of interfacial water. It is an example of coherent emission from superradiance (Li, 1992) and it has been referred to by Popp and colleagues as coherent biophotons (Popp *et al.*, 1984; Popp *et al*, 1994; Popp & Chang, 1998). Popp *et al.* (2002) suggest that coherent biophotons function like a biological "laser" light (i.e., collective motion of local distortions of quasifree electron densities in large number of interfacial water molecules interacting with the quantized EM field) which has been suggested to be a biophoton field (cf. Hagan *et al.*, 1994; Jibu *et al.*, 1994; Cacha & Poznanski, 2014).

If each neuron is emitting a biophoton field, then the whole brain is, in effect, a resonating biophoton field — a ubiquitous energy field. And since biophotons are the entities through which the entire brain can readily communicate information, there is near instantaneous intercommunication throughout the whole brain that provide an information character of the electric coherence (Van Wijk *et al.*, 2007) as a consequence of global quantum coherence or quantum long-range order between the collective modes of oscillating quantum fields and dipole moments throughout the human brain. The frequencies of biophoton fields are comparable to coherent order of interfacial water molecules and this allows for the biophoton field to propagate along the filament of protein molecules by "piercing" the shield from which dipolar molecules in interfacial watter are isolated from the surrounding dipole disorder of the solvent molecules. This is referred to as filamentary propagation (Del Guidence *et al.*, 1983).

If the EM biophoton field theory is correct then external self-propagating EM fields that release photons at much higher frequency then the proposed biophoton fields would not be expected to have any impact. Furthermore, putatively EM patterns could be capable of acting as patterns not only in the same neurons that generated them, but throughout the whole brain since quantum coherence within the ionized milieu of proteins produces a biological "laser" light inside and outside of neurons interconnecting neurons through the exchange of energy quanta. However, there is yet no reliable proof for biological 'laser' light[6] and coherent emission or reliable proof for coherence of biophotons (i.e., a biophoton field) in neuronal systems. An electronic excited state in a neuron, which gives rise to biophotons per second, seems to be far too low to exert any effects

[6]A biological "laser" light carries different meaning from a wave of "virtual" particles denoted as a "virtual" laser.

and since biophotons are ultra-weak emissions of photons (incoherent) they are unlikely to interact with charged subatomic particles because their wavelength is inversely proportional to their energy.

Quantum interactions in interfacial water entrained by the brain's endogenous electric field result in the exchange "virtual" photons — particles that exist for only a short time determined by Heisenberg's uncertainty principle. Living organisms in fact possess a living aura, a "virtual" wave that pervades the entire organism with a 'virtual' photonic flux. Popp (1979) erroneously concluded that the wave was a "laser" light of real photons (termed biophotons) that continually receives inputs from the environment and is continually outputting coherent biophotons, particularly near the ULF range. The problem is that there are no laser lights in the brain as pointed out by the famous neurologist Karl Pribram (2013). The misconception made by Popp was that coherent biophotons are real photons. We have retained the use of biophoton to mean "virtual" photons as such "lasing" properties of interfacial water, resonating biophoton field as an ubiquitous energy field is replaced with "photon" behavior in interfacial water.

### 11.3.2. *Instantiation of "photon" behavior as a macroergic effect*

Neural activity involves the transfer of macroscopic energy that leads to breakup of hydrogen bonds leading to conformational changes in neuronal DNA (Suberbielle *et al.*, 2013). This is because it leaves changes in protein expression and synthesis implying polymerases which then becomes exposed to DNA to transcribe to RNA resulting in conformational changes. Breaking of hydrogen bonds within DNA will result in energy transfer/release which can lead to macroergic effects such as exciplex states of DNA in the cell nucleus of neurons that trigger chemical reactions, which produces chemical energy that can result in the release of biophotons in neurons (Isojima *et al.*, 1995; Kataoka *et al.*, 2001; Kobayashi *et al.*, 1999; Rahnama *et al.*, 2011). The emission of biophotons originates from oxidative metabolism (chemical reactions) of mitochondria and lipid peroxidation (Thar & Kuhl, 2004).

The basic structure of DNA is composed of two parts: the external sugar-phosphate backbone, and the internal bases. The sugar-phosphate backbone of DNA is polar, and therefore hydrophillic. The interior portions of DNA, the bases, are relatively non-polar and therefore hydrophobic. The hydrophobic core of the DNA molecule hidden inside the sugar-phosphate backbone acts to isolate it from the polar water molecules. Polar molecules, because of their charge, can interact with water without disrupting the ubiquitous lattice of hydrogen bonds that the water molecules naturally form. Non-polar molecules, however, cannot form electrostatic bonds with the hydrogen and oxygen atoms in water. Non-polar molecules, to be immersed in water, must break potential electrostatic hydrogen bonds between water molecules. Breaking potential bonds represents a net increase in the free energy of the system of water molecules and this has a destabilizing effect. The result of which is that non-polar molecules tend to get pushed out of water. As a result, DNA 'breathing' (see Peyrad *et al.*, 2009) occurs, which is localized excitations that result in the exposure of some bases to the hydrophobic environment due to rupture of chemical hydrogen bonds causing electronic transitions of quantum events to take place. These localized quantum events will aid the transfer of energy across proteins in macromolecules as a result of protein-ligand binding (Turton *et al.*, 2014) causing DNA chromosomes acquiring unlimited information from metabolic activity that in turn produce regulative electric fields along the chain of polarized proteins as "photon" behavior in interfacial water.

Nagl & Popp (1983) developed a model where they assumed biophoton transfer from cytoplasmic metabolism which resulted in changes of the DNA conformation via exciplex/excimer formation. Biophotons are absorbed by DNA producing exciplex states in DNA molecules when moving in-between proteins of macromolecules in histones encased by DNA in the cell nucleus (see Fig. 4) that provide chemical energy and charge transport during metabolic activity (Tuszynski *et al.*, 1984). This provides an instance of a claim that a macroergic effect by way of reabsorption of biophotons by exciplex states of DNA serves as a "feed-back" mechanism to release "virtual" biophotons forming a "virtual" wave. In this sense, the electric dipole oscillations at the molecular level constitute a wave of "virtual" particles resulting from patterns of dipolar waves due to biophoton excitation.

Macroergic molecules can store chemical energy produced by oxidization of glucose and they are adenosine triphosphate (ATP) that provide neurons with chemical energy needed to perform physicochemical activities. ATP provides most chemical energies stored within cells for metabolism. The chemical energy is stored in phosphate bonds of ATP molecules. The largest producer of ATP is the mitochondria. At the molecular level, hydrolysis of ATP is the major exchanger of chemical

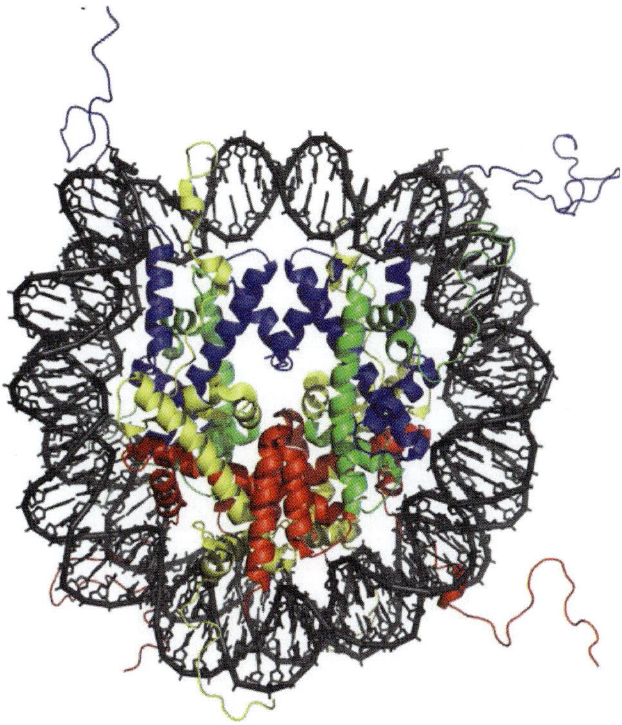

**Fig. 4.** Electric field of DNA due to mostly negative charges of phosphate backbone interacting with positively charged histones encased by the DNA in the cell nucleus, results in the electric field around DNA to be changed by conformational changes to proteins in histones and ionic bonds in interfacial water. This occurs at the quantum level when an electric charge in collision with quasifree electrons changes its momentum. The loss of momentum appears as a "virtual" biophoton which can be generated in the decay of unstable elementary particles.

energy within neurons. External feeding of energy through ATP reactions may trigger the formation and dissipative propagation of dipolar waves on the protein chains propagating into the spatial region occupied by interfacial water molecules (Jibu & Yasue, 2004).

Subsequent decay of dipolar waves propagating through a medium such as the intracellular space of neurons, transfers energy via interactions with DNA in the nucleus to induce conformational changes contributing to the creation and annihilation dynamics of energy quanta at the quantum level of the brain. In the presence of an electric field, polar protein molecules can undergo coherent conformational excitations in the presence of continuous supply of biochemical energy from metabolic activity in the DNA. Biophoton emission due to DNA reabsorption and energy transfer due to continuous supply of energy through metabolic activity (Popp *et al.*, 1981; Rattemeyer *et al.*, 1981; Chwirot, 1986; Li, 1992) that is stored as a form of collective mode of excitation (Del Giudice *et al.*, 1985).

### 11.3.3 *EM zeitgeber as the synchronizing mechanism*

The "Two-Brains" hypothesis (Goodman *et al.*, 2014 and Chapter 9) postulates that the macrolevel neural activity associated with cognitive functioning in the cerebral networks is tangible, known and accessible referred to as the "electroionic brain" is inseparable (spatiotemporally), yet neurophenomenologically segregated by the microlevel quantum activity which is intangible and difficult to access experimentally referred to as the "electromagnetic brain." The "Two-Brains" hypothesis according to which an "electroionic brain" interacts by way of resonance with an "electromagnetic" brain, thereby enhancing higher processes of cognitive functioning. The EM quantum interactions originate in the midbrain reticular formation (Parvizi & Damasio, 2001) from where consciousness moves upstream through the claustrum in the forebrain and throughout the cerebral hemispheres for further processing by cognition.

The question pertaining to the "Two-Brains" hypothesis (Goodman *et al.*, 2014) is the elucidation of the origin of the endogenous EM fields in brains (see Hales, 2014). It has been shown mostly at the cellular level, that effects of electric field exposure can be blocked by calcium channel blockers, demonstrating that activation of voltage-gated calcium channels (VGCCs) appears to be essential for many and perhaps all electric field responses (Pall, 2013). Is the source of endogenous electric fields in the brain caused by the effects of voltage-dependent ionic channels, in particular VGCCs? Electric current flow in neurons at the macrolevel is not carried by electrons, but ions in electrolytic solutions. Therefore electrolytic conduction of ionic charge produces an endogenous electric field in the entire brain. Electromagnetic fields are generated by the net effect of electrolytic current flow from a classical perspective. While the magnitude of fields associated with an individual neuron is negligible, the effect of multiple neurons excited together in a specific area generates a measureable magnetic field that is nevertheless extremely small — a billionth of the strength of the earth's magnetic field. In the brain the magnetic field is considered negligible and only an endogenous electric field dominates through neuroelectrodynamics (Aur & Jog, 2010). This is of course from a classical perspective of the origins of endogenous electric field.

Brain magnetic field intensity due to spontaneous currents is less than thermal noise energies and cannot be expected to affect brain's processes. However, this quantum interference effect does not require a strong magnetic field to exist endogenously because it is not the intensity of the magnetic field, but the interference pattern shifts that describe the rhythm of electric dipole oscillations of the field expressed in terms of *periodicities of discrete energies.* Therefore the negligibility of a magnetic field in the brain likely remains immaterial for a "quantum force" as waves of "virtual" particles in "quantum water" can pilot brain dynamics similar way to the Aharonov & Bohm (1959) quantum effect can pilot the magnetic vector potential interpreted as "action-at-a-distance" quantum dynamic effect between quasifree electrons in interfacial water and the distant magnetic field.

This is caused by the magnetic dipole moments sensitivity to the phase shifts acting upon the molecular dipoles in interfacial water and EM *zeitgeber* as the environmental entrainment agent which aligns them generating a coherent domain. The coherent domain has a decreased frequency of oscillation resulting in the entanglement of quasiparticles (self-trapped biophotons) due to a change in mass of the quasiparticle. The coherent domains containing trapped biophotons can be induced by a magnetic vector potential when the immediate surrounding space has negligible magnetic field. This coupling between the magnetic dipole moments of the quasifree electrons in interfacial water and the magnetic field originating in the environment gives rise to a quantum dynamic effect, similar to the Aharonov-Bohm effect. The phase of the system is changed by the magnetic vector potential meaning that *periodicities of discrete energies* are conveying quantum information. The phase of every coherent system becomes a "memory" of the events occuring in the whole environment (Brizhik *et al.*, 2009).

We can therefore point to the quantum dynamic effect as elemental consciousness in living organisms, since through the magnetic vector potential, the organism begins to be self-aware of the environment through the exchange of information (Trukhan & Anosov, 2007) and generates more self-awareness through quantum information spanning the spatial volume of the brain. The consequence of this effect is that

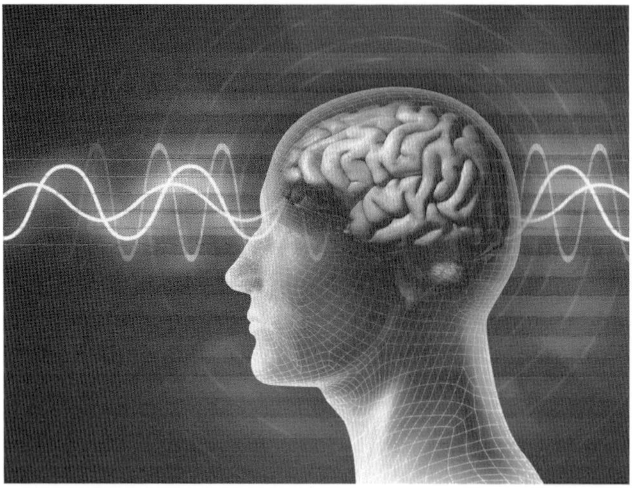

**Fig. 5.** The geomagnetic field induces molecular magneto-sensitivity along the filaments of protein molecules in interfacial water through a quantum dynamic effect. It is discussed that the geomagnetic field (which cannot be completely shielded) acts as an EM *zeitgeber*, triggering self-awareness through quantum information spanning the spatial volume of the brain.

interfacial water molecular dipole field can harbor causality through the magnetic vector potential (Del Giudice *et al.*, 2010). Then is the earth's magnetic field the *zeitgeber* of brain-based consciousness? According to Vitiello (2001) consciousness emerges from the interaction of the brain oscillators with external ensemble of oscillators in the environment such as the magnetic field of the earth (Persinger *et al.*, 2013) and as a result the "Double" of the brain refers to the EM *zeitgeber* that induces elemental consciousness upon interaction with the brain (see Figure 5).

### 11.4. The Subjective Aspect of Consciousness

Subjectivity seems to originate at a different interface — between quantum field interactions and the residual complex interactions resulting from such events, and how these events bring about subjectivity in the aconscious where it is recalled to consciousness in a way similar to the role of potentialities describing the passage of potentiality to an actualization.

Discrete quantum processes act as one whole informational 'conscious field' providing an informational character of quantum coherence that instantaneously is unified through protein conformation changes due to metabolic activity and 'virtual' wave propagation in the cerebral cortex (cortico-thalamic system) and subcortical areas (e.g., midbrain and hindbrain). The resultant informational "conscious field" is attributable to the facilitation of qualia throughout the brain and in specific networks of neurons emerging from the midbrain reticulum. Our hypothesis is based on the source of intrinsic information arising from a novel property of DNA molecular sequences instantiating waves of "virtual" particles in interfacial water and the resultant quantum dynamic effect.

The hard problem is how any objective explanation can explain subjective phenomena given that objectivity seems not to have direct access to such phenomena. Such an indirect access that links the ontologically subjective experience with epistemologically objective concepts in the aconsciousis brought about by the exchange of energy quanta and transfer of quantum information into more stable associable representations is a mechanism for the selection of subjective experiences relayed higher up through cognitive functions. Subjectivity as the interplay between "virtual" wave propagation and electric dipole oscillations in macromolecules forming dipolar waves in the aconscious.

### 11.4.1. *Bridging the explanatory gap*

Subjective experiences involve the transfer of quantum information (i.e., aconscious subjective experiences) that is conveyed through an informational "conscious field" that is encoded in terms of *periodicities of discrete energies* where energy driven by information-physical interactions of the quantum field oscillations throughout the cerebral cortex collides with Shannon information coming from the environment via cognition. Subjectivity gets facilitated through informational field arising from quantum field interactions between composite quantum fields, consisting of composite quasiparticle fields. Quantum fields are made up of charged subatomic particles and energy quanta where energy is understood to have a fundamentally subjective aspect in nature. Energy is intrinsically qualitative so, for example, the qualitative feeling of seeing "red" is a form of physical energy.

The subjective aspect of consciousness arises from the informational field which is a quantum field that gives rise to the quantum potential and given that internal motion associated with quantum spin has been shown to be responsible for the quantum potential (Salesi & Recami, 1998) it is possible that by invoking spin-mediated theory of consciousness (see Hu & Wu, 2004 and Chapter 12) the mechanism by which the explanatory gap is bridged exists through proton nuclear spin ensembles in neural membranes and molecular proteins through the Grotthuss mechanism (Hassanali *et al.*, 2013). This could arise when composite quantum field oscillations with *periodicities of discrete energies* are tethered through a common quantum potential as a "quantum force" manifesting upscale at the molecular level through collective modes of excitations as electric dipole oscillations or dipolar waves in interfacial water. The *periodicities of discrete energies* induced by the interaction of quasiparticle fields with the endogenous electric field can be related to absorption bands of interfacial water molecules in the brain. Nevertheless, it appears the uniqueness of the anharmonic *periodicities of discrete energies* that link together the elements into a whole subjective experience are yet to be theoretically determined with precision.

The difference between contemporary neuroscientific understanding of the brain and elemental consciousness theories is the added component of quantum informational processes at the most fundamental level. Neurons are there to contain and help amplify the quantum information or quantum informational processing that is occuring within and between those cells. However, consciousness arises not from the separation in space and time between quantum processing and neural information processing in the explicate order, but in the non-segregation of *periodicities of discrete energies* in the implicate order where causative "quantum force" induced dynamic effect appears as correlations of entangled quantum information associated with difference in the phase (S) of the coherent system encoded as *periodicities of discrete energies* detectable at specific frequency domains.

The following de Broglie equation represents quantum field oscillations due to quasiparticles interacting with an endogenous electric field:

$$f = \frac{\mathrm{mc}^2}{\mathrm{h}},$$

where $f$ is the frequency of composite quantum field oscillation (periodicity), m is the mass of quasiparticles, c is the speed of energy quanta in the ionized milieu of interfacial water and $\mathrm{h} = 6.626 \times 10^{-34}$ Js is Planck's constant. The speed of energy quanta in a vacuum is constant, but is reduced significantly depending on the medium. For example, in a coherent medium like water the photons acquire a finite rest mass, a longitudinal polarization and a frequency that is proportional to their velocity based on non-Maxwellian propagation (Smith, 1994).

There needs to be sufficient number of quasiparticles to enable the composite quantum fields to oscillate at a comparable frequency to the coherent order for consciousness to emerge in the informational field. Different composite quantum field oscillations depend on the mass of the quasiparticles. Our hypothesis is that consciousness to be reportable higher up; the periodicities of discrete energies need to be monotonically increasing from the microlevel to macrolevel invoking spatiotemporal inseparability across any level, but neurophenomenological segregate through sensitivity of the phase shifts of the coherent system induced by distant magnetic field as a quantum dynamic effect. A genuine case of loss of consciousness is correlated with the brain's loss of energy due to loss of autopoietic processes in death. However, under anesthesia, anesthetic gases weaken quantum mechanical interactions and supress proteins from oscillating at a specific frequency for energy transfer between micro-level and macrolevel systems thus erasing consciousness.

Energy scales in physics decrease as length scales increase because of the quantum relation:

$$\iota = \frac{hc}{E}$$

where $\iota$ is the length scale and E is the energy scale in units of Planck's constant. A useful approximate value is $hc = 1240$ (MeVfm) $= 1.986 \times 10^{-25}$ (Jm). Moreover, the quantum dynamic effect to be causative requires a monotonically increasing energy scale (see Fig. 6). The causative, "quantum force" induced dynamic effect applies to low energy levels where the usual framework of quantum field theory applies given that the detection by a molecule of the earth's magnetic field on a single quantum spin would be the order $10^{-8}$eV (Leask, 1977). At higher energies approaching or exceeding the Planck scale which occur at a length scale $10^{-32}$ cm, the effects of quantum gravity can no longer be ignored and a new theory of consciousness such as the Orch-OR theory is required (see Chapter 14).

The conscious and aconscious are different in energy level of brain quantum field oscillation frequency. The brain quantum field oscillations can form from two-level atom system. The oscillation frequency of

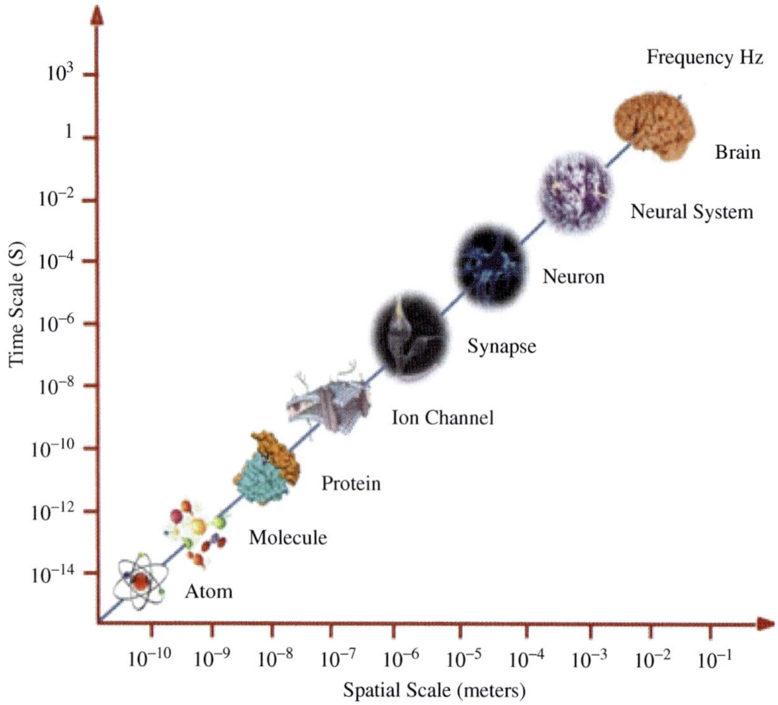


**Fig. 6.** This diagram shows phase-coherence in which the informational "conscious" field is not spatiotemporally confined but resides in the frequency domain expressed in terms of energy as $E = hf$ having a monotonically increasing variation in the possibilities for the *periodicities of discrete energies* to fathom an almost infinite number of subjective experiences across the informational field.

two-level atom is known as Rabi oscillation frequency, which is defined as (Pantian & Yupapin, 2013):

$$\omega_{i,j} = \mathbf{E}_o \cdot \mathbf{p}_{i,j} / \hbar,$$

where $\mathbf{E}_o = \hat{\varepsilon}\, E_o$ denotes the amplitude of electric field $\mathbf{E}$, which includes the polarization $\hat{\varepsilon}$ and $\mathbf{p}_{i,j}$ is the transition dipole moment for a transition from initial state i to final state j. In principle, the individual brain frequency function cannot be the same, for conscious experiences and subjective experiences in the aconscious. Any two-state system can also be used to model a "qubit." However, the equivalence between "quantum force" ($f$) in Section 11.2.4 and the *periodicities of discrete energies* is that quantum causality through protonic conduction results in the transfer of energy across the informational "conscious" field.

### 11.4.2. *Unpredictability of spontaneous breaking of symmetry of subjective experiences*

The biological aspect which constitutes function that potentially leads to the capability of expressing subjectivity is latent in inanimate matter. (This excludes some forms of energy in Nature). Living organisms such as mammals and in particular primates have hierarchically driven informational-functional organization that is sufficiently complex to sustain subjective experiences. Subjectivity is hidden from the external world in the aconscious to be recalled to consciousness as *qualia*. The unconscious transfer of information in the brain is uncognized, but becomes cognizable when *qualia* transcends cognition. Subjective experiences are cognizable through the subtle aspect of physical phenomena which are information based undermines the subjective raw qualities of conscious experience — the *qualia*.[7]

In the absence of highly complex organization among neurons it would be unlikely that aconscious subjective experiences could self-organize into conscious experiences, since self-referential quantum information would be seriously underdeveloped, most likely latent and limited to quantum-generalized processes. Therefore subjective experiences in complex adaptive systems emerge as a quantum phenomenon due to spontaneous breaking of symmetry (Del Giudice *et al.*, 1985; Freeman & Vitiello, 2008). This suggests that the resolution of subjective experiences in brains cannot be emulated without an understanding of the unpredictability of spontaneous breaking of symmetry. What this means is that some unpredictability allows spontaneous breaking of symmetry to act upon the structure. Therefore consciousness as a quantum dynamic effect can impact changes in brain function by acting as a facilitator of *qualia*.

Spontaneous breaking of symmetry in quantum field theory generates Goldstone bosons as energy quanta in the cortical quantum field that manifests itself in a collective mode of excitation in brain matter (Del Giudice

[7]*Qualia* would seem to be the intrinsic quantum information that gets to be facilitated by the quantum dynamic effect in response to external cues and collision with extrinsic information from the environment.

*et al.*, 1985). A special case of spontaneous breaking of symmetry is dynamical symmetry breaking. Freeman & Vitiello (2008) had shown spontaneous breaking of symmetry in brain dynamics at the molecular level. The symmetry which gets broken is the rotational symmetry of the spin magnetic dipole moment. Spontaneous breaking of symmetry is a mechanism for change of scale from microlevel to the macrolevel is dynamically achieved through spontaneous breaking of symmetry leading to coherent order of energy quanta. In quantum field theory, the symmetry in the intrinsic quantum information is broken down by extrinsic information ("qubits"). Information is carried by ordered patterns maintained by long-range correlations mediated by energy quanta. The mechanism of spontaneous breaking of symmetry by which long-range correlations (made up of Goldstone bosons) are dynamically generated is expected to be a major facilitator or constraint on quantum brain dynamics. Brain processes are dependent on random fluctuations. These continuous perturbations of the dynamics involving the cortex give rise to myriads of local phase transitions, which are quenched to maintain the cortex in a state of conditional stability known as metastability (Freeman & Vitiello, 2008).

The metastable dynamic interactions invoking associable representations in brain processes are uncognizable in space and time which means not repressed and not recalled to consciousness, i.e., in the aconscious. The dynamical interactions metastabilize (cf. Kelso, 1995; Fingelkurts & Fingelkurts, 2004; Freeman & Holmes, 2005; Fingelkurts *et al.*, 2009) and cognize in the process of spontaneous breaking of symmetry of subjective experiences. The realization of subjective experiences into cognized higher brain functions follows the process of resonance (Golant, 1989; Grossberg, 1999, 2007). Resonance is an organizational reservoir of potential changes to neural dynamics without any specific structural reorganization. The idea behind resonance in neural systems concerns the topology and includes the tendency for maximum potential of organization. Resonance occurs when bottom-up patterns of microscopic dynamic activity as quantum information collide with top-down expectations of macroscopic dynamics that is influenced by the environment as "qubits." This bottom-up / top-down resonance lasts longer and is more amplified than any individual activation with resonance. A schematic diagram of the process whereby resonance acts upon an associable representation arising from the aconscious to form *qualia* is shown in Fig. 7.

The collision process involves a mechanism for attaining a transient coherence in a very large assembly of neurons is based on quantum

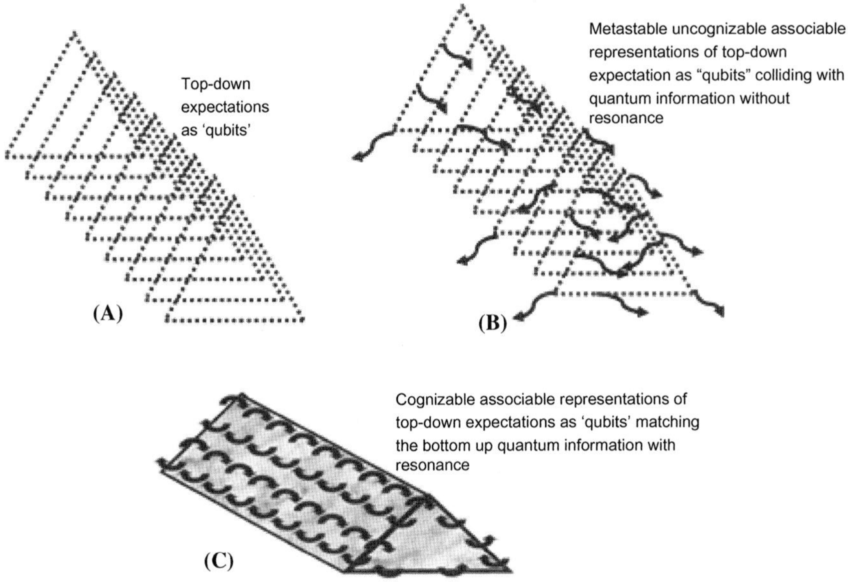


**Fig. 7.** A schematic illustration of the emergence of subjective experience as a resonant phenomenon whereupon the collision between top-down expectations of quantized information influenced by the environment as "qubits" and bottom-up quantum information. (A) Top-down expectations in dendritic neural systems representing is associable representations of subjective experiences. (B) Without resonance, there is inadequate matching of the top-down quantized information and the bottom-up quantum information. This results in a highly metastable and uncognizable associable representation of a subjective experience as indicated by the arrows. (C) Feedback relationships among neurons by way of resonance unifying top-down expectations of quantized information with bottom-up quantum information that result in associable representation of subjective experiences. The triangle is not a representation, but an ontological neurophenomenon that unifies the epistemological aspects as resonance develops. The dipolar wave is a collective vibration of molecular origin that can transport energy without transporting electric charge — it is quasiparticle fields that forms when biophotons are re-absorbed by DNA.

theory where the act of observation causes a system to be in one type of unpredictable state: a quantum phenomenon called "subjective reduction." It is subjective because it requires an observer. In the brain, however, there is no outside observer hence quantum events like the collapse of the wave function are not downgraded subjectively, but occur spontaneously without a homunculus or observer being present in the brain. Further subjective experiences that emerge from quantum field oscillations and become conscious experiences as collective modes of excitations. This coherent order through the process of resonance leads to higher-order cognitive functions associated with conscious experiences that are reportable through language and are causally communicated by wave packets (Freeman, 2003) to the language areas in the cerebral cortex.

This resonance phenomenon is capable of transmitting patterns of DNA molecular sequences by waves of "virtual" particles in interfacial water (Montagnier, *et al.*, 2009). The quantum field theory behind such a "virtual" laser has been reviewed in Montagnier *et al.* (2011). This is based on the theory of diphasic water where dipolar molecules organize (given low energy input) into quantum coherent domains capable of carrying quantum information. DNA in water solution forms short-lived associations through larger number of water molecules which can form long polymers of electric dipoles associated with hydrogen bonds (Ruan *et al.*, 2004). DNA and proteins are polyelectrolytes surrounded by a cloud of positive counter-ions. A resonant alternating magnetic field with a frequency of the counter-ions extracts these ions and thus producing a counter-rotation of quasifree electrons in the interfacial water coherent domain which become energetically excited to produce the observed waves of "virtual" particles in interfacial water (Montagnier *et al.*, 2009).

## 11.5. The Self-Organizing Neural Processes of Conscious Experiences

Conscious experiences can be attributed to both cognitive aspects of consciousness such as core consciousness (Damasio, 1999) as a spontaneous self-organizing process of consciousness (Görner & Combs, 1998) and aconscious subjective experiences that have arisen from quantum events. Cognition is the processing of perception, memory, learning, attention, thought, and language disclosed by behaviors that exhibit self-awareness and awareness of an entity's surroundings. By this definition, a sleeping person is clearly unconscious. If we are trying to measure consciousness in the clinical sense, then we would use this definition. If someone is anesthetized they are unconscious. If they are awake then they are conscious. Subjective experience is the direct experience in the aconscious. For example experiencing a lucid dream, one is "conscious" within that dream despite not having any actual awareness of all surroundings. This phenomenological definition requires understanding of the subtle states of consciousness on a deeper level. Common accounts of subjective experiences as an emergent property of the brain constitute a number of elemental stages before reaching cognition to report feelings and conscious experiences through language (see Chapter 3).

### 11.5.1. *Selectional self-organization of dynamically involving neuronal groups*

The self-referential character of the brain facilitated by degeneracy in biological systems (Edelman & Gally, 2001) represents a conceptual idea in understanding how subjective experiences evolved by natural selection. The hallmark of natural selection in biological systems is information that is stored in a genetic database and self-replicated with occasional variations. Evolution's sole mechanism by natural selection in the context of the brain has been referred to as "selectionism" (Sporns & Tononi, 1994). Selectionism transcends between "developing" and "mature" neural activities (Sporns, 1997). The idea is that selection through variance drives self-organization. It is through variance in neuroanatomy that essential features related to conscious experiences are established. Selectionism forms an integral feature of neuronal group selection theory or neural Darwinism (Edelman 1978, 1981, 1987, 1993) and is a dynamic process that is predominant during brain development (Frank & Wenner, 1993).

Based on neural Darwinism, two distinct processes of selection through variation have been postulated (Edelman, 1987): (i) during embryonic development, neural populations form groups with strong synaptic connections and the environment exerts selectional pressure as a result of diversity, variance, and imprecision of the synaptic connections. The majority of anatomical connections are not functionally expressed via so-called developmental selection processes. (ii) After ontogenesis in which both structural and functional modules are continually modified, new groups form through the so-called experiential selection process (Edelman, 1981). A corollary of this latter selectionism is that selection applied to neuronal populations in which selection processes shape synaptic connectivity and hence dynamic connectivity (Reeke & Sporns, 1990; Reeke, 1994). The mechanism by which changes to synaptic efficiency en route for synaptic connections bring about dynamic connectivity involves a selection process that invokes excitatory changes through synaptic plasticity. Post-ontogenic pressures governing dynamical connectivity could in essence reflect the coordinated firing patterns observed in neural assemblies. It is equivalent in some sense to dynamic core as neuronal groups of strongly connected neuronal assemblies (Tononi & Edelman, 1998). However, dynamical connectivity alone is insufficient to establish a physical process as advanced by neural Darwinism (Crick, 1990).

What physical processes govern selectionist theories? If synapses transfer signals indifferently by a similar chemical modulation, then what determines a weak synapse as opposed to a strong synapse? This question was not fully explored in Edelman's work. If selectionist theories do not require a rule for selecting the appropriate "re-entry" path, then such a process must be assumed to depend on some adaptive behavior. Indeed, Calvin (1998) defined a "Darwinian process" as one "...capable of recursively bootstrapping random novelties into something of quality." Although from the neural Darwinian perspective, the output of the system as a whole is of an adaptive value to the organism (Reeke *et al.*, 1990), in the population approach it is difficult to generalize at the individual neuron level (Edelman, 1995).

The process whereby selectional self-organization installs "neural" connectivity has been discussed by Willshaw & von Malsburg (1976). Gierer (1988) has suggested the possibility of a genetic contribution toward "fine tuning" of neuronal connections, although this is a misconception because DNA encoding through explicit specification is associated with formation of structural changes and not the dynamical flows that ultimately govern conscious experiences. Genetic evolution is itself the result of selectionist processes involving millions of years (i.e., phylogenetics) and not the life-span of an individual organism (i.e., epigenetics). For instance, DNA duplication errors in a gene SRGAP2 have led to structural changes that may have influenced cognition (Dennis *et al.*, 2012). Cognitive semantics that are fluidly engrained within the neural structure as information is entangled with self-referential qualities through quantum causality and are realized as conscious experiences.

### 11.5.2. *Self-organization of textured neural activity*

The rate of firing at synapses can invoke dynamic connectivity as a result of changes to the electrochemical signature that is modulated globally by self-organization of textured neural activity. These changes to electrochemical signature are broadly tuned in areas of cerebral cortex, both are changing the electrochemical signature, which is local and fast in addition to global and slow to guide the dynamics associated with self-organization of textured neural activity that supports cognition (Freeman, 2005). Analyzing cognition requires a good grasp of how the brain integrates separated tasks into a coherent function. Higher cognitive functions (e.g., language, thinking, planning, reasoning, problem solving, and free will) are all influenced by *qualia* to be harnessed by the electrochemical signaling through the midbrain reticulum to various cortical connections (e.g., associative, commissural, and projection fibers) and neural pathways defined on the basis of the neurotransmitter used (e.g., cholinergic, serotonergic, dopaminergic, and noradrenergic pathways).

Neural correlates of given subjective experiences and their qualities can be understood in ontological process, but not in an epistemological sense. Therefore cognitive brain functions allow for conscious experience to be actualized and a function can have neural correlates as determined by fMRI analysis. Putative continuity of dynamical processes across spatiotemporal scales is relevant for projecting cognitive function to bring about physical integration due to their recurrent interconnections within neural networks environment. This dynamical process undergoes a continuous change via chemical modulation and serves as an informational field for quantized information. Cognition cannot be confined solely to the thalamocortical complex alone, but also is present in the forebrain or the upper brainstem system, which extends from the midbrain to the basal diencephalon in the absence of cortical input. The cerebral cortex serves as a medium for the elaboration of conscious content or cognition (Merker, 2007).

The formation of a relational organization relies on associative memory (i.e., new signals invoking old responses) and in particular on the ability to integrate concepts. In other words, distantly related information is organized in long-term memory by integrating concepts to related concepts stored in short-term memory. A relational organization is the information that has been stored and retrieved associated with physical concepts involving cognitive semantics derived from the process of information integration. Cognitive semantics from the viewpoint of relational organization is semantic entailment in the brain of information-physical interactions. In living organisms, such information-physical interactions are hierarchically integrated. Hierarchical integration of quantum information with quantized information facilitates first-person ontology. The subtle aspect enables the realization of subjective experience to be cognizable through information-physical interactions as neural correlates disclosing objective third-person perspective (3pp).

Brain dynamics allow monitoring neural activities thus providing a mechanistic understanding of brain functions (i.e., cognitive concepts and uncognized concepts) to be adaptive to the environment and other pressures specifically tuned to the individual organism thereby allowing for subjectivity to emerge. The dynamical interactions in the brain remain predominantly dependent on physical interactions as macroscopic manifestation of underlying microscopic phenomena (Freeman & Vitiello, 2006) and are influenced by the environment, which is the resultant effect of subjective experience in the aconscious attributed to such dynamical interactions.

To understand subjective experiences as cognized concepts, a hierarchically driven informational-hierarchical organization actualizes the subjective experiences through the metastability of an evolving dynamical information flow aided by selective and adaptive processes from the self-referential organization. The self-referential qualities that are expressed in terms of self-referential quantum information are actualizations of subjective experiences that belong to the first-person perspective (1pp). Subjective experiences are uncognized manifestations of hierarchical integration leading to their cognizability. These subjective experiences reconcile the epistemological qualities with ontological considerations within a unified informational-hierarchical organization that comes about as a causal account of a relational organization that derives from informational integration in which dynamical interactions involving information transfer form associable representations (Cacha & Poznanski, 2011).

Conscious experiences are cognizable concepts. Cognitive concepts require both segregation and integration of Shannon information (such as in a main complex in thalamocortical neural networks for consciousness (Tononi, 2008, 2012) from across spatiotemporal scales, resulting in changes to synaptic connectivity over broad areas of cerebral cortex leading to the formation of neural activity that form neuronal groups or cell assemblies. Neural activity in assemblies of neural networks and from which subjective experiences as metastable and uncognized phenomenon are self-recognized as a cognizable neurophenomenon (Orpwood, 2007). Aconscious subjective experiences are the self-referential qualities that we directly experience, and this is regardless of whether we are conscious or unconscious during sleep. This suggests that subjective experiences can evoke either cognized or uncognized concepts through access consciousness. For example, the subjective experience of pain is a cognizable neurophenomenon, but the process of transduction, transmission, and modulation associated with perception of pain, are based on empirical methods that reveal objective physical aspects such as neural correlates of various conscious experiences that do not directly reveal epistemological, self-referential qualities, such as *qualia*.

## 11.6. Discussion

There is no way of resolving the hard problem of consciousness unless it is approached through non-reductive physicalism. Set against the philosophical views of "eliminativism," "reductive physicalism," "dualism," we have shown how a new view "anomalous monism" — "non-reductive physicalism" can unfold consciousness. Non-reductive physicalism is more like biological naturalism (see Chapter 1) than anomalous monism (Davidson, 1980) since the condition that there are no "natural laws" connecting mental phenomena with physical phenomena are replaced with mental phenomena as being putatively physical with first person ontology. The approach is integrative beginning at the quantum level as a foundational conscious act, which contributes to higher levels of cognitive aspects of consciousness such as core consciousness (Damasio, 1999). This integrative approach enforces to close the gap between the quantum world and the classical world. How this is done is of vital importance as both worlds are independently computational, yet their integration seems to be non-computational requiring new physics as suggested by Sir Roger Penrose or it seems to be understood in other ways, for example through de Broglie–Bohm pilot wave theory discussed throughout this chapter.

The computational doctrine that the human brain is an information processing system and that thinking is a form of computing has been largely discredited (Hales, 2011; Manzotti, 2012; Cicurel & Nicolelis, 2015). Conscious sentient awareness as discrete computation (Dehaene *et al.*, 2014) is just as fallible. A machine that deploys its experience of being, then it is sentient, but it cannot be self-aware for it lacks the salient features of subjective experiences associated with quantum phenomena. In other words, for subjective experiences to be actualized there should be an ingredient that such sentient artifacts must possess — characterizing the topological nature of autopoietic processes. The causative interactions define the topology of the autopoietic process broken down into components that have space and time scales, yet are tethered *to periodicities of discrete energies* as they interact and collectively represent consciousness. According to Reeke & Edelman (1995) sentient artifacts may be possible, but they will certainly not be computers that process information. Self-referential quantum information distinguishes an organism from an artificial system. In an artificial system, like a robotic machine with a supercomputer brain, every function is entailed by another function, thus an infinite regress exists, but the infinite regress does not exist when there is intrinsic information (Rosen, 1991; Cicurel & Nicolelis, 2015).

Consciousness is not a fundamental property intrinsic to all matter in the universe. It is not reproducible in inorganic matter or any other nonliving matter, since the ecology for consciousness invokes biological laws which are specifically tuned physical laws applicable only to organic matter. In the fabrication of sentience and sapience in machines, assuming the detailed complexity of brain operations are known a priori, the machine would also be required to implement biological laws in a process that is called biological computation. At present this is of course problematic because

(a) Computational models of brain functioning (Izhikevich & Edelman, 2008; Eliasmith *et al.*, 2012; Stewart & Eliasmith, 2014) lack subjective experiences and therefore the possibility for sentient awareness (MacGregor, 2006);
(b) Modeling subjectivity through computational neuroscience (Taylor, 2009; Reggia, 2013) is problematic since it defies the existence of *qualia*;
(c) There are conceptual problems of highly conceivable and realistic single neuron models of consciousness (Edwards, 2005; Sevush, 2006; LaBerge & Kasevich, 2007);
(d) The creation of brain models that reconcile all possible knowledge, at present and in the future, en route for continuing improvement in accuracy until eventually the "brain in a supercomputer" model will encounter series problems with reductionism (Cicurel & Nicolelis, 2015).
(e) Connectionist or human connectome approaches to brain connectivity (Sporns *et al.*, 2005) assume both segregation and integration processes exist in the brain and reconcile with discontinuous consciousness (Bodovitz, 2004).

In this regard, understanding the role that physical processes have in subjective experiences and the extent to which these processes act in conscious experiences requires an understanding of the nature of subjectivity in terms of energy exchange and quantum information transfer. In particular, subjectivity instantiates as a resonant phenomenon arising from the quasiparticle interactions between biophotons and exciplex states of DNA that result in quanta energy transfer in and between macromolecular proteins. The quasiparticle interactions between biophotons and exciplex states of DNA are inclusively occurring in organic matter and there is no guarantee that such can be effectively reproduced in man-made quantum computing machines. As such any theory that suggests consciousness is integration of Shannon information (Tononi, 2004) based on identity theory (i.e., mind-body problem) undermines the tenets of consciousness as an emergent neurophenomenon that cannot be integrated or be a physical entailment of an integration process. Moreover, the framework of the integrated information theory of consciousness (Balduzzi & Tononi, 2008) is based on discrete dynamical systems, which is totally incompatible with the spontaneous breaking of symmetry process that was shown to lead to the unfolding of subjectivity through resonance.

Consciousness is an emergent neurophenomenon residing in the frequency domain of the informational field. Through the transfer of quantum information that is influenced by an EM *zeitgeber* giving rise to a "quantum force" induced quantum dynamic effect makes it possible for the collision of quantum information with the quantized information to facilitate *qualia*. The consequence of this effect is that interfacial water dipolar molecules become sensitive detectors of geomagnetic fields allowing the organism to be self-aware of the environment through the exchange of information and the facilitation of *qualia*. Further empirical evidence is sought using high voltage holography electron microscopy, specifically adapted to capture this quantum dynamic effect within interfacial water of DNA and biomolecules.

The human brain foster macroergic effects with sufficiently strong electric energy fields for collective modes of excitations on the surface of alpha-helical integral proteins as dipolar waves. The transfer of energy via interactions with DNA molecules, to induce conformational changes to proteins in histones and ionic bonds resulting from chemical (hydrogen) bonds breakage in interfacial water, causing molecular electronic transitions that emit an ultra-weak non-thermal spontaneous emission of biophotons. The instantiation of biophotons confined to structures of DNA molecules within histones facilitates "photon" behavior along interfacial water via dipole–dipole interactions, providing quantized informational character of collective quantum dynamics.

It was hypothesized the transition involving quantum-spin enables objectivity to have access to subjectivity through neurophenomenological segregation and manifests as subjectivity in conscious experiences mediated by molecular quantum mechanics and an EM *zeitgeber.*

## Appendix A — Quantum Field Interactions

At the quantum level, endogenous EM fields in the brain are carried by subatomic particles or sub-particles which are force particles referred to as gauge bosons. Fermions (e.g., quarks, leptons) on the other hand are subatomic particles and the electron is one such elementary particle that makes interatomic bonding and chemical reactions and hence life possible are building blocks of brain matter, which do not aggregate into atoms are found in interfacial water. The most familiar gauge boson is the photon, which is the energy quanta of the EM field. Gauge bosons of electromagnetism are force particles that carry energy and momentum and in the brain also convey information. In quantum field theory, subatomic particles like both fermions and gauge bosons are expressed in terms of fields as well as their interactions. The composite quasiparticle fields contain fermions quanta or other elementary charged subatomic particles of matter that as a result of their interactions produce "cortical" quantum field (see Fig. A.1). The composite quasiparticles can be either bosonic quasiparticles or fermionic quasiparticles depending on their constituents. More precisely, because of the relation between spin and statistics, composite quasiparticles containing an even number of fermions are bosonic quasiparticles.

It is a daunting task to understand brain environment in terms of elementary particles (e.g., matter particles like fermions or force particles like bosons) so in quantum field theory physical processes are described in terms of energy exchange. The emission of energy quanta by an endogenous EM field causes absorption of energy in composite quasiparticle fields and vice versa. These force particles scatter the energy, but they are not pure energy quanta particles, but quasiparticles that are the result of quantum field interaction with charged subatomic particles (e.g., quasi-free electrons). Quasiparticles convey a way to represent a large number of charged subatomic particle interactions in a simpler way compared to directly describing every subatomic particle. Assuming that quantum field oscillations are anharmonic, only a few elementary excitations (or low-lying excited states) which are approximately independent are needed to be described.

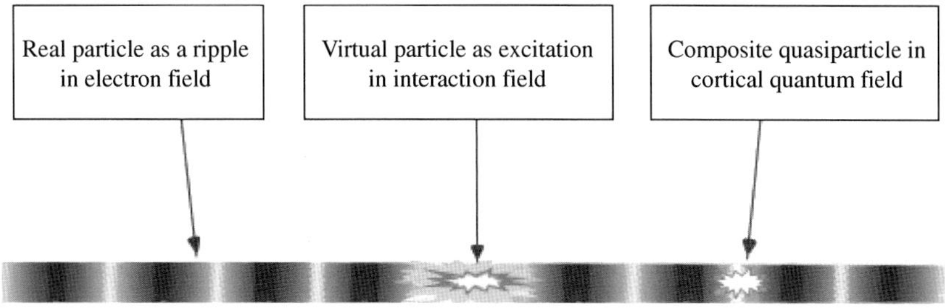


**Fig. A.1.** An electron field as in polarized molecular proteins found in the interfacial water of neurons contains a narrow zone of protons, as positively charged hydrogen ions, which can move very rapidly in water from one water molecule to the next, but the electron interacts with the endogenous EM field by creating an excitation or pattern and ceases to be a normal particle and becomes "virtual" particle in the "interaction" field or pattern of quasiparticles in an "interaction" field. A combination of two disturbances (i.e., the two "virtual" particles) remains a quasiparticle with the energy, momentum and mass of the incoming electron that is a composite quasiparticle incorporating different patterns of quasiparticles in the "cortical" quantum field. The ripple in the electron field is an oversimplification in situations where there are no interactions among particles at all. The disturbance in the electron field is caused by a real particle, and the disturbance in the "interaction" field is a "virtual" particle (e.g., gauge boson). The pattern of quasiparticles is the "cortical" quantum field.

## Appendix B — Quantum Coherence

All binding forces between atoms and molecules are of EM nature, and as a consequence, all chemical reactions are EM processes. In particular, the release of photons from chemical reactions during metabolic activity plays a fundamental role in EM couplings to molecular proteins and macromolecular structures. These EM quantum interactions produce dynamic order or quantum coherence in systems that are far from thermodynamic equilibrium. Global order emerges in the collective mode of the emission and absorption of energy quanta (photons) within the composite quantum field (i.e., composite quasiparticle fields). Due to the presence of this global order, the emission and absorption dynamics produces a macroscopic effect resulting in a collective excitation as a reflection of the aggregate behavior of the quantum biological system with no real particles only "virtual" particles. Under certain conditions quasiparticles containing an even number of fermions can lose their independence and coalesce into a collective quantum dynamic state.

Collective excitations are related to force-transmitting gauge bosons or "force" particles that cause a disturbance in the composite quasiparticle "cortical" fields in many-body systems like biomolecules. They are a reflection of the aggregate behavior of the system with no real particle at its core. A perturbation of a dipole propagates to other dipoles causing an array of dipoles to be excited as quantum field oscillations. Dipolar molecules are arranged to give an array of dipoles, that is, linear electric dipolar chains corresponding to long chain of proteins or DNA molecules. Interaction between groups of charges of the dipoles in macromolecule manifests in vibrational motion with frequency. Each dipole interacts with other dipoles via intermolecular forces, e.g., dispersion forces; result of these interactions causes the dipolar oscillations to spread with frequency into a narrow range and to provide energy exchange between dipoles.

Fröhlich (1975) argued that as the energy of metabolic drive is long enough and the dielectric properties of the materials are sufficiently extreme, then there is the possibility of large scale quantum coherence that are present within biological systems (Engel *et al.*, 2007; Collini *et al.*, 2010; Lloyd, 2011; Bischof & Del Giudice, 2013). It turns out that not only is the metabolic energy high enough and the dielectric properties unusually extreme there is now direct evidence for the microwave range oscillation within cells (Grundler & Keilmann. 1983). In biological systems that are far from thermodynamic equilibrium and where metabolic energy is continuously supplied into the system a dynamic order is formed in interfacial water surrounding polarized proteins. It is these collective excitations that are considered responsible for generating and maintaining global order via protonic conduction of the quantum field interactions which give access to electric polarization. Electric polarization in macroscopic systems occurs through Fröhlich-style condensation of energy quanta (photons) (cf. Del Giudice *et al.*, 1985).

The suggested problem with Fröhlich (1968) theory that incoherence due to thermal environment cannot be ruled out is a half truth without taking into account interfacial water. Tegmark (2000) estimated that, at normal brain temperature (300 K), individual neurons have decoherence timescales of $10^{-13}$ s yet no estimates were done specifically for quantum decoherence of interfacial water. While research conducted by undergraduate student, Laura McKemmish and colleagues (see Reimers *et al.*, 2009) shown that Fröhlich condensates involving protein amide-I vibrations are possible at temperatures of order $10^5$–$10^7$K. Nevertheless, they investigated bulk water (e.g., equilibrium thermal bath) and not interfacial water such as those close to the surfaces of molecular structures, which mimic the water in biological systems (Del Giudice *et al.*, 2010). Thus, Fröhlich-style condensates are most likely tenable in biological environments.

Interfacial water enables quantum information to be carried in neurons and between neurons as relevant frequencies of the quantum coherence domains (approximately 0.1 microns in diameter including millions of molecules) and is a quasifree electron supplier giving rise to energy transfer and long-lasting excitation (Szent-Gyorgyi, 1960). The frequency of the coherent domains is based on Szent-Gyorgyi (1957) concept of frequency changes when energy is stored in a coherent domain and if at the boundary other molecules oscillate with the same frequency then they will become part of the coherent domain. In this way, harnessing of stored energy of the coherent domain and propagating information across the brain. Intrinsic non-Shannon information originating as a quantum phenomenon needs quantum coherence to propagate information upscale to the molecular level, so quantum properties of interfacial water have a pivotal role in protecting from decoherence and allowing for periodicities of discrete energies of coherent excitations to serve as a messenger of quantum "intrinsic" information.

Interfacial water where water molecules are bound to membranes and proteins, the attraction between water molecules and the surface could protect the domain from thermal noise by way of strong and weak forces that may be associated with hydrogen isotopes of interfacial water molecules, giving rise to a stabilization of the collective excitations. Semiheavy water, HDO, is water in which hydrogen atom has been replaced with deuterium, the isotope of hydrogen containing one proton and one neutron. It is present naturally in water, but in only small amounts (~0.03%). Deuteration has been shown to raise the temperature necessary for the thermal denaturation of proteins. Indeed the disruptive role of thermal collision is neutralized and the whole interfacial water is allowed to stay coherent (Buzzacchi *et al.*, 2002).

## Acknowledgement

The first author is grateful to Michal Cifra, Huping Hu, Michael Persinger and Giuseppe Vitiello for valuable discussions.

## References

Aharonov, Y. & Bohm, D. (1959) Significance of electromagnetic potentials in quantum theory. *Phys. Rev.*, **115**, 485–491.

Alkire, M. T. Hudetz, A. G. Tononi, G. (2008) Consciousness and anesthesia. *Science*, **322**, 876–880.

Aur, D. & Jog, M. S. (2010) *Neuroelectrodynamics: Understanding the Brain Language.* IOS Press, Amsterdam.

Aur, D., Jog, M. S. & Poznanski, R.R. (2011) Computing by physical interaction in neurons. *J. Integr. Neurosci.* **10**, 413–422.

Baars, B. J. (1988) *A Cognitive Theory of Consciousness.* Cambridge, MA: Cambridge University Press.

Baars, B. J. (1997) *In the Theater of Consciousness: The Workspace of the Mind.* New York: Oxford University Press.

Baars, B. J. & Edelman, D. B. (2012) Consciousness, biology and quantum hypotheses. *Phys. Life Rev.*, **9**, 285–294.

Balduzzi D., & Tononi, G. (2008) Integrated information in discrete dynamical systems: Motivation and theoretical framework. *PLoS Comput. Biol.*, **4**, e1000091.

Balduzzi, D. & Tononi, G. (2009) Qualia: The geometry of integrated information. *PLoS Comput. Biol.*, **5**, e1000462.

Bazyan, A. S. (2010) Chailakhyan's views of the origin of consciousness and mentality. *Biophysics*, **55**, 508–512.

Beck, F. & Eccles, J. C. (1992) Quantum aspects of brain activity and the role of consciousness. *Proc. Natl. Acad. Sci. USA*, **89**, 11357–11361.

Bischof, M. & Del Giudice, E. (2013) Communication and the emergence of collective behavior in living organisms: A quantum approach. *Molec. Biol. Inter.* 987549.

Block, N. (1995). On a confusion about a function of consciousness. *Behav. Brain Sci.* 18, 227–287.

Bodovitz, S. (2004) Consciousness is discontinuous. *Med Hypotheses* **62**, 1003–1005.

Bohm, D. (1952) A suggested interpretation of the quantum theory in terms of hidden variables. *Phys. Rev.*, **85**, 166–193.

Bohm, D. (1980) *Wholeness and the Implicit Order.* London: Routledge.

Bohm, D. & Hiley, B. J. (1993) *The Undivided Universe: An Ontological Interpretation of Quantum Mechanics.* London: Routledge.

Brizhik, L., Del Guidice, E., Jorgensen, S. E., Marchettini, N. & Tiezzi, E. (2009) The role of electromagnetic potentials in evolutionary dynamics of ecosystems. *Ecol. Mod.*, **220**, 1865–1869.

Buzsaki, G., Anastassiou, C. A. & Koch, C. (2012) The origin of extracellular fields and currents-EEG, ECoG, LFP and spikes. *Nat. Rev. Neurosci.*, **13**, 407–420.

Buzzacchi, M., Del Guidice, E. & Preparata, G. (2002) Coherence of the glassy state. *Int. J. Mod. Phys. B.* **16**, 3771–3786.

Cacha, L. A. & Poznanski, R. R. (2011) Associable representations as a field of influence for dynamic cognitive processes. *J. Integr. Neurosci.*, **10**, 423–437.

Cacha, L. A. & Poznanski, R. R. (2014) Genomic instantiation of consciousness in neurons through a biophoton field theory. *J. Integr. Neurosci.*, **13**, 253–292.

Calvin, W. H. (1998) Competing for consciousness: A Darwinian mechanism at an appropriate level of explanation. *J. Conscious. Stud.*, **5**, 389–404.

Carruthers, P. & Veillet, B. (2007). The phenomenal concept strategy. *J. Conscious. Stud.*, **14**, 212–236.

Chalmers, D. J. (1993) Toward a theory of consciousness. Unpublished Ph.D. thesis. Indiana University, Bloomington.

Chalmers, D. J. (1996) *The Conscious Mind: In Search of a Fundamental Theory*. NY: Oxford University Press.

Chauvet, G. A. (1993) Non-locality in biological systems results from hierarchy: Application to the nervous system. *J. Math. Biol.*, **31**, 475–486.

Chauvet, G. A. (1996) *Theoretical Systems in Biology: Hierarchical & Functional Integration*. Oxford: Pergamon Press.

Chauvet, G. A. (2002) On the mathematical integration of the nervous tissue based on the S-propagator formalism. *J. Integr. Neurosci.*, **1**, 31–68.

Chwirot, W. B. (1986) New indication of possible role of DNA in ultraweak photon emission from biological systems. *J. Plant. Physiol.* **122**, 81–86.

Cicurel, R. & Nicolelis, M. A. L. (2015) *The Relativistic Brain: How It Works and Why It Cannot be Simulated by a Turing Machine*. Kios Press, Sao Paulo.

Cifra, M., Fields, J. Z. & Farhadi, A. (2011) Electromagnetic cellular interactions. *Prog. Biophys. Mol. Biol.* **105**, 223–246.

Collini, E., Wong, C. Y., Wilk, K. E., Curmi, P. M. G. Brumer, P. & Scholes, G. D. (2010) Coherently wired light-havesting in photosynthetic marine algae at ambient temperature. *Nature* **463**, 644–647.

Crick, F. (1990) Neural edelmanism: Reply. *Trends Neurosci.*, **13**, 13–14.

Damasio, A. R. (1999) *The Feeling of What Happens: Body and Emotion in the Making of Consciousness*. New York: Harcourt Brace.

Davidson, D. (1980) Mental Events. In *Essays on Actions & Events*. Clarendon Press, Oxford.

Dehaene, S. (2014) *Consciousness and the Brain: Deciphering How the Brain Codes Our Thoughts*. New York: Viking.

Dehaene, S. & Naccache, L. (2001) Towards a cognitive neuroscience of consciousness: Basic evidence and a workspace framework. *Cognition*, **79**, 1–37.

Dehaene, S., Charles, L., King, J.-R. & Marti, S. (2014) Toward a computational theory of conscious processing. *Curr. Opin. Neurobiol.*, **25**, 76–84.

Del Giudice, E., Doglia, S. & Milani, M. (1983) Self-focusing and ponderomotive forces of coherent electric waves: a mechanism for cytoskeleton formation and dynamics. In, H. Fröhlich & F. Kremer (eds) *Coherent Excitations in Biological Systems*. Berlin: Springer-Verlag.

Del Giudice, E., Doglia, S., Milani, M. & Vitiello, G. (1985) A quantum field theoretical approach to the collective behaviour of biological systems. *Nucl. Phys. B*, **251**, 375–400.

Del Giudice, E., Preparata, G. & Vitiello, G. (1988) Water as a free electric dipole laser. *Phys. Rev. Lett.*, **61**, 1085–1088.

Del Giudice, E., De Ninno, A. & Fleischmann, M. (2005) Coherent quantum electrodynamics in living matter. *Electromag. Biol. Med.*, **24**, 199–210.

Del Giudice, E., Spinetti, P. R. & Tedeschi, A. (2010) Water dynamics at the root of metamorphosis in living organisms. *Water* **2**, 566–586.

Dennis, M. Y., Nuttle, X., Sudmant, P. H., Antonacci, F., Graves, T. A., Nefedov, M., Rosenfeld, J. A, Sajjadian, S., Malig, M., Kotkiewicz, H., Curry, C. J., Shafer, S, Shaffer, L. G., deJong, P. J., Wilson, R. & Eichier, E. E (2012) Evolution of human-specific neural SRGAP2 genes by incomplete segemental duplication. *Cell*, **149**, 912–922.

Dicke, R. H. (1954) Coherence in spontaneous radiation processes. *Phys. Rev.*, **93**, 99–110.

Eccles, J. C. (1992) Evolution of consciousness. *Proc. Natl. Acad. Sci. USA*, **89**, 7320–7324.

Edelman, G. M. (1978) Group selection and phasic re-entrant signaling: A theory of higher brain function. In: G. M. Edelman and V. B. Mountcastle, eds. *The Mindful Brain: Cortical Organization and the Group-Selective Theory of Higher Brain Function*. Cambridge, MA: The MIT Press.

Edelman, G. M. (1981) Group selection as the basis for higher brain function. In: F. O. Schmitt, F. G. Worden, G. Adelman and S. G. Dennis, eds. *The Organization of the Cerebral Cortex*. Cambridge, MA: The MIT Press.

Edelman, G. M. (1987) *Neural Darwinism: The Theory of Neuronal Group Selection*. New York: Basic Books.

Edelman, G. M. (1993) Neural Darwinism: Selection & re-entrant signaling in higher brain function. *Neuron*, **10**, 115–125.

Edelman, G. M. (1995) Memory and the individual soul: Against silly reductionism. In: J. Cornwell, ed. *Nature's Imagination: The Frontiers of Scientific Vision*. Oxford: Oxford University Press.

Edelman, G. M. & Gally, J. A. (2001) Degeneracy and complexity in biological systems. *Proc. Natl. Aced. Sci. USA*, **98**, 13763–13768.

Edelman, G. M. & Tononi, G. (1995) *A Universe of Consciousness: How Matter Becomes Imagination*. New York: Basic Books.

Eliasmith, C., Stewart, T. C., Choo, X., Bekolay, T., De Wolf, T. Tang, Y. & Rasmussen D. (2012) A large-scale model of the functioning brain. *Science*, **338**, 1202–1205.

Edwards, J. C. W. (2005) Is consciousness only a property of individual cells. *J. Conscious. Stud.*, **12**, 60–76.

Engel, G. S., Calhoun, T. R. Read, E. L. Ahn, T. K. Mancal, T., Cheng, Y. C. Blankenship, R. E. & Fleming, G. R. (2007) Evidence for wavelike energy transfer through quantum coherence in photosynthetic systems. *Nature* **446**, 782–786.

Esposito, S. (1999) On the role of spin in quantum mechanics. *Found. Phys. Lett.* **12**, 165–177.

Feinberg, T. E. (2012) Neuroontology, neurobiological naturalism, and consciousness: A challenge to scientific reduction and a solution. *Phys. Life. Rev.*, **9**, 13–34.

Feinberg, T. E. & Mallatt, J. (2013) The evolutionary and genetic origins of consciousness in the Cambrian period over 500 million years ago. *Front. Psychol.* **4**, 667.

Fingelkurts, A. A. & Fingelkurts, A. A. (2004) Making complexity simpler: Multivariability and mestability in the brain. *Int. J. Neurosci.*, **114**, 843–862.

Fingelkurts, A. A., Fingelkurts, A. A., & Neves, C. F. M. (2009). Phenomenological architecture of a mind and operational architectonics of the brain: The unified metastable continuum. *New Math.Nat.Comput*, **5**, 221–244.

Frank, E. & Wenner, P. (1993) Environmental specification of neuronal connectivity. *Neuron*, **10**, 779–785.

Freeman, W. J (2003) The wave packet: An action potential for the 21st century. *J. Integr. Neurosci.*, **2**, 3–30.

Freeman, W. J. (2005) NDN, volume transmission, and self-organization in brain dynamics. *J. Integr. Neurosci.*, **4**, 407–421.

Freeman, W. J. (2007) Indirect biological measures of consciousness from field studies of brains as dynamical systems. *Neural Networks* **20**, 1021–1031.

Freeman, W. J. & Holmes, M. D. (2005) Metastability, instability, and state transitions in neocortex. *Neural Networks*, **18**, 497–504.

Freeman, W. J. & Vitiello, G. (2006) Nonlinear brain dynamics as a macroscopic manifestation of underlying many-body field dynamics. *Phys. Life. Rev.* **3**, 93–118.

Freeman, W. J. & Vitiello, G. (2008) Dissipation and spontaneous symmetry breaking in brain dynamics. *J. Physics A: Math. Theor.* **41**, 304042.

Freud, S. (1915) The Unconscious. *The Standard Edition of the Complete Psychological Works of Sigmund Freud.*, Vol. 14. London: Hogarth Press, pp. 166–204.

Fröhlich, H. (1968) Long-range coherence and energy storage in biological systems. *J. Quantum Chem.*, **2**, 641–649.

Fröhlich, H. (1975) Evidence for Bose condensation-like excitation of coherent modes in biological systems. *Phys. Lett.* A **51**, 21–22.

Gelbard-Sagiv, H., Mukamel, R., Harel, M. & Fried, I. (2008) Internally generated reactivation of single neurons in human hippocampus during free recall. *Science*, **322**, 96–101.

Gierer, A. (1988) Spatial organization and genetic information in brain development. *Biol. Cybern.* **59**, 13–21.

Görner, S. & Combs, A. (1998) Consciousness as a self-organizing processes. *Biosystems*, **46**, 123–127.

Golant, M. B. (1989) Resonance effect of coherent electromagnetic radiation of millimetre range of waves on the living organisms. *Biofizika* **34**, 1004–1014.

Goodman, G, Gershwin, M. E. & Bercovich, D. (2014) Mars can wait: Facing the challenges of our civilization. *IMAJ* **16**, 744–747.

Greenfield, S. A. (2001) *The Private Life of the Brain: Emotions, Consciousness, and the Secret of the Self*. New York: Wiley.

Grossberg, S. (1999) The link between brain learning, attention, and consciousness. *Conscious. Cognit.*, **8**, 1–44.

Grossberg, S. (2007) Consciousness CLEARS the mind. *Neural Networks*, **20**, 1040–1053.

Grundler, W. & Keilmann, F. (1983) Sharp resonances in growth proved non thermal sensitivity to microwaves. *Phys. Rev. Lett.*, **51**, 1214–1216.

Hagan, S., Jibu, M. & Yasue, K. (1994) Consciousness and anesthesia: an hypothesis involving biophoton emissions in the microtubular cytoskeleton of the brain. In: K. H. Pribram, ed. *Origins: Brain and Self-Organization*. Hillsdale: Lawrence Erlbaum.

Hales, C. G. (2011) On the status of computationalism as a law of nature. *Int. J. Machine Conscious.*, **3**, 55–90.

Hales, C. G. (2014) The origins of the brain's endogenous electromagnetic field and its relationship to provision of consciousness. *J. Integr. Neurosci.*, **13**, 313–361.

Hameroff, S. & Penrose, R. (2003) Conscious events as orchestrated space time selection. *NeuroQuantology*, **1**, 10–35.

Hameroff, S. & Penrose, R. (2014) Consciousness in the universe: A review of the 'Orch OR' theory. *Phys. Life Rev.*, **11**, 39–78.

Hassanali, A., Gilberti, F., Cuny, J., Kuhne, J. D. & Parrinello, M. (2013) Protons transfer through the water gossamer. *Proc. Natl. Acad. Sci. USA*, **110**, 13723–13728.

Hiley, B. J. (1995) Nonlocality in microsystems. In: J. King and K. H. Pribram, eds. *Scale in Conscious Experience: Is the Brain Too Important to be Left to the Specialists to Study*? Hillsdale: Lawrence Erlbaum.

Hodgkin, A. L. & Huxley, A. F. (1952) A quantitative description of membrane current and its application to conduction and excitation in nerve. *J. Physiol. Lond.*, **177**, 500–544.

Hu, H. & Wu, M. (2004) Spin-mediated consciousness theory: Possible roles of neural membrane nuclear spin ensembles and paramagnetic oxygen. *Med. Hypotheses*, **63**, 633–646.

Isojima, Y., Isoshima, T., Nagai, K., Kikuchi, K. & Nakagawa, H. (1995) Ultraweak biochemiluminescence detected from rat hippocampal slices. *Neuroreport*, **6**, 658–660.

Izhikevich, E. M. & Edelman, G. M. (2008) Large-scale model of mammalian thalamocortical systems. *Proc. Natl. Acad. Sci. USA*, **105**, 3593–3598.

Jibu, M., Hagan, S., Hameroff, S. R., Pribram, K. H. & Yasue, K. (1994) Quantum optical coherence in cytoskeletal microtubules: implications for brain function. *Biosystems*, **32**, 195–209.

Jibu, M., Pribram, K. H. & Yasue, K. (1996) From conscious experience to memory storage and retrieval: The role of quantum brain dynamics and boson condensation of evanescent photons. *Int. J. Mod. Phys. B*, **10**, 1735–1754.

Jibu, M. & Yasue, K. (1995) *Quantum Brain Dynamics & Consciousness: An Introduction*. John Benjamins, Amsterdem.

Jibu M., Yasue, K. (1997). What is mind? Quantum field theory of evanescent photons in brain as quantum theory of consciousness. *Informatica*, **21**, 471–490.

Jibu M., Yasue, K. (2004). Quantum brain dynamics and quantum field theory. In: G. G. Globus, K. H. Pribram and G. Vitiello, eds. *Brain and Being*. John Benjamins, Amsterdam.

John, E. R. (2001) A field theory of consciousness. Conscious. Cognit., **10**, 184–213.

John, E. R. (2005) From synchronous neuronal discharges to subjective awareness? *Prog. Brain Res.*, **150**, 143–171.

Kataoka, Y., Cui, Y., Yamagata, A., Niigaki, M., Hirohata, T., Oishi, N. & Watanabe, Y. (2001) Activity-dependent neural tissue oxidation emits intrinsic ultraweak photons. *Biochem. Biophys. Res. Commun.*, **285**, 1007–1011.

Kelso, J. A. S. (1995) *Dynamic Patterns: the Self-organization of Brain and Behaviour*. Cambridge, MA: MIT Press.

Kobayashi, M., Takeda, M., Sato, T., Yamazaki, Y., Kaneko, K., Ito, K., Kato, H. & Inaba, H. (1999) In vivo imaging of spontaneous ultraweak photon emissions from a rat's brain correlated with cerebral energy metabolism and oxidative stress. *Neurosci. Res.*, **34**, 103–113.

Kohler, W. (1940) *Dynamics in Psychology*. New York: Liveright.

Korzeniewski, B. (2010) *From Neurons to Self-Consciousness*. Amherst: Prometheus Books.

Korzeniewski, B. (2013) Formal similarities between cybernetic definition of life and cybernetic model of self-consciousness: universal definition/model of individual. *Open J. Philos.*, **3**, 314–328.

LaBerge, D. & Kasevich, R. (2007) The apical dendrite theory of consciousness. *Neural Networks*, **20**, 1004–1020.

Leask, M.J.M. (1977). A physiochemical mechanism for magnetic field detection by migratory birds and homing pigeons. *Nature* **267**, 144–145.

Levin, J. (2006). What is a Phenomenal Concept? In: T. Alter and S. Walter, eds. *Phenomenal Concepts and Phenomenal Knowledge. New essays on Consciousness and Physicalism*, Oxford: Oxford University Press, pp. 87–110.

Levin, J. (2008). Taking Type-B Materialism Seriously. *Mind Lang.*, **23**, 402–425.
Levine, J. (1983). Materialism and qualia: The explanatory gap. *Pac. Philos. Quart.*, **64**, 354–361.
Li, K. H. (1992) Coherent radiation from DNA molecules. In: F. A. Popp, K. H. Li and Q. Gu, eds. *Recent Advances in Biophoton Research and its Applications.* Singapore: World Scientific.
Libet, B. J. (1996) Conscious mind as a field. *J. Theor. Biol.*, **178**, 223–243.
Lindahl, B. I. B. & Arhem, P. (1994) Mind as a force field: Comments on a new interactionistic hypothesis. *J. Theor. Biol.*, **181**, 95–96.
Lipkind, M. (2003) "Virtual" photons as carriers of consciousness: A critical comment. *Ind. J. Exp. Biol.*, **41**, 539–543.
Llinas, R., Ribary, V., Contreras, D. & Pedroarena, C. (1998) The neuronal basis for consciousness. *Philos. Trans. R. Soc. B*, **353**, 1841–1849.
Lloyd, S. (2011) Quantum coherence in biological systems. *J. Phys. Conf. Ser.* **302**: 012037.
MacGregor, R. J. (2006) *On the Contexts of Things Human: An Integrative View of Brain, Consciousness, and Freedom of Will.* Singapore: World Scientific.
Maddox, J. B. & Bittner, E. R. (2003) Estimating Bohm's quantum force using Bayesian statistics. *J. Chem. Phys.* **119**, 6465–6474.
Manzotti, R. (2012) The computational stance is unfit for consciousness. *Int. J. Mach. Conscious.* **4**, 401–420.
Matsuno, K. & Paton, R. C. (2000) Is there a biology of quantum information? *Biosystems*, **55**, 39–46.
McFadden, J. (2002a) Synchronous firing and its influence on the brain's electromagnetic field: Evidence for an electromagnetic theory of consciousness. *J. Conscious. Stud.*, **9**, 23–50.
McFadden, J. (2002b) The conscious electromagnetic information (CEMI) field theory: The hard problem made easy? *J. Conscious. Stud.*, **9**, 45–60.
McFadden, J. (2013) The CEMI field theory closing the loop. *J. Conscious. Stud.*, **20**, 153–168.
McGinn, C. (1995) Consciousness and space. J. Conscious. Stud., **2**, 220–230.
Mentre, P. (2012) Water in the orchestration of the cell machinery. Some misunderstandings: A short review. *J. Biol. Phys.*, **38**, 13–26.
Merker, B. (2007). Consciousness without a cerebral cortex: A challenge for neuroscience and medicine. *Behav. Brain Sci.*, **30**, 63–81.
Montagnier, L., Aissa, J., Ferris, S., Montagnier, J.-L. & Larllee, C. (2009) Electromagnetic signals are produced by aqueous nanostructures derived from bacterial DNA sequences. *Interdiscip. Sci. Comput. Life Sci.* **1**, 245–253.
Montagnier, L., Aissa, J., Del Giudice, E., Lavallee, C., Tedeschi, A. & Vitiello, G. (2011) DNA waves and water. *J. Phys. Conf. Ser.* **306**, 012007.
Nagl, W. & Popp, F. A. (1983) A physical model of differentiation. *Cytobios*, **37**, 45–62.

Orpwood, R. D. (2007). Neurobiological mechanisms underlying qualia. *J. Integr. Neurosci.*, **6**, 523–540.

Orsucci, F. (2009) *Mind Force. On Human Attractions.* Singapore: World Scientific Publishers.

Pall, M. L. (2013) Electromagnetic fields act via activation of voltage-gated calcium channels to produce beneficial or adverse effects. *J. Cell. Mol. Med.* **17**, 958–965.

Pantian, S. & Yupapin, P. P. (2013) THz rabi frequency oscillation for human consciousness/sub-consciousness detection probe use. *J. Biosens. Bioelectron.*, **4**, e126.

Parvizi, J. & Damasio, A. (2001) Consciousness and the brainstem. *Cognition*, **79**, 135–159.

Patel, A. (2001) Why genetic information processing could have a quantum basis. *J. Biosci.*, **26**, 145–151.

Peressini, A. (2013). Consciousness as integrated information: A provisional philosophical critique. *J. Conscious. Stud.*, **20**, 180–206.

Persinger, M.A., Dott, B.T., Saroka, K.S. & Scott, M.A. (2013). Congruence of energies for cerebral photon emissions, quantitative EEG activities and ~5nT changes in the proximal geomagnetic field support spin-based hypothesis of consciousness. *J. Consciousness Exp. & Res*, **4**, 1–24.

Peyrard, M., Cuesta-Lopez, S. & James, G. (2009) Nonlinear analysis of the dynamics of DNA breathing. *J. Biol. Phys.*, **35**, 73–89.

Popp, F. A. (1979) Photon storage in biological systems. In: F. A. Popp, G. Becker, H. L. König, & W. Peschka, eds. *Electromagnetic Bioinformation.* Baltimore: Urban and Schwarenberg.

Popp, F. A. & Chang, J. J. (1998) The physical background and the informational character of biophoton emission. In: J. J. Chang, J. Fish and F. A. Popp, eds. Biophotons. Dordrecht: Kluwer.

Popp, F. A., Chang, J. J., Herzog, A., Yan, Z. and Yan, Y. (2002) Evidence of nonclassical (squeezed) light in biological systems. *Phys. Lett. A*, **293**, 98–102.

Popp, F. A., Gu, Q. & Li, K.-H. (1994) Biophoton emission: experimental background and theoretical approaches. *Mod. Phys. Lett. B*, **8**, 1269–1296.

Popp, F. A., Nagl, W., Li, K. H., Scholz, W., Weingartner, O. & Wolf, R. (1984) Biophoton emission: New evidence for coherence and DNA as source. *Cell Biophys.*, **6**, 33–52.

Popp, F. A., Ruth, B. & Bahr, W. (1981) Emission of visible and ultraviolet radiation by active biological systems. *Collective Phenomena*, **3**, 187–213.

Popper, K. & Eccles, J. C. (1977) *The Self and its Brain: An Argument for Interactionism.* New York: Routledge.

Popper, K., Lindahl, B. I. B. & Arhem, P. (1993) A discussion on the mind-body problem. *Theor. Med.*, **14**, 167–180.

Pribram, K. H. (2013) *The Form Within: My Point of View.* Westport, CT: Prospecta Press.

Rahnama, M., Tuszynski, J. A., Bokkon, I., Cifra, M., Sardar, P. & Salari, V. (2011) Emission of mitochondrial biophotons and their effect on electrical activity of membrane via microtubules. *J. Integr. Neurosci.*, **10**, 65–88.

Rattemeyer, M., Popp, F. A. & Nagl, W. (1981) Evidence of photon emission from DNA in living systems. *Naturwissen*, **68**, 572–580.

Reeke, Jr., G. N. (1994) Selection versus instruction: Use of computer models to compare brain theories. *Int. Rev. Neurobiol.*, **37**, 211–241.

Reeke, Jr. G. N. & Edelman, G. M. (1995) A Darwinist view of the prospects for conscious artifacts. In: G. Trautteur, ed. *Consciousness: Distinction and Reflection*. Naples: Bibliopolis.

Reeke, Jr., G. N. & Sporns, O. (1990) Selectionist models of perceptual and motor systems and implications for functionalist theories of brain function. *Physica D*, **42**, 347–364.

Reeke, Jr., G. N., Sporns, O. & Edelman, G. M. (1990) Synthetic neural modeling: the "Darwin" series of recognition automata. *Proc. IEEE*, **78**, 1498–1530.

Reggia, J. A. (2013). The rise of machine consciousness: Studying consciousness with computational models. *Neural Ntwks* 44, 112–131.

Reimers, J. R. McKemmish, L. K., McKenzie, R. H. Mark, A. E. & Hush, N. S. (2009) Weak, strong and coherent regimes of Fröhlich Condensation and their application to terahertz medicine & quantum consciousness *Proc. Natl. Acad. Sci. USA* **106**, 4219–4224.

Revonsuo, A. and Newman, J. (1999) Binding and consciousness. *Conscious. Cogn.*, **8**, 123–127.

Romijn, H. (2002) Are virtual photons the elementary carriers of consciousness? *J. Consc. Studies* **9**, 61–81.

Rosen, R. (1991) *Life Itself: A Comprehensive Inquiry into the Nature, Origin, and Fabrication of Life*. New York: Columbia University Press.

Rouleau, N. & Dotta, B. T. (2014) Electromagnetic fields as structure-function zeitgebers in biological systems: environmental orchestrations of morpogenesis and consciousness. *Front. Integr. Neurosci.*, **8**, 84.

Ruan, C. Y., Lobastov, V. A. Vigliotti, F., Chen, S. & Zewall, A. H. (2004) Ultra fast electron crystallography of interfacial water. *Science* **304**, 80–84.

Rudrauf, D., Lutz, A., Cosmelli, D., Lachaux, J.-P. and Le Van Quyen, M. (2003) From autopoiesis to neurophenomenology: Francisco Varela's exploration of the biophysics of being. *Biol. Res.*, **36**, 21–59.

Salesi, G. & Recami, E. (1998) Hydrodynamics of spinning particles. *Phys. Rev. A*, **57**, 98–105.

Searle, J. R. (2000) Consciousness. *Ann. Rev. Neurosci.*, **23**, 557–578.

Searle, J. R. (2007) Biological naturalism. In: M. Velmans and S. Schneider, eds. *The Blackwell Companion to Consciousness*. Oxford: Blackwell.

Sejnowski, T. J. (2015) Consciousness. *Daedalus* 144, 123–132.

Seth, A. K., Izikevich, E., Reeke, G. N. & Edelman, G. M. (2006) Theories and measures of consciousness: An extended framework. *Proc. Natl. Acad. Sci. USA*, **103**, 10799–10804.

Sevush, S. (2006) Single-neuron theory of consciousness. *J. Theor. Biol.*, **238**, 704–725.

Smith, C. W. (1994) Electromagnetic and magnetic vector potential bio-information and water. In: P. C. Endler and J. Schulte, eds. *Ultra High Dilution-Physiology and Physics*. Dordrecht: Kluwer Academic Publishers.

Solms, M. & Turnbull, O. (2002) *Brain and the Inner World: An Introduction to the Neuroscience of Subjective Experience*. New York: Other Press.

Sperry, R. W. (1980) Mind-brain interactions: Mentalism, yes; dualism, no. *Neuroscience*, **5**, 195–206.

Sporns, O. (1997) Deconstructing neural constructivism. *Behav. Brain Sci.*, **20**, 576–577.

Sporns, O. & Tononi, G., eds. (1994) *Selectionism and the Brain*. New York: Academic Press.

Sporns O., Tononi, G., Kötter, R. (2005) The human connectome: A structural description of the human brain. *PLoS Comput. Biol.* **1**, e42.

Stewart, T. C. & Eliasmith, C. (2014) Large-scale synthesis of functional spiking neural circuits. *Proc. IEEE*, **102**, 881–898.

Stuart, C. I. J. M., Takahashi, Y. & Umezawa, H. (1979) Mixed-system brain dynamics. *Foundations of Physics*. **9**, 301–327.

Suberbielle, E., Sanchez, P. E., Kravitz, A. V., Wang, X., Ho, K. Eilertson, K., Devidze, N., Kreitzer, A. C. & Mucke, L. (2013) Physiologic neuron activity causes DNA double strand breaks in neurons with exacerbation by amyloid — beta. *Nat. Neurosci.*, **16**, 613–621.

Szasz, A., Vincze, G. Andocs, G. & Szasz, O. (2009) Do field-free electromagnetic potentials play a role in biology? *Electromagn. Biol. Med.* **28**, 135–147.

Szent-Gyorgyi, A. (1957) *Bioenergetics*. Academic Press, NY.

Szent-Gyorgyi, A. (1960) *Introduction to a Supramolecular Biology*. Academic Press, NY.

Taylor, J. G. (2009) Modeling consciousness. *New Math. Nat. Comp.*, **5**, 31–46.

Tegmark, M. (2000). Importance of quantum decoherence in brain processes. *Phys. Rev. E*, **61**, 4194–4206.

Tegmark, M. (2015) Consciousness as a state of matter. *Chaos, Solitons & Fractals*, **76**, 238–270.

Thar, R. and Kuhl, M. (2004) Propagation of electromagnetic radiation in mitochondria? *J. Theor. Biol.*, **230**, 261–270.

Tononi, G. (2004). An information integration theory of consciousness. *BMC Neurosci.*, **5**, 42.

Tononi, G. (2008). Consciousness as integrated information: A provisional manifesto. *Biol. Bull.*, **215**, 216–242.

Tononi, G. (2010) Information integration: Its relevance to brain function and consciousness. *Arch. Ital. Biol.*, **148**, 299–322.

Tononi, G. (2012) Integrated information theory of consciousness: An updated account. *Arch. Ital. Biol.*, **150**, 293–329.

Tononi, G. & Sporns, O. (2003). Measuring information integration. *BMC Neurosci.*, **4**, 31.

Tononi, G. & Edelman, G. M. (1998) Consciousness and complexity. *Science*, **282**, 1846–1851.

Trukhan, E. M. & Anosov, V. N. (2007) Vector potential as a channel of informational effect on living objects. *Biofizika* **52**, 376–381.

Turton, D. A., Senn, H. M., Harwood, T., Lapthorn, A. J., Ellis, E. M., & Wynne, K. (2014) Terahertz underdamped vibrational motion governs protein-ligand binding in solution. *Nat. Commun.*, **5**, 3999.

Tuszynski, J. A, Paul, R, Chatterjee, R. & Sreenivasa, S. R. (1984) Relationship between Fröhlich and Davydov models of biological order. *Phys. Rev. A*, **30**, 2666–2675.

van Gaal, S. & Lamme, V. A. F. (2012) Unconscious high-level information processing: Implications for neurobiological theories of consciousness. *Neuroscientist*, **18**, 287–301.

Van Wijk, R., Godaerts, G. & Van Wijk, E. P. A. (2007) Human ultra weak light emission in consciousness research. In: J. T. Locks, ed. *New Research on Consciousness*. New York: Nova Science Publishers.

Varela, F. J. (1971) Self-consciousness: Adaptation or epiphenomenon? *Stud. Gen.*, **24**, 426–439.

Varela, F. J. (1979) *Principles of Biological Autonomy*. New York: North-Holland.

Varela, F. J. (1996) Neurophenomenology: A methodological remedy to the hard problem. *J. Consc. Stud.*, **3**, 330–350.

Varela, F. J. (1999) The specious present: The neurophenomenology of time consciousness. In: J. Petitot, F. J. Varela, B. Pachoud and J. M. Roy, eds. *Naturalizing Phenomenology*. Stanford, CA: Stanford University Press.

Vitiello, G. (2001) *My Double Unveiled*. John Benjamins, Amsterdam.

Walker, E. H. (1970) The nature of consciousness. *Math. Biosci.*, **7**, 131–178.

Wallstrom, T. C. (1994) Inequivalence between the Schrödinger equation and the Madelung hydrodynamic equations. *Phys. Rev. A*, **49**, 1613–1617.

Willshaw, D. J. & von Malsburg, C (1976) How patterned neural connections can be set-up by self-organization. *Proc. R. Soc. Lond B*, **194**, 431–445.

Wissner-Gross, A. D. & Freer, C. E. (2013) Causal entropic forces. *Phys. Rev. Lett.*, **110**, 168702.

Biophysics of Consciousness: A Foundational Approach
R. R. Poznanski, J. A. Tuszynski and T. E. Feinberg
Copyright © 2017 World Scientific, Singapore.

# 12

# QUANTUM SPIN FORMALISM ON CONSCIOUSNESS

*Huping Hu and Maoxin Wu*

*Research & Development, Quantum Dream Inc.,*
*Stony Brook, NY 11790, USA*

## Abstract

Within the context of brain-based consciousness, the spin-mediated consciousness theory claims that quantum spin is a cog in the wheel of consciousness of consciousness and the linchpin between mind and the brain, that is, spin is the "mind-pixel," and the unity of mind is achieved by quantum entanglement of these mind-pixels. Here, mind is understood as the cognitive faculty in the brain which enables subjectivity to arise from consciousness. The theory is based on the fact that quantum spin is the most basic quantum bit ('qubit') for encoding information and a fundamental quantum process associated with the structure of neural membranes and proteins that are saturated with quantum spin carrying nuclei and these nuclear spins have relatively long quantum coherence time. Furthermore, in both Bohm and Hestenes quantum formalisms, the origin of quantum effects is due to internal motion associated with quantum spin. In this chapter, we first ask three questions: (1) What is the mind-pixel? (2) How does mind achieve binding? (3) How mind-pixels are connected with the classical neural activities in the brain? We then offer possible solutions to each of these three key questions. We then proceed to illustrate the spin-mediated consciousness theory insofar as the mechanism by which subjectivity is sourced from "qubits" through the internal motion associated with the quantum spin.

*Keywords*: Quantum spin, consciousness, mind-pixel, subjectivity, nuclear spin, electron spin.

## 12.1. Introduction

There is no coherent view as to what is and causes consciousness. Some neuroscientists would say that it is the connections between the neurons and the coherent firing patterns thereof (e.g., Edelman, 1989; Crick, 1994; Poznanski, 2001). Some physicists would propose that it is connected to the measurement problem in quantum theory and thus the solution lies there (e.g., Donald, 1990; Stapp, 1993; Penrose, 1989, 1994). A few philosophers would suggest that it is an emergent property of the complex brain (e.g., Searle, 1992; Freeman, 2001) or new kinds of properties and laws are required (Chalmers, 1996). Philosophically, Searle (1992) argues that consciousness is an emergent biological phenomenon thus cannot be reduced to physical states in the brain. Chalmers (1996) argues that consciousness cannot be explained through reduction, because mind does not belong to the realm of matter.

On the theoretical front, there are quite a few quantum theories of mind (e.g., Beck & Eccles, 1992; Donald, 1990; Stapp, 1993; Penrose, 1994; Walker, 2000; Hameroff & Penrose, 1996). Among these, Penrose (1994)'s Objective Reduction ("OR") together with Hameroff's microtubule computation is perhaps the most popular, and the combination of the two produced the Orchestrated Objective Reduction ("Orch OR") in microtubules (Hameroff & Penrose, 1996). According to Penrose (1989, 1994), each quantum state has its own space–time geometry, thus superposition of quantum states entails superposition of different space–time geometries. Under certain conditions, such space–time geometric superpositions would separate under its own "weight" through a non-computable process, which in turn would collapse said quantum state superposition. Hameroff suggested that such self-organized OR could occur in microtubules because of their particular structures, thus, born the Orch OR. According Orch OR, each collapse of macroscopic space–time geometry superposition corresponds to a discrete conscious event (Hameroff & Penrose, 1996). In addition, it seems that Penrose (1989, 1994) accepts a separate mental world so called "platonic" world with grounding in the physical world.

There are also a number of theories based on conventional neuroscience (e.g., Edelman, 1989, Crick, 1994; Poznanski, 2001; LeDoux, 2002).

Our view on these is that whatever the final accepted version based on neuroscience ("classical physics"), it could be accepted as classically correct. The reason is that we must rely on the classical parts of the brain working according to conventional neuroscience to provide us the necessary neural components and dynamics such as coherent neural firings, neurotransmitter releases and neural plasticity to support any realistic quantum activities of the brain. The situation is much like that in quantum computation where classical components form the supporting system of a quantum computer. Without these classical components, quantum computation could not be implemented at all.

In comparison, our working philosophy in the context of a brain-based (physicalist's) theory of consciousness is that consciousness is grounded at the bottom of physical reality and emerges from the collective dynamics of known physical candidates inside the brain (Hu & Wu, 2002, 2004a). Next we ask "where" and "how." To answer these, we take the reductionist approach both down to the end of physics to see what is left there and to the microscopic domain of a neuron to see what may be really important for the functioning of a conscious brain.

What we found is that there is almost nothing left at the end of physics except the fundamental ideas of quantized space–time and spin. On the other hand, we found that what may be really important in the microscopic domain of a neuron are the nuclear spin ensembles and the fluxes of biologically available unpaired electrons carried by small molecules such as $O_2$ and NO. Naturally, we draw the conclusion that quantum spin together with its connection to space–time geometry is needed to ground consciousness in physical reality such that conscious experience emerges from the successive dynamics of various entangled neural nuclear spin states.

Specifically, we are searching answers to the following fundamental questions: (1) What are the mind-pixels or neural substrates of consciousness and what physical processes are involved in conscious experience? (2) What binding "glue" or binding mechanism allows the mind to achieve unity? (3) How mind-pixels are connected to the classical electrical activities in the brain and what physical and/or chemical process allows consciousness to influence the brain?

## 12.2. Is Quantum Spin the Mind-Pixel?

In this section, we express our fundamental view based on Hu & Wu (2002, 2004b) that spin is a primordial self-referential process driving quantum mechanics, space–time dynamics and consciousness. Further, it is the source of *zitterbewegung* ("internal motion"). To justify such a view, we will draw support from a reductionist perspective the essential properties spin should possess as mind-pixel and explore further the nature of spin to see whether such properties are present. Our conclusion is that these properties are indeed endowed to spin by Nature. Thus, we have postulated that spin is a cog in the wheel of consciousness and the linchpin between mind and the brain, that is, spin is the mind-pixel (Hu & Wu, 2002, 2004a).

For clarity, we specifically make the following basic assumptions so as to avoid complication from potentially intractable arguments so that we can focus on manageable, but fundamental questions related to consciousness: (1) that consciousness exists (not an illusion); and (2) that consciousness is quantum mechanical. It follows from (2) that the process generating quantum effects also generates consciousness.

Let us assume that spin is the fundamental process driving quantum mechanics, that is, it is spin that generates all the quantum effects. Do we have any supporting evidence? The answer indeed is "Yes."

First, spin is uniquely quantum mechanical often being said to have no classical counterpart (e.g., Tomonaga, 1997). The definition and mathematics of quantum spin are treated in the Appendix. As shown there, unlike mass and charge that enter a dynamic equation as arbitrary parameters, spin reveals itself through the structure of the relativistic quantum equation for fermions, the Dirac equation, which combines quantum mechanics with special relativity (Dirac, 1928). Indeed, modern physics leads us right down to the microscopic domain of space–time where many models of elementary particles and even space–time itself are built with spinors (e.g., Penrose 1960, 1967; Budinich, 2002) that were first used by Pauli (1927) and Dirac (1928) to describe the spin ½ electron.

Second, in Hestenes (1973, 1983) formulism of Dirac equation using real Clifford algebra (space–time algebra), the *zitterbewegung* (*zbw*) associated with the spin is qualitatively shown to be responsible for all known quantum effects of said electron and the imagery complex number "*i*" in the Dirac equation is said to be due to spin tensor in space–time algebra. The concept of *zbw* was introduced by Schrödinger (1930) to interpret high-frequency oscillations in free-particle wave packets of the Dirac electron. These oscillations arise from interference between positive and negative energy components of a wave packet. Hestenes (1983) concluded that literal interpretation of *zbw* implies that the electron is the seat of an oscillating bound electromagnetic field similar to de Broglie's pilot wave. On this basis, Hestenes (1983) gave qualitative explanations for electron diffraction, the existence of quantized radiationless states, the Pauli principle, and other features of quantum mechanics.

Third, in de Broglie–Bohm formulation of non-relativistic quantum mechanics, also known as the pilot-wave theory or Bohmian mechanics, the quantum potential is a term within the Schrödinger equation which acts to guide the movement of quantum particles, thus, responsible for quantum effects (Bohm, 1952). This approach is a more complete exposition of the idea presented by de Broglie. Recami and Salesi (1998) showed that quantum potential is a pure consequence of "internal motion" associated with spin and *zbw* that the quantum behavior is a direct consequence of the fundamental existence of spin. Esposito (1999) expanded this result by showing that "internal motion" is due to the spin of the particle, whatever its value.

Let us further assume that spin is the primordial process driving space–time dynamics such that it generates our perceptions of space–time. Do we have any supporting arguments? The answer is also "Yes."

As already discussed earlier, spin is deeply connected to the microscopic structure of space–time as reflected by the Dirac equation for fermions (Dirac, 1928; also see the Appendix). Indeed, Penrose (1960, 1967) had considered early on that spin might be more fundamental than space–time and invented spinor and twistor algebras for a combinatorial description of space–time geometry. Bohm and Hiley (1984) generalized the twistor idea to Clifford algebra as a possible basis for describing Bohm's "implicit order." Further, various spin foams have been formulated as extensions to Penrose's spin networks for the purpose of constructing a consistent theory of quantum gravity (e.g., Baez, 1998; Smolin, 2001). In the spin foam picture, the quantum transitions of spin networks represent the time evolution of said spin networks (Smolin, 2001).

Many others have also studied the nature of spin from both classical and quantum-mechanical perspectives. For example, Newman (2002) showed that spin might have a classical geometric origin. By treating the real Maxwell field and real linearized Einstein equations as being embedded in complex Minkowski space, he was able to interpret spin-angular momentum as arising from a charge and "mass monopole" source moving along a complex world line. Galiautdinov (2002) has considered a theory of space–time quanta and suggested that spin might manifest the atomic structure of space–time.

Furthermore, Sidharth (2001a,b) had discussed the nature of spin within the context of quantized fractal space–time and showed that spin is symptomatic of the non-commutative geometry of space–time at the Compton scale of a fermion and the three dimensionality of the space result from the spinorial behavior of fermions. He showed that mathematically an imaginary shift of the space–time coordinate in the Compton scale of a fermion introduces spin ½ into general relativity and curvature to the fermion theory. The reason why an imaginary shift is associated with spin is to be found in the quantum mechanical *zbw* within the Compton scale and the consequent quantized fractal space–time.

Now, we discuss from a reductionist perspective the essential properties spin should possess as mind-pixel and further explore the nature of spin to see whether these properties are endowed to spin by Nature (Hu & Wu, 2004b).

Most importantly, spin as a "protopsychic" process, should be self-referential. Such requirement of spin is well supported by Hofstadter's view of what is at the crux of consciousness. According to Hofstadter (1979), consciousness is based on a kind of self-reference he termed as a "strange loop" and further explained as an interaction between levels in which the top level reaches back down towards the bottom level influencing it, while at the same time being itself determined by the bottom level.

Now, do we have any evidence supporting the existence of this property? The answer is "Yes." Indeed, the spin, being associated with *zbw* by definition is connected to self-interaction. Hestenes (1983) has suggested the possibility of *zbw* producing quantum mechanics through such self-interaction. Further, quantum potential which is responsible for quantum effects in Bohmian mechanics is produced by "internal motion" associated with spin (Recami & Salesi, 1988; Esposito, 1999).

Penrose–Hameroff's self-organized OR model of space–time geometry also implies that space–time dynamics is driven by a certain self-referential process (Hameroff & Penrose, 1996). In addition, Cahill's work on a self-referentially limited to neural-network model of reality supports the view of a primordial self-referential network underlying reality (Cahill, 2002).

Penrose (1989, 1994) argued forcibly that human thought involves non-computable processes, as Gödel's theorem of incompleteness would suggest. According to Gödel, any consistent system of axioms beyond a certain basic level of complexity yields statements that cannot be proved or disproved with said axioms. Yet humans can decide whether those statements are true, thus human thought cannot be reduced to a set of rules or computations. Penrose had further argued that it is these non-computable processes that drive the collapse of superposed quantum state to produce our conscious experience of a classical world.

The experience of classical world involves what Hofstadter (1979) would call a "level-crossing" from the quantum to the classical. So where can one find non-computable processes? We suggest here the collapse of spin quantum state due to self-reference of spin manifested as self-interaction (*zbw* or "internal motion") (Hu & Wu, 2004b).

Is there any indication that spin is the "linchpin" between what is quantum and what is classical. The answer is "Yes." Bogan's results indicate that spin is responsible for "level-crossing," because he derived a spin-dependent gauge transformation between the Hamilton–Jacobi equation of classical mechanics and the time-dependent Shrödinger equation of quantum mechanics, which is a function of the quantum potential of Bohmian mechanics (Bogan, 2002). The unitary evolution of quantum state is dictated by the requirement of probability conservation. It is our view that, in the collapsing scheme, the appearance of a classical world is dictated by the primordial self-referential collapse of spin state (Hu & Wu, 2004b). Furthermore, if space–time is indeed built with Penrose's spin network or its variations such as the spin foams, the appearance of classical world is necessarily mediated by a spin process.

As we all know, conscious experience is highly contextual. Indeed, Aerts *et al.* (2000) has proposed that intrinsic contextuality as the crux of consciousness. Now, is the collapse of spin quantum state also contextual? The answer is "Yes."

Aerts *et al.* (2000) showed that quantum system and cognitive function share similar context-based probabilistic structure, cognitive function can be represented by the simplest quantum state — the quantum state of a single spin ½, and, indeed, even paradoxes can be represented by a quantum state. Further, Khrennikov (2001) had also shown the similarities between quantum system and cognitive function and pointed out that the key feature of quantum mechanics is its context-dependent probability distribution.

As we also all know, conscious experience is highly subjective. Now, does the collapse of spin quantum state (qubit) provide an explanation for subjectivity? The answer is most likely "Yes."

Indeed, the subjectivity of experience is shaped by the context of the cognitive function in the brain, and, on the other hand, the cognitive function can be represented by the simplest quantum state — the quantum state of a single spin ½ — as shown by Aerts *et al.* (2000). Thus, the contextual collapse of spin quantum state (qubit) can provide an explanation for subjectivity of conscious experience which is influenced by the information stored in the neural nuclear spin ensembles and/or made available by quantum entanglement of these nuclear spins with the environment.

Penrose (1989) once said that "central to our feelings of awareness is the sensation of the progression of time." Can the primordial self-referential spin process provide a mechanism to this phenomenon? The answer is most likely "Yes."

Two modes of time have been tentatively suggested (Hu & Wu, 2004b). A local closed/circular time attached to each self-referential spin process and a perceptual time connected to the sequential collapses of entangled spin states. Since the self-referential collapse process is irreversible so is the perceptual time, thus, the rise of the time's arrow. Indeed, a classical model of spin process also has two modes of time attached: a local time is attached to the spin rest frame and the global time is attached to the center-of-mass reference frame (Barut & Zanghi, 1984).

Finally, central to our feelings of self is the indescribable unity. Can the primordial self-referential spin process provide an explanation? We answer this question affirmatively (Hu & Wu, 2004b). Indeed, the collective dynamics of these self-referential spin processes, that is, these mind-pixels, can form "tangled hierarchy" as Hofstadter (1979) would call it through quantum entanglement, thus, forming unity of self. "The self comes into being at the moment it has the power to reflect itself" through self-referential collapse of primordial spin process.

## 12.3. Is Quantum Entanglement the Binding "Glue"?

Many experiments have shown that quantum entanglement is physically real since its demonstration by Aspect *et al.* (1982). Further, it is ubiquitous, appears everywhere in the microscopic world and under some circumstances manifests itself macroscopically (See, e.g., Arnesen *et al.*, 2001; Julsgaard *et al.*, 2001; Ghosh *et al.*, 2003). Indeed, it is currently the most intensely studied subject in physics. Speculations abound as to its nature and implications (see, e.g., Josephson & Pallikari-Viras, 1991; Clarke, 2004; Radin, 2004). We have outlined our propositions on the implications and applications of quantum entanglement as follows (Hu & Wu, 2006a):

(1) It is the quantum "glue" that holds together interacting quantum entities together in non-local domain, which implies genuine interconnectedness and inseparableness of the quantum entities that can be directly sensed and utilized by the entangled quantum entities.
(2) It can influence chemical/biochemical reactions, other physical processes and micro- and macroscopic properties of different states of matter, thus, playing vital role in many biological processes and consciousness.

Shrödinger (1935) showed that entanglement arises from the interactions of two particles through the evolution of Shrödinger equation and called this phenomenon the characteristic trait of quantum theory. Einstein called quantum entanglement "spooky action at a distance" in the famous EPR debate (see, e.g., Einstein *et al.*, 1935). We have argued that quantum entanglement arises from the primordial self-referential spin processes which are the driving force behind quantum effects (Hu & Wu, 2004b, 2006a). It could be said that the driving force behind the evolution of Shrödinger equation is quantum spin and, since quantum entanglement arises from the evolution of Shrödinger equation, the said spin is the genuine cause of quantum entanglement (Hu & Wu, 2006a).

It is often said that instantaneous signaling through quantum entanglement is impossible because of Eberhard's theorem that basically says that since local measurements always produce random results no information can be sent through quantum entanglement alone (Eberhard, 1978). However, there are at least two ways to circumvent this impossibility. The first is to assume that the statistical distribution can be modified and utilized to transmit information. Quite a few authors have expressed this view (Walker, 1974; Stapp, 1982; Josephson, 1991), especially when discussing the role of consciousness in parapsychology such as telepathy. The second is to assume that each quantum entity can directly sense and utilize quantum entanglement as already mentioned before (Hu & Wu, 2006a). This latter view is the view we subscribe to and it is permissible in the Bohmian picture.

The implication of the second view is far-reaching. It means that quantum entanglement can influence chemical and biochemical reactions and other physical processes (Hu & Wu, 2006b, 2006c). Thus, it plays a vital role in many biological processes including consciousness and is the genuine cause of many anomalous effects. It can affect the micro- and macroscopic properties of all states of matter.

## 12.4. Modulation of Neural Spin Networks as the Connection with Neural Electrical Activities

We will show in this section that nuclear spin networks in neural membranes and proteins are modulated by action potentials through *J*-coupling, dipolar coupling and chemical shielding tensors and perturbed by microscopically strong and fluctuating internal magnetic fields produced largely by paramagnetic oxygen (Hu & Wu, 2004c). We have suggested that these spin networks could be involved in brain functions since said modulation inputs information carried by the neural spike trains into them, said perturbation activates various dynamics within them and the combination of the two likely produce stochastic resonance thus synchronizing said dynamics to the neural firings. Although quantum coherence is desirable and may indeed exist, it is not required for these spin networks to serve as the subatomic components for the conventional neural networks.

The range of electric field strength $E_m$ inside the neural membranes during a typical action potential as calculated from $E_m = V_m/d$ where $V_m$ and $d$ are respectively the membrane voltage and thickness. It oscillates during the course of each action potential with strengths comparable to those causing electroporation of cell membranes and dielectric breakdown of many materials at which the covalent bonds of the constituent molecules are torn apart (see, e.g., Barnet & Weaver, 1991). So it significantly affects the conformations and collective dynamics of the neural membrane components such as phospholipids, cholesterols and proteins. Indeed, voltage-dependent ion channels perform their function through electric field induced conformation changes of the constituent protein (see, e.g., Morais-Cabral *et al.*, 2001) and studies on the effect of electric fields on lipids support the above conclusion (see, e.g., Sargent, 1975; Saux *et al.*, 2001).

The spins carried by the nuclei such as $^1$H, $^{13}$C and $^{31}$P inside the neural membranes form complex intra- and inter-molecular spin networks through various intramolecular J- and dipolar couplings and both short- and long-range intermolecular dipolar couplings (Hu & Wu, 2004c). Since J-coupling is the indirect interaction between two nuclear spins through covalent bonds and dipolar coupling is the direct interaction of two nuclear spins through space, their strengths and anisotropies strongly depend on the conformations of the neural membrane components (see, e.g., Peshkovsky & McDermott, 2000; Grayson, 2003). Further, the chemical shielding of each nuclear spin also depends on the conformations of surrounding covalent bonds (see, e.g., Buckingham, 1960). Thus, when these spin networks are subjected to the enormous changing electric field produced during each action potential, the J-coupling, dipolar coupling and chemical shielding tensors oscillate with it, although nuclear spins themselves do not directly interact with electric fields (Hu & Wu, 2004c). Studies on the effect of electric fields on these tensors (see, e.g., Buckingham, 1960; Peshkovsky & McDermott, 2000; Grayson, 2003) also support this conclusion.

In the simple case of two ½-spin system inside neural membranes coupled to each other through isotropic J-coupling $J = J_{zz} = J_{xx} = J_{yy} = J_R + J_A$ the Hamiltonian of the system is

$$\hat{H} = h\left(J_R + J_A\right)\left(\hat{I}_{1z}\hat{I}_{2z} + \hat{I}_{1x}\hat{I}_{2x} + \hat{I}_{1y}\hat{I}_{2y}\right) \tag{12.1}$$

where $J_R$ is the J-coupling at resting potential and $J_A$ is the first-order contribution to $J$ from action potential modulation — thus it is a function of membrane voltage $V_m$ (Hu & Wu, 2004c).

For a given value of $V_m$ the two ½-spins form a triplet consisting of

$$|1\rangle = |\uparrow\uparrow\rangle,\ |3\rangle = \frac{1}{\sqrt{2}}\left(|\uparrow\downarrow\rangle + |\downarrow\uparrow\rangle\right)\ \&\ |4\rangle = |\downarrow\downarrow\rangle \tag{12.2}$$

and a singlet

$$|2\rangle = \frac{1}{\sqrt{2}}\left(|\uparrow\downarrow\rangle - |\downarrow\uparrow\rangle\right) \tag{12.3}$$

with energies

$$E_1 = E_3 = E_4 = \frac{1}{4}h(J_R + J_A),\ E_2 = -\frac{3}{4}h(J_R + J_A) \tag{12.4}$$

thus an energy gap $J = h(J_R + J_A)$.

The J-coupling strengths between $^1$H and $^1$H are typically in the range of 5–25 Hz. Further, J-couplings among biologically available nuclear spins such as $^1$H, $^{13}$C, and $^{31}$P are in the range of 5–250 Hz that are also the frequency spectra of various brain activities associated with different functional states.

In the principal axes system of dipolar coupling tensor $\boldsymbol{D}$ for the two ½-spin system,

$$\hat{H} = h(J_R + J_A + D_R + D_A)\hat{I}_{1z}\hat{I}_{2z} + h\left(J_R + J_A - \tfrac{1}{2}D_R - \tfrac{1}{2}D_A\right)\left(\hat{I}_{1x}\hat{I}_{2x} + \hat{I}_{1y}\hat{I}_{2y}\right) \tag{12.5}$$

is the Hamiltonian with both isotropic J-coupling

$$J = J_R + J_A \tag{12.6}$$

and dipolar coupling

$$D = D_{zz} = -\frac{1}{2}D_{xx} = -\frac{1}{2}D_{yy} = D_R + D_A \tag{12.7}$$

where $D_R$ is the dipolar coupling at resting potential and $D_A$ is the first-order contribution to $D$ from action potential modulation thus it is also a function of membrane voltage $V_m$. $D$ is typically in the range of 100 Hz to 10 kHz.

It can be verified that $|1\rangle$, $|3\rangle|4\rangle$ and $|2\rangle$ are also the eigenstates of the above Hamiltonian with energies

$$E_1 = E_4 = \frac{1}{4}h(J_R + J_A + D_R + D_A),\ E_3 = \frac{1}{4}h(J_R + J_A) - \frac{1}{2}h(D_R + D_A),\quad E_2 = -\frac{3}{4}h(J_R + J_A) \tag{12.8}$$

Thus, dipolar coupling has no effect on the singlet state, but partially removes the energy degeneracy of the triplet states thus producing zero-field splitting.

Further, the chemical shielding tensor $\boldsymbol{\sigma}$ of each nuclear spin also contains contribution from action potential modulation of its surrounding covalent bonds. That is, for the first spin $\boldsymbol{\sigma}_1 = \boldsymbol{\sigma}_{1R} + \boldsymbol{\sigma}_{1A}$ and for the second spin $\boldsymbol{\sigma}_2 = \boldsymbol{\sigma}_{2R} + \boldsymbol{\sigma}_{2A}$ where $\boldsymbol{\sigma}_{1R}$ and $\boldsymbol{\sigma}_{2R}$ are the chemical shielding tensors at resting potential and, $\boldsymbol{\sigma}_{1R}$ and $\boldsymbol{\sigma}_{2R}$ are the first-order contribution to $\boldsymbol{\sigma}_1$ and $\boldsymbol{\sigma}_2$ respectively from action potential modulations thus they are functions of membrane voltage $V_m$. So when the effects of both internal and external magnetic fields $\mathbf{B}_i$ and $\mathbf{B}_e$ are taken into accounts the total Hamiltonian for the two ½-spin system in neural membranes is

$$\hat{H} = -\hbar\gamma_1 \hat{\mathbf{I}}_1 \cdot (1 - \boldsymbol{\sigma}_{1R} - \boldsymbol{\sigma}_{1A}) \cdot (\mathbf{B}_{1i} + \mathbf{B}_{1e}) - \hbar\gamma_2 \hat{\mathbf{I}}_2 \cdot (1 - \boldsymbol{\sigma}_{2R} - \boldsymbol{\sigma}_{2A}) \cdot (\mathbf{B}_{2i} + \mathbf{B}_{2e}) + h\hat{\mathbf{I}}_1 \cdot (\mathbf{J}_R + \mathbf{J}_A) \cdot \hat{\mathbf{I}}_2 + h\hat{\mathbf{I}}_1 \cdot (\mathbf{D}_R + \mathbf{D}_A) \cdot \hat{\mathbf{I}}_2 \tag{12.9}$$

where $\mathbf{B}_{1i}$, $\mathbf{B}_{1e}$, $\mathbf{B}_{2i}$ and $\mathbf{B}_{2e}$ are respectively the internal and external magnetic fields at the locations of first and second ½-spins without chemical shielding and, $\gamma_1$ and $\gamma_2$ are respectively the gyromagnetic ratios of the said first and second ½- spins.

In general, microscopically

$$\frac{|\mathbf{B}_i|}{|\mathbf{B}_e|} >> 1 \tag{12.10}$$

at each spin location as shown later but macroscopically

$$\langle \mathbf{B}_i \rangle_{\mathbf{r}} = 0,\ \langle \mathbf{B}_e \rangle_{\mathbf{r}} \neq 0,\ \langle \mathbf{B}_i \rangle_t = 0 \text{ and } \langle \mathbf{B}_e \rangle_t \neq 0 \tag{12.11}$$

where $\mathbf{r}$ and $t$, respectively denote spatial and time average. So in many cases the effects of $\mathbf{B}_e$ on these spin networks are small.

These results from consideration of a simple two-1/2-spin system in neural membranes demonstrate that the large neural spin networks inside the membranes can form complex modulated structures through action potential driven oscillations of *J*-coupling, dipolar coupling and chemical shielding tensors. Thus, the neural spike trains of various frequencies can directly input information carried by them into these spin networks.

The fluctuating internal magnetic fields are produced by the paramagnetic species such as $O_2$ and NO and spin-carrying nuclei themselves such as $^{1}H$, $^{13}C$ and $^{31}P$. Table 1 shows the maximal magnetic field strengths produced by the magnetic dipoles of the unpaired electrons of $O_2$ and NO and the nucleus of $^{1}H$ along the axes of said dipoles at given distances. Because the magnetic dipole moment of an unpaired electron is 658 times larger than that of the $^{1}H$ nucleus, $O_2$ and NO can respectively produce magnetic fields 1,316 and 658 times larger than $^{1}H$.

**Table 1.** Magnetic Fields Produced by $O_2$, NO and $^1H$.

| Distance (Å) | $O_2$ (Tesla) | NO (Tesla) | $^1H$ (Tesla) |
|---|---|---|---|
| 1.0 | 3.713940 | 1.856970 | 0.002821 |
| 2.0 | 0.464243 | 0.232122 | 0.000353 |
| 3.0 | 0.137553 | 0.068777 | 0.000104 |
| 4.0 | 0.058030 | 0.029015 | 0.000044 |
| 5.0 | 0.029712 | 0.014856 | 0.000023 |
| 10.0 | 0.003714 | 0.001857 | 0.000003 |

As distance $r$ increases, the strength of the magnetic dipole field quickly attenuate according to

$$B = \frac{\mu_0 m}{4\pi r^3} \tag{12.12}$$

where $\mu_0$ is the permeability of free space and $m$ is the magnetic dipole moment. In addition, $O_2$ and NO are hydrophobic small molecules so their concentrations in neural membranes are much higher than in aqueous solutions such as cytoplasma (Marsh, 2001). As they rapidly tumble and diffuse, they produce microscopically strong and fluctuating magnetic fields. Indeed, $O_2$ are the predominant sources of internal magnetic fields in neural membranes as evidenced by the strong effect of $O_2$ on spin–spin and spin–lattice relaxation rates (Marsh, 2001; Prosser *et al.*, 2001).

These fluctuating internal magnetic fields continuously perturb the neural spin networks. The intensities of said perturbations depend on the concentrations of $O_2$ and NO that are highly regulated in the brain. Thus, these perturbations not only activate various modulated dynamics within the neural spin networks but also are likely capable of enhancing the synchronization of these dynamics to the neural spike trains through non-linear processes such as stochastic resonance that is known to occur in the brain (Bezrukov & Vodyanoy, 1995; Simonotto *et al.*, 1997). So, stochastic resonance of dipolar splitting transitions and spin-forbidden singlet-triplet transitions are possible inside the neural membranes under said modulations and perturbations. Stochastic resonance in two-state nuclear spin system was demonstrated by NMR spectroscopy (Viola *et al.*, 2000).

It is therefore suggested that the collective dynamics of the neural spin networks under modulations by action potentials and perturbations by fluctuating internal magnetic fields represent meaningful information to the brain. An analogy to this suggestion is the mechanism of liquid crystal display (LCD) where information-carrying electric voltages applied to the pixel cells change the optical properties of the constituent molecules such that when lights pass through these cells their phases rotate differently which in turn represent different information to the viewer of the LCD screen (Bryan-Brown *et al.*, 1998). According to this suggestion, significant $^1$H replacements by $^2$H and large external disturbances of the collective dynamics of the neural spin networks will affect the functional states of the brain to certain extent. Further, drug-induced large changes to membrane structures and $O_2$ pathways in neural membranes have similar adverse effects. These predications are testable and provide alternative interpretations to the causes of neural effects produced by some drugs and external stimulations. For example, the effect of transcranial magnetic stimulations (TMS) on cognitive functions (see, e.g., Walsh & Cowey, 2000) can be partly attributed to the direct disturbances of the dynamics of the said spin networks by TMS and the cause of unconsciousness by general anaesthetics can be explained as the direct consequence of their effects on neural membrane structures and $O_2$ pathways inside (Hu & Wu, 2001).

However, how can we explain based on the above suggestion that cognitive functions seem in general insensitive to environmental and even medical strength external magnetic fields such as those generated by the power lines and the ones used in MRI? First, the strengths of environmental magnetic fields are in the range of $10^{-4}$–$10^{-6}$ Tesla (see, e.g., Marino, 1988). For example, the magnetic field strength of the earth is about $5 \times 10^{-5}$ Tesla. In comparison, the internal fluctuating magnetic fields can be as high as several Tesla as indicated by Table 1. Thus, the microscopically strong and fluctuating internal magnetic fields overshadow them. But the strengths of magnetic fields used in clinical and research MRI systems are in the range of 0.064 to 8.0 Tesla (see, e.g., Shellock, 2002) that is comparable to or even higher than the strengths of said internal magnetic fields. So, additional explanations are called for.

Indeed, the net magnetization of nuclear spins even by magnetic field of several Tesla is only about a few parts per million at room temperature (see, e.g., Gershenfeld & Chuang, 1997) which shows that even strong static magnetic fields have only small effects on the thermal dynamics of the neural spin networks. Third, to the extent that spin networks are disturbed by external magnetic fields, it is argued that most of these disturbances do not represent meaningful information to the brain and, further, the brain likely has developed other mechanisms through evolution to counter the effects of external magnetic fields. In the case where external magnetic disturbances were reported to have observable effects on cognition, the above suggestion provides a basis for interpreting these effects as said disturbances contain meaningful information to the brain (Hu & Wu, 2004c).

Although quantum coherence is not required for the neural spin networks to serve as the subatomic components for the conventional neural network according to the above suggestion, it likely exists within some parts of spin networks as studies in other fields suggest (see., e.g., Hu & Wu, 2002; Vedral, 2011; Saeedi, *et al.*, 2013; Turanov & Khitrin, 2013).

## 12.5. Explanation of Consciousness with Quantum Spin Formulism

In reference to preceding sections, we postulate the following (Hu & Wu, 2002, 2004a): (a) Consciousness is intrinsically connected to quantum spin; (b) The mind-pixels of the brain are comprised of the nuclear spins distributed within the neural membranes and proteins, the pixel-activating agents are comprised of biologically available paramagnetic species such as $O_2$ and NO, and memory is comprised of all possible entangled quantum states of the mind-pixels; (c) Action potential modulations of nuclear spin interactions input information to the mind-pixels and spin chemistry is the output circuit to classical neural activities; and (d) Consciousness emerges from the collapse of those entangled quantum states which are able to survive decoherence: collapse is contextual, irreversible and non-computable and the unity of consciousness is achieved through quantum entanglement of the mind-pixels.

In postulate (a), the relationship between quantum spin and consciousness is defined based on the fact that spin is the origin of quantum effects in both Bohm quantum formulism (Recami & Salesi, 1998; Esposito, 1999) and Hestenes quantum formulism (Hestenes, 1975, 1983) and a fundamental quantum process associated with the structure of space–time (Dirac, 1928). Combining this fundamental idea with those stated in postulates (b), (c) and (d) allows us to build a qualitatively detailed working model of consciousness.

In postulate (b), we specify that the nuclear spins in both neural membranes and neural proteins serve as the mind-pixels and propose that biologically available paramagnetic species such as $O_2$ and NO are the mind-pixel activating agents. We also propose that memory is comprised of all possible entangled quantum states of mind-pixels. This concept of memory is an extension to the associative memory in neuroscience.

In postulate (c), we propose the input and output circuits for the mind-pixels. As shown earlier, the strength and anisotropies of nuclear spin interactions through J-couplings and dipolar couplings are modulated by action potentials. Thus, the neural spike trains can directly input information into the mind-pixels made of neural membrane nuclear spins (Hu & Wu, 2004c). Further, spin chemistry can serve as the bridge to the classical neural activity since biochemical reactions mediated by free radicals are very sensitive to small changes of magnetic energies (Minaev, 1996; Nagakura, 1998; Hayashi, 2002).

In postulate (d), we propose how conscious experience emerges. Since there are several interpretations of the measurement problem in quantum mechanics, we choose to accept the collapsing view (see, e.g., Penrose, 1989, 1994). Thus, we adopt a quantum state collapsing from which conscious experience emerges as a set of such collapses of the decoherence-resistant entangled spin quantum states. We further theorize that the unity of consciousness is achieved through quantum entanglements of the mind-pixels (Stapp, 1993).

The subjectivity of experience is explained by the contextual collapse of entangled spin quantum states (qubits) which is influenced by the information stored in the neural nuclear spin ensembles and/or made available by quantum entanglement of the nuclear spins with the environment.

Figure 1 is a highly schematic drawing of the overall picture of a spin-mediated consciousness model proposed herein (Hu & Wu, 2002, 2004a). At the top of Fig. 1, a two-neuron network is shown. The connections are self-explanatory. The neural activities of the postsynaptic membrane are immediately shown below the neurons in Fig. 1. These activities include biochemical reactions immediately following the release of neurotransmitters into the synaptic cleft, the ensuing collective activities of multiple ion channels and the action potentials and their propagations thereof, and other enzymatic activities.

The present model is mainly concerned with the dynamics of the nuclear spin ensembles within neural membranes and proteins such as those on the dendrites and soma under modulations by action potentials

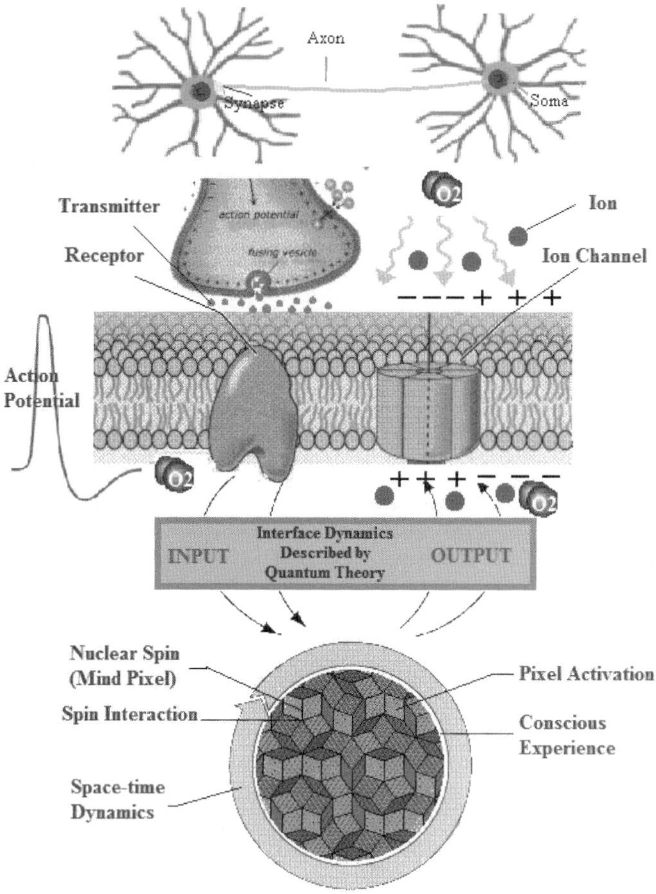


**Fig. 1.** Illustration of spin-mediated consciousness theory. The drawing is self-explanatory except the part dealing with conscious experience, see text for detailed explanations.

and activations by paramagnetic species such as rapidly tumbling and diffusing $O_2$ and neural transmitter NO, and the connections of such dynamics to conscious experience. The input and out interface of the neural nuclear spin ensembles are schematically shown in the middle of Fig. 1. On the bottom of Fig. 1, what the conscious brain perceives is schematically shown. The neural substrates and mechanism of the spin-mediated consciousness are described below.

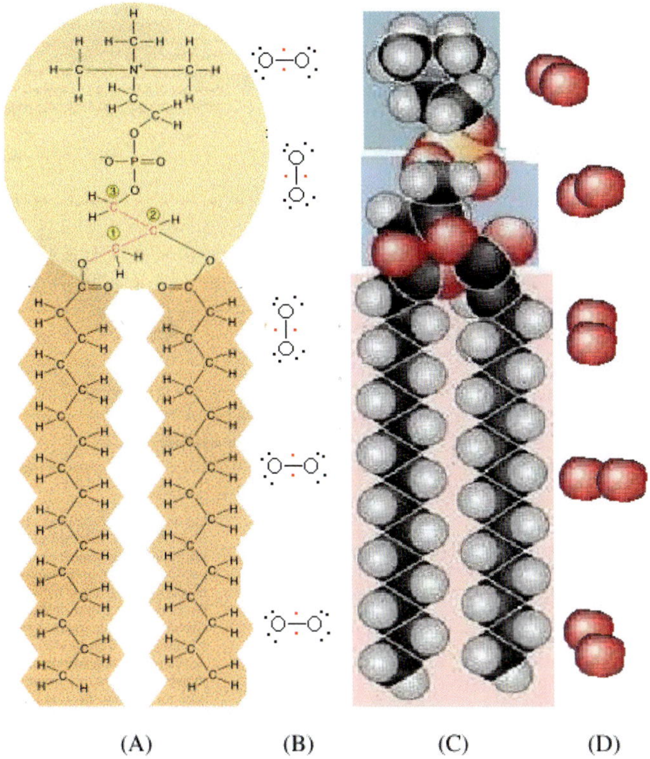


**Fig. 2.** Schematic drawings of chemical structure (A) and atomic model (C) of a typical phospholipid with diffusing $O_2$ in Lewis structure (B) and atomic model (D) shown along the side. Two unpaired electronic electrons of $O_2$ are shown in Lewis structure as two red dots (B). In the atomic model, the white, black, purple and orange balls respectively represent hydrogen, carbon, oxygen and phosphonium atoms. $^{1}$H, $^{13}$C and $^{31}$P are spin-carrying nuclei.

Figure 2 shows side-by-side a typical phospholipid found in neural membranes together with the diffusing $O_2$. A similar but much complex picture can be drawn for a neural protein. As can be seen, each phospholipid molecule contains 1 $^{31}$P, less than 1 $^{13}$C (1.1%) and more than 60 $^{1}$H on its two lipid chains. Because the small mobility of nuclei and weak interactions of nuclear spins with their environments, nuclear spins have long relaxation time after excitations (see, e.g., Morris, 1986). This property of nuclear spins is ideal for them to serve as the mind-pixels. Importantly, these nuclear spins can form various intra/inter-molecularly entangled quantum states under different external activations through J-couplings and dipolar coupling (see, e.g., Khitrin & Ermakov, 2002a, b, c).

The unpaired electrons attached to the paramagnetic species such as $O_2$ and NO can interact with nuclear spins through their large magnetic dipoles and collision-induced Fermi-contact mechanism (see, e.g., Wertz & Bolton, 1972) thus activating the neural nuclear spin ensembles. Indeed, because the magnetic dipole moment of an unpaired electron is 658 times larger than that of the $^1H$ nucleus, $O_2$ and NO can respectively produce magnetic fields 1,316 and 658 times larger than $^1H$. In addition, $O_2$ are the predominant sources of internal magnetic fields in neural membranes as evidenced by the strong effect of $O_2$ on spin–spin and spin–lattice relaxation rates (Marsh, 2001, Prosser, *et al.*, 2001).

The mechanism of spin-mediated consciousness is concisely stated here and the related issues such as decoherence effect will be treated next (see also Hu & Wu, 2002, 2004a). Through action potential modulated nuclear spin interactions and paramagnetic $O_2$/NO driven activations the nuclear spins inside neural membranes and proteins form various entangled spin states some of which survive rapid decoherence through quantum Zeno effects or in decoherence-free subspaces and then collapse contextually through non-computable and irreversible means thus producing consciousness and, in turn, the collective spin dynamics associated with collapses have effects through spin chemistry on classical neural activities thus influencing neural networks of the brain. Our proposal calls for extension of associative encoding of memory to the dynamical structures of neural membranes and proteins.

Therefore, according to the quantum spin mediated theory of consciousness (Hu & Wu, 2002, 2004a), (a) the dynamical nuclear spin ensembles are the "mind-screen" with nuclear spins as its pixels, (b) the neural membranes and proteins are the mind-screen and memory matrices; and (c) the fluxes of biologically available paramagnetic $O_2$/NO are the beam for pixel-activations. Together, they form the neural substrates of consciousness. An analogy to this is the mechanism of LCD where information-carrying electric voltages applied to the pixel cells change the optical properties of the constituent molecules such that when lights pass through these cells their phases rotate differently which in turn represent different information to the viewer of the LCD screen (see, e.g., Bryan-Brown *et al.*, 1998).

The decoherence effect which causes a quantum system to lose quantum coherence through interactions with its environment is a major concern for any quantum theory of the brain (see, e.g., Tegmark, 2000; Hagan, *et al.*, 2002). However, nuclear spins only have weak interactions with their environments thus long relaxation time after excitation (see, e.g., Morris, 1986). Indeed, there are both theoretical and experimental studies (see, e.g., Khitrin & Ermakov, 2002a,b,c; Warren *et al.*, 1993, 1998; Julsgaard *et al.*, 2001) indicating the possibility of large-scale quantum coherence with entanglement in the nuclear spin ensembles distributed within the neural membranes and proteins. Paradoxally, the interactions of the neural membrane nuclear spin ensembles with their noisy environments may even enhance quantum coherence through quantum Zeno effect which prevents a quantum system to evolve/decohere through repeated collisions with their environments (Stapp, 1993). Further, studies show that decoherence-free subspaces can exist theoretically within the Hilbert space of a complex quantum system (see, e.g., Lidar & Whaley, 2003).

In a series of experiments, Khitrin *et al.* (2002a,b,c) have demonstrated that a cluster of dipolar coupled $^1$H nuclear spins in the molecules of a nematic liquid crystal at room temperature can be manipulated to achieve long-lived intra-molecular quantum coherence with entanglement such that a large amount of information may be stored in said cluster. In particular, they have succeeded in storing at room temperature a 2D pattern consisting of 1025 bits of information in the proton nuclear spin states of a molecular system of nematic liquid crystal and then retrieved the same as a stack of NMR spectra.

Second, Warren *et al.* (1993, 1998) discovered long-ranged intermolecular multiple-quantum coherence in NMR spectroscopy and imaging and have since successfully applied said coherence as MRI contrast agents. Indeed, they found that even the intermolecular dipolar couplings of the nuclear spins at distances larger than 10 microns are not averaged away by diffusion.

Third, Julsgaard *et al.* (2001) have theoretically predicted and then experimentally demonstrated at room temperature a long-lived entanglement of two macroscopic spin ensembles formed by two caesium gas samples each of which contains about $10^{12}$ atoms. The entangled spin-state can be maintained for 0.5 milliseconds and was generated via interactions of the samples with a pulse of light. The state they demonstrated is not a maximally entangled 'Schrödinger cat' state but a state similar to a two-mode squeezed state; thus, it is an example of a non-maximally entangled state. In addition, Kun *et al.* (2003) have theoretically predicted a "Schrödinger cat" state to be found in highly-excited and strongly-interacting many-body system.

These results apparently contradict the claim that there is no large-scale quantum coherence in the noisy brain (Tegmark, 2000). At least, this claim does not seem to apply to the nuclear spin ensembles within neural membranes and proteins. Further, in the case of Penrose–Hameroff microtubule model, it was strongly argued by Hagan *et al.* (2002) that Tegmark's theoretical calculations also do not apply there. Indeed, even if we assume this claim to be true then, consciousness can still emerge from the statistical mixtures of coherent and incoherent quantum states of the mind-pixels, as long as we accept some kind of emergence theory (Searle, 1992).

In accordance with our spin-mediated theory is schematically shown at the bottom of Fig. 1 the geometry inside the spinning circle represents conscious experience and is part of a Penrose tiling (Penrose, 1989). It symbolizes that consciousness emerges from the non-computable collapse of entangled quantum states of the mind-pixels under the influence of space–time dynamics schematically shown as the spinning circle. The edges in the Penrose tiling represent the nuclear spins within neural membranes and proteins as mind-pixels, the nodes represent interactions between these nuclear spins through J-couplings and dipolar couplings and the colors represent activations of mind-pixels by biologically available paramagnetic species such as $O_2$ and NO. The whole tiling pattern in Fig. 1 represents conscious experience and the underlying space–time geometry. This pattern successively evolves under repeated activations by the paramagnetic species representing successive collapse of the entangled quantum states of the mind-pixels that have survived decoherence as a stream of conscious experience.

We adopt Penrose's long-standing view that human thought may involve non-computable processes, as Gödel's incompleteness theorem (Penrose, 1989, 1994). According to Gödel, any consistent system of axioms beyond a certain basic level of complexity yields statements that cannot be proved or disproved with said axioms. Yet human can decide whether those statements are true, thus human thought cannot be reduced to a set of rules or computations. So where can one find non-computable process in physics? Obviously it cannot be found in classical physics because classical physics is deterministic so, in principle, can be simulated by a computer. Thus, Penrose reasoned that some kind of non-computable quantum process must be involved in consciousness and further suggested gravity-induced reduction ("R") process of quantum state superposition to be the candidate. One may recall that, according to Einstein's theory of general relativity, gravity is space–time geometry and, further, as we have discussed before quantum mechanical spin is associated with the structure of space–time. Therefore, the quantum state of spin must be connected to the underlying space–time geometry. However, we still have the task of working out the details in future research. This will be especially difficult because at the present we do not have a satisfactory theory of quantum gravity.

If we assume that there is no large-scale quantum coherence in the noisy brain because of decoherence (Tegmark, 2000), how can consciousness still emerge from the statistically mixed quantum states of the nuclear spin ensembles within neural membranes and proteins? There are indeed at least two ways. The first is to adopt an emergence theory (Searle, 1992) and the second is to take a dualistic approach (Beck & Eccles, 1992; Chalmers, 1996). Here, we will focus our discussion on the dualistic approach.

In such approach, we can propose that mind has its own independent existence and resides in a non-local domain. Then, the question becomes how does mind process and harness the information from the brain so that it can have conscious experience? We can theorize that conscious experience emerges from those quantum states of the mind-pixels in the statistical mixtures that have encapsulated the attention of the mind through quantum Zeno effect (Stapp, 1993) or through entanglement in a non-local domain. Indeed, the many-mind interpretation of quantum theory as proposed by Donald (1990) supports this type of formulation. Thus, in this scenario, mind does not depend on large-scale quantum coherence.

Furthermore, each nuclear spin is a quantum bit (qubit), each $O_2$ spin triplet is tightly entangled two-qubit and there exist coherent and incoherent intra-molecular superpositions of multiple nuclear spins under external stimulations (Khitrin & Ermakov, 2002a,b,c). So, it is plausible that mind could utilize quantum statistical computations similar to those proposed in by Castagnoli & Finkelstein (2001). In their model, a triode network made of triplet pairs of spin ½ fermions and their quantum statistical relations due to particle indistinguishability was utilized to implement the computation, together with a randomly-varying magnetic field as heat bath for annealing. The scheme develops quantum parallelism through the incoherent superposition of parallel computation paths (i.e., the mixtures). It replaces the superposition of coherent parallel computation paths with the almost indestructible superposition of different permutations of identical particles subject to a given statistics thus surviving decoherence.

We have proposed that memory is comprised of all possible entangled quantum states of the nuclear spins inside neural membranes and proteins (Hu & Wu, 2002, 2004a). This proposal calls for extension of the existing associative memory concept in neuroscience to include all possible conformations of neural membranes and proteins in a single neuron (see, e.g., Marder *et al.*, 1996). A few illustrations are given here. Figure 3(a) schematically shows a patch of neural membrane containing the same phospholipids. Such a membrane is much like a blank tape. Figure 3(b) shows the same neural membrane after cholesterols are added. The changes in membrane configuration are quite noticeable (Raffy & Teissie, 1999; Smondyrev & Berkowitz, 1999). These changes can represent memory or information. Figure 3(c) shows the chemical structure and atomic model of a stearic acid molecule — a saturated fatty acid. Figure 3(d) shows the chemical structure and atomic model of oleic acid molecule — an unsaturated fatty acid. The only difference between the two fatty acids is that the latter contains a double bond in the middle that causes its kink formation when the double bond is in the cis isomer form. When the double bond is in the transform, the chain is doubly bent so there is no kink. Certainly insert either one of the fatty acids into the membrane shown in Fig. 3(b) would further increase its complexity thus information content. Furthermore, insertions of proteins to neural membranes also significantly change their conformation and dynamics surrounding the inserted proteins (Woolf & Roux, 1996). Thus, inserting different proteins into neural membranes both in numbers and types can significantly increase the information content of the neural membranes.

We have suggested that action potential modulations of nuclear spin interactions input information to the mind pixels and spin chemistry is the output circuit to classical neural activities (Hu & Wu, 2004b). With respect to the input circuit, published studies show that the dynamics of membrane proteins significantly affect the dynamics of surrounding membranes (see, e.g., Marder *et al.*, 1996; Woolf & Roux, 1996). Secondly, due to the very small thickness of the neural membrane

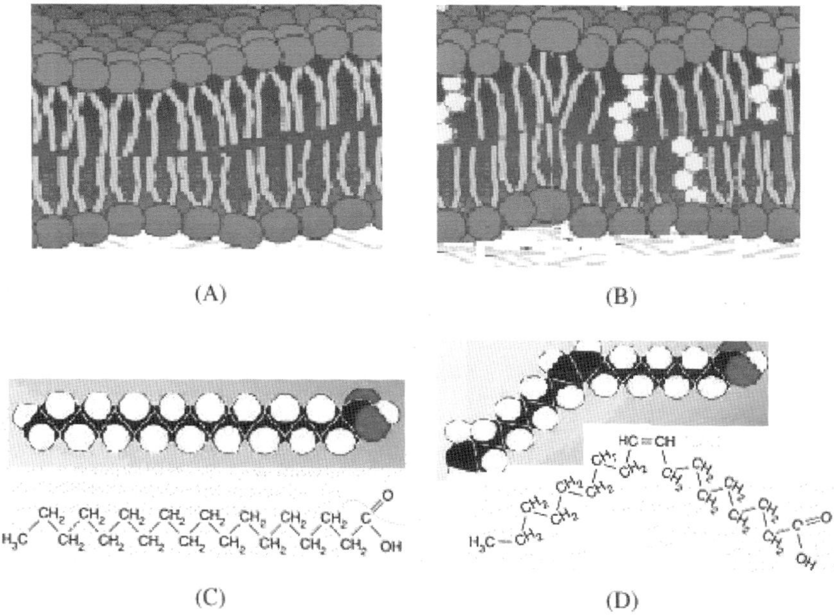


**Fig. 3.** Illustration of neural memory. (A) a neural membrane containing only the same phospholipids. (B) the neural membrane after cholesterols are added. (C) the chemical structure and atomic model of a stearic acid molecule and (D) the chemical structure and atomic model of oleic acid molecule.

(~10 nm), even voltage of an action potential can create enormous oscillating electric field inside the neural membranes that in turn can affect the conformations and dynamics of the membrane components such as phospholipid and proteins (see e.g., Akinlaja & Sachs, 1998; Sens & Isambert, 2002) . Indeed, the strength and anisotropies of nuclear spin interactions through J-couplings and dipolar couplings are modulated by action potentials. Thus, the neural spike trains can directly input information into the mind-pixels made of neural membrane nuclear spins (Hu & Wu, 2004b).

Secondly, the weak magnetic field produced collectively by all neural activities may also directly serve as the input. However, the magnitude of magnetic field is only in the order of $10^{-12}$ Tesla (Wikswo, 1990). In comparison, $O_2$ and NO can produce a fluctuating local magnetic field as high as a few Tesla (unpublished calculations). Thus, the effect of said weak magnetic field on the dynamics of mind-pixels is probably small unless non-linear processes such as stochastic resonance are involved (Hu & Wu, 2004b).

Further, we have already pointed out earlier that spin chemistry can serve as the output circuit to classical neural activities because biochemical reactions mediated by free radicals are very sensitive to small changes of magnetic energies as discussed previously. Indeed, many biochemical reactions mediated by radical pairs and biradicals, such as those dual path radical reactions driven/initiated by NO and active oxygen species, have been found to be influenced by the magnetic field in their local environment (Nagakura, 1998; Hayashi, 2002). Thus, the functional output of the mind-pixels, being the varying local magnetic field generated by the dynamics of the nuclear spin ensembles as mind-pixels, can directly affect classical neural activities. Further, there may be other mechanisms through which the mind-pixels can influence the classical neural activities of the brain.

Several experimentally verifiable predictions can be drawn from our spin-mediated consciousness theory: (1) Significant replacement of $^1$H with $^2$H will affect or disrupt consciousness; (2) Significant external disturbances to the dynamics of the nuclear spin ensembles in neural membranes and proteins will interfere with normal conscious functions; (3) Significant drug-induced disturbances to the structure and dynamics of the neural membranes and protein themselves will affect or disrupt consciousness; (4) Significant drug-induced disturbances to the $O_2$ pathways inside the neural membranes will diminish or block consciousness; and (5) Significant lack of $O_2$ in neural membranes will directly affect or disrupt consciousness even if everything else in the brain functions normally. Of course, other predictions and inferences can also be drawn from the present theory. But we will focus our discussions on the above listed a few to see whether there are any experimental evidence supporting these predictions.

With respect to prediction (1), there are published results concerning the biological effects of heavy water ($^2H_2O$ or $D_2O$) that lend indirect support to our proposals (Kushner *et al.*, 1999). Indeed, the only observable physical and chemical difference between regular and heavy water is that the latter has a slightly higher viscosity. It was reported that rats restricted to heavy water (99.8%) would drink it freely the first day, then drank progressively less and died within 14 days (Richter, 1977). It was also reported that the incremental increase of the amount of heavy water in regular drinking water lengthened the 24-hour clock of the Circadian rhythm in blind hamsters or normal hamsters kept in constant darkness.

In principle, these observations can be explained as animals literally losing their minds because of the loss of their mind-pixels due to gradual $^{2}H$ substitutions of $^{1}H$ in neural membranes and proteins. The complication is that the higher viscosity of heavy water might have also contributed to the observed effects. It is possible to design an experimental procedure to either account for or exclude those effects attributable to the higher viscosity.

With respect to prediction (2), there are also quite a few published studies (see e.g., Limoniemi, 1997; Chicurei, 2002) based on "TMS" that can be at least partially explained based on our theory, although common wisdom is that TMS induces electrical currents in the brain, causing depolarization of cellular membranes and thereby neural activation (Limoniemi, 1997). It has been found that depending on the locations of stimulation TMS affects the test subject's verbal ability, visualization and other conscious functions (Chicurei, 2002). According to our theory, TMS can directly affect the dynamics of nuclear spin ensembles within neural membranes and proteins which in turn result in altered, diminished and/or disrupted conscious functions of the brain (Hu & Wu, 2002, 2004a).

With respect to predictions (3) and (4), many general anesthetics have been found to disturb the structures and dynamics of neural membranes (see, e.g., Cantor, 1997; Tu, 1998; Koubi, 2000). Thus, the mechanism of their action can be interpreted, according our spin-mediated theory, as caused by their direct effects on $O_2$ pathways and nuclear spin ensembles inside the neural membranes and proteins. Our earlier work discusses a detailed treatment on anesthetic perturbations of oxygen pathways and membranes themselves (Hu & Wu, 2001). Here, we will focus on one particularly small anesthetic agent, the nitrous oxide ($N_2O$), also known as laughing gas. Indeed, the size of $N_2O$ is similar to that of $O_2$ but it does not contain unpaired electrons and is not reactive. It has low polarity that makes it soluble in both water and lipid. Thus, it can be carried to the brain through blood stream and accumulate in the neural membranes. Inhalation of $N_2O$ will cause disorientation, euphoria, numbness and ultimately loss of consciousness if the inhalation dosage is high. The cellular mechanism of these actions by $N_2O$ is so far unknown, but seems confined to postsynaptic targets (Mennerick, 1998). On the other hand, its closely related 'cousin' NO contains one unpaired electron and has been discovered as the first small and highly diffusive neural transmitter produced in the brain through enzymatic reactions (see, e.g., Philippides *et al.*, 2000). According to our theory, there indeed exists a natural and straightforward explanation. By dissolving into neural membranes in an inhalation-dose-dependent fashion, $N_2O$ gradually displace $O_2$ in the neural membranes thus diminish or disrupt the activating function of $O_2$ (Hu & Wu, 2001, 2002, 2004a).

With respect to prediction (5), it is probably very hard to deprive brain $O_2$ and yet at the same time require its neurons to keep their metabolic functions normal since $O_2$ is an essential component of brain energy production. However, according to our theory in the case of temporary-hypoxia-induced unconsciousness such as that due to sudden loss of air pressure in an airplane, the actually cause may not be the depletion of brain energy resources because of the lack of $O_2$, but the direct loss of $O_2$ as the activating agents.

Now, we briefly turn our attention to the associative memory model proposed herein. There are tens of thousands of research papers on the subject of synaptic plasticity/modification (see, e.g., Fu *et al.*, 2002). The commonly accepted assumption in neuroscience is that synaptic efficacy is both necessary and sufficient to account for learning and memory (see, e.g., Marder *et al.*, 1996). Our, associative memory model does not conflict with the synaptic efficacy view, but extends it to the sub-neural and microscopic domain. Studies show that neural activities modify not only the synaptic efficacy, but also the intrinsic properties of the neuron.

More recently, we have also considered the possible roles of neural electron networks in memory and consciousness (Hu & Wu, 2004d). Besides free $O_2$ and NO, the main sources of unpaired electron spins within neural membranes and proteins are transition metal ions and $O_2$ and NO bound/absorbed to large molecules, free radicals produced through biochemical reactions and excited molecular triplet states induced by fluctuating internal magnetic fields. We have shown that unpaired electron spin networks inside neural membranes and proteins are also modulated by action potentials through exchange and dipolar coupling tensors and spin-orbital coupling and g-factor tensors and perturbed by microscopically strong and fluctuating internal magnetic fields produced largely by diffusing $O_2$. Thus, we have argued that these spin networks could also be involved in brain functions since said modulation inputs information carried by the neural spike trains into them, said perturbation activates various dynamics within them and the combination of the two likely produce stochastic resonance thus synchronizing said dynamics to the neural firings. Although quantum coherence is desirable, it is not required for these spin networks to serve as the microscopic components for the classical neural networks.

On the quantum aspect, we speculate that human brain works with unpaired electron spins being the mind-pixels. Through action potential modulated electron spin interactions and fluctuating internal magnetic field driven activations, the neural electron spin networks inside neural membranes and proteins form various entangled quantum states some of which survive decoherence through quantum Zeno effects or in decoherence-free subspaces and then collapse contextually via irreversible and non-computable means producing consciousness and, in turn, the collective spin dynamics associated with these collapses have effects through spin chemistry on classical neural activities thus influencing the neural networks of the brain. Thus, according to this alternative model, the unpaired electron spin networks are the "mind-screen," the neural membranes and proteins are the mind-screen and memory matrices, and diffusing $O_2$ and NO are pixel-activating agents. Together, they form the neural substrates of consciousness.

## 12.6. Evidence Supporting Quantum-spin Formulism of Consciousness

In this section, we discuss some experimental evidence supporting spin-mediated consciousness theory from various sources.

Experimentally, quantum spins of nuclei, electrons and photons have now been successfully manipulated and entangled in a variety of ways for the purposes of quantum computation, memory and communication. By way of examples, nuclear spins have relatively long relaxation times after excitations (see, e.g., Khitrin & Ermakov, 2002a,b,c; Warren, *et al.*, 1993, 1998; Julsgaard *et al.*, 2001). Julsgaard *et al.* (2001) achieved long-lived (~0.05ms) entanglement of two macroscopic electron spin ensembles in room temperature. Khitrin *et al.* (2002a,b,c) showed that when a nematic liquid crystal is irradiated with multi-frequency pulse magnetic fields, its $^1H$ spins can form long-lived intra-molecular quantum coherence with entanglement for information storage.

Saeedi *et al.* (2013) have observed a room-temperature coherence time of over 39 minutes for an ensemble of phosphorous-31 donors in isotopically purified silicon-28. They further showed that a coherent spin superposition can be cycled from 4.2K to room temperature and back with a cryogenic coherence time of 3 hours in the same system. In another experiment, Turanov & Khitrin (2013) observed extremely long-lived spin states in systems of dipolar-coupled nuclear spins in solids which allow large amounts of information to be encoded in such long-lived states, stored in a form of multi-spin correlations, and subsequently retrieved by an application of a single "reading" pulse. They reported that nuclear spin in $^{1}$H of Naphthalene has a ~1700 sec (28.3 minutes) spin-lattice relaxation time (T1).

The above facts suggest that the quantum phenomena responsible for brain functions such as self-awareness are likely to be nuclear and/or electronic spins. Indeed, neural membranes and proteins contain vast numbers of nuclear spins such as $^{1}$H, $^{13}$C, $^{31}$P and $^{15}$N. These nuclear spins and unpaired electronic spins are the natural targets of interaction with action potentials in the brain through their motions and the photons of the magnetic pulses or other sources (Hu & Wu, 2002, 2004a). These spins form complex intra- and inter-molecular networks through various intra-molecular J- and dipolar couplings and both short- and long-range intermolecular dipolar couplings (Hu & Wu, 2004b).

Persinger *et al.* (2013) has shown that spin energies can accommodate the interactions among protons, electrons, photons and action potentials associated with intention and consciousness.

Recent studies in spintronics have shown that an electric field will exert a transverse torque/force on a moving spin (see, e.g., Sun *et al.*, 2004; Shen, 2005). This is actually not hard to understand since according to special theory of relativity a moving spin in an electric field "sees" a magnetic field. Sun *et al.* (2004) has shown that a moving spin is affected by an external electric field and "feels" a force/torque as $m \times [(v/c^2) \times E]$ where $m$ and $v$ are respectively the magnetic moment and the velocity of the moving spin and $E$ is the external electric field. Shen (2005) has shown that, as a relativistic quantum mechanical effect, an external electric field exerts a transverse force on an electron spin 1/2 if the electron is moving. The said spin force, analogue to the Lorentz force on an electron charge in a magnetic field, is perpendicular to the electric field and the spin motion when the spin polarization is projected along the electric field.

Therefore, the interactions between the moving nuclear/electronic spins in neural membranes and proteins and the varying high-voltage electric fields there directly feed information into the brain (Hu & Wu, 2008). To illustrate this particular mechanism, we now consider the spin transverse force exerted on a proton spin of a hydrogen atom connected to the carbon chain of a phosphate lipid located inside the neural membrane as shown in Fig. 4. As the carbon chain rotates in parallel to the intense electric field **E** across the neural membranes, the vertical proton

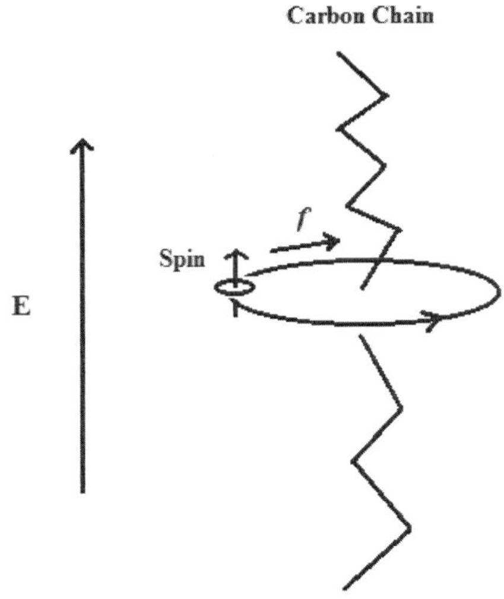


**Fig. 4.** Illustration of spin transverse torque/force $f$ exerted on a nuclear/electronic spin on a molecular chain or fragment inside the neural membrane. Upper: Excitation starting on tryptophan 2 (TRP 2). Lower: Excitation starting on tryptophan 4 (TRP 4). E denotes an external electric field.

spin moving in a circle perpendicular to the carbon chain "sees" a magnetic field in the rotating frame of reference thus feels a transverse torque/force $f$ toward the rotating plane. This spin transverse torque/force enables the neural spike trains to directly influence the nuclear/electronic spin networks within neural membranes and proteins thus inputting information into mind in the dualistic approach.

It has been long known that in an external electric field, the Dirac particle such as an electron or nuclear sub-entity acts as if it has an imaginary electric moment $i\,|\,\mathbf{d}\,|\, = ie\hbar/2$ mc (Dirac, 1928, 1983). It was Hestenes who showed that Dirac magnetic and electric dipole moments have same origin associated with spin and magnetization (for a review, see Hestenes, 2003). In Hestenes' formulism, magnetic moment density is not directly proportional to the spin but "dually proportional." The duality factor $e^{i\beta}$ has the effect of generating an effective electric dipole moment for the Dirac particle. Hestenes commented that "this seems to conflict with experimental evidence that the electron has no detectable electric moment, but the issue is subtle".

Other researchers have also shown that the magnetic and electric dipole moments of a fermion are closely related because they determine the real and imaginary part of the same physical quantity (Feng *et al.*, 2001; Graesser & Thomas, 2002). Further, Silenko has shown in the Foldy–Wouthuysen representation that although the influence of the electric dipole moment on the Dirac particle motion is negligibly small in an external electric field, it influences significantly the spin motion of the said particle (Silenko, 2006).

Furthermore, in the classical models of the Dirac particle, fast oscillating electric dipole moments also appear (Rivas, 2002; Gauthier 2006). These findings coincide with earlier finding that a moving magnetic dipole induces an electric dipole $d = (v/c^2) \times m$, where $m$ and $v$ are respectively the magnetic moment and the velocity of the moving spin, as a relativistic effect (Rosser, 1964). Rivas (2002) believes that what is lacking in the typical quantum mechanical wave equation is this oscillating electric dipole. He states that "*in general, the average value of this term in an electric field of smooth variation is zero, [but] in high intensity fields or in intergranular areas in which the effective potentials are low, but their gradients could be very high, this average value should not be negligible.*" Rivas further showed that the electric moment of the classical Dirac electron could lead to interesting physical effects.

In the context of spin-mediated consciousness theory, the interactions between the Dirac-Hestenes electric dipoles of nuclei and/or electrons with the varying electric fields within neural membranes and proteins may directly feed information carried by the neural spike trains into the quantum realm through the varying action potentials (Hu & Wu, 2008). Even if the Dirac electric dipole is purely imaginary with no known physical consequence, we argue that it may serve as an information receiver.

## 12.7. Conclusions

The spin-mediated consciousness theory claims that quantum spin is a cog in the wheel of consciousness and the linchpin between mind and the brain, that is, spin is the mind-pixel (Hu & Wu, 2002, 2004a). The starting point is the fact that spin is basic quantum bit ('qubit') for encoding information and, moreover, neural membranes and proteins are saturated with nuclear spin carrying nuclei and form the matrices of brain electrical activities. Indeed, spin is embedded in the microscopic structure of space–time as reflected by Dirac equation (Dirac, 1928) and is likely more fundamental than space–time itself as implicated by

Penrose's work (Penrose, 1960, 1967). In the Hestenes picture, the *zbw* associated with spin was shown to be responsible for the quantum effects of the fermion (Hestenes, 1975, 1983). Further, in the Bohm picture, the internal motion associated with spin has been shown to be responsible for the quantum potential which, in turn, is responsible for quantum dynamic effects (see Chapter 11). Thus, nuclear spins and possibly unpaired electron spins become natural candidates for mind-pixels (Hu & Wu, 2002, 2004a).

Applying these ideas to particular structures and dynamics of the brain, we have argued that human brain operates through action potential modulated nuclear spin interactions and paramagnetic $O_2$/NO driven activations; the nuclear spins within neural membranes and proteins form various entangled quantum states and, in turn, the collective dynamics of the entangled quantum states produces consciousness through contextual, irreversible and non-computable means that influences the classical neural activities through spin chemistry.

Therefore, according to our theory, consciousness is intrinsically connected to the spin process and emerges from the collective dynamics of various entangled spin states and the unity of consciousness is achieved by entanglement of these mind-pixels. As with other quantum consciousness theories, decoherence is a major concern as pointed out by Tegmark (2000), but may not be insurmountable (Hu & Wu, 2002, 2004a,b,c, 2008).

We strongly believe that the solutions to the binding problem and decoherence can be solved by quantum entanglement. According to our theory, quantum spins are the mind-pixels, consciousness emerges from the collective dynamics of various entangled spin states and the unity of consciousness is achieved by entanglement of these mind-pixels (Hu & Wu, 2002, 2004a).

The subjectivity of conscious experience is explained by the contextual collapse of entangled spin quantum states (qubits) under the influence of the information stored in the neural nuclear spin ensembles as memory and/or made available by quantum entanglement of these nuclear spins with the environment.

Quantum entanglement as understood herein allows brains to achieve binding and overcome decoherence (Hu & Wu, 2004c). The essential question is, then, how does the brain process and harness the information from the mind-pixels which form various entangled spin states so that conscious experience emerge. We have argued that contextual, irreversible and non-computable means are utilized by brain to do this (Hu & Wu, 2002, 2004a). Furthermore, we recognize that there may not be any large-scale quantum coherence in the warm and wet brain to support those quantum theories of consciousness that require macroscopic quantum effects. However, our theory does not depend on such a coherence to work in the dualistic approach (Hu & Wu, 2002, 2004a).

## Appendix A — What Is Quantum Spin?

Quantum spin refers to angular momentum of a particle (e.g., electron or proton) due to self-rotation or internal motion (see e.g., Tomonaga, 1997). Although unexplained at the time of its discovery, space quantization (quantization of orientation) due to quantum spin was first indirectly observed by Zeeman (1897) (known as anomalous Zeeman Effect), and then directly observed in the Stern–Gerlach experiment in 1922.

The two-valued quantum degree of freedom was introduced by the Nobel Laureate Wolfgang Pauli when he was formulating his exclusion principle. The non-relativistic quantum equation of a spin ½ particle (e.g., an electron) was formulated by Pauli (1927), which can be written in SI units as:

$$i\hbar\partial_t \begin{pmatrix} \psi_1(\mathbf{x},t) \\ \psi_2(\mathbf{x},t) \end{pmatrix} = \left( \frac{1}{2m}(\mathbf{p}-e\mathbf{A})^2 - \frac{e\hbar}{2m}\boldsymbol{\sigma}\cdot\mathbf{B} + e\phi \right) \begin{pmatrix} \psi_1(\mathbf{x},t) \\ \psi_2(\mathbf{x},t) \end{pmatrix} \tag{A.1}$$

where $\psi = \begin{pmatrix} \psi_1(\mathbf{x},t) \\ \psi_2(\mathbf{x},t) \end{pmatrix}$ is the Pauli 2-spinor wave function, $\mathbf{p}$ is momentum operator, $A^\mu = (A, \phi)$ is the electromagnetic four-potential, $B = (\nabla \times A)$ is the magnetic field, and $\boldsymbol{\sigma} = (\sigma_x, \sigma_y, \sigma_z) = (\sigma_1, \sigma_2, \sigma_3)$ are Pauli matrices representing spin ½ operators:

$$\sigma_1 = \begin{pmatrix} 0 & 1 \\ 1 & 0 \end{pmatrix} \quad \sigma_2 = \begin{pmatrix} 0 & -i \\ i & 0 \end{pmatrix} \quad \sigma_3 = \begin{pmatrix} 1 & 0 \\ 0 & -1 \end{pmatrix} \tag{A.2}$$

Thus, the Hamiltonian term $-\frac{e\hbar}{2m}\boldsymbol{\sigma}\cdot\mathbf{B}$ in the above Pauli equation together with the Pauli bi-spinor wave function (quantum state) explains the space quantization observed in the anomalous Zeeman Effect (Zeeman, 1897) and Stern–Gerlach experiment (Stern & Gerlach, 1922).

It turns out that spin ½ operator $\boldsymbol{\sigma}$, thus, quantum spin ½ of a fermion, solely originates from the relativistic quantum mechanics of fermions, that is, the fundamental equation in modern physics, known as the Dirac equation, discovered by Dirac (1928).

$$i\hbar\partial_t\psi(\mathrm{x},t) = \left(c(\alpha_1 p_1 + \alpha_2 p_2 + \alpha_3 p_3) + \beta mc^2\right)\psi(\mathrm{x},t) \tag{A.3}$$

where $\boldsymbol{\alpha} = (\alpha_1, \alpha_2, \alpha_3)$ and $\beta$ are Dirac matrices and $H = \boldsymbol{\alpha} \bullet c\mathbf{p} + \beta mc^2$ is the Dirac Hamiltonian, and

$$\psi(\mathbf{x},t) = \begin{pmatrix} \psi_1(\mathbf{x},t) \\ \psi_2(\mathbf{x},t) \\ \psi_3(\mathbf{x},t) \\ \psi_4(\mathbf{x},t) \end{pmatrix} \tag{A.4}$$

is the Dirac four-spinor wave function.

Dirac equation (A.3) can be recast in the following form (Hu & Wu, 2010):

$$\begin{pmatrix} E-mc^2 & -\boldsymbol{\sigma}\cdot c\mathbf{p} \\ -\boldsymbol{\sigma}\cdot c\mathbf{p} & E+mc^2 \end{pmatrix}\begin{pmatrix} \psi_+(\mathbf{x},t) \\ \psi_-(\mathbf{x},t) \end{pmatrix} = 0 \text{ or}$$
$$\begin{pmatrix} i\hbar\partial_t - mc^2 & i\hbar c\boldsymbol{\sigma}\cdot\nabla \\ i\hbar c\boldsymbol{\sigma}\cdot\nabla & i\hbar\partial_t + mc^2 \end{pmatrix}\begin{pmatrix} \psi_+(\mathbf{x},t) \\ \psi_-(\mathbf{x},t) \end{pmatrix} = 0 \tag{A.5}$$

where $\psi_+(\mathbf{x},t) = \begin{pmatrix} \psi_1(\mathbf{x},t) \\ \psi_2(\mathbf{x},t) \end{pmatrix}$ and $\psi_-(\mathbf{x},t) = \begin{pmatrix} \psi_3(\mathbf{x},t) \\ \psi_4(\mathbf{x},t) \end{pmatrix}$. With the Gordon–Klein equation for a scalar (spin-less) particle as the starting point, Eq. (A.5) can be derived from the following exercise (Hu & Wu, 2010):

First, the Gordon–Klein equation

$$\left(E^2 - m^2c^4 - \mathbf{p}^2c^2\right)\phi = 0 \tag{A.6}$$

can be recast as

$$\left(Det\begin{pmatrix} E & 0 \\ 0 & E \end{pmatrix} + Det\begin{pmatrix} -mc^2 & 0 \\ 0 & mc^2 \end{pmatrix} + Det\begin{pmatrix} 0 & -c|\mathbf{p}| \\ -c|\mathbf{p}| & 0 \end{pmatrix}\right)\phi = 0 \tag{A.7}$$

Now, if we remove the determinant sign *Det*, we may write:

$$\begin{pmatrix} E & 0 \\ 0 & E \end{pmatrix} + \begin{pmatrix} -mc^2 & 0 \\ 0 & mc^2 \end{pmatrix} + \begin{pmatrix} 0 & -c|\mathbf{p}| \\ -c|\mathbf{p}| & 0 \end{pmatrix} \begin{pmatrix} \phi_+ \\ \phi_- \end{pmatrix}$$
$$= \begin{pmatrix} E-mc^2 & -c|\mathbf{p}| \\ -c|\mathbf{p}| & E+mc^2 \end{pmatrix} \begin{pmatrix} \phi_+ \\ \phi_- \end{pmatrix} = 0 \tag{A.8}$$

Thus, after the substitution

$$|\mathbf{p}| = \sqrt{\mathbf{p}^2} = \sqrt{-\mathrm{Det}(\boldsymbol{\sigma}\cdot\mathbf{p})} \rightarrow \boldsymbol{\sigma}\cdot\mathbf{p} \tag{A.9}$$

we arrive at the Dirac equation (A.5). From the above exercise, we see that

$$|\mathbf{p}| \rightarrow \boldsymbol{\sigma}\cdot\mathbf{p} = -i\hbar\boldsymbol{\sigma}\cdot\nabla = -i\hbar\left(\sigma_x\partial_x + \sigma_y\partial_y + \sigma_z\partial_z\right) \tag{A.10}$$

gives rise to quantum spin ½ operator in relativistic quantum mechanics of fermion and, further, since in special theory of relativity

$$s^2 = (ct)^2 - \mathbf{x}^2 = \mathrm{Det}(I_2 ct) + \mathrm{Det}(\boldsymbol{\sigma}\cdot\mathbf{x}) \tag{A.11}$$

we see that quantum spin ½ operator $\boldsymbol{\sigma}$ is associated with the space–time structure in theory of relativity.

In an external electromagnetic field $A^\mu = (\phi, \mathbf{A})$, the Dirac equation becomes:

$$\begin{pmatrix} E+e\phi-mc^2 & -\boldsymbol{\sigma}\cdot c(\mathbf{p}-e\mathbf{A}) \\ -\boldsymbol{\sigma}\cdot c(\mathbf{p}-e\mathbf{A}) & E+e\phi+mc^2 \end{pmatrix} \begin{pmatrix} \psi_+(\mathbf{x},t) \\ \psi_-(\mathbf{x},t) \end{pmatrix} = 0 \tag{A.12}$$

or

$$\begin{pmatrix} i\hbar\partial_t - e\phi - mc^2 & i\hbar c\boldsymbol{\sigma}\cdot\left(\nabla - \frac{ie\mathbf{A}}{\hbar}\right) \\ i\hbar c\boldsymbol{\sigma}\cdot\left(\nabla - \frac{ie\mathbf{A}}{\hbar}\right) & i\hbar\partial_t - e\phi + mc^2 \end{pmatrix} \begin{pmatrix} \psi_+(\mathbf{x},t) \\ \psi_-(\mathbf{x},t) \end{pmatrix} = 0 \tag{A.13}$$

It can be shown that the Pauli Equation (A.1) is the non-relativistic approximation of the above Dirac equation (A.13) in the electromagnetic field $A^\mu = (\phi, \mathbf{A})$.

The action potentials in the brain are driven by voltage-gated ion channels on neural membranes as embodied by the well-known Hodgkin & Huxley (1952) model.

$$\partial_t V_m = -\frac{1}{C_m}\left(\sum_i (V_m - E_i) g_i\right), \tag{A.14}$$

where $V_m$ is the electric potential across the neural membranes, $C_m$ is the capacitance of the membranes, $g_i$ is the $i$th voltage-gated or constant-leak ion channel .

The overall effect of the action potentials and other surrounding factors, especially the magnetic dipoles carried by oxygen molecules due to their two unpaired electrons, is that inside the neural membranes and proteins, there exist varying strong electric field $E$ and fluctuating magnetic field $B$ that are governed by the Maxwell equation:

$$\begin{pmatrix} \partial_t \mathbf{E} = \nabla \times \mathbf{B} - \mathbf{j} \\ \partial_t \mathbf{B} = -\nabla \times \mathbf{E} \\ \nabla \cdot \mathbf{E} = \rho \\ \nabla \cdot \mathbf{B} = 0 \end{pmatrix} \quad \text{or} \quad \begin{pmatrix} E & -\boldsymbol{\sigma}\cdot\mathbf{p} \\ -\boldsymbol{\sigma}\cdot\mathbf{p} & E \end{pmatrix} \begin{pmatrix} \boldsymbol{\sigma}\cdot\mathbf{E} \\ i\boldsymbol{\sigma}\cdot\mathbf{B} \end{pmatrix} = \begin{pmatrix} -i\boldsymbol{\sigma}\cdot\mathbf{j} \\ -i\rho \end{pmatrix}, \tag{A.15}$$

where $\rho$ is electric density and $j$ is electric current inside the neural membranes. For simplicity, we have not considered the medium effect of the membranes, that is, we have treated the membranes as a vacuum.

In turn, electromagnetic fields **E** and **B** interact with a particle with spin ½ and charge $e$ (e.g., an electron or proton) inside neural membranes or proteins through electromagnetic potential representation $A^\mu = (\phi, \mathbf{A})$ as follows (assuming the particle is otherwise free for simplicity):

$$\begin{pmatrix} E - e\phi - m & -\boldsymbol{\sigma}\cdot(\mathbf{p} - e\mathbf{A}) \\ -\boldsymbol{\sigma}\cdot(\mathbf{p} - e\mathbf{A}) & E - e\phi + m \end{pmatrix} \begin{pmatrix} \psi_+ \\ \psi_- \end{pmatrix} = 0 \tag{A.16}$$

In the above equation, the interactions of **B** and/or **E** with proton and/or electron spin operator $\boldsymbol{\sigma}$ are hidden. These two couplings are explicitly shown during the process of non-relativistic approximation of the Dirac equation.

In eqns (A.15) and (A.16), natural units $c = 1$ and $\hbar = 1$ are used.

## Acknowledgment

The authors would like to thank Prof. Roman R. Poznanski for his suggestions during the preparation of this chapter.

## References

Aerts, D., Broekaert, J. & Gabora, L. (2000) Intrinsic contextuality as the crux of consciousness. In, K. Yasue, ed. *Fundamental Approaches to Consciousness*, John Benjamins Publishing Company.

Akinlaja, J. & Sachs, F. (1998) The breakdown of cell membranes by electrical and mechanical stress. *Biophys. J.*, **75**, 247–254.

Arnesen M. C., Bose, S. & Vedral, V. (2001) Natural thermal and magnetic entanglement in the 1D Heisenberg model. *Phys. Rev. Lett.*, **87**, 017901.

Aspect, A., Dalibard, J. & Roger, G. (1982) Experimental test of Bell's inequalities using time-varying analyzers. *Phys. Rev. Lett.*, **49**, 1804–1806.

Baez, J. C. (1984) Spin foam models. *Class. Quant. Grav.*, **15**, 1827–1858.

Barnet, A. & Weaver, J. C. (1991) Electroporation: a unified, quantitative theory of reversible electrical breakdown and mechanical rupture in artificial planar bilayer membranes. *Bioelectrochem. Bioenerg.*, **25**, 163–182.

Barut, A. O. & Zanghi, N. (1984) Classical model of the Dirac electron. *Phys. Rev. Lett.*, **52**, 2009–2012.

Beck, F. & Eccles, J. C. (1992) Quantum aspects of brain activity and the role of consciousness. *Proc. Natl. Acad. Sci. USA*, **89**, 11357–11361.

Bezrukov, S. M. & Vodyanoy, I. (1995) Noise-induced enhancement of signal transduction across voltage-dependent ion channels. *Nature*, **378**, 362–364.

Bogan, J. R. (2002) Spin: The classical to quantum connection. Available at http://www.arxiv.org/pdf/quant-ph/0212110.

Bohm, D. (1952) A Suggested Interpretation of the Quantum Theory in Terms of 'Hidden Variables' I. *Phys. Rev.*, **85**, 166–179.

Bohm D. & Hiley B. J. (1984) Generalisation of the twistor to Clifford algebras as a basis for geometry. *Revista Brasilera de Fisica*, **70**, 1–26.

Bryan-Brown, G. P. Brown, C. V., Sage, I. C & Hui, V. C. (1998) Voltage-dependent anchoring of a nematic liquid crystal on a grating surface. *Nature*, **392**, 365–367.

Buckingham, A. D. (1960) Chemical shifts in the nuclear magnetic resonance spectra of molecules containing polar groups. *Can. J. Chem.*, **38**, 300–307.

Budinich, P. (2002) From the Geometry of Pure Spinors with their Division Algebras to Fermion's Physics. *Found. Phys.* **32**, 1347–1398.

Castagnoli, G. & Finkelstein, D. R. (2001) Quantum statistical computation. http://www.arxiv.org/pdf/hep-th/0111120 .

Cahill, R. T. (2002) Process physics: from quantum foam to general relativity. Available at http://www.arxiv.org/pdf/gr-qc/0203015 .

Chalmers, D. (1996) *The Conscious Mind*. Oxford: Oxford University Press.

Chicurei, M. (2002) Magnetic mind games. *Nature*, **417**, 114–116.

Clarke, C. (2004) Quantum mechanics, consciousness and the self. In: D. Lorimer, ed. *Science, Consciousness and Ultimate Reality*. Exeter: Imprint Academic.

Crick, F. (1994) *The Astonishing Hypothesis*. New York: Simon & Schuster.

Dirac, P. A. M. (1928) The quantum theory of the electron. *Proc. R. Soc. A*, **117**, 610–624.

Dirac, P. A. M. (1983) *The Principles of Quantum Mechanics*, 4th edn. Oxford: Clarendon Press.

Donald, M. J. (1990) Quantum theory and the brain. *Proc. R. Soc. A*, **427**, 43–93.

Eberhard, P. (1978) Bell's theorem and the different concepts of locality. *Nuovo Cimento B*, **46**, 392–419.

Edelman, G. M. (1989) *The Remembered Present: A Biological Theory of Consciousness*. New York: Basic Books.

Einstein, A., Podolsky, B. & Rosen, N. (1935) Can quantum-mechanical description of physical reality be considered complete? *Phys. Rev.*, **47**, 777–780.

Esposito, S. (1999) On the role of spin in quantum mechanics. *Found. Phys. Lett.*, **12**, 165–177.

Feng, *et al.* (2001) Theoretical expectations for the muon's electric dipole moment. *Nucl. Phys. B*, **613**, 366–381.

Freeman, A. (2001) *The Emergence of Consciousness*. Imprint Academic.

Fu, Y. X. *et al.* (2002) Temporal specificity in the cortical plasticity of visual space representation. *Science*, **296**, 1999–2003.

Galiautdinov A. A. (2002) Quantum theory of elementary process. *Inter J. Theor. Phys.* **41**, 1423–1480.

Gerlach, W. & Stern, O. (1922) Der experimentelle Nachweis der Richtung squantelung im Magnetfeld. *Zeitschrift für Physik*, **9**, 349–352.

Gershenfeld, N. & Chuang, I. L. (1997) Bulk spin resonance quantum computation. *Science*, **275**, 350–356.

Ghosh, S. *et al.* (2003) Entangled quantum state of magnetic dipoles. *Nature*, **425**, 48–51.

Grayson, M. (2003) Electric field effects on 2JHH spin–spin coupling constants. *Int. J. Mol. Sci.*, **4**, 218–230.

Hagan, S., Hameroff, S. R. & Tuszynski, J. A. (2002) Quantum computation in brain microtubules: Decoherence and biological feasibility. *Phys. Rev. E.*, **65**, 061901.

Hameroff, S. & Penrose, R. (1996) Conscious events as orchestrated spacetime selections. *J. Conscious. Stud.*, **3**, 36–53.

Hayashi, H. (2002) Advent of spin chemistry. *RIKEN Rev.*, **44**, 7–10.

Hestenes, D. (1975) Observables, operators, and complex numbers in the Dirac theory. *J. Math. Phys.*, **16**, 556–572.

Hestenes, D. (1983) Quantum mechanics from self-interaction. *Found. Phys.*, **15**, 63–87.

Hofstadter, D. R. (1979) *Gödel, Escher, Bach*. New York: Basic Books.

Hu, H. & Wu, M. (2001) Mechanism of Anesthetic Action: Oxygen pathway perturbation hypothesis. *Med. Hypotheses*, **57**, 619–627.

Hu, H. & Wu, M. (2002) Spin-mediated consciousness theory: Possible roles of oxygen unpaired electronic spins and neural membrane nuclear spin ensemble in memory and consciousness. Available at http://arxiv.org/abs/quant-ph/0208068v1.

Hu, H. & Wu, M. (2004a) Spin-mediated consciousness theory: Possible roles of neural membrane nuclear spin ensembles and paramagnetic oxygen. *Med. Hypotheses*, **63**, 633–646.

Hu, H. & Wu, M. (2004b) Spin as primordial self-referential process driving quantum mechanics, spacetime dynamics and consciousness. *NeuroQuantology*, **2**, 41–49.

Hu, H. & Wu, M. (2004c) Action potential modulation of neural spin networks suggests possible role of spin. *NeuroQuantology*, **2**, 309–317.

Hu, H. & Wu, M. (2004d) Possible roles of neural electron spin networks in memory and consciousness. Available at http://cogprints.org/3544/.

Hu, H. & Wu, M. (2006a) Thinking outside the box: The essence and implications of quantum entanglement and the story of spin-mediated consciousness theory. *NeuroQuantology*, **4**, 5–16.

Hu, H. & Wu, M. (2006b) Photon induced non-local effects of general anesthetics on the brain. *NeuroQuantology*, **4**, 17–31.

Hu, H. & Wu, M. (2006c) Evidence of non-local physical, chemical and biological effects support quantum brain. *NeuroQuantology*, **4**, 291–306.

Hu, H. & Wu, M. (2008) Concerning spin as mind-pixel: How mind interacts with the brain through electric spin effects. *NeuroQuantology*, **6**, 26–31.

Hu, H. & Wu, M. (2010) Prespacetime model of elementary particles, four forces & consciousness. *Prespacetime J.*, **1**, 77–146.

Hunt, S. P. & Mantyh, P. W. (2001) The Molecular dynamics of pain control. *Nature Rev. Neurosci.*, **2**, 83–91.

Gauthier R. (2006) The Dirac equation and the superluminal electron model. http://superluminalquantum.org.

Graesser, M. & Thomas, S. (2002) Supersymmetric relations among electromagnetic dipole operators. *Phys. Rev. D*, **65**, 075012.

Hodgkin, A. L. & Huxley, A. F. (1952) A quantitative description of membrane current and its application to conduction and excitation in nerve. *J. Physiol.*, **177**, 500–544.

Josephson, B. D. & Pallikari-Viras, F. (1991) Biological utilisation of quantum nonlocality. *Found. Phys.*, **21**, 197–207.

Julsgaard, B., Kozhekin, A. & Polzik, E. S. (2001) Experimentally long-lived entanglement of two macroscopic objects. *Nature*, **413**, 400–403.

Khitrin, A. K. & Ermakov, V. L. (2002a) Spin Processor. Available at http://www.arxiv.org/pdf/quant-ph/0205040.

Khitrin, A. K., Ermakov, V. L. & Fung, B. M. (2002b) Information storage using a cluster of dipolar-coupled spins. *Chem. Phys. Lett.*, **360**, 161–166.

Khitrin, A. K., Ermakov, V. L. & Fung, B. M. (2002c) NMR molecular photography. *J. Chem. Phys.*, **117**, 6903–6906.

Khrennikov, A. (2001) 'Quantum probabilities' as context depending probabilities. Available at http://arxiv.org/abs/quant-ph/0106073 .

Kun, S. Y. Benet, L., Chadderton, L. T., Greiner, W. & Haas, F. (2003) Macroscopic quantum superpositions in highly-excited strongly-interacting many-body system. *Phys. Rev. C* **67**, 011604R.

Kushner, D. J., Baker, A. & Dunstall, T. G. (1999) Pharmacological uses and perspectives of heavy water and denatured compound. *Can. J. Physiol. Pharmacol.*, **77**, 79–88.

LeDoux, J. (2002) *Synaptic Self.* New York: Viking — a member of Penguin Putnam Inc.

Lidar, D. A. & Whaley, K. B. (2003) Decoherence-free subspaces and subsystems. In: F. Benatti and R. Floreanini, eds. *Irreversible Quantum Dynamics.* 83–120. Berlin: Springer Lecture Notes in Physics, Vol. 622.

Limoniemi, R. J. (1997) Neuronal responses to magnetic stimulation reveal cortical reactivity and connectivity. *NeuroReport*, **8**, 3537–3540.

Marder, E. Abbott, L. F, Turrigiano, G. G., Liu, Z. & Golowasch, J. (1996) Memory from the dynamics of intrinsic membrane currents. *Proc. Natl. Acad. Sci. USA*, **93**, 13481–13486.

Marino, A. A. (1983) Environmental electromagnetic fields and public health. In: A. Marino, eds. *Foundations of Modern Bioelectricity.* New York: Marcel Dekker.

Marsh, D. (2001) Polarity and permeation profiles in lipid membranes. *Proc. Natl. Acad. Sci. USA*, **98**, 7777–7782.

Mennerick, S. (1998) Effect of nitrous oxide on excitatory and inhibitory synaptic transmission in Hippocampal cultures. *J. Neurosci.*, **18**, 9716–9726.

Minaev, B. F. (1996) Intermolecular exchange in the system O2 + H2 as a model of spin-catalysis in radical recombination reaction. *Theor. Exp. Chem.*, **32**, 229–233.

Morais-Cabral, J. H., Zhou, Y. & MacKinnon, R. (2001) Energy optimisation of ion conduction rate by the K selectivity filter. *Nature*, **414**, 37–42.

Morris, P. G. (1986) *Nuclear Magnetic Resonance Imaging in Medicine and Biology.* Oxford: Clarendon Press.

Nagakura, S., Hayashi, H. & Azumi, T. (1998) *Dynamic Spin Chemistry.* New York: Wiley.

Newman, T. E. (2002) On a classical, geometric origin of magnetic moments, spin-angular momentum and the Dirac gyromagnetic ratio. *Phys. Rev. D*, **65**, 104005.

Pauli, W. (1927) Zur quantenmechanik des magnetischen electrons. (in German)*Z. Phys.*, **43**, 601–623.

Penrose, R. (1960) A spinor approach to general relativity. *Ann. Phys.*, **10**, 171–201.

Penrose, R. (1967) Twistor algebra. *J. Math. Phys.*, **8**, 345–366.

Penrose, R. (1989) *The Emperor's New Mind*. Oxford: Oxford University Press.

Penrose, R. (1994) *Shadows of the Mind*. Oxford: Oxford University Press.

Persinger, M. A., Dotta, B. T., Saroka, K. S. & Scott, M. A. (2013) Congruence of energies for cerebral photon emissions, quantitative EEG activities and –5nT changes in the geomagnetic field support spin-based hypothesis of consciousness. *Journal of Consciousness Research & Exploration*, **4**, 1–24.

Peshkovsky, A. & McDermott, A. E. (2000) Dipolar interactions in molecules aligned by strong AC electric fields. *J. Magn. Reson.*, **147**, 104–109.

Philippides, A., Husbands, P. & O'Shea, M. (2000) Four-dimensional neural signaling by nitric oxide: A computer analysis. *J. Neurosci.*, **20**, 1199–1207.

Poznanski, R. R. (2001) Biophysical Neural Networks: Foundations of Integrative Neuroscience. In: R. R. Poznanski, ed. *Dendritic Spike-like Potentials in a Neural Network*. New York: Mary Ann Liebert.

Prosser, R. S. Luchette, P. A., Westerman, P. W., Rozek, A. & Hancock, R. E. W. (2001) Determination of membrane immersion depth with O2: A high-pressure 19F NMR study. *Biophys. J.*, **80**, 1406–1416.

Radin, D. (2004) Entangled minds. *Shift*, **5**, 10–14.

Raffy, S. & Teissie, J. (1999) Control of membrane stability by cholesterol content. *Biophys. J.*, **76**, 2072–2080.

Recami, E. & Salesi, G. (1998) Kinematics and hydrodynamics of spinning particles. *Phys. Rev. A*, **57**, 98–105.

Richter, C. P. (1976) A study of taste and smell of heavy water (99.8%) in rats. *Proc. Soc. Exp. Biol. Med.*, **152**, 677–684.

Richter, C. P. (1977) Heavy water as a tool for study of the forces that control length of period of the 24-hour clock of the hamster. *Proc. Natl. Acad. Sci. USA*, **74**, 1295–1299.

Rivas, M. (2002) *Kinematical Formalism of Elementary Spinning Particles*. New York, Springer.

Rosser, W. G. V. (1964) *An Introduction to the Theory of Relativity*. London: Butterworths Press.

Sargent, D. F. (1975) Voltage jump/capacitance relaxation studies of bilayer structure and dynamics. *J. Membr. Biol.*, **23**, 227–247.

Saux, A. L., Ruysschaert, J. M. & Goormaghtigh, E. (2001) Membrane molecule reorientation in an electric field recorded by attenuated total reflection Fourier-transform infrared spectroscopy. *Biophys. J.*, **80**, 324–330.

Saeedi, K., Simmons, K, Salvail, J. Z, Dluhy, P., Riemann, H., Abrosimov, N. V., Becker, P. Pohl, H. J., Morton, J. J. L. & Thewalt, M. L. W. (2013) Room-temperature quantum bit storage exceeding 39 minutes using ionized donors in silicon-28. *Science*, **342**, 830–833.

Searle, J. (1992) *The Rediscovery of the Mind*. Cambridge, MA: MIT Press.

Schrödinger, E. (1930) Über die kräftefreie Bewegung in der relativistischen Quantenmechanik ('On the free movement in relativistic quantum mechanics'), Berliner Ber. pp. 418–428.

Schrödinger, E. (1935) Discussion of probability relations between separated systems. *Proc. Cambridge Philos. Soc.*, **31**, 555–563.

Sens, P. & Isambert, H. (2002) Undulation instability of lipid membranes under electric field. *Phys. Rev. Lett.* **88**, 128102.

Shellock, F. G. (2002) Magnetic Resonance Safety Update 2002: Implants and devices. *J. Magn. Resonan. Imaging*, **16**, 485–496.

Shen, S. Q. (2005) Spin transverse force on spin current in an electric field. *Phys. Rev. Lett.*, **95**, 187–203.

Sidharth, B. G. (2001a) Issues and ramifications in quantized fractal space-time: an interface with quantum superstrings. *Chaos Sol. Fract.*, **12**, 1449–1457.

Sidharth, B. G. (2001b) *Chaotic Universe*. New York: Nova Science.

Simonotto, E. Riani, M., Seife, C., Roberts, M., Twitty, J. & Moss, F. (1997) Visual perception of stochastic resonance. *Phys. Rev. Lett.*, **78**, 1186–1189.

Silenko, A. J. (2005) Quantum-mechanical description of the electromagnetic interaction of relativistic particles with electric and magnetic dipole moments. *Russ. Phys. J.*, **48**, 788–792.

Smolin, L. (2001) *Roads to Quantum Gravity*. New York: Basic Books.

Smondyrev, A. M. & Berkowitz, M. L. (1999) Structure of Dipalmitoyl phosphatidylcholine-cholesterol bilayer at law and high cholesterol concentrations: molecular dynamics simulation. *Biophys. J.*, **77**, 2075–2089.

Stapp, H. E. (1982) Mind, matter and quantum mechanics. *Found. Phys.*, **12**, 363–399.

Stapp, H. P. (1993) *Mind, Matter and Quantum Mechanics*. New York: Springer-Verlag.

Sun, Q. F. Guo, H. & Wang, J. (2004) Spin-current-induced electric field. *Phys. Rev. B*, **69**, 054409.

Tegmark, M. (2000) The importance of quantum decoherence in brain processes. *Phys. Rev. E*, **61**, 4194–4206.

Tomonaga, S. (1997) *The Story of Spin* — Translated by T. Oka. Chicago: The University Press of Chicago.

Turanov, A. & Khitrin, A. K. (2013) Long-lived states of nuclear spins in solids. *J. Chem. Phys.* **139**, 091102.

Uhlenbeck, G. E. & Goudsmit, S. (1925) Ersetzung der Hypothese vom ... jeden einzelnen Elektrons. *Naturwissenschaften*, **13**, 953–954.

Vedral, V. (2011) Living in a quantum world. *Sci. Am.*, **304**, 38–43.

Viola, L. *et al.* (2000) Stochastic resonance and nonlinear response using NMR spectroscopy. *Phys. Rev. Lett.*, **84**, 5466–5469.

Walker, E. H. (1974) Consciousness and quantum theory. In: J. White, ed. *Psychic Exploration*. New York: Putnam, pp. 544–568.

Walker, E. H. (2000) *The Physics of Consciousness*. Cambridge, MA: Perseus Publishing.

Walsh, V. & Cowey, A. (2000) Transcranial magnetic stimulation and cognitive neuroscience. *Nature Rev. Neurosci.*, **1**, 73–79.

Warren, W. S. *et al.* (1993) Generation of impossible correlation peaks between bulk water and biomolecules in solution NMR. *Science*, **262**, 2005–2009.

Warren, W. S. *et al.* (1998) MR imaging contrast based on intermolecular zero-quantum coherence. *Science*, **281**, 274–250.

Wertz, J. E. & Bolton J. R. (1972) *Electron Spin Resonance: Elementary Theory and Practical Application*. New York: McGraw-Hill Book Company.

Wikswo, J. P. (1990) Biomagnetic sources and their models. In: S. J. Williamson *et al.* eds. *Advances in Biomagnetism*. New York: Plenum.

Woolf, T. B. & Roux, B. (1996) Structure, energetics, and dynamics of lipid-protein interactions: A molecular dynamics study of the gramicidin a channel in a DMPC bilayer. *Proteins: Struct. Funct. Gen.*, **24**, 92–114.

Zeeman, P. (1897) Doubles and triplets in the spectrum produced by external magnetic forces. *Philos. Mag.*, **44**, 55–60.

Biophysics of Consciousness: A Foundational Approach
R. R. Poznanski, J. A. Tuszynski and T. E. Feinberg
Copyright © 2017 World Scientific, Singapore.

# 13

# THE "QUANTUM UNDERGROUND": WHERE LIFE AND CONSCIOUSNESS ORIGINATE

*Travis J. A. Craddock*[*,†], *Stuart R. Hameroff*[‡] *and Jack A. Tuszynski*[§,‖]

*[*]College of Psychology, College of Engineering and Computing, and College of Osteopathic Medicine, Nova Southeastern University, Fort Lauderdale, Florida, USA*

*[†]Clinical Systems Biology Group, Institute for Neuro-Immune Medicine Nova Southeastern University, Fort Lauderdale, Florida, USA*

*[‡]Departments of Anesthesiology and Psychology, Center for Consciousness Studies, Banner-University Medical Center, The University of Arizona, Tucson, Arizona*

*[§]Department of Physics, University of Alberta, Edmonton, Alberta, Canada*

*[‖]Department of Experimental Oncology, Cross Cancer Institute, Edmonton, Alberta, Canada*

## Abstract

As our knowledge of reductionist details of living systems continues to grow, the gap in understanding life and consciousness remains wide. Progress has been made, e.g., biological organisms are seen as complex hierarchical amalgamations of elements interacting in self-similar, "scale-invariant" patterns within and across spatio-temporal scales ("$1/f$"). But the mechanism for trans-scalar communication is unknown, as are the origin and foundation, i.e., the "bottom floor" of scale-invariant systems in biology. Here, we describe scale-invariant hierarchies in brain and living organisms in general, originating in a biomolecular "quantum underground" pervading neurons, glia and all living cells, most specifically within cytoskeletal microtubules. The quantum underground is a non-polar solubility phase composed largely of π-electron resonance clouds of aromatic amino acids, similar to pi-resonance arrays mediating quantum coherence in photosynthesis proteins. In the brain, the quantum underground is identified as the origin of consciousness by the Meyer–Overton correlation, showing where anesthetics act to erase consciousness while sparing non-conscious brain activities. Evidence points to anesthetics acting to dampen quantum dipole oscillations in the "Meyer–Overton quantum underground" within brain neuronal microtubules. These quantum dipole oscillations are seen as the "inward apex," the origin of scale-invariant processes in consciousness and life.

*Keywords*: Consciousness; emergence; life; scale-invariance; $1/f$; tubulin; microtubules; quantum coherence; tryptophan; anesthesia; Meyer–Overton; hierarchy.

## 13.1. Reductionism: The "Road to Nowhere"

Once considered isolated entities, biomolecular components of living systems are now regarded as being embedded in scale-free communicative networks. From gene regulation and intracellular signaling, to coordinated patterns of neuronal activity, hierarchically connected systems lead to purposeful behavior, intelligence and consciousness. Systems biology, the rapidly emerging discipline positioned as a translational catalyst at the interface of the life, physical, and computational sciences, attempts to integrate these various networks into multi-scale interlocking hierarchies with information and instruction flowing in both directions among spatio-temporal scales. Yet, while our knowledge of these components and their behaviors grows at a dizzying pace, the nature and origin of trans-scalar interactions (Pribram, 1991) leading to purposeful life and consciousness remains unexplained, as does the origin, the "bottom floor" of life's hierarchical systems.

In the biological sciences, the rise of "omic" research (genomic, proteomic, transcriptomic, metabolomic, etc.) has led to a rapid increase in our ability to collect and store large data sets involving comprehensive snapshots of biological elements and their interconnections. Indeed more data can now be collected regarding a single constitutive element in a year, than has been gathered over the course of previous scientific history (Chuang *et al.*, 2010). Science, from its inception, has been largely based in reductionism: breaking entities into smaller and smaller elements, categorizing, labeling and cataloging. This has led to marvelous breakthroughs in medicine, technology, and the understanding of our world in general. It is these very advances in our ability to manipulate the external world, which gives us confidence that this age-old method of understanding the universe around us is correct. However, life is not composed of isolated elements. Rather, it appears that the nature of biological systems and processes is more properly explained within the framework of a holistic paradigm.

The behavior of biological elements depends on the context within which they occur, analogous to words in a novel or notes in a song. At its bare minimum, the field of systems biology attempts to integrate these elements and synthesize their interactions into a comprehensive picture in order to identify the underlying organizing principles of biological behavior (Broderick & Craddock, 2013; Ideker *et al.*, 2001). This is an enormous task as the challenge is to retain, organize and analyze this wealth of information to find meaningful and consistent patterns. A pile of books is not a library, nor does an organized library mean that all the books have been read. Consider this, for 100 genes simplistically described as either in an "on/active" or "off/inactive" state it would take $2^{100}$ bits of information to describe all the different possible functions afforded to this rudimentary genome. This is more than a trillion petabytes of information. This is an astronomical amount of memory within the current computer technology framework. Now consider also that the human genome contains roughly 35,000 genes, and the problem rapidly becomes computationally insurmountable easily exceeding the immense number of $10^{110}$, which represents a real computational barrier. Continuing along the same line of logic, consider further that the human brain contains roughly 100 billion ($10^{11}$) neurons (connected with up to 100,000 other neurons via synapses) that may again be very crudely simplified as either being in an "on" or an "off" state, and one can begin to understand the hopelessness in the attempt to model the entire brain using brute force approaches (Markram, 2006). The size, depth and breadth of this problem require not only the state-of-the-art computing resources, but especially some incredibly ingenious ways of synthesizing the data into a coherent picture describing brain function. Yet even if this is accomplished, will the resulting simulation explain consciousness? Will the simulation itself be conscious? We think not.

Currently, our understanding suggests that the organizing principles of life are conserved across all scales of biology. For example, biological networks such as protein–protein interaction networks and the metabolic networks for eukaryotes and bacteria, follow scale-free or small-world topology in which the number of highly interacting elements decreases according to a universal power law (Goh, 2002; Albert & Barabási, 2002). This unique, but universal pattern of interactivity defines network architecture, reaching from molecular interactions within a cell to social interactions between individuals (Barabási, 2009). Specifically, this topology results in a small number of highly influential governing elements within an overall relatively sparsely connected network. This interactivity pervades all of biology and it is from this very interactivity that function arises. A small number of interacting elements can thus result in numerous governing behaviors and complex modes of regulation. This is seen in gene regulation (Flintoft, 2010), the dynamics of immune cells (Almeida *et al.*, 2012), the overall immune response (Reynolds *et al.*, 2006), and cell fate selection (Hanel *et al.*, 2012). Ultimately, this means that overall system behavior cannot be fully understood in terms of isolated biological elements alone or even their pair-wise interactions, but rather must be realized in the context of the entire system.

Arguably, more complex and intractable still is the concept of consciousness (Greenfield & Collins, 2005). There have been numerous attempts to describe the nature of consciousness through spiritual, philosophical (Eastern and Western), and even scientific terms, however consciousness continues to be the most difficult and enigmatic problem for empirical science. The mechanisms and origins of this phenomenon have eluded scientific discourse, as even its definition remains problematic. At the dawn of the 20th century the American philosopher and psychologist William James popularized the concept of consciousness, however this popularity diminished with the rise of behaviorism, the measurable actions of people and animals, relegating the status of consciousness studies to a taboo subject. It was not until the late 20th century that consciousness returned to scientific discourse, yet some of this same bias and prejudice still remains today.

Currently, the standard accepted working definition of consciousness is the "condition of being aware of one's surroundings and one's own existence, or self-awareness" (Tuszynski & Woolf, 2006). From the neuroscientific viewpoint it is agreed that consciousness is a function of the brain (Koch, 2004; Walker, 1970), and it is therefore attributable to complex synaptic computation among brain cells. This current view in neuroscience states that consciousness is a highly non-linear, *emergent* property arising from neuronal features of the brain that are fully compatible with the laws of classical physics (Koch, 2004; Walker, 1970; Gazzaniga, 2004; Scott, 1995), and that when a critical level of complexity is reached, interacting neurons form a conscious experience.

However, let us examine this more closely. Classical physics envisions objects such as the brain, as an aggregate of logically independent, local, functional units, which only interact with their direct neighbors forming larger systems. Yet the question remains, how do spatially distributed brain activities bind together to produce the unity of conscious perception? Known broadly as the binding problem, and more specifically as the combination problem (Revonsuo, 1999), it seeks a description of how objects, their background, as well as abstract or emotional features are combined into a single coherent experience. The classical physics-based description explains this in terms of non-linear and deterministic chaotic behavior leading to non-computable results, stating that it is not easy to predict the property of consciousness. This interpretation leaves the combination problem largely unanswered. To seek insights that may lead to possible solutions of this problem we need to consider the structural and functional organization of the brain in more detail.

The brain is a fascinating and marvelous organ. It is approximately 1.5 kg of soft tissue that is composed of a collection of highly organized electrically excitable cells, mainly neurons and glia. The approximately 100 billion ($10^{11}$) neurons, comprising 50% of the brain, connect and communicate which each other via 1000 trillion ($10^{15}$) synaptic connections to control the central nervous system, regulating autonomic functions, cognition and behavior. Consciousness correlates strongly with coherent neural oscillations, however the mechanisms by which neurons synchronize are not fully understood. Thus, somewhere in this magnificent and seemingly deterministic clockwork, consciousness appears as a mere byproduct of the machine, an afterthought. Some researchers have gone so far as to argue that consciousness is a mere illusion, something with no influence on the world at large (Dennett, 1991; Wegner, 2002). However, this line of reasoning contributes nothing to our understanding of this truly remarkable phenomenon that everyone intrinsically comprehends without being able to precisely define, analyze or describe.

One attempt to reconcile these discrepancies is the Integrated Information Theory (IIT) of consciousness (Tononi, 2004; Oizumi *et al.*, 2014). IIT provides a mathematical characterization of the amount of information in a system when it transitions to a specific state through causal, deterministic interactions among its elements, above and beyond the information generated independently by its individual elements. Such a description captures key properties of consciousness, namely that there exists a large repertoire of conscious experiences so that, when one particular experience occurs, it generates a large amount of information by ruling out all others, and this information is integrated, in that each experience appears as a whole that cannot be decomposed into independent parts. IIT specifies which conditions must be satisfied by physical mechanisms, such as neurons and their connections, to account for the phenomenology of consciousness. However, for this uniform distribution of perturbations to be imposed at the macro-level, the probability of the underlying micro perturbations must be modified by averaging the micro-states that map into the same macro-state (Hoel *et al.*, 2013). That is to say, the mapping must be such that no individual micro-level information is available at the macro-level (the identity of the micro-elements grouped into a macro-element is lost). Currently, it is not clear if this is the case. If it is, it is a cry for the strong emergence of consciousness, if not then it is weakly emergent (Chalmers *et al.*, 2006). Here, we argue the latter is the case.

The integration and binding of information into a whole implies that classical mechanics has to become holistic or non-separable in some way. Averaging the effects of micro-level elements is well, good and fine for particles behaving randomly (like a gas in a box), but, as we have discussed above, this is not the case for highly organized complex biological systems such as the human brain. It has been shown that the dynamics of scale-free networks are significantly affected by the fine details of the network (i.e., both by the local neighborhood interactions and the network's topology), and are thus important in determining the synchronization properties of the system (DeVille & Peskin, 2012). On the other hand, quantum theory, the most accurate and thoroughly tested theory of physical systems in existence, is fundamentally a non-separable theory and circumvents the logical dilemma produced by a classical separable theory. In regards to IIT there is only one known physical condition where information can exist in a single state, namely by quantum superposition. If consciousness arises from the complexity of the collective action of 100's of millions of neurons acting cooperatively/synchronously, then the collective action of 100's of millions of sub-neural components (including for example those comprising the cytoskeleton) can only add to the richness and complexity of this process. Here, we argue that the underlying phenomena necessary for life and consciousness represent non-trivial quantum level effects amplified from the micro- to the macro-level, and therefore cannot be averaged away. This amplification process while not as mysterious as, say qualia, is a major challenge to the present-day neuroscience. For example, how does the brain perceive a flash of light that was caused only by a few photons at the threshold of visual perception? These photons struck receptors of the retina and eventually resulted in an action potential that traveled along an axon of the optic nerve to the visual perception area of the brain, but what was the mechanism of amplification of quantum events so they became classical effects at the macro-level?

Since Schrödinger asked "What is Life?" (Schrödinger, 1943) over 70 years ago, a possible link between the "weird" world of quantum phenomena and biology has been hypothesized but not yet conclusively demonstrated. It has often been argued that living systems are too "warm, wet and noisy" to support quantum effects, however recent cutting edge research suggests this is not necessarily always the case. From quantum coherence in photosynthesis and magneto-reception in bird navigation to quantum olfaction (Lambert *et al.*, 2013) and individual photon effects in vision (Fleming *et al.*, 2011), the field of quantum biology is leaping into the mainstream. This leads us to inquire if the brain also makes use of quantum effects and if so how? We believe that it does, and that it occurs in a "quantum underground" pervading the brain and all living systems.

## 13.2. The Neuroscience of Consciousness

At the level of measurable observations, consciousness is somehow related to synchronized activities of groups of neurons (Craddock *et al.*, 2014; Plankar *et al.*, 2013; Uhlhaas *et al.*, 2009; Tononi & Koch, 2008; Crick & Koch, 1990). Each neuron is considered an integrate-and-fire threshold device consisting of multiple dendrites and a cell body (soma), which receive and "integrate" synaptic inputs to a firing threshold, and a single axon along which "firings" occur. Synchronous activity as measured from scalp or cortical surface as electro-encephalography ("EEG") reflects local field potentials due to dendritic-somatic "integration." Implanted electrodes can detect firings along axons, including synchronized firings from groups of neurons. Synchronized axonal firings result from synchronized dendritic-somatic integration, e.g., in gamma synchrony EEG.

Higher cognitive functions including perception, attention, decision-making, learning, memory, and consciousness are associated with synchronized oscillations of large neuronal groups, known as neuronal

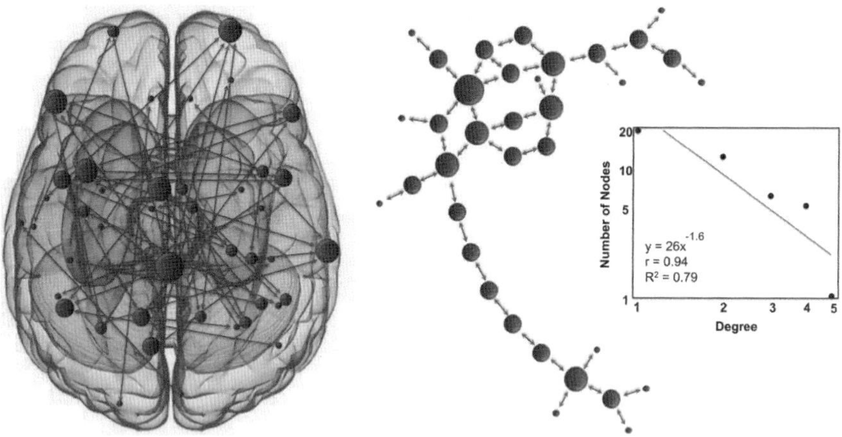


**Fig. 1.** "Small world network" representation of scale-free cortico–cortical coherence indicating synchronized oscillations. Nodes represent prescribed anatomical locations. Line segments represent reliable coherence between anatomical locations during a cognitive task. Node size represents node degree. Plotting number of nodes vs. node degree reveals a power law distribution, indicating that the network of synchronized oscillations possesses a scale-free topology. This image and caption are a modification of the original found in the Open Access publication (Bardouille & Boe, 2012) and used under the PLoS Creative Commons Attribution license.

synchrony, and reflecting mainly dendritic-somatic synchrony. Interruptions of these patterned oscillations are observed in various diseases of the brain such as epilepsy or Parkinson's disease (Uhlhaas & Singer, 2006; Spencer *et al.*, 2003; Uhlhaas & Singer, 2010; Hammond *et al.*, 2007). Upon the conscious perception of a stimulus, enhanced global synchronous oscillations in the gamma frequency range (~40 Hz) of the EEG occur in short bursts between occipital, parietal and frontal cortices across brain hemispheres (see Fig. 1). However, when the stimulus is not consciously perceived, no such pattern is observed (Senkowski *et al.*, 2008; Melloni, 2007). This large-scale coherent synchronization induces a cascade of cognitive processes related to perceptual binding and consciousness including the maintenance of working memory, and perceptual stabilization (Uhlhaas *et al.*, 2009). These coherent neuronal oscillations thus cannot be considered a mere byproduct of brain function, and are likely to have a specific functional role in the organization of signaling among neurons (Uhlhaas & Singer, 2010; Fell & Axmacher, 2011; Fries, 2009; Fries *et al.*, 2007; Jensen, 2007; Singer, 2009).

Yet, it is unclear how consciousness and cognition result from synchronized activities of brain cells. On their own, oscillating electrical patterns appear to serve as a temporal reference frame for neuronal signaling (Axmacher *et al.*, 2006). Rapid signaling events conducted via gap junctions coupled with postsynaptic potential inhibition can create coherence, or temporal gaps, allowing time for the generation of action potentials via excitatory neurons (Bartos *et al.*, 2007; Cardin, 2009). However, what remains unknown is how groups of neurons function to produce brief periods of synchronization, while rapidly switching between coherent and incoherent states that correlate with cognitive function (Melloni *et al.*, 2007; Breakspear *et al.*, 2004; Taylor *et al.*, 2005). How do these states relate to scale-invariant hierarchies?

Such neuronal excitation patterns may be regulated by background activity (Matej *et al.*, 2013). Bombardment of synaptic connections via noise in classical highly recurrent local and distant networks can cause increased levels of neuron activity causing a rapid neuron response to a variety of inputs (Matej *et al.*, 2013). The balancing of synaptic excitation and inhibition in neural network dynamics naturally generates a background of signals (Barbieri & Brunel, 2008; Compte *et al.*, 2003; Mattia & Del Giudice, 2004; Renart *et al.*, 2007). This fine balance is required to regulate the response to synchronized synaptic signals (Atallah & Scanziani, 2009; Haider & McCormick, 2009) indicating that rapid control of neuronal responsiveness via synaptic bombardment is a fundamental property of cortical dynamics (Haider & McCormick, 2009). Such oscillations are extremely responsive to phase perturbations and high-frequency oscillations (Wang *et al.*, 2010) indicating that without an underlying organization in this cacophony, specific information would be lost to synaptic noise. While most often this background is considered as "white" (random) noise, this may not be the case, as "colored" noise (Guo, 2011), i.e., patterns with frequency-based information, like $1/f$ "pink noise" can modulate synchronization patterns. Ongoing "background" activity is correlated over wide regions of the brain, and at different locations within single pyramidal neurons. Where can this organization be found? What colors the noise? One key aspect of the rapid transitions in cortical activity across various spatio-temporal scales is the scale-invariant cascade of neural activity measured in the cortex of higher-vertebrate animals (Van De Ville *et al.*, 2010; Petermann *et al.*, 2009; Stam & De Bruin, 2004).

## 13.3. Scale Invariant Hierarchies

Many views of how the brain produces consciousness are hierarchical, "scale invariant," layered systems with self-similar patterns repeating at different spatiotemporal scales (He *et al.*, 2010; Kitzbichler *et al.*, 2009; Van De Ville *et al.*, 2010; Eguiluz *et al.*, 2005). Various systems throughout nature exhibit this scale-invariant property, sometimes referred to as '$1/f$' (inverse frequency) "pink noise," "fractal-like," or "holographic." Here, $1/f$ implies the intensity and pattern of information are roughly the same at different frequency ranges and spatial scales. In living systems, scale-invariance is seen in heart beats, neuronal activities, microtubule fluctuations, DNA sequences, psychological reaction times, protein-protein interactions, metabolic networks, gene regulation, immune response, intracellular signaling, neuronal and social interactions, and EEG patterns which can be self-similar over several orders of magnitude.

Structurally, brain network architectures are described as "small world networks" (Bullmore & Sporns, 2009) and "nested hierarchies" (Feinberg, 2011) with self-similarity at different scales. One example of this type of architecture is the distribution of airports — a few, large hubs with many connections, and many small hubs with few connections, linking any two places in a minimal number of steps. In the brain, nodes, or clusters of neuronal cell bodies with multiple incoming synapses ("brain nuclei") are arranged in such "small world networks" (Bullmore & Sporns, 2009). Systems theory integrates this small-world network information which flows both upward and downward ("inward") in scale.

Memory storage has been considered distributed and holographic, i.e., encoded at various scales and locations (Lashley, 1950; Pribram, 1991), and spatial location uses $1/f$ representations. Moser, Moser and O'Keefe were awarded the 2014 Nobel Prize in Medicine and Physiology for their discovery of fractal-like "grid cells" mapping spatial location in the brain (O'Keefe & Dostrovsky, 1971; O'Keefe, 1976; Fyhn *et al.*, 2004; Hafting *et al.*, 2005; Sargolini *et al.*, 2006). They found that spatial grids of an animal's immediate and larger scale surroundings are represented in several layers of hexagonal grid patterns in entorhinal cortex with each layer at a different scale. Zooming into finer scales involves changing anatomical layers.

Why $1/f$ scale-invariance abounds in nature is unknown, although Barabasi and Albert proposed that large networks self-organize to be scale-invariant (Barabási & Albert, 1999). Some theories accommodate specific examples, e.g., $1/f$ in semiconductor materials, while others attempt to be universal, e.g., the "resonance chain" theory of Ghosh and Bandyopadhyay (Ghosh and Bandyopadhyay). Scale-invariant $1/f$ pink noise is seen in gravity wave detectors suggesting $1/f$ processes may be embedded in the very structure of "holographic" space-time geometry (Hogan, 2008a,b; Hooft, 1999).

Like holograms or "fractals," scale-free architectures are inherently robust and resistant to disruption, arising most commonly in nature and technology. In the brain, evidence suggests neural network structures, temporal dynamics, and representation of mental states are all "scale-free," with self-similar patterns repeating at multiple spatio-temporal levels and locations spatiotemporal levels and locations (He *et al.*, 2010; Kitzbichler *et al.*, 2009; Van De Ville *et al.*, 2010; Eguiluz *et al.*, 2005). Taking the synaptic background activity as a scale-free dynamics maintaining itself close to criticality suggests that a richer, deeper level of sub-neural information processing may contribute to the complex spatio-temporal patterns of neural oscillations, and ultimately consciousness.

### 13.4. Consciousness in the Brain — Hierarchies within Hierarchies

Conscious perceptions of the external world are driven by sensory inputs. We do not know how accurately our conscious perceptions portray the external world (nor if the external world actually exists at all). Moreover, internally generated conscious states also arise, e.g., in default mode networks, but sensory-based conscious processes are reasonably well characterized.

Most inputs from sensory organs are relayed to the thalamus, a walnut-sized region in the middle of the brain, and from the thalamus to the cortex and other regions, e.g., for visual, auditory and somato-sensory perception. In the case of vision, (see Fig. 2) sensory inputs from the retina to the thalamus result in (1) projections to the primary sensory cortex, e.g., visual area 1 (V1) in the occipital cortex in the back of the brain. From there (2) secondary feed-forward projections go to associative and "executive" areas e.g., the pre-frontal cortex (PFC). Finally, (3) tertiary projections go to other regions. It is in this third wave that consciousness occurs.

The "third wave" correlation with consciousness is consistent with philosophical approaches called "higher order thought" ("HOT"). In HOT theories (Gennaro, 2004, 2014; Lau & Rosenthal, 2011), "bottom-up" activities (e.g., the first two waves) reach the pre-frontal "executive" cortex, which then sends third wave projections to other cortical regions.

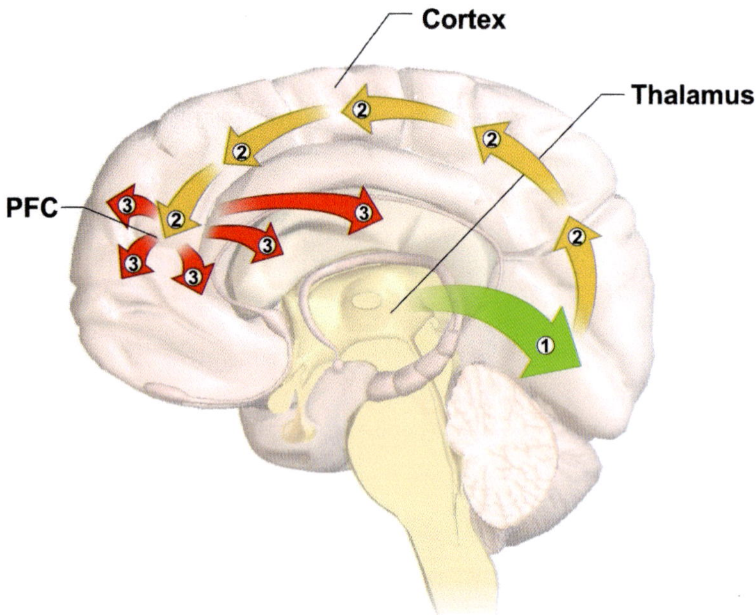


**Fig. 2.** Three waves in sensory processing. Sensory inputs from the spinal cord and cranial nerves to the thalamus result in primary projections (1) to the primary sensory cortex, e.g., visual area 1 (V1) in the occipital cortex in the back of the brain. From these areas, feed-forward projections (2) go to secondary associative and "executive" areas, e.g., the pre-frontal cortex (PFC) from which tertiary projections (3) go to other brain regions whose content then becomes conscious.

These third wave activities correlate with consciousness. In neuroscience, recurrent feedback models (Edelman & Gally, 2013) propose similar ideas.

Experimentally, only third wave activity is inhibited by anesthesia, which selectively prevents consciousness, sparing non-conscious processes. Lee *et al.* (2013) showed that all three types of general anesthetics (volatile gas anesthetics, propofol and ketamine) selectively inhibit third wave activity, sparing primary and secondary projections. Moreover, third wave feedback seems to correlate with gamma synchrony EEG, which is currently considered to be the best measurable correlate of consciousness.

Third wave processes themselves seem to be composed of three waves, successively integrating information (see Fig. 3). The cortex is arranged in six horizontal layers, which cut across vertical, hexagonal columns. Sensory inputs from the thalamus go (1) to layer IV, and thence

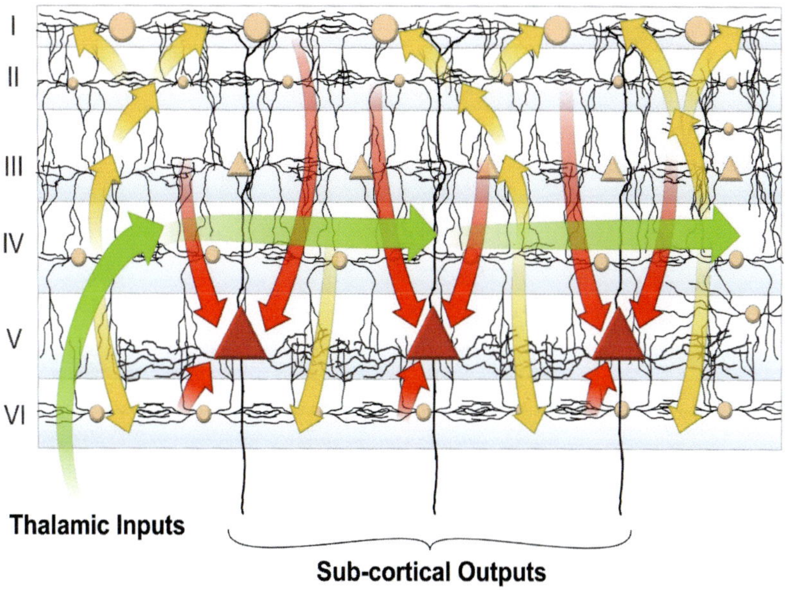


**Fig. 3.** Three waves of sensory processing in the cerebral cortex, a thin mantle on the very top of the brain composed of six hierarchical cellular layers. Primary sensory projections from the thalamus (1) arrive in layer IV which projects secondary activity (2) to layers I, II, III and VI. These areas then project tertiary (3) activity to giant pyramidal neurons in layer V, where consciousness is most likely to occur. Outputs from layer V pyramidal neurons project sub-cortically, e.g., to manifest "conscious" behavioral actions. Activity in apical dendrites from pyramidal neurons which ascend to the cortical surface are most directly responsible for measurable EEG.

(2) from layer IV to layers I, II, III, and VI. Finally, (3) projections from these layers converge on layer V giant pyramidal neurons, the most likely site for consciousness in the brain (see Fig. 4).

Basilar dendrites from layer V pyramidal neurons extend horizontally and interact (via gap junctions) with basilar dendrites from other pyramidal neurons, and cortical inter-neurons, in a planar sheet of pyramidal neurons. Apical dendrites from pyramidal neurons ascend vertically to layer V to the cortical surface, and are most directly responsible for measurable EEG, e.g., "40 Hz" gamma synchrony, the best neural correlate of consciousness. Axonal firing outputs from layer V pyramidal neurons descend sub-cortically, e.g., to implement behavior, exerting

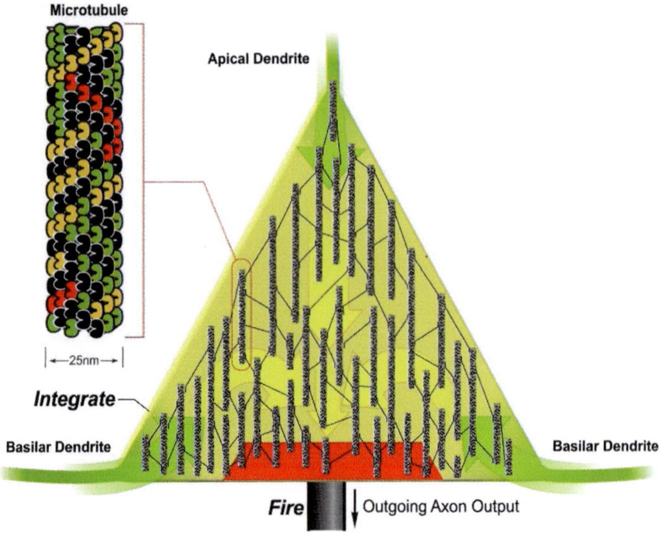


**Fig. 4.** Layer V pyramidal neuron with internal networks of microtubules connected by microtubule-associated proteins ("MAPs"). Inputs from apical and basilar dendrites are integrated in pyramidal neuronal membranes and cytoskeletal microtubules. On the left, a single microtubule is shown comprised of individual tubulin proteins, each in three possible states.

causal efficacy in the world. Third wave integration in cortical layer V pyramidal neurons therefore appears to be the most likely site for consciousness in the brain.

Like other neurons, layer V pyramidal cells are considered "integrate-and-fire" units according to the Hodgkin–Huxley (HH) standard model (Scott, 1995). In HH models, each neuron is a threshold logic device in which dendrites and cell body (soma) receive and integrate synaptic inputs via excitatory and inhibitory membrane potentials to a threshold at the proximal axon (axon hillock, or axon initiation segment (AIS)). When the AIS membrane potential reaches a critical threshold, the axon "fires" to convey signals to the next synapse and layer of neurons.

Integration implies merging and consolidation of multiple disparate information sources (e.g., Tononi's "Integrated Information Theory" (Tononi, 2004). At the level of individual neurons integration is approximated as a linear summation of synaptic membrane potentials. However, integration in branching dendrites and the soma requires logic, amplification of distal inputs, branch point effects, and signaling in dendritic spines and local dendritic regions. Nonetheless, according to HH, all such factors are reflected in membrane potentials, and thus the HH neuron is completely algorithmic and deterministic. For a given set of inputs, synaptic strengths and firing threshold, a fixed output in the form of axonal firings, or spikes will occur. Networks of integrate-and-fire neurons regulated by synaptic strengths and firing thresholds can integrate at various anatomical scales, providing highly non-linear functional processing. However, in the end, such processes are algorithmic and deterministic, leaving no place for conscious causal action or free will.

Real neurons significantly differ from idealized HH neurons. For example Naundorf *et al.* (2006) showed that firing thresholds in the cortical neurons of animals which are awake vary from firing to firing, or "spike to spike". Some factors other than inputs, synaptic strengths and the integrated membrane potential at the AIS must therefore contribute to firing, or not firing, thus controlling behavior. This integration "x-factor" deviation from HH behavior, of modulating integration and adjusting firing threshold, e.g., in layer V pyramidal neurons, is perfectly positioned for consciousness, causal action and free will, yet is in some way divorced from membrane potentials. The most logical source for this deviation is a continuation of the cortical hierarchy downward, inward, to smaller, faster scales within the neuronal cell interior.

## 13.5. Networks Inside the Neuron

Every neuron is a highly structured matrix of nucleic acids, proteins (channels and receptors included), ions and water molecules that work in cooperation to produce normal cellular function. Thus, it seems prudent to examine this machinery as a whole.

Conventional signaling mechanisms within and between cells include autocrine, paracrine, and endocrine chemical signaling, binding to and activating receptors and electrical signals (mediated by ions) passing through synapses and gap junctions. In basic terms, the traditional view of the cell interior has been a bag of randomly diffusing ions, proteins and enzymes (Spitzer & Poolman, 2005), which in the course of their diffusive transport, bind to and unbind from other molecules. More bluntly, the conventional, reductionist view in cell biology is that intracellular signaling and information processing occur in a "minestrone soup" of randomly diffusing ions, proteins and enzymes. However, the cell's interior, the "cytoplasm," is partially in solid state due to its major constituents: the cytoskeleton, internal membranes, organelles and the ordered water surrounding these structures.

Indeed, the cell is an extremely crowded environment vastly different from the dilute, idealized conditions found in most biophysical experiments (Wang *et al.*, 2012). The sophisticated arrangement of internal membranes and cytoskeletal networks causes the space available for the diffusion of proteins and ions in cytoplasm to be significantly complex, directly affecting diffusivity. This has led to a view of the cell as a highly ordered collection of molecular machines operating in unison to give rise to cell function (Alberts, 1998; Vale, 2003).

The cytoskeleton, in specific, extends throughout cells from the centrosome ("centriole") near the nucleus to the cell membrane, defining cell architecture and filling much of the cell interior. Microtubules and other cytoskeletal components self-assemble to guide and define cell shape and function. Accordingly, these processes are intimately linked to neuronal development, differentiation, synaptic generation, and neuronal plasticity. Active transport of intracellular material, particularly in asymmetric cells like neurons, occurs by motor proteins conveying synaptic components along microtubules to their appropriate destination. Delivery at particular synaptic locations appears to be guided by placement of microtubule-associated proteins ("MAPs") at specific microtubule lattice locations. In cognitive dysfunction associated with Alzheimer's disease, mild Traumatic Brain Injury, and Chronic Traumatic Encephalopathy (Goldstein *et al.*, 2012; Faden & Loane, 2015), the MAP tau is displaced from microtubules, causing destabilization and depolymerization of their aggregate structure. Importantly, microtubule disruption occurs in nearly all neurodegenerative diseases (Woolf *et al.*, 2010). Supporting this view, the cytoskeletal matrix has been shown to act as a target for neurotransmitters via post-translational modifications mediated by signal transduction pathways (Craddock *et al.*, 2012; Janke & Kneussel, 2010; Hameroff *et al.*, 2010).

As the cytoskeleton extends from the cell membrane to the cell nuclei it forms a continuous mechanical linkage throughout the neuron and may be viewed as an interconnected "tensegrity" matrix that allows for long-range mechano-transduction (Ingber *et al.*, 2014; Wang *et al.*, 1993; Wang *et al.*, 2009; Friesen *et al.*, 2015) through the neuron. Ionic and neurotransmitter signals acting to modify and reorganize cytoskeletal protein interconnections in one region of the cell can thus coordinate the overall architecture of neuron morphology (Gardiner *et al.*, 2011; Woolf *et al.*,

2009). This influences the spacing of synaptic contacts, the placement of ion channels and receptors, and the overall plasticity of the neuron. The resulting changes in synaptic strength regulate neural firing patterns and alter cognitive functions including learning and memory. In essence the cytoskeleton exerts immediate control over neural signal propagation. While the exact mechanism through which this occurs remains to be determined, disturbances in cytoskeletal cohesion have been shown to directly reduce the transmission of action potentials (Gardiner *et al.*, 2011). Moreover, ion channels and receptors exchange information in a bi-directional manner with the cytoskeleton that is integral for cognitive activity (Craddock *et al.*, 2010).

Alterations of cytoskeletal stability reliably regulate the gating properties of voltage- and ligand-gated ion channels of the major ionic currents ($Na^+$, $K^+$ and $Cl^-$, and $Ca^{2+}$) (Janmey, 1998; Sun *et al.*, 2008; Shcherbatko *et al.*, 1999; Schubert & Akopian, 2004; Strege *et al.*, 2003). This regulation appears to be independent of filament transport roles. Drug induced alterations of microtubules modulate the inactivation of calcium channels in isolated rat hippocampal neurons influencing calcium signaling (Furukawa *et al.*, 2003; Furukawa & Mattson, 1995; Johnson & Byerly, 1993, 1994). The $GABA_A$ receptor, known to bind anesthetics, is also dependent on the microtubule cytoskeleton for proper function. Disintegration of the microtubule lattice has been shown to cause a 78% decrease in $GABA_A$ signaling (Meyer *et al.*, 2000). GABA is the prime inhibitory neurotransmitter in the brain and has the potential to significantly alter neural firing patterns.

Nonetheless, while mechanical changes in the cytoskeleton can account for slower neuromodulation, it does not adequately account for the rapid dynamic changes required for functional connectivity. The cytoskeletal matrix, however, is not just a mechanical structure. Consider that this protein network is ultimately composed of electrically responsive microtubules, and actin filaments. Experimental, computational and theoretical evidence indicates that both ionic and electronic conduction through these protein filaments offers the possibility of signal propagation and processing in these protein structures (Freedman *et al.*, 2010; Chelminiak *et al.*, 2009; Sataric *et al.*, 2009; Dixon *et al.*, 2008; Priel & Tuszynski, 2008; Priel *et al.*, 2005a, 2005b; Tuszynski *et al.*, 2004; Lin & Cantiello, 1993; Cantiello *et al.*, 1991). Effectively then, the cytoskeleton forms an electromechanical signaling network allowing local electrical and/or mechanical perturbations to be propagated throughout the cell's interior (Friesen *et al.*, 2015). Such a matrix within the crowded cellular environment would allow for more rapid communication than allowed for by conventional diffusion based signaling and actin filaments.

The highly negative surface charge of microtubules induces positively charged ions in the cytosol (such as $K^+$, $Na^+$, $Ca^{2+}$) to condense forming a cloud of positive counter-ions around the filament. In theory a voltage gradient (i.e., action potential) can cause this cloud to move along the filament in a non-linear, soliton-like ionic wave (Sataric *et al.*, 2009; Priel & Tuszynski, 2008; Priel *et al.*, 2005a; Tuszynski *et al.*, 2004; Lin & Cantiello, 1993; Sekulić *et al.*, 2011). These properties depend on the adsorbed counter-ions (Priel *et al.*, 2005a; Lin & Cantiello, 1993; Cantiello *et al.*, 1991; Friesen *et al.*, 2015; Sekulić *et al.*, 2011; Wong & Pollack, 2010), which in the case of actin filaments have been shown to display extremely coordinated dynamics (Angelini *et al.*, 2006; Angelini *et al.*, 2003). Due to the charged nature of microtubules (Tuszynski *et al.*, 2005), actin filaments and their ionic clouds, these filaments can be manipulated by external electric fields (Tuszynski *et al.*, 2004) indicating that electric fields generated by firing neurons can directly and indirectly affect the neural cytoskeleton offering a rapid means of intercellular and intracellular communication. Such a mechanism may explain the rapid dynamic changes required for functional connectivity.

As the neuronal cytoskeletal network effectively connects synapses throughout the cell, electric perturbation (e.g., ion influx at the postsynaptic density) can feasibly propagate throughout the network responding to multiple inputs. The resulting signal, composed and integrated from these multiple inputs, could then affect the neuron response through electrical control of voltage-gated ion channels (Craddock *et al.*, 2010; Priel *et al.*, 2010) affecting neural signaling and higher cognitive function. The temporal focusing of synaptic inputs by coincidence detection, can allow for communication via synchrony (Fell & Axmacher, 2011; Fries, 2009). Additionally, the interaction between cytoskeletal electric signaling and electromagnetic fields generated by firing neurons would effectively make the cytoskeleton an antennae network working to enhance, and amplify the neural response to exogenous or endogenous electromagnetic fields (Funk *et al.*, 2009; Gartzke & Lange, 2002). Overall, this conceptual framework suggest that the electrically active neuronal cytoskeleton is integral to neuron function and cognitive processes (Craddock *et al.*, 2014). An alternative or perhaps parallel channel of information storage and processing could involve intrinsic electronic conductivity through the protein itself via electron hopping based on the hypothesized semiconducting bands of tubulin within the microtubule periodic lattices (Tuszynski *et al.*, 1998).

## 13.6. The Neuronal Cytoskeleton

Major components of the cytoskeleton, microtubules are self-assembling cylindrical polymers of the protein tubulin. 25 nm (nm = $10^{-9}$ meter) in diameter, microtubules are comprised usually of 13 longitudinal protofilament chains of tubulin. Each tubulin is a dimer composed of alpha and beta monomers. Tubulin dimers self-assemble and arrange in two types of twisted hexagonal lattices (A-lattice and B-lattice), which are illustrated in Fig. 5.

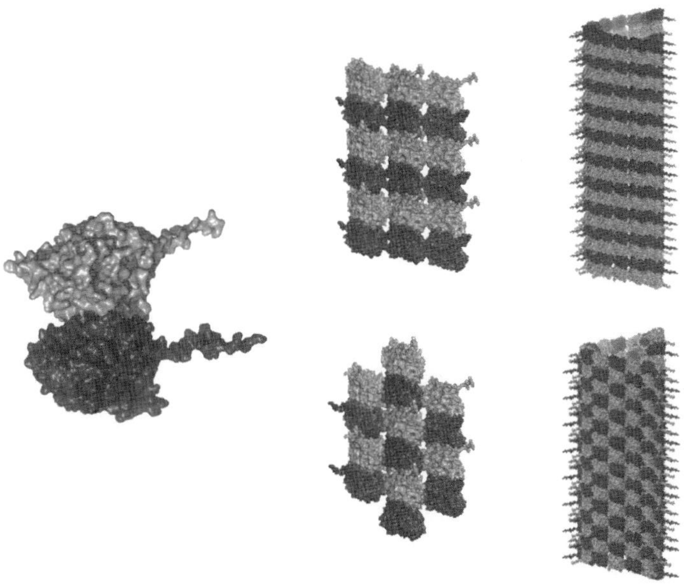

**Fig. 5.** Schematic illustration of a microtubule and its hexagonal lattices. Left: Tubulin dimer, alpha chain — light gray, beta chain — dark gray. Top: Type B lattice structure 9 dimer neighborhood, and microtubule structure. Bottom: Type A lattice structure 7 dimer neighborhood, and microtubule structure.

Within microtubules, each tubulin may differ from its neighbors by genetic variability, post-translational modifications, phosphorylation states, binding of ligands and MAPs, and transient dipole orientation, so the information capacity in such a mosaic-like microtubule is enormous. Microtubules are particularly prevalent in neurons ($10^9$ tubulins/neuron), and uniquely suited for information processing, encoding and memory.

In cell division ("mitosis"), microtubules disassemble, and then reassemble as mitotic spindles, which separate chromosomes, establish daughter cell polarity, and rearrange for cellular structure and function. However unlike other cells, neurons once formed, do not divide, and so neuronal microtubules remain assembled indefinitely, providing a stable medium for memory encoding.

Microtubules in neuronal soma and dendrites are unique in other ways. In axons, and all non-neuronal cells throughout biology, microtubules are arrayed radially, like spokes in a wheel, extending continuously from the hub-like centriole outward toward the cell membrane, all with the same polarity. Microtubules in dendrites and cell bodies/soma are short and arrayed in mixed polarity networks interconnected by MAPs, and stabilized by MAP capping proteins. Dendritic-somatic microtubules are ideal for memory encoding, and well situated for consciousness, e.g., in dendrites and soma of layer 5 cortical pyramidal neurons.

Due to their key roles in cell organization, and their "computer-like lattice structure," microtubules and the cytoskeleton have long been considered as possible information processing devices. Early models suggested each tubulin in a microtubule lattice could represent a particular state, e.g., binary "bits" of either 1 or 0 (Hameroff & Watt, 1982) and interact with neighbor tubulin states by dipole coupling in molecular microtubule "automata" to generate interactive patterns and process information (Hameroff *et al.*, 1992). Subsequently, tubulin "bits" were also considered as quantum bits ("qubits") in quantum computational microtubule models (Hameroff, 2007; Hameroff *et al.*, 2014; Craddock *et al.*, 2009) and microtubule lattice pathways of tubulins, rather than specific tubulin states, were considered as topological bits and qubits (Hameroff *et al.*, 2002). Such information processing capabilities can be modified in the presence of post-translational modifications, or interaction with MAPs, and may be used to store information in a form of molecular memory.

Memory is generally attributed to synaptic plasticity, however synaptic proteins are transient, lasting hours to days, and yet memories can last lifetimes. The best experimental model for memory is long term potentiation

(LTP) in which synaptic activation results in calcium influx, and activation of the hexagonal holoenzyme calcium-calmodulin kinase II (CaMKII). Each activated CaMKII extends two sets of 6 kinase domains, each set able to encode 6 bits of synaptic information by phosphorylation in an appropriate hexagonal lattice for storage and functional processing. In Craddock *et al.* (2012) we showed how CaMKII can encode 6 bits (a "byte") of memory in hexagonal lattices in microtubules, major components of the cell cytoskeleton inside neurons. Could memory be encoded in hexagonal hierarchies extending inward to, and emanating from microtubules? It would be very efficient for processes mediating consciousness to occur in the same medium in which memory is embedded.

Assuming these forms of information processing stroage and taking tubulin states to act simply as binary bits switching, at 10 megahertz ($10^7$ Hz), $10^9$ tubulins per neuron gives $10^{16}$ operations per second per neuron, and $10^{11}$ neurons per brain gives roughly $10^{27}$ operations per second for the entire brain. However, reduction to this lower level of the hierarchy appears to move further away from global unity and binding, essential features in life and consciousness. Something is missing.

### 13.7. Order All the Way Down

At least until the present time, consciousness appears to be purely a biological phenomenon, and therefore a property of living systems. To understand consciousness, essential features of life must first be considered.

Historically, life has been described in two types of approaches, "functionalism" and "vitalism." Functionalism characterizes life by its behaviors (Margulis *et al.*, 1995), including self-organization, homeostasis, metabolism, growth, adaption, response to stimuli, replication/reproduction and evolution. But these can also occur in non-biological computer programs, e.g., as "artificial life" (Langton & Christopher, 1997), and video games.

On the other hand, 19th-century "vitalists" (or "animists") saw a unifying energy field, or life force pervading living systems — *"elan vital."* However as reductionist biology developed, vitalism fell from favor, with the notion of a "life force" becoming taboo and considered "unscientific." 19th-century vitalism was based either on electromagnetism, or on forces outside science, as quantum mechanics had yet to be discovered.

But in classical descriptions of biology, the motion of molecules in a liquid or gaseous medium are believed to be stochastic (i.e., truly random), negating long-range ordering effects in molecular living systems.

In his famous book *What is Life?*, Erwin Schrödinger remarked that, on its own, classical physics seems to be insufficient to explain the robust organization observed in biology (Schrödinger, 1943) suggesting that life's unitary oneness derived from quantum coherence in biomolecular lattices. In other words, life is "order arising from order" rather than "order from chaos." Long-range macroscopic coherence mediated by quantum dipole oscillations within the hydrophobic core of proteins was initially proposed as a means to regulate protein conformation and function (Fröhlich, 1968, 1975, 1978). However, living systems were considered too "warm, wet and noisy" for seemingly delicate quantum coherence.

That is until recently. Evidence in the past decade has shown several examples of functional quantum mechanisms in living systems, for example bird navigation, our sense of smell and photosynthesis, all at ambient, or warm temperatures (Lambert *et al.*, 2013). Engel and colleagues have provided the most rigorous example of experimentally demonstrated quantum effects in biology (Engel *et al.*, 2007; Panitchayangkoon *et al.*, 2010). Their direct observation of quantum beats in the Fenna–Matthews–Olsen (FMO) photosynthetic light-harvesting complex is a direct indication of electronic quantum coherence in biology. Following this initial discovery, similar coherent quantum phenomena have been shown to operate in the light harvesting complexes of plants (LHCII) (Schlau-Cohen *et al.*, 2012; Calhoun *et al.*, 2009; Schlau-Cohen *et al.*, 2009), bacteria (LH2) (Harel & Engel, 2012; Ostroumov *et al.*, 2013) and phycobiliproteins (Collini *et al.*, 2010; Turner *et al.*, 2012).

It is hard to imagine that the use of quantum processes in plants and bacteria would not be retained or even perfected through evolutionary achievements into higher-developed life forms such as mammals. Energy production mechanisms in plants involve light harvesting through chlorophyll while animals predominantly utilize glucose metabolism in mitochondria. It is, therefore, expected that at least some of the many complex steps involving oxidative phosphorylation in mitochondria, such as electron transfer in the mitochondrial protein complex I, employ quantum tunneling. Thus, there is potential for quantum effects to occur within the unique biological environment of the neuronal cytoskeleton.

The theory describing coherent energy transfer in plants is the capture, and transfer of light (i.e., photons) via the interaction of transition dipoles in chlorophyll molecules. Light harvesting complexes in plants are a unique and elegant geometrical arrangement of chromophoric chlorophyll pigments. This arrangement allows a photon to be efficiently captured and funneled from the external environment to a reaction center where the energy is processed and stored in chemical components. This biological light-harvesting operates at nearly 100% efficiency. The remarkable performance is achieved through quantum mechanisms, which allow the photon to be shared across the chlorophyll network (Strümpfer *et al.*, 2012). The initially excited donor chromophore transfers its electronic energy to acceptor chromophores via electrodynamic coupling of their transition electric dipole moments. This energy is shared due to the close relation, or 'resonance', between their energy levels. Originally, proposed by Forster (Forster, 1948) this resonant energy transfer (RET) is the dominant theory applied in electronic energy transport (Andrews *et al.*, 2011; Beljonne *et al.*, 2009). The core of this mechanism is the optimal packing of light sensitive chromophores that possess a significant transition dipole upon excitation

Chlorophyll, is one of a number of biologically available chromophores. Other bio-chromophores include pyrroles, prophyrins, cytochromes, phycobiliproteins, carotenoids, ferredoxins, flavins, heme (forming the red pigment in blood) and melanin (responsible for skin pigmentation). Additionally, three, naturally occurring, amino acids possess fluorescent chromophoric properties: tryptophan, tyrosine and phenylalanine. Tryptophan, tyrosine and phenylalanine possess resonant ring structures in which electrons are mobile and delocalizable and thus, due to their high polarizability, they are convenient sites for electronic energy transfer. Tryptophan is the most highly suited amino acid for transferring electrons and exchanging photons, as it has the greatest electron resonance and is thus the most fluorescent. Such intra-molecular singlet excitation energy transfer among and between tryptophan, tyrosine and phenylalanine has been shown to occur in protein environments (Visser *et al.*, 2008; Eisinger *et al.*, 1969; Karreman *et al.*, 1958), and crystals of globular proteins have been shown to efficiently transfer energy between chromophoric residues (Desie *et al.*, 1986). While these crystal structures are artificial constructs, and are therefore not expected to be biologically relevant, similar processes may take place in other natural proteins, including the polymers forming the cell cytoskeleton.

## 13.8. Life in the Quantum Underground

Living organisms including the human body may be structurally subdivided in several ways, e.g., into various organs, tissues, cells, organelles, molecules, ions, atoms etc. However pharmacologists divide the body differently, into various solubility phases to predict where certain types of drugs will bind. Liquid regions and surfaces of solid-state biomolecules are "polar" due to electric charges on water, hydrogen bonds, ions and protein outer surfaces. Most drugs, neurotransmitters and hormones are charged, soluble in water (and blood) and bind by charge interactions in the polar, aqueous medium inside cells and on surfaces, e.g., to membrane receptors and surfaces of the solid state cytoskeleton.

However cells, and the body and brain in general, also include non-polar, oil-like solubility phases ("oil and water do not mix!"). Such non-polar, "hydrophobic" regions are found within membranes, cytoskeleton and other biomolecules shielded from the charged, polar regions. This distinction is important in the formation of biomolecules, for example protein folding. Amino acids which comprise proteins have residues which are either polar and water soluble ("hydrophilic") or non-polar and soluble in oil-like media ("hydrophobic"). Non-polar, hydrophobic amino acid residues coalesce and cluster together by van der Waals attraction, excluding water, forming non-polar "hydrophobic pockets" within proteins. Similar non-polar interiors are found in membranes and nucleic acids, indeed within molecular structures throughout biology.

Non-polar regions are comprised mainly of pi-electron resonance clouds, e.g., (in proteins) of the aromatic rings of amino acids phenylalanine, tyrosine and tryptophan. Similar to chromophores mediating quantum coherence in photosynthesis proteins, these pi-resonance clouds stem from the benzene ring, the basis for organic chemistry.

Life is based on carbon chemistry. Carbon atoms each have four electrons to bond other atoms, e.g., hydrogens and others to form "hydrocarbon chains" called alkanes, with the general formula $C_nH_{2n+2}$ where $n$ is the number of carbons, and length of the chain. Carbons can also share two electrons, a "double bond," or "pi-resonance" bond, and hydrocarbon chains with a single double bond are called alkenes, with the general formula $C_nH_{2n}$. The "pi-resonance" implies a free electron (from carbon's outermost "pi-orbital") shared between carbon atoms, either oscillating between the two (molecular orbital theory), or delocalized, i.e., in quantum superposition within a "pi-electron resonance cloud" covering both carbons (resonance theory).

The 19th century chemists were puzzled by the structure of benzene, an oily, flammable hydrocarbon with the formula $C_6H_6$, fitting neither alkanes nor alkenes. Then, the German chemist Kekule had a dream about

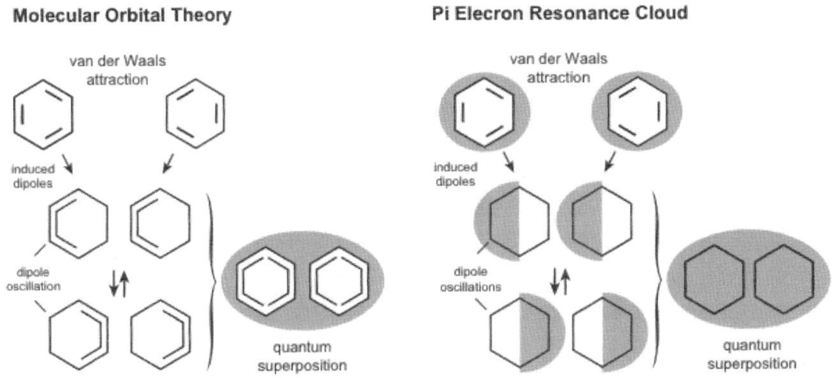


**Fig. 6.** "Molecular orbital theory" (left) and "pi-electron resonance cloud" (right) descriptions of van der Waals attraction and dipole coupling of mobile electrons in phenyl (benzene) rings, the basis for organic chemistry. In pi-resonance, electrons are viewed as dipole clouds, which can couple and entangle.

snakes of varying lengths, which he later recognized as alkanes and alkenes. One snake in Kekule's dream swallowed its tail, forming a ring. Kekule concluded it was benzene, $C_6H_6$, a hexagonal ring with three extra pi-electron resonance bonds either oscillating among possible locations (molecular orbital theory), or superpositioned in an electron resonance cloud tethered to the six carbon ring (resonance theory, Fig. 6).

Replacing hydrogens in benzene makes an attachable "phenyl" ring which, along with "indole" (6 + 5) rings are called "aromatic" because of generally sweet, pleasant smells. Adding one or more charged, polar ends to aromatic rings makes "amphipathic" biomolecules, non-polar on one end, polar on another. They include aromatic amino acids phenylalanine, tyrosine and tryptophan, and the psychoactives dopamine, serotonin, LSD, DMT, etc.

When these amphipathic molecules bind to proteins such as membrane receptors or microtubules, non-polar groups insert into non-polar interior regions, and their charged, polar ends stick outward toward the exterior. Thus, within the protein interior non-polar ring structures stack and align. In bulk form, benzene is flammable and oily, but when geometrically constrained, e.g., in flat sheets as graphene, or cylindrical fullerene nanotubes, pi-electron resonance clouds couple in extended cooperative systems with interesting quantum properties.

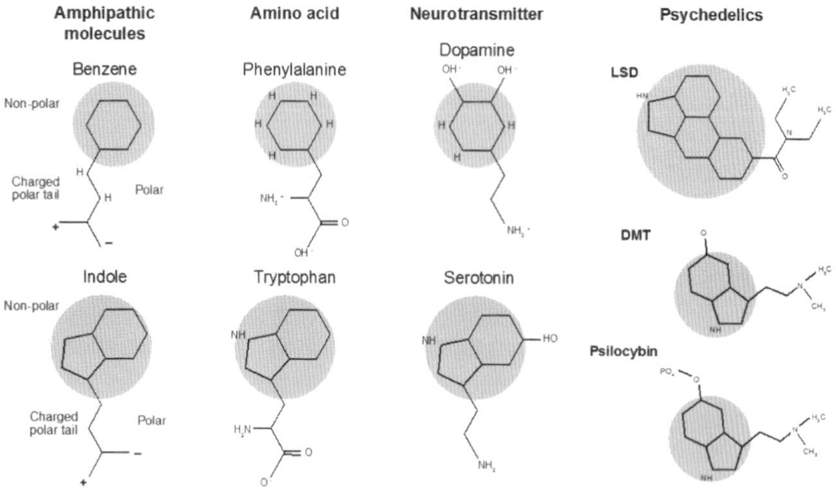


**Fig. 7.** Amphipathic biomolecules with both polar and non-polar regions (gray circles) based on benzene (phenyl) and indole rings. Psychoactives bind to membrane receptors and also inside neurons in the microtubule "quantum underground."

Though electrically neutral and non-polar, pi-electron resonance clouds are polarizable, and, when close (angstrom level van der Waals radius) attract and couple by induced quantum dipoles (van der Waals London forces, quadrupoles, magnetic dipoles). Such coupling drives protein folding and results in "pi-stacking" in which resonance rings are arrayed at specific spacings and geometric orientations.

The pi-resonance clouds are optically active, able to fluoresce, absorbing photons, which induce excited electron states, which then drop to lower states, emitting lower energy photons. This quantum-level energy can also be efficiently transferred non-radiatively between excited resonance clouds separated up to 2 nm by Forster electron resonance transfer (FRET), as occurs in quantum coherent energy transfer in photosynthesis proteins.

## 13.9. Binding it All Together

Taken together, a quantum-friendly, non-polar solubility phase of pi-electron resonance is shielded inside biomolecules, distributed throughout living systems. Geometric arrays of pi-resonance clouds ("pi-stacks")

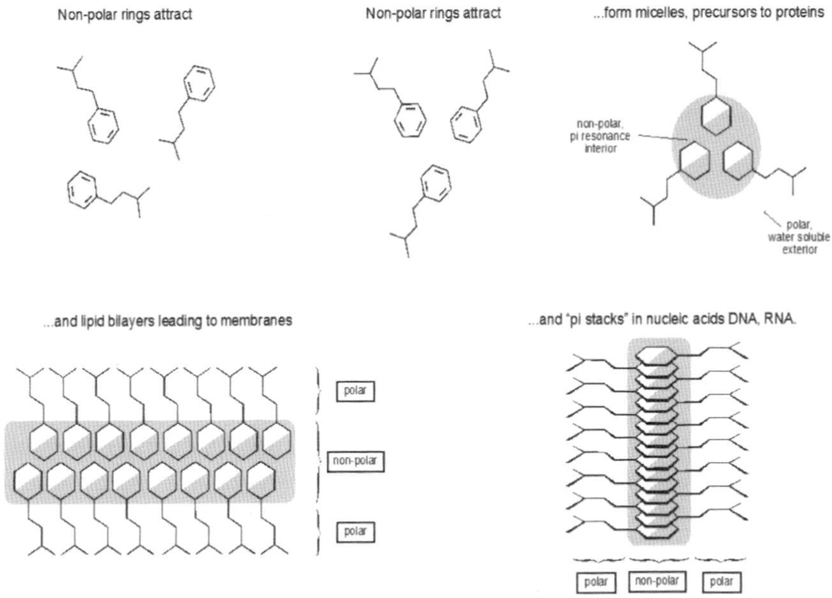


**Fig. 8.** The "quantum underground." Amphipathic biomolecules self-assemble by non-polar van der Waals attractions to form (top row) micelles leading to proteins, (lower left) lipid bilayers in membranes, and (lower right) nucleic acids. Non-polar regions (gray) can host quantum coherence and constitute the "quantum underground."

in non-polar regions within biomolecules, shielded from the charged, polar environment, are capable of sustaining quantum coherence and constitute a "quantum underground." Interiors of DNA and membranes have similar arrays and stacks of pi-resonance clouds and non-polar regions (Fig. 8). Vattay *et al.* showed that pi-resonance clouds in biomolecules are poised at positions of "quantum criticality," at the boundary between conductivity and insulation (Vattay *et al.*, 2014, 2015). We suggest microtubule mechanical vibrations (e.g., in the megahertz frequency range) can regulate quantum properties, vibrating in and out of quantum criticality.

In the highly structured neuron, various MAPs interconnect microtubules into networks, forming the scaffolding for the neuronal and synaptic architecture. In axons, microtubules form long continuous structures orienting with the same polarity, while in dendrites they are formed as short interrupted filaments of mixed polarity. These structures in neurons are relatively stable, as opposed to the dynamic instability observed in all other cell types. The microtubule constituent protein tubulin possesses a unique network of chromophoric amino acid residues (Fig. 9), similar to the chlorophyll architectures observed in photosynthetic light harvesting complexes.

Tubulin proteins comprising microtubules each have eight tryptophan indole rings arrayed in a geometry similar to that in photosynthesis proteins. The fluorescence quantum yield for pure tryptophan is comparable to the room temperature fluorescence quantum yields of bacteriochlorophyll (Craddock *et al.*, 2014). Tryptophan to tryptophan energy transfer has also been observed in tubulin (Guha *et al.*, 1996; Weber & Shinitzky, 1970). Due to the symmetrical structure of microtubules, the tryptophan networks in microtubules are continuous along the protofilament length, effectively forming 13 parallel conduction pathways (Hameroff *et al.*, 2002), and may also wrap around helical pathways. Could these aromatic chromophores support coherent quantum effects within tubulin and microtubules?

Our group has applied structure-based simulations similar to current studies of photosynthetic light harvesting complexes to probe potential energy transfer mechanisms in tubulin, and by extension microtubules (Craddock *et al.*, 2014). Accounting for the warm and wet thermal environment, our results found the existence of coherent quantum beatings lasting for upwards of 600 sampled frequency, with the excitation traveling the length of the tubulin dimer, Fig. 10. Therefore, it is feasible that the tryptophan arrangement in tubulin could effectively transfer energy along the length of the microtubule. The unique cylindrical lattice symmetries found in the tubulin lattices of microtubules may effectively serve to enhance transfer rates and distances, and potentially with the involvement of tyrosine and phenylalanine enable energy transfer along helical pathways.

These aromatic chromophores can thus support coherent quantum oscillations within microtubules and allow for the emergence of quantum states in protein filaments of the cytoskeleton. The stability, length and oriented polarity of these networks are expected to play a role in both ionic and electronic signal conduction. Such rapid signaling through aromatic conduction pathways may coordinate the complex organization of the microtubule cytoskeleton required for the tasks of cell division, motor protein trafficking, and motility.

Arrays of aromatics, behaving as cellular automata, have been shown to be capable of solving a variety of computationally intense problems in a manner that could potentially far exceed the capabilities of today's

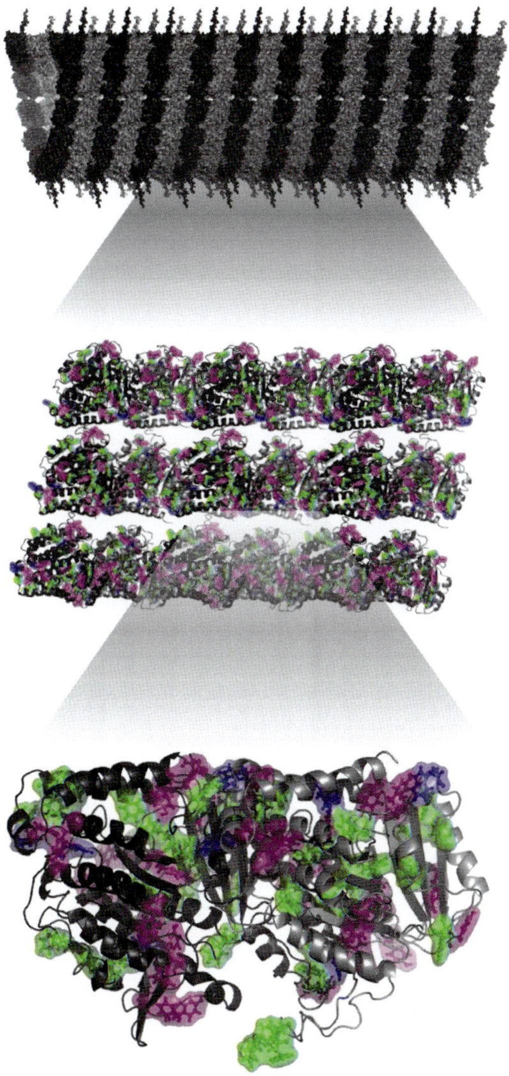

**Fig. 9.** Unique network of chromophoric amino acid in microtubules. Light gray — alpha tubulin; Dark gray — beta tubulin; Blue — tryptophan residues; Purple — tyrosine residues; Green — phenylalanine residues. (A) Microtubule fragment. (B) A patch of microtubule showing nine tubulin dimer lattice. (C) Microtubule constituent protein dimer Tubulin.

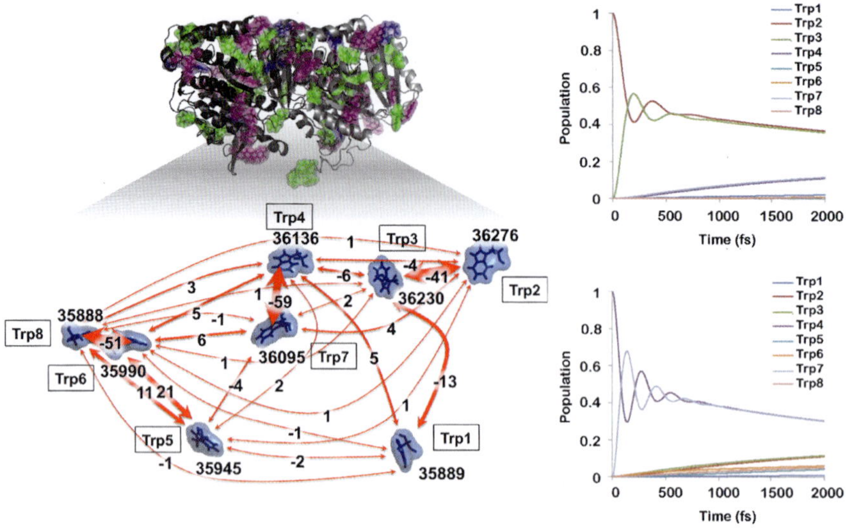


**Fig. 10.** Quantum Coherent Energy Transfer in Tubulin. Left: Eight tryptophans in a tubulin within a microtubule are coupled via dipole oscillations and capable of quantum coherent excitations as occur in photosynthesis proteins. Right: Coherent dynamics of a single photon excitation moving through tryptophan residues in tubulin. Upper: Excitation starting on tryptophan 2 (TRP2). Lower: Excitation starting on tryptophan 4 (TRP 4).

computers (Adamatzky, 2010; Bandyopadhyay *et al.*, 2010). Cellular automata models of dipole fluctuations in microtubules have shown that their geometry is conducive to such computations (Hameroff *et al.*, 1986; Rasmussen *et al.*, 1990; Craddock *et al.*, 2009), which may underlie the coordination of microtubule architecture and intracellular transport required for proper neuron development and function.

While aromatic excitation dynamics in microtubules have not been experimentally investigated, there is a high likelihood that they would be able to support this form of computation. Supportive of this notion, microtubules have been shown to reorganize in a dose-dependent manner after exposure to UV light (Krasylenko *et al.*, 2013; Staxén *et al.*, 1993; Zamansky & Chou, 1987; Zaremba *et al.*, 1984). Feasible mechanisms for these changes include the reduction of disulfide or peptide bonds induced by photo-excitation of tryptophan groups (Petersen *et al.*, 2002; Vanhooren *et al.*, 2002; Wu *et al.*, 2008; Weisenborn *et al.*, 1996), or subtle protein structural changes due to photo-induced alterations in tryptophan flexibility

(Weisenborn *et al.*, 1996). Aromatic amino acids like tryptophan, tyrosine and phenylalanine can also serve as acceptors for weak hydrogen bonds that may be perturbed upon excitation, playing a pivotal role in determining the conformations and motion of proteins (Sandorfy, 2004). Such signaling mechanisms may explain the observations of an apparent UV mediated cell-to-cell influence on cell division (Fels, 2009).

At the cellular level living systems emit ultra-weak spontaneous photons without external excitation (Yoon *et al.*, 2005; van Wijk *et al.*, 2006; Takeda *et al.*, 2004; Chang, 2008; Cohen & Popp, 1997; Devaraj *et al.*, 1991; Quickenden & Que Hee, 1974; Kobayashi & Inaba, 2000; Kobayashi *et al.*, 2009). These photons are produced via various biochemical reactions, but principally from bioluminescent radical reactions of reactive oxygen and nitrogen species and the relaxation of excited states. The oxidative metabolism of mitochondria and lipid peroxidation appear to be a primary source for this activity (Nakano, 2005; Thar & Kuhl, 2004). Neurons also continuously produce photons during their ordinary metabolism (Kataoka *et al.*, 2001; Isojima *et al.*, 1995), and it has been shown *in vivo* that the intensity of photon emission from rat brain correlates with cerebral energy metabolism, electrical activity, blood flow, and oxidative stress (Kobayashi *et al.*, 1999a, 1999b). Furthermore, Sun *et al.* (2010) have revealed that ultraweak bioluminescent photons can propagate along neural fibers and can be considered a means of neural communication. Radical recombination within mitochondria can emit photons in the UV range required to excite the chromophoric network within microtubules (Voeikov, 2001). Additionally, biophotons emitted in the infrared or visible spectrum (Tafur *et al.*, 2010), may excite the UV aromatic chromopores in microtubules in a multiphoton excitation manner (Kaushalya & Maiti, 2007). Such multiphoton excitation exploits nonlinear quantum effects to allow a molecule to be photo-excited by light that has much lower energy, and longer wavelength, compared to what would be predicted from its conventional single-photon excitation spectrum. Thus, when the intensity of the excitation light is very high, a chromophore molecule can absorb multiple low-energy photons simultaneously, whose total energy equals or exceeds its one-photon excitation energy, causing it to excite.

These biophotonic and bioelectronic activities are not independent of biological events in the nervous system, and their synergistic action can play an important role in cytoskeletal ionic signaling, neural signal transduction, and further to neural synchrony.

Alteration of these activities via physical disruption (i.e., mechanical vibration, ultrasound), or the chemical effect of drugs would be expected to alter this signaling cascade. The cytoskeleton is a target for drugs, which alter mood and subjective experience. For example, the antidepressant effects of fluoxetine (Prozac) take several weeks, apparently because of the need for cytoskeletal reconfiguration. Anesthetics also appear to act in microtubules to prevent consciousness, not exclusively membrane proteins as has been previously assumed (see below), however the effects of anesthesia cannot be explained via a simple mechanistic mode of action.

## 13.10. Anesthesia — Preventing Quantum Dipole Oscillations in the Microtubule Quantum Underground

Arguably the greatest medical discovery of all time, anesthesia remains enigmatic. Despite a century of research, the mechanism by which anesthetic molecules cause reversible loss of consciousness and memory is still unknown. The initial discovery of these miracle drugs, which enable all of modern surgery sparing the patient the agony of pain, was the result of serendipity, and while rational drug design is increasingly the standard practice in pharmacology, serendipitous discovery remains the only way to find novel anesthetics. Shedding light on the mechanism of anesthetic action will settle not only philosophical and existential conundrums surrounding consciousness, but will also aid the design and development of novel anesthetics and conversely psychoactive mood-altering drugs.

In the mid-19th century, gases like diethyl ether and nitrous oxide ("laughing gas") were discovered to have "anesthetic" properties, i.e., inhaling these gases caused humans and animals to lose consciousness, or some observable correlate, like mice falling over, or animals not moving in response to stimulation. When the anesthetic gas was exhaled away, the subjects and animals woke up, snapping back to consciousness, essentially unchanged. Potencies were determined for each anesthetic, i.e., the gas concentration at which half of a group of mice will fall over, and half remain upright. Amazingly, each anesthetic acted at the same concentration in all animals, identical concentrations of any particular gas anesthetizing a mouse, salamander or human. By the turn of the 20th century, dozens of anesthetic gases had been discovered, curiously with various dissimilar chemical structures, e.g., ethers, halogenated hydrocarbons, nitrous oxide and the inert gas xenon. Yet despite their chemical differences, each gas had a particular potency for all animals. Scientists sought a common molecular property independent of structure, which correlated with anesthetic action. The answer was solubility in a particular non-polar medium.

The earliest glimpse into the mechanism of anesthetic action was the observation associating the potency of an anesthetic with its lipid solubility. Working separately, Germany's Hans Meyer and Britain's Charles Overton ranked anesthetic potency for many anesthetics in various animal models, and tested anesthetic solubility in many types of solvents, each finding the same result (Meyer, 1901; Overton, 1901; Missner, 2009). Over many orders of magnitude, among many chemically disparate structures, potency of all anesthetics correlated near-perfectly with solubility in a particular non-polar, lipid-like, hydrophobic environment, characterized by (what is now recognized as) a low Hildebrand coefficient lambda (15.2–19.3 SI Units), closely resembling olive oil and benzene.

This naturally led to the notion that anesthetics act in a unitary fashion in lipid regions of brain cell surface membranes to disrupt membrane fluidity and neuronal signaling, a notion that persisted through much of the 20th century. However, this notion is wrought with inconsistencies (Lugli *et al.*, 2009). As a general example, an increase in body temperature will disrupt membrane fluidity comparable to an anesthetic, yet this does not result in anesthesia, as incorrectly predicted by the lipid-theory (Eger *et al.*, 1965). More specifically, there exists a class of compounds[1] whose structure and lipid solubility are predicted by the Meyer–Overton correlation to act as general anesthetics, but rather exert convulsive effects (Koblin *et al.*, 1994; Fang *et al.*, 1996). Chemically homologous anesthetics, such as n-alkanes, show increasing potency with increasing molecular size consistent with the Meyer–Overton correlation; yet, above a critical molecular size the series loses anesthetic potency despite increasing lipid solubility (Raines & Miller, 1994).

Finally, enantiomers of many general anesthetics can differ greatly in their anesthetic potency despite the similar oil/gas partition coefficients (Nau & Strichartz, 2002; Franks & Lieb, 1991). This is problematic for the lipid-theory as enantiomers, which are chemically identical compounds with non-superposable mirror image structures, exert identical physiochemical effects in an achiral environment, such as the lipid bilayer. Hyrdophobic pockets within proteins, however are chiral environments.

[1]These compounds are referred to as non-immobilizers.

As neuronal firing and membrane excitability are attributed to protein receptors and ion channel proteins, these became the next logical target. In the 1980s, Franks and Lieb showed Meyer–Overton correlations with anesthetic effects on isolated protein functions, rather than lipid membranes (Franks & Lieb, 1984). Further, anesthetics are relatively small, and non-polar molecules that are consistent with Meyer–Overton solubility, but too large (greater than eight carbon alkanes) lack anesthetic properties, presumably because they cannot "fit" into non-polar hydrophobic regions within proteins.

Which proteins mediate anesthetic action? Anesthetics were shown to impair post-synaptic dendritic-somatic integration, with little or no effects on axonal firing. A search began for one or more post-synaptic membrane protein receptors or channels to account for anesthetic action, and anesthetics were shown to bind with relatively high affinity to receptors for GABA, nicotinic acetylcholine, glycine, serotonin and certain ion channels. Yet results for anesthetic effects on isolated functions of these proteins were mixed and confusing. Eger and colleagues concluded there was no evidence that anesthetics exert their effects on any particular membrane protein, or a set of membrane proteins, and despite Meyer–Overton they all must act differently (Eger *et al.*, 2008). If a unitary anesthetic mechanism was to be found, it was suggested to return to membrane lipid-based theories. Yet, the problems stated above, still remain. Cytoskeletal proteins were not considered.

Anatomically, general anesthetics have been shown to bind in, and affect various brain regions including the posterior cingulate cortex, orbitofrontal cortex, the right angular gyri, and the thalamus (Fiset *et al.*, 1999). Different effects of anesthesia may be differentiated. Loss of consciousness appears to be associated with the cortex (Imas *et al.*, 2005a, 2005b; Plourde *et al.*, 2006), loss of memory (amnesia) with the limbic system (Uchida *et al.*, 2000; Simon *et al.*, 2001; Alkire & Nathan, 2005), and loss of movement and pain (immobility and analgesia, respectively) with the spinal cord (Grasshoff & Antkowiak, 2004; Haseneder *et al.*, 2004; Rehberg *et al.*, 2004; Wakai *et al.*, 2005) Current theories of anesthetic action use functional and/or anatomical processes related to neural correlates of consciousness, e.g., John and Prichep's anesthetic cascade (John & Prichep, 2005), Mashour's cognitive unbinding theory (Mashour, 2004), Flohr's information processing theory (Flohr, 1995), Alkire and colleagues unified narcosis theory (Alkire *et al.*, 2000). However, these network-level approaches fail to account for the Meyer–Overton effect, and offer no molecular-level mechanisms for pharmacological approach.

Thus, modern mainstream anesthesia, neuroscience and pharmacology offer neither functional targets nor mechanisms of action for direct effects of anesthetics on consciousness or memory. Yet deciphering the mechanisms of anesthesia is still the best possible approach to understanding consciousness and memory encoding. During anesthesia many non-conscious brain activities continue, and no new memories are generated, yet old memories are still retained. Despite the effect of anesthesia non-conscious brain activities including sensory-evoked potentials continue, indicating that anesthetics are quite selective for consciousness. Understanding consciousness might require an understanding of anesthesia, and *vice versa*. Anesthesia and consciousness may be part of the same mystery, making anesthetics natural probes for consciousness.

What do anesthetics share in common to allow them to produce such a unitary effect? Urban *et al.* (2006) make it explicitly clear that the empirical observation that the Meyer–Overton rule is held so closely in the interactions between proteins and anesthetics underscores the importance of hydrophobic effects in anesthetic action. The high selectivity and stereo-specificity of ligand and protein interactions indicates that the involvement of weak intermolecular forces (i.e., van der Waals interactions) is substantial (Urban, 2008). Consistent with this, quantum chemical and statistical thermodynamic investigations of anesthetic activity have shown that anesthetics must possess a large polarizability (Hobza *et al.*, 1981, 1982; Hobza & Sandorfy, 1984). Additionally, these same studies indicate that anesthetics must contain an acidic hydrogen. Such an inclusion is important for the polar interactions and breakage of hydrogen bonds for the halogenated hydrocarbon anesthetics (isoflurane, enflurane, sevoflurane, desflurane, halothane and chloroform) (Abraham *et al.*, 1991; Urban & Haydon, 1987; Davies *et al.*, 1976), however, hydrogen bonds may be broken by compounds that themselves do not form hydrogen bonds (i.e., n-alkanes with gramicidin A) (Elliott *et al.*, 1983; Hendry *et al.*, 1978). Thus, while hydrogen bond breaking appears to be important to anesthetic function the existence of acidic hydrogen may not be a necessity.

Thus, anesthetic action appears to depend on a molecule's hydrophobicity, dipolarity/polarizability, and potential to disrupt hydrogen bonds. Attempts have been made to identify the contribution of each of these factors with results indicating that they all contribute to the overall interaction (Abraham *et al.*, 1991; Davies *et al.*, 1976) albeit these analyses did not examine the hydrophobic and weak polar interactions directly.

Physical chemistry research clearly shows that anesthetics bind in non-polar regions via the weak forces of van der Waals', London dipole dispersion' interactions, disturbing pi-electron resonance clouds, hydrogen bonds and hydrophobic interactions. Therefore, the "how" of anesthetic binding is well understood. The main questions to be addressed then are "where" this binding takes place, and "why" does disruption of these weak forces result in loss of consciousness.

Allison & Nunn (1968) showed that the anesthetic halothane caused complete depolymerization of microtubules, but at about five times the amount needed to cause loss of consciousness. Binding studies revealed anesthetics bind only weakly in tubulin, the component protein of microtubules, roughly 1000 times weaker than anesthetic binding to $GABA_A$ receptors. But there are 1000 to 10,000 times more tubulins per neuron than there are $GABA_A$ receptors, and about 10 potential binding sites per tubulin (Craddock *et al.*, 2012). So at clinically relevant concentrations, anesthetics clearly bind to microtubules, though without causing depolymerization.

Is anesthetic binding to tubulin the functional mechanism for loss of consciousness? Eckenhoff's group (Xi *et al.*, 2004; Pan *et al.*, 2007) systematically surveyed anesthetic-binding proteins, measuring radiolabeled halothane bound to proteins in brain neurons from mice and human brain samples. Using chromatography they found halothane binding to 23 membrane proteins and 34 cytoplasmic proteins (Xi *et al.*, 2004; Pan *et al.*, 2007). Among the anesthetic-binding cytoplasmic proteins were cytoskeletal actin, and tubulin, the subunit component of microtubules.

Which anesthetic-binding proteins mediate anesthesia and consciousness? Techniques which address functional mechanisms involve genomics and proteomics, e.g., measuring changes in gene expression following exposure. Proteomic analysis of genetic expression following halothane anesthesia in animals points to protein networks involved in "neuronal growth, proliferation, division and communication" (Pan *et al.*, 2008), all microtubule-dependent functions. In rodent brain cortical neurons, genetic expression of seven proteins changed following either halothane or isoflurane anesthetic exposure, with only three proteins affected by both anesthetics. These three included tubulin, and two others, a heat shock protein, and an acetyltransferase. Genetic expression of membrane proteins did not change (Pan *et al.*, 2008). Other studies show alterations in tubulin gene expression for three days after desflurane (Futterer *et al.*, 2004), and 28 days following sevoflurane exposure (Kalenka *et al.*, 2007). Using optogenetics, Emerson *et al.* (2013) found that the fluorescent anesthetic 1-azidoanthracene also acts through tubulin and microtubules at the non-polar, hydrophobic colchicine site. The results of Emerson *et al.* (2013) are consistent with the Meyer–Overton correlation. The cumulative evidence suggests anesthetics act on microtubules to prevent consciousness. What precisely do they do there?

The "Meyer–Overton correlation" shows that anesthetics act (by quantum London "dipole dispersion" forces), in non-polar regions involving pi-electron resonance clouds. As anesthesia is fairly selective (e.g., sparing non-conscious brain activities which continue during anesthesia), consciousness apparently originates in non-polar pi-resonance regions in the "quantum underground." In Craddock *et al.* (2015) we simulated tubulin three-dimensional structure and showed that pi-resonance clouds of aromatic amino acid rings are clustered together near the van der Waals radius separation. This enables quantum dipole couplings and coherence among rings, and creates a non-polar solubility region consistent with Meyer–Overton. The pi-resonance clusters were geometrically arrayed with seven other clusters within tubulin, each separation less than 2 nm. As we know from photosynthesis proteins, this is sufficient for quantum resonance transfer through the entire 8-nm length of a tubulin dimer, a "quantum underground" (Craddock *et al.*, 2014). The tubulin quantum underground can align its neighboring tubulins in microtubule lattices, forming pathways of a quantum underground spanning mesoscopic lengths of microtubules, e.g., within pyramidal neuron dendrites and soma (Fig. 11).

In Craddock *et al.* (2012) we also showed that halothane and other anesthetics bind in tubulin's quantum underground. Coupled with genomic, proteomic and optogenetic results from the Eckenhoff group, it seems likely anesthetic action in the microtubule quantum underground mediates anesthesia and loss of consciousness, but what precisely do they do there? What critical function does anesthesia prevent?

As described earlier, several approaches suggest consciousness depends on collective quantum dipole oscillations of pi-resonance clouds in microtubules, and that anesthetics act to prevent the coupled oscillations. Penrose–Hameroff Orch OR suggests quantum dipoles in microtubules relate to processes in fundamental space-time geometry, through quantum gravity and general relativity (Hameroff & Penrose, 1996). Quantum mobility theory places consciousness in quantum oscillations in microtubules, without quantum gravity (Hameroff & Stuart, 2002; Craddock *et al.*, 2014).

In Craddock *et al.* (2015), we showed the anesthetic halothane binds a mere 4.5 Angstroms from a phenyl ring pair in the "Meyer–Overton

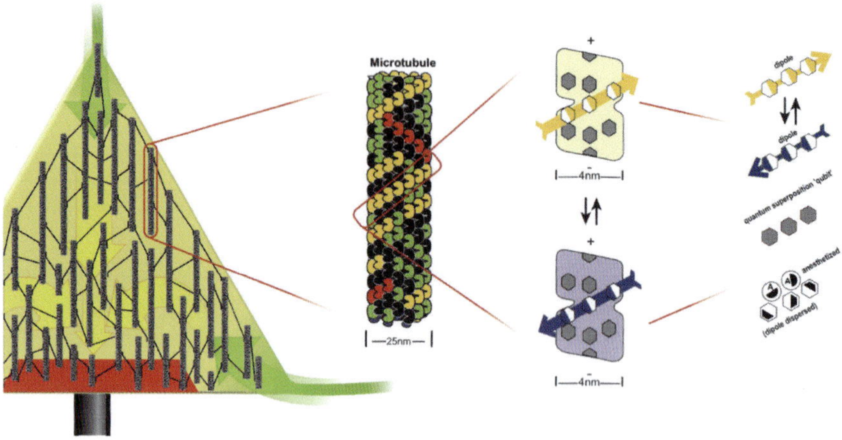


**Fig. 11.** Inward to the quantum underground. From left, pyramidal neuron cell body with microtubules, single microtubule, row of three tubulins with aligned pi-resonance dipoles, dipole oscillations and anesthetic dampening.

quantum underground." Dipole-coupled switching for a single phenyl (benzene) ring has an energy barrier of $4.5 \times 10^{-20}$ J, with an average frequency of 68 Terahertz (THz). In the presence of halothane 4.5 Angstroms away the energy barrier for coupled oscillations increases 20% to $5.1 \times 10^{-20}$ J. Thus, anesthetics block normally-occurring Terahertz quantum dipole oscillations, an activity we propose occurs in the microtubule quantum underground, e.g., in dendrites and soma of layer V cortical pyramidal neurons, and is critical to consciousness (Fig. 11).

Terahertz beating in microtubules may be part of a self-similar scale-invariant hierarchy leading to EEG and important for consciousness. Bandyopadhyay's group appears to have shown self-similar spectra for electrical resonances with similar spectra in gigahertz and megahertz ranges for tubulin and microtubules, and kilohertz for bundles of microtubules (Hameroff *et al.*, 2013) (Fig. 12). Another self-similar spectral pattern at the next slowest frequency band gives, essentially, EEG rhythms. Terahertz in oscillations microtubule quantum underground may provide the fine scale clocking mechanism for a scale-invariant brain hierarchy, one in which consciousness occurs.

Moving downward, or inward in the brain's scale-invariant hierarchy, and upward in frequency to terahertz in pi-resonance rings is seemingly

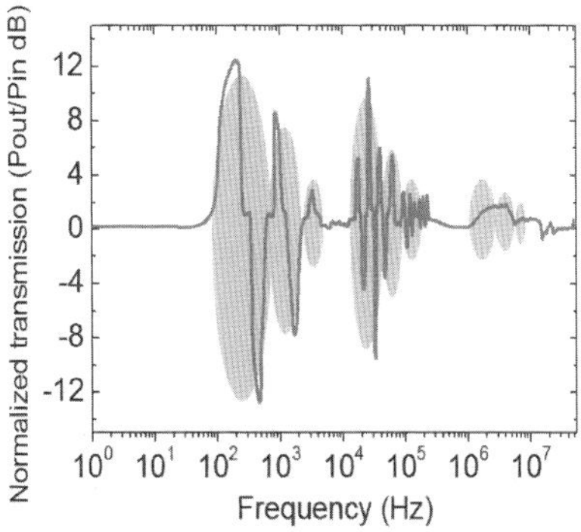


**Fig. 12.** Self similar electrical resonance spectral pattern for microtubules.

reductionist, but reaching the quantum underground provides non-locality and global coherence. Mind and life originate in the quantum underground.

## 13.11. Conclusions

In this chapter, we have discussed the hypothesis that quantum coherent oscillations may be playing a roles in life and consciousness and outlined a molecular basis for this possibility. In view of recent experimental evidence of quantum beats in light-harvesting complexes of plants and bacteria, it is logical to seek similar effects in important structures found in animal cells, especially in neurons. After all, it is common in nature to preserve evolutionary advances. If bacteria and plants found a way to harness electromagnetic energy using quantum processes, why would animals not find a use for this method of energy and information transmission. We argue that microtubules, which play critical roles in all eukaryotic cells, possess structural and functional characteristics that are consistent with quantum coherent excitations in the aromatic groups of their aromatic residues. We further outline the consequences of these findings on neuronal processes including the emergence of consciousness. It is speculative at this stage but not far-fetched to expect quantum mechanisms to provide important clues regarding such long-lasting puzzles as the efficiency of bioenergetic conversion processes and the enigma of the unitary sense of self.

Scientific understanding of the conscious mind may well depend on understanding what life is, as both have common, ineffable qualities of unitary oneness and global unity. Two promising approaches for both mind and life are scale-invariant hierarchies, and quantum coherence. Here we combine these, extend the brain's scale-invariant hierarchy inward, within the neuron, and further within cytoskeletal microtubules, specifically in an oil-like non-polar solubility phase (the "quantum underground") where quantum dipole oscillations occur which are necessary for consciousness.

The brain's network structure and electrodynamics both exhibit scale-invariant hierarchical organization, with "fractal-like" self-similar information patterns following $1/f$ power law distributions, repeating at different spatiotemporal scales. For example neuronal "nuclei," or clusters of cell bodies and synapses are arranged in the brain as "small world networks," like airport hubs, with relatively few large hubs having many connections, and many smaller hubs with few connections. Electrodynamically, brain EEG shows self-similar patterns at scales from roughly $10^2$ hertz ("high gamma"), through 10 hertz (alpha, beta), hertz (delta, theta) and 0.1 hertz (default mode oscillations) (He *et al.*, 2010; Chialvo, 2010). But to date, hierarchies stop at the neuronal membrane. We extend structural and dynamical scale-invariant brain hierarchies to faster, deeper levels inside neurons, and specifically to microtubules, major components of the cytoskeleton, which organize cellular activities.

Microtubules are cylindrical lattice protein polymers apparently shown to have self-similar resonance vibrations at gigahertz ($10^9$ Hz), megahertz ($10^6$ Hz) and kilohertz ($10^3$ Hz) scales (Ghosh *et al.*, 2014). EEG rhythms in 1 to 100 Hz are suggested to derive as 'beat frequencies' of faster microtubule vibrations (e.g., megahertz $10^6$ Hz) inside neurons (Hameroff & Penrose, 2014). In microtubules, mechanical and electromagnetic vibrations couple in megahertz frequencies (Sahu *et al.*, 2011; Sahu *et al.*, 2013).

Smaller scales run into the quantum world. Recent experimental evidence of quantum beats in light-harvesting proteins of plants and bacteria suggest quantum mechanisms would be useful for energy and information processing in animal cells, especially brain neurons. After all, it is common in nature to preserve evolutionary advances, and if bacteria and plants are able to harness electromagnetic energy using quantum processes, would not animals be so able? But the biochemical milieu at quantum-level scales is generally considered too warm, wet and noisy for useful quantum effects. And indeed polar, aqueous phases in neuronal cytoplasm are seemingly inhospitable. But oil and water do not mix, and oil-like solubility phases in living systems, unlike aqueous phases, are conducive to quantum coherence.

Biomolecules tend to have non-polar, "hydrophobic" (water-excluding) interiors comprised largely of pi-electron resonance clouds of aromatic amino acid rings. These regions constitute a particular "olive oil-like" solubility phase in which anesthetic molecules bind and act to selectively erase consciousness (the "Meyer–Overton correlation"). Unlike aqueous phases, non-polar regions of pi-resonance are conducive to quantum mechanisms, such as seen in plant photosynthesis. In plant proteins involved in photosynthesis, arrays of eight pi-resonance "chromophores" mediate quantum coherence necessary for food production, with "excitons," or quantum resonance transfers propagating among all chromophores simultaneously. We have found within "tubulin," the protein subunit of microtubules, clusters and channels of pi-resonance clouds of aromatic amino acid rings, including an array of eight tryptophans which can feasibly support excitons and resonance transfers. As tubulins adjoin neighboring tubulins in microtubule lattices, quantum excitons, or resonance transfers can extend in quantum channels through entire lengths of microtubules. Evidence indicates anesthetics act in microtubule quantum channels to specifically erase consciousness. Along with similar regions inside other proteins, membranes and nucleic acids, non-polar quantum channels constitute a "quantum underground" pervading, and regulating living systems.

The quantum underground in microtubules looks like the bottom floor, the faster, deeper limit of the brain's hierarchy, the origin of consciousness. The pi-resonance clouds have natural resonance vibrations in terahertz, $10^{12}$ hertz. Using pairs of phenyl (benzene) pi-resonance rings with natural dipole oscillations of 68 terahertz ($68 \times 10^{12}$ Hz) as a model system, Craddock *et al.* (2015) showed that the energy barrier for these oscillations is increased by 20% when an anesthetic molecule is nearby. We suggest this effect prevents cooperative global coherence of quantum states among pi-resonance rings in the quantum underground in microtubules (particularly within dendrites and soma of cortical layer V pyramidal neurons). Consciousness is prevented, while non-quantum, non-conscious brain activities continue. And as quantum coherence continues in non-polar regions too small for anesthetic binding, during anesthesia, "life goes on."

Are coupled terahertz oscillations of pi-resonance rings in the microtubule quantum underground truly the limit, the small, fast end of a scale-invariant brain hierarchy in which consciousness occurs? Or beyond terahertz, do quantum brain vibrations engage the structure of the universe? In any case, the quantum underground in brain microtubules should be the focus of therapeutic approaches to mental states and cognitive disorders.

## Acknowledgments

TJAC would like to acknowledge funding support for this work from Nova Southeastern University's College of Psychology. JAT acknowledges funding support received from the Natural Sciences and Engineering Council of Canada (NSERC).

## References

Abraham, M. H., Lieb, W. R. & Franks, N. P. (1991) Role of hydrogen bonding in general anesthesia. *J. Pharm. Sci.*, **80**, 719–724.

Adamatzky, A. (2010) Aromatic arithmetic. *Nat. Phys.*, **6**, 325–326.

Albert, R. & Barabási, A. L. (2002) Statistical mechanics of complex networks. *Rev. Mod. Phys.*, **74**, 47–97.

Alberts, B. (1998) The cell as a collection of protein machines: Preparing the next generation of molecular biologists. *Cell*, **92**, 291–293.

Alkire, M. T. & Nathan, S. V. (2005) Does the amygdala mediate anesthetic-induced amnesia? Basolateral amygdala lesions block sevoflurane-induced amnesia. *Anesthesiology*, **102**, 754–760.

Alkire, M. T., Haier, R. J. & Fallon, J. H. (2000) Toward a unified theory of narcosis: brain imaging evidence for a thalamocortical switch as the neurophysiologic basis of anesthetic-induced unconsciousness. *Conscious. Cogn.*, **9**, 370–386.

Allison, A. C. & Nunn, J. F. (1968) Effects of general anesthetics on microtubules: A possible mechanism of anesthesia. *Lancet*, **2**, 1329–1396.

Almeida, A. R., Amado, I. F., Reynolds, J., Berges, J., Lythe, G., Molina-París, C. & Freitas, A. A. (2012) Quorum-sensing in $CD4^+$ T cell homeostasis: A hypothesis and a model. *Front. Immunol.*, **3**, 125.

Andrews, D. L., Curutchet, C. & Scholes, G. D. (2011) Resonance energy transfer: Beyond the limits. *Laser Photon. Rev.*, **5**, 114–123.

Angelini, T. E., Golestanian, R., Coridan, R. H., Butler, J. C., Beraud, A., Krisch, M., Sinn, H., Schweizer, K. S. & Wong, G. C. L. (2006) Counterions between charged polymers exhibit liquid-like organization and dynamics. *Proc. Nat. Acad. Sci. USA*, **103**, 7962–7967.

Angelini, T. E., Liang, H., Wriggers, W. & Wong, G. C. L. (2003) Like-charge attraction between polyelectrolytes induced by counterion charge density waves. *Proc. Nat. Acad. Sci. USA*, **100**, 8634–8637.

Atallah, B. V. & Scanziani, M. (2009) Instantaneous modulation of gamma oscillation frequency by balancing excitation with inhibition. *Neuron,* **62,** 566–577.

Axmacher, N., Mormann, F., Fernández, G., Elger, C. E. & Fell, J. (2006) Memory formation by neuronal synchronization. *Brain Res. Rev.,* **52,** 170–182.

Bandyopadhyay, A., Pati, R., Sahu, S., Peper, F. & Fujita, D. (2010) Massively parallel computing on an organic molecular layer. *Nat. Phys,* **6,** 369–375.

Barabási, A. L. & Albert, R. (1999) Emergence of scaling in random networks. *Science,* **286,** 509–512.

Barabási, A. L. (2009) Scale-free networks: A decade and beyond. *Science,* **325,** 412–413.

Barbieri, F. & Brunel, N., (2008) Can attractor network models account for the statistics of firing during persistent activity in prefrontal cortex? *Front. Neurosci.,* **2,** 114–122.

Bardouille, T. & Boe, S. (2012) State-related changes in MEG functional connectivity reveal the task-positive sensorimotor network. *PLoS One,* **7,** e48682.

Bartos, M., Vida, I. & Jonas, P. (2007) Synaptic mechanisms of synchronized gamma oscillations in inhibitory interneuron networks. *Nat. Rev. Neurosci.,* **8,** 45–56.

Beljonne, D., Curutchet, C., Scholes, G. D., Silbey, R. J. (2009) Beyond Forster resonance energy transfer in biological and nanoscale systems. *J. Phys. Chem. B,* **113,** 6583–6599.

Breakspear, M., Williams, L. M. & Stam, C. J. (2004) A novel method for the topographic analysis of neural activity reveals formation and dissolution of 'dynamic cell assemblies'. *J. Comput. Neurosci.,* **16,** 49–68.

Broderick, G. & Craddock, T. J. A. (2013) Systems biology of complex symptom profiles: capturing interactivity across behavior, brain and immune regulation. *Brain Behav. Immunity,* **29,** 1–8.

Bullmore, E. & Sporns, O. (2009) Complex brain networks: Graph theoretical analysis of structural and functional systems. *Nat. Rev. Neurosci.,* **10,** 186–198.

Calhoun, T. R., Ginsberg, N. S., Schlau-Cohen, G. S., Cheng, Y.-C., Ballottari, M., Bassi, R. & Fleming, G. R. (2009) Quantum coherence enabled determination of the energy landscape in light-harvesting complex II. *J. Phys. Chem. B,* **113,** 16291–16295.

Cantiello, H. F., Patenaude, C. & Zaner, K. (1991) Osmotically induced electrical signals from actin Filaments. *Biophys. J.,* **59,** 1284–1289.

Cardin, J. A., Carlén, M., Meletis, K., Knoblich, U., Zhang, F., Deisseroth, K., Tsai, L. H. & Moore, C. I. (2009) Driving fast-spiking cells induces gamma rhythm and controls sensory responses. *Nature,* **459,** 663–667.

Chalmers, D. J. (2006) Strong and weak emergence. In: *The Reemergence of Emergence,* (P. Davies & P. Clayton, eds.) Oxford University Press, New York. pp. 244–256.

Chang, J. J. (2008) Physical properties of biophotons and their biological functions, *Ind. J. Exp. Biol.*, **46**, 371–377.

Chelminiak, P. W., Tuszynski, J. A. & Dixon, J. M. (2009) Effects of C-termini distribution on ionic transport along microtubules. *Interdiscip. Sci. Comput. Life Sci.*, **1**, 108–112.

Chialvo, D. (2010) Emergent complex neural dynamics. *Nat. Phys.*, **6**, 744–750.

Chuang, H. Y., Hofree, M. & Ideker, T. (2010) A decade of systems biology. *Ann. Rev. Cell Develop. Biol.*, **26**, 721–744.

Cohen, S, & Popp, F. A. (1997) Biophoton emission of the human body. *J. Photochem. Photobiol. B*, **40**, 187–189.

Collini, E., Wong, C. Y., Wilk, K. E., Curmi, P. M. G., Brumer, P. & Scholes, G. D. (2010) Coherently wired light-harvesting in photosynthetic marine algae at ambient temperature. *Nature*, **463**, 644–647.

Compte, A., Constantinidis, C., Tegnér, J., Raghavachari, S., Chafee, M. V., Goldman-Rakic, P. S. & Wang, X. J. (2003) Temporally irregular mnemonic persistent activity in prefrontal neurons of monkeys during a delayed response task. *J. Neurophysiol.*, **90**, 3441–3454.

Craddock, T. J. A., Hameroff, S. R., Ayoub, A. T., Klobukowski, M. & Tuszynski, J. A. (2015) Anesthetics act in quantum channels in brain microtubules to prevent consciousness. *Curr. Topics Medic. Chem.*, **15**, 523–533.

Craddock, T. J. A., Tuszynski, J. A. & Hameroff, S. (2012) Cytoskeletal signaling: Is memory encoded in microtubule lattices by CaMKII phosphorylation? *PLoS Comput. Biol.*, **8**, e1002421.

Craddock, T. J. A., Tuszynski, J. A., Priel, A. & Freedman, H. (2010) Microtubule ionic conduction and its implication for higher cognitive functions. *J. Integr. Neurosci.*, **9**, 103–122.

Craddock, T. J., Beauchemin, C. & Tuszynski, J. A. (2009) Information processing mechanisms in microtubules at physiological temperature: Model predictions for experimental tests. *Biosystems*, **97**, 28–34.

Craddock, T. J., George, M. S., Freedman, H., Barakat, K. H., Damaraju, S., Hameroff, S. & Tuszynski, J.A. (2012) Computational predictions of volatile anesthetic interactions with the microtubule cytoskeleton: implications for side effects of general anesthesia. *PLoS One*, **7**, e37251.

Craddock, T. J., Priel, A. & Tuszynski, J. A. (2014) Keeping time: Could quantum beating in microtubules be the basis for the neural synchrony related to consciousness? *J. Integr. Neurosci.*, **13**, 293–311.

Craddock, T.J.A., Friesen, D., Mane, J., Hameroff, S. & Tuszynski, J.A. (2014) The feasibility of coherent energy transfer in microtubules. *J. Roy. Soc. Interf.*, **11**, 20140677.

Crick, F. & Koch, C. (1990) Towards a neurobiological theory of consciousness. In: *Seminars in the Neurosciences*, Vol. 2. Saunders Scientific Publications, pp. 263–275.

Davies, R. H., Bagnall, R. D., Bell, W. & Jones, W. G. M. (1976) The hydrogen bond proton donor properties of volatile halogenated hydrocarbons and ethers and their mode of action in anaesthesia. *Int. J. Quant. Chem.*, **10**, 171–185.

Dennett, D. C. (1991) *Consciousness Explained*. Boston, MA: Little Brown.

Desie, G., Boens, N. & De Schryver, F. C. (1986) Study of the time-resolved tryptophan fluorescence of crystalline alpha-chymotrypsin. *Biochemistry*, **25**, 8301–8308.

Devaraj, B, Scott, R. Q., Roschger, P. & Inaba, H. (1991) Ultraweak light emission from rat liver nuclei, *Photochem. Photobiol.*, **54**, 289–293.

DeVille, R. L. & Peskin, C. S. (2012) Synchrony and asynchrony for neuronal dynamics defined on complex networks. *Bull. Math. Biol.*, **74**, 769–802.

Dixon, J. M., Chelminiak, P. & Tuszynski, J. A. (2008) Surface conductance in seamless microtubules. *Physica A*, **387**, 4183–4194.

Edelman, G. M. & Gally, J. A. (2013) Reentry: A key mechanism for integration of brain function. *Front. Integr. Neurosci.*, **7**, 63.

Eger, E. I., Raines, D. E., Shafer, S. L., Hemmings, H. C. Sonner, J. M. (2008). Is a new paradigm needed to explain how inhaled anesthetics produce immobility? *Anesth. Analg.*, **107**, 832–848.

Eger, E. I., Saidman, L. J. & Brandstater, B (1965). Temperature dependence of halothane and cyclopropane anesthesia in dogs: Correlation with some theories of anesthetic action. *Anesthesiology*, **26**, 764–770.

Eguiluz V. M., Chialvo D. R., Cecchi G. A., Baliki M. & Apkarian A. V. (2005) Scale-free brain functional networks. *Phys. Rev. Lett.*, **94**, 018102.

Eisinger, J., Feuer, B. & Lamola, A. A. (1969) Intramolecular singlet excitation transfer. Applications to polypeptides. *Biochemistry*, **8**, 3908–3915.

Elliott, J. R., Needham, D., Dilger, J. P. & Haydon, D. A. (1983) The effects of bilayer thickness and tension on gramicidin single-channel lifetime. *Biochim. Biophys. Acta (BBA)-Biomembranes*, **735**, 95–103.

Emerson, D. J., Weiser, B. P., Psonis, J., Liao, Z., Taratula, O., Fiamengo, A., Wang, X., Sugasawa, K., Smith, A. B., Eckenhoff, R. G. & Dmochowski, I. J. (2013) *J. Am. Chem. Soc.*, **135**, 5389–5398.

Engel, G. S., Calhoun, T. R., Read, E. L., Ahn, T. K., Mancal, T., Cheng, Y. C., Blankenship, R. E. & Fleming, G.R. (2007) Evidence for wavelike energy transfer through quantum coherence in photosynthetic systems. *Nature*, **446**, 782–786.

Faden, A. I. & Loane, D. J. (2015) Chronic neurodegeneration after traumatic brain injury: Alzheimer disease, chronic traumatic encephalopathy, or persistent neuroinflammation? *Neurotherapeutics*, **12**, 143–150.

Fang, Z., Sonner, J., Laster, M. J., Ionescu, P., Kandel, L., Koblin, D. D., Eger, E. I. & Halsey, M. J. (1996) Anesthetic and convulsant properties of aromatic compounds and cycloalkanes: Implications for mechanisms of narcosis. *Anesth. Analg.*, **83**, 1097–1104.

Feinberg, T. E. (2011) The nested neural hierarchy and the self. *Conscious. Cogn.*, **20**, 4–15.

Fell, J. & Axmacher, N. (2011) The role of phase synchronization in memory processes. *Nat. Rev. Neurosci.*, **12**, 105–118.

Fels, D. (2009) Cellular communication through light. *PLoS One*, **4**, e5086.

Fiset, P., Paus, T., Daloze, T., Plourde, G., Meuret, P., Bonhomme, V., Hajj-Ali, N., Backman, S. B. & Evans, A. C. (1999) *J. Neurosci.*, **19**, 5506–5513.

Fleming, G. R., Scholes, G. D. & Cheng, Y. C. (2011) Quantum effects in biology. *Proc. Chem.*, **3**, 38–57.

Flintoft, L. (2010) Systems biology: Small genome, complex regulation. *Nat. Rev. Gen.*, **11**, 3.

Flohr, H. (1995) An information processing theory of anaesthesia. *Neuropsychologia*, **33**, 1169–1180.

Forster, T. (1948) Excitation transfer. *Ann. Phys.*, **2**, 12–19.

Franks, N. P. & Lieb, W. R. (1984) Do general anaesthetics act by competitive binding to specific receptors? *Nature*, **310**, 599–601.

Franks, N. P. & Lieb, W. R. (1991) Stereoselective effects of inhalational general anesthetic optical isomers on nerve ion channels. *Science*, **30**, 427–30.

Freedman, H., Rezania, V., Priel, A., Carpenter, E., Noskov, S. Y. & Tuszynski, J.A. (2010) Model of ionic currents through microtubule nanopores and the lumen. *Phys. Rev. E*, **81**, 051912.

Fries, P. (2009) Neuronal gamma-band synchronization as a fundamental process in cortical computation. *Ann. Rev. Neurosci.*, **32**, 209–224.

Fries, P., Nikolic, D. & Singer, W. (2007) The gamma cycle. *Trends Neurosci.*, **30**, 309–316.

Friesen, D. E., Craddock, T. J., Kalra, A. P. & Tuszynski, J. A. (2015) Biological wires, communication systems, and implications for disease, *Biosystems*, **127**, 14–27.

Fröhlich, H. (1968) Long-range coherence and energy storage in biological systems. *Int. J. Quant. Chem.*, **2**, 641–649.

Fröhlich, H. (1975) The extraordinary dielectric properties of biological materials and the action of enzymes. *Proc. Nat. Acad. Sci. USA*, **72**, 4211–4215.

Fröhlich, H. (1978) Coherent electric vibrations in biological systems and the cancer problem. *IEEE Trans. Microwave Theory and Tech.*, **26**, 613–618.

Funk, R. H. W., Monsees, T. & Özkucur, N. (2009) Electromagnetic effects — from cell biology to medicine. *Prog. Histochem. Cytochem.*, **43**, 177–264.

Furukawa, K. Y., Wang, P. J., Yao, W., Fu, W., Mattson, M. P., Itoyama, Y., Onodera, H., D'Souza, P. H., Poorkaj, P. H., Bird, T. D. & Schellenberg, G. D., (2003) Alteration in calcium channel properties is responsible for the neurotoxic action of a familial frontotemporal dementia tau mutation. *J. Neurochem.*, **87**, 427–436.

Furukawa, K., Mattson, M. P., (1995) Taxol stabilizes $[Ca^{2+}]_i$ and protects hippocampal neurons against excitotoxicity. *Brain Res.*, **689**, 141–146.

Fütterer, C. D., Maurer, M. H., Schmitt, A., Feldmann, R. E., Kuschinsky, W. & Waschke, K. F. (2004) *Anesthesiology*, **100**, 302–308.

Fyhn, M., Molden, S., Witter, M. P., Moser, E. I. & Moser, M. B. (2004) Spatial representation in the entorhinal cortex. *Science,* **305**, 1258–1264.

Gardiner, J., Overall, R. & Marc, J. (2011) The microtubule cytoskeleton acts as a key downstream effector of neurotransmitter signaling. *Synapse,* **65**, 249–256.

Gartzke, J. & Lange, K. (2002) Cellular target of weak magnetic fields: ionic conduction along actin filaments of microvilli. *Amer. J. Physiol. Cell Physiol.,* **283**, 1333–1346.

Gazzaniga, M. S., ed. (2004) *The Cognitive Neurosciences*. MIT Press, Camb., MA.

Gennaro, R. J. (2014) *In Defense of the HOT Thesis: A BIT of The Consciousness Paradox*. MIT Press, Camber., MA.

Gennaro, R. J. ed. (2004) *Higher-order Theories of Consciousness: An Anthology,* Vol. 56. John Benjamins Publishing, Amsterdam.

Ghosh, S. & Bandyopadhyay, A. Ecplising & replacing the two most fundamental religions of science: Turing world of computation and Darwin's world of evolution. Function space: Social learning network for science blog.

Ghosh, S., Aswani, K., Singh, S., Sahu, S., Fujita, D. & Bandyopadhyay, A. (2014) Design and construction of a brain-like computer: A new class of frequency-fractal computing using wireless communication in a supramolecular organic, inorganic system. *Information,* **5**, 28–100.

Goh, K. I., Oh, E., Jeong, H., Kahng, B. & Kim, D. (2002) Classification of scale-free networks. *Proc. Natl. Acad. Sci. USA,* **99**, 12583–12588.

Goldstein, L. E., Fisher, A. M., Tagge, C. A., Zhang, X. L., Velisek, L., Sullivan, J. A. & McKee, A. C. (2012) Chronic traumatic encephalopathy in blast-exposed military veterans and a blast neurotrauma mouse model. *Sci. Trans. Med.,* **4**(134), 134ra60–134ra60.

Grasshoff, C. & Antkowiak, B. (2004) Propofol and sevoflurane depress spinal neurons in vitro via different molecular targets. *Anesthesiology* **101**, 1167–1176.

Greenfield, S. A. & Collins, T. F. (2005) A neuroscientific approach to consciousness. *Prog. Brain Res.,* **150**, 11–587.

Guha, S., Rawat, S. S., Chattopadhyay, A. & Bhattacharyya, B. (1996) Tubulin conformation and dynamics: A red edge excitation shift study. *Biochemistry,* **35**, 13426–13433.

Guo, D. (2011) Inhibition of rhythmic spiking by colored noise in neural systems. *Cogn. Neurodyn.,* **5**, 293–300.

Hafting, T., Fyhn, M., Molden, S., Moser, M. B. & Moser, E. I. (2005) Microstructure of spatial map in the entorhinal cortex. *Nature,* **436**, 801–806.

Haider, B. & McCormick, D. A. (2009) Rapid neocortical dynamics: cellular and network mechanisms. *Neuron,* **62**, 171–189.

Hameroff, S. & Penrose, R. (2014) Consciousness in the universe — A review of the 'Orch OR'theory. *Phys. Life Rev.,* **11**, 39–78.

Hameroff, S. & R. Watt (1982) Information processing in microtubules. *J. Theor. Biol.,* **98**, 549–561.

Hameroff, S. (2007) The brain is both neurocomputer and quantum computer. *Cogn. Sci.*, **31**, 1035–1045.

Hameroff, S. R., Dayhoff, J. E., Beltra-Lahoz, R., Samsonovich, A. V. & Rasmussen, S. (1992) Models for molecular computation: Conformational automata in the cytoskeleton. *Computer*, **25**, 30–39.

Hameroff, S. R., Nip, A., Porter, M. & Tuszynski, J. A. (2002) Conduction pathways in microtubules, biological quantum computation, and consciousness. *Biosystems*, **64**, 149–168.

Hameroff, S. R., Craddock, T. J. A. & Tuszynski, J. A. (2010) Memory bytes molecular match for CaMKII phosphorylation encoding of microtubule lattices. *J. Integr. Neurosci.*, **9**, 253–267.

Hameroff, S. R., Smith, S. A. & Watt, R. C. (1986) Automaton model of dynamic organization in microtubules. *Ann. NY Acad. Sci.*, **466**, 949–952.

Hameroff, S. & Penrose, R. (1996) Orchestrated reduction of quantum coherence in brain microtubules: A model for consciousness. *Math. Comput. Simul.*, **40**, 453–480.

Hameroff, S., Nip, A., Porter, M. & Tuszynski, J. (2002) Conduction pathways in microtubules, biological quantum computation, and consciousness. *Biosystems*, **64**, 149–168.

Hameroff, S., Penrose R., Ghosh, S., Sahu, S. & Bandyopadhyay, A. (2014) Evidence of massive global synchronization and the consciousness Comment on "Consciousness in the universe: A review of the 'Orch OR' theory". *Phys. Lett. Rev.*, **11**, 83–84.

Hammond, C., Bergman, H. & Brown, P. (2007) Pathological synchronization in Parkinson's disease: Networks, models and treatments. *Trends Neurosci.*, **30**, 357–364.

Hanel, R., Pöchacker, M., Schölling, M. & Thurner, S. (2012) A self-organized model for cell-differentiation based on variations of molecular decay rates. *PloS one*, **7**, e36679.

Harel, E. & Engel, G. S. (2012) Quantum coherence spectroscopy reveals complex dynamics in bacterial light-harvesting complex 2 (LH2) *Proc. Nat. Acad. Sci. USA*, **109**, 706–711.

Harris, B. D., Moody, E. J., Basile, A. S. & Skolnick, P. (1994) Volatile anesthetics bidirectionally and stereospecifically modulate ligand binding to GABA receptors. *Eur. J. Pharmacol.*, **267**, 269–274.

Harris, B., Moody, E. & Skolnick, P. (1992) Isoflurane anesthesia is stereoselective. *Eur. J. Pharmacol.*, **217**, 215–216.

Haseneder, R., Kurz, J., Dodt, H.-U. U., Kochs, E., Zieglgänsberger, W., Scheller, M., Rammes, G. & Hapfelmeier, G. (2004) Isoflurane reduces glutamatergic transmission in neurons in the spinal cord superficial dorsal horn: evidence for a presynaptic site of an analgesic action. *Anesth. Analg.*, **98**, 1718–1723.

He, B. J., Zempel J. M, Snyder A. Z. & Raichle, M. E. (2010) The temporal structure and functional significance of scale-free brain activity. *Neuron*, **66**, 353–369.

Hendry, B. M., Urban, B. W. & Haydon, D. A. (1978) The blockage of the electrical conductance in a pore-containing membrane by the n-alkanes. *Biochim. Biophys. Acta (BBA)-Biomembranes*, **513**, 106–116.

Hobza, P. & Sandorfy, C. (1984) Quantum chemical and statistical thermodynamic investigations of anesthetic activity. 3. The interaction between CH4, CH3Cl, $CH_2Cl_2$, $CHCl_3$, $CCl_4$, and an O–H… O hydrogen bond. *Can. J. Chem.*, **62**, 606–609.

Hobza, P., Mulder, F. & Sandorfy, C. (1981) Quantum chemical and statistical thermodynamic investigations of anesthetic activity. 1. The interaction between chloroform, fluoroform, cyclopropane and an OH. O hydrogen bond. *J. Amer. Chem. Soc.*, **103**, 1360–1366.

Hobza, P., Mulder, F. & Sandorfy, C. (1982) Quantum chemical and statistical thermodynamic investigations of anesthetic activity. 2. The interaction between chloroform, fluoroform, and an NH... O: C hydrogen bond. *J. Amer. Chem. Soc.*, **104**, 925–928.

Hoel, E. P., Albantakis, L. & Tononi, G. (2013) Quantifying causal emergence shows that macro can beat micro. *Proc. Natl. Acad. Sci. USA*, **110**, 19790–19795.

Hogan, C. J. (2008a) Measurement of quantum fluctuations in geometry. *Phys. Rev. D*, **77**, 104031.

Hogan, C. J. (2008b) Indeterminacy of holographic quantum geometry. *Phys. Rev. D*, **78**, 087501.

Hooft, G. T. (1999) Quantum gravity as a dissipative deterministic system. *Class. Quantum Grav.*, **16**, 3263.

Ideker, T., Galitski, T. & Hood, L. (2001). A new approach to decoding life: Systems biology. *Ann. Rev. Genom. Human Gen.*, **2**, 343–372.

Imas, O. A., Ropella, K. M., Ward, B. D., Wood, J. D. & Hudetz, A. G. (2005a) Volatile anesthetics disrupt frontal — posterior recurrent information transfer at gamma frequencies in rat. *Neurosci. Lett.*, **387**, 145–150.

Imas, O. A., Ropella, K. M., Ward, B. D., Wood, J. D. & Hudetz, A. G. (2005b) Volatile anesthetics enhance flash-induced $\gamma$ oscillations in rat visual cortex. *Anesthesiology*, **102**, 937–947.

Ingber, D. E., Wang, N., Stamenovic, D. (2014) Tensegrity, cellular biophysics, and the mechanics of living systems. *Rep. Progr. Phys.*, **77**, 046603.

Isojima, Y., Isoshima, T., Nagai, K., Kikuchi, K. & Nakagawa, H. (1995), Ultraweak biochemiluminescence detected from rat hippocampal slices, *NeuroReport*, **6**, 658–660.

Jan Spitzer J. & Poolman, B. (2005), Electrochemical structure of the crowded cytoplasm, *Trends in Biochem. Sci.*, **30**, 536–541.

Janke, C. & Kneussel, M. (2010) Tubulin post-translational modifications: Encoding functions on the neuronal microtubule cytoskeleton. *Trend Neurosci.*, **33**, 362–372.

Janmey, P. A. (1998) The cytoskeleton and cell signaling: Component localization and mechanical coupling. *Physiol. Rev.*, **78**, 763–781.

Jensen, O., Kaiser, J. & Lachaux, J. P., (2007) Human gamma-frequency oscillations associated with attention and memory. *Trends Neurosci.*, **30**, 317–324.

John, E. R. & Prichep, L. S. (2005) The anesthetic cascade: a theory of how anesthesia suppresses consciousness. *Anesthesiology*, **102**, 447–471.

Johnson, B. D., Byerly, L. (1993) A cytoskeletal mechanism for calcium channel metabolic dependence and inactivation by intracellular calcium. *Neuron*, **10**, 797–804.

Johnson, B. D., Byerly, L., (1994) $Ca^{2+}$ channel $Ca^{2+}$ -dependent inactivation in a mammalian central neuron involves the cytoskeleton. *Pfluegers Arch.*, **429**, 14–21.

Kalenka, A., Hinkelbein, J., Feldmann, R. E., Kuschinsky, W., Waschke, K. F. & Maurer, M. H. (2007) *Anesth. Analg.*, **104**, 1129–1135.

Karreman, G., Steele, R. H. & Szent-Gyorgi, A. (1958) On resonance transfer of excitation energy between aromatic amino acids in proteins. *Proc. Nat. Acad. Sci. USA*, **44**, 140–143.

Kataoka, Y., Cui, Y., Yamagata, A., Niigaki, M., Hirohata, T., Oishi, N. & Watanabe, Y. (2001) Activity-dependent neural tissue oxidation emits intrinsic ultraweak photons. *Biochem. Biophys. Res. Commun.*, **285**, 1007–1011.

Kaushalya, S. K. & Maiti, S. (2007) Quantitative imaging of serotonin autofluorescence with multiphoton microscopy. In: A. Chattopadhyay, ed. *Serotonin Receptors in Neurobiology.*, Boca Raton, FL: CRC Press, pp. 1–18.

Kitzbichler M. G., Smith M. L., Christensen S. R. & Bullmore E. (2009) Broadband criticality of human brain network synchronization. *PLoS Comput. Biol.*, **5**, e1000314.

Kobayashi, M. & Inaba, H. (2000) Photon statistics and correlation analysis of ultraweak light originating from living organisms for extraction of biological information. *Appl. Opt.*, **39**, 183–192.

Kobayashi, M., Kikuchi, D. & Okamura, H. (2009) Imaging of ultraweak spontaneous photon emission from human body displaying diurnal rhythm. *PLoS One*, **4**, e6256.

Kobayashi, M., Takeda, M., Ito, K. I., Kato, H. & Inaba, H. (1999a) Two-dimensional photon counting imaging and spatiotemporal characterization of ultraweak photon emission from a rat's brain *in vivo*. *J. Neurosci. Meth.*, **93**, 163–168.

Kobayashi, M., Takeda, M., Sato, T., Yamazaki, Y., Kaneko, K., Ito, K., Kato, H. & Inaba, H. (1999b) *In vivo* imaging of spontaneous ultraweak photon emission from a rat's brain correlated with cerebral energy metabolism and oxidative stress. *Neurosci. Res.*, **34**, 103–113.

Koblin, D. D., Chortkoff, B. S., Laster, M. J., Eger, E. I., Halsey, M. J. & Ionescu, P. (1994) Polyhalogenated and perfluorinated compounds that disobey the Meyer–Overton hypothesis. *Anesth. Analg.*, **79**, 1043–1048.

Koch, C. (2004) *The Quest for Consciousness*. Roberts & Co. Publishers, Englewood, Co.

Krasylenko Y., Yemets A. & Blume Y. (2013) Plant microtubules reorganization under the indirect UV-B exposure and during UV-B-induced programmed cell death. *Plant Signal Behav.*, **8**, e24031.

Lambert, N., Chen, Y. N., Cheng, Y. C., Li, C. M., Chen, G. Y. & Nori, F. (2013) Quantum biology. *Nat. Phys.*, **9**, 10–18.

Langton, C. G., ed. (1997) *Artificial Life: An Overview*. MIT Press.

Lashley, K. (1950) In search of the engram. In: *Physiological Mechanisms in Animal Behavior*. New York: Academic Press, pp. 454–482.

Lau, H. & Rosenthal, D. (2011) Empirical support for higher-order theories of conscious awareness. *Trends in Cogn. Sci.*, **15**, 365–373.

Lee, U., Ku, S., Noh, G., Baek, S., Choi, B. & Mashour, G. A. (2013) Disruption of frontal-parietal communication by ketamine, propofol, and sevoflurane. *Anesthesiology*, **118**, 1264–1275.

Lin, E. C., Cantiello, H. F. (1993) A novel method to study the electrodynamic behavior of actin filaments: Evidence for cable-like properties of actin. *Biophys. J.*, **65**, 1371–1378.

Lugli, A. K., Yost, C. S. & Kindler, C. H. (2009) Anaesthetic mechanisms: update on the challenge of unravelling the mystery of anaesthesia. *Eur. J. Anaesthesiol.*, **26**, 807–820.

Lysko, G. S., Robinson, J. L., Casto, R. & Ferrone, R. A. (1994) The stereospecific effects of isoflurane isomers *in vivo*. *Eur. J. Pharmacol.*, **263**, 25–29.

Margulis, L. & Sagan, D. (1995) *What is Life?* University of California Press.

Markram, H. (2006) The blue brain project. *Nat. Rev. Neurosci.*, **7**, 153–160.

Mashour, G. A. (2004) Consciousness unbound: toward a paradigm of general anesthesia. *Anesthesiology*, **100**, 428–433.

Matej P., Brežan S., Jerman I. (2013) The principle of coherence in multi-level brain information processing. *Progr. Biophys. Mol. Biol.*, **111**, 8–29.

Mattia, M. & Del Giudice, P. (2004) Finite-size dynamics of inhibitory and excitatory interacting spiking neurons. *Phys. Rev. E*, **70**, 052903.

Melloni, L., Molina, C., Pena, M., Torres, D., Singer, W. & E. Rodriguez (2007) Synchronization of neural activity across cortical areas correlates with conscious perception. *J. Neurosci.*, **27**, 2858–2865.

Meyer, D. K., Olenik, C., Hoffmann, F., Barth, H, Leemhuis, J., Brunig, I., Aktories, K. & Norenberg, W. (2000) Regulation of somatodendritic $GABA_A$ receptor channels in rat hippocampal neurons: Evidence for a role of the small GTPase Rac1. *J. Neurosci.*, **20**, 6743–6751.

Meyer, H (1901) Zur theorie der alkolnarkose: der einfuss wehselnder temperatur auf wirkungsstarke und theilungscofficient der narcotica (in German). *Naunyn Schmied Arch. Exp. Path Pharmakol.*, **46**, 338–346.

Mihic, S. J., Ye, Q., Wick, M. J., Koltchine, V. V., Krasowski, M. D. & Finn, S. E. *et al.* (1997) Sites of alcohol and volatile anaesthetic action on $GABA_A$ and glycine receptors. *Nature*, **389**, 385–389.

Missner, A. & Pohl, P. (2009) 110 years of the Meyer-Overton rule: predicting membrane permeability of gases and other small compounds. *Chemphyschem. Eur. J. Chem. Phys. Phys. Chem.*, **10**, 1405–1414.

Nakano, M. (2005) Low-level chemiluminescence during lipid peroxidations and enzymatic reactions. *J. Biolumin. Chemilum.*, **4**, 231–240.

Nau, C. & Strichartz, G. R. (2002) Drug chirality in anesthesia. *Anesthesiology*, **97**, 497–502.

Naundorf, B., Wolf, F. & Volgushev, M. (2006) Unique features of action potential initiation in cortical neurons. *Nature*, **440**, 1060–1063.

Neves-Petersen, M. T., Gryczynski, Z., Lakowicz, J., Fojan, P., Pedersen, S., Petersen, E. & Petersen, S. B. (2002) High probability of disrupting a disulphide bridge mediated by an endogenous excited tryptophan residue. *Protein Sci.*, **11**, 588–600.

O'Keefe, J. & Dostrovsky, J. (1971) The hippocampus as a spatial map. Preliminary evidence from unit activity in the freely moving rat. *Brain Res.*, **34**, 171–175.

O'Keefe, J. (1976) Place units in the hippocampus of the freely moving rat. *Exp. Neurol.*, 51, 78–109.

Oizumi, M., Albantakis, L. & Tononi, G. (2014) From the phenomenology to the mechanisms of consciousness: Integrated information theory 3.0. *PLoS Comput. Biol.*, **10**, e1003588.

Ostroumov, E. E., Mulvaney, R. M., Cogdell, R. J. & Scholes, G. D. (2013) Broadband 2D electronic spectroscopy reveals a carotenoid dark state in purple bacteria. *Science*, **340**, 52–56.

Overton, E. (1901) *Studien ueber die Narkose*. Jena, Switzerland: Gustav Fischer.

Pan, J. Z., Xi, J., Eckenhoff, M. F. & Eckenhoff, R. G. (2008) Inhaled anesthetics elicit region-specific changes in protein expression in mammalian brain. *Proteomics*, **8**, 983–2992.

Pan, J. Z., Xi, J., Tobias, J. W., Eckenhoff, M. F., Eckenhoff, R. G. (2007) Halothane binding proteome in human brain cortex. *J. Proteome Res.*, **6**, 582–592.

Panitchayangkoon, G., Hayes, D. Fransted, K. A., Caram, J. R., Harel, E., Wen, J., Blankenship, R. E. & Engel, G. S. (2010) Long-lived quantum coherence in photosynthetic complexes at physiological temperature. *Proc. Nat. Acad. Sci. USA*, **107**, 12766–12770.

Petermann, T., Thiagarajan, T. C., Lebedev, M. A., Nicolelis, M. A. L., Chialvo, D. R. & Plenz, D. (2009) Spontaneous cortical activity in awake monkeys composed of neuronal avalanches. *Proc. Natl. Acad. Sci. USA*, **106**, 15921–15926.

Plankar, M., Brežan, S. & Jerman, I. (2013) The principle of coherence in multi-level brain information processing. *Prog. Biophys. Mol. Biol.*, **111**, 8–29.

Plourde, G., Belin, P., Chartrand, D., Fiset, P., Backman, S. B., Xie, G. & Zatorre, R. J. (2006) Cortical processing of complex auditory stimuli alterations of consciousness with the general anesthetic propofol. *Anesthesiology*, **104**, 448–457.

Pribram, K. H. (1991) *Brain and Perception: Holonomy and Structure in Figural Processing*. Psychology Press.

Priel, A., Tuszynski, J. A. (2008) A nonlinear cable-like model of amplified ionic wave propagation along microtubules. *Eur. Lett.*, **83**, 68004.

Priel, A., Tuszynski, J. A. & Cantiello, H. F. (2005a) Electrodynamic signaling by the dendritic cytoskeleton: Toward an intracellular information processing model. *Electromag. Biol. Med.*, **24**, 221–231.

Priel, A., Tuszynski, J. A. & Woolf, N. J. (2010) Neural cytoskeleton capabilities for learning and memory. *J. Biol. Phys.*, **36**, 3–21.

Priel, A., Tuszynski, J. A. & Woolf, N. J. (2005b) Transitions in microtubule C-termini conformations as a possible dendritic signaling phenomenon. *Eur. Biophys. J.*, **35**, 40–52.

Quickenden, T. I. & Que Hee, S. S. (1974) Weak luminescence from the yeast *Saccharomyces cerevisiae, Biochem. Biophys. Res. Commun.*, **60**, 764–770.

Raines, D. E. & Miller, K. W. (1994) On the importance of volatile agents devoid of anesthetic action. *Anesth. Analg.*, **79**, 1031–1033.

Rasmussen, S., Karampurwala, H., Vaidyanath, R., Jensen, K. S. & Hameroff, S. (1990) Computational connectionism within neurons: A model of cytoskeletal automata subserving neural networks. *Phys. D: Nonlin. Phenom.*, **42**, 428–449.

Rehberg, B., Grünewald, M., Baars, J., Fuegener, K., Urban, B. W. & Kox, W. J. (2004) Monitoring of immobility to noxious stimulation during sevoflurane anesthesia using the spinal H-reflex. *Anesthesiology*, **100**, 44–50.

Renart, A., Moreno-Bote, R., Wang, X. J. & Parga, N. (2007) Mean-driven and fluctuation-driven persistent activity in recurrent networks. *Neural Comput.*, **19**, 1–46.

Revonsuo, A. (1999) Binding and the phenomenal unity of consciousness. *Conscious. Cogn.*, **8**, 173–185.

Reynolds, A., Rubin, J., Clermont, G., Day, J., Vodovotz, Y. & Bard Ermentrout, G., (2006) A reduced mathematical model of the acute inflammatory response: I. Derivation of model and analysis of anti-inflammation. *J. Theor. Biol.*, 242, 220–236.

Sahu, S., Ghosh, S., Hirata, K., Fujita, D. & Bandyopadhyay, A. (2011) Computational myths and mysteries that have grown around microtubule in the last half a century and their possible verification. *J. Comput. Theor. Nanosci.*, **8**, 509–515.

Sahu, S., Ghosh, S., Hirata, K., Fujita, D. & Bandyopadhyay, A. (2013) Multi-level memory-switching properties of a single brain microtubule. *Appl. Phys. Lett.*, **102**, 123701.

Sandorfy, C. (2004) Hydrogen bonding and anaesthesia. *J. Mol. Struct.*, **708**, 3–5.

Sargolini, F., Fyhn, M., Hafting, T., McNaughton, B. L., Witter, M. P., Moser, M. B. & Moser, E. I. (2006) Conjunctive representation of position, direction, and velocity in the entorhinal cortex. *Science*, **312**, 758–762.

Sataric, M. V., Ilic, D. I, Ralevic, N., Tuszynski, J. A. (2009) A nonlinear model of ionic wave propagation along microtubules, *Eur. Biophys. J.*, **38**, 637–647.

Schlau-Cohen, G. S., Calhoun, T. R., Ginsberg, N. S., Read, E. L., Ballottari, M., Bassi, R. van Grondelle, R. & Fleming, G. R. (2009) Pathways of energy flow in LHCII from two-dimensional electronic spectroscopy. *J. Phys. Chem. B*, **113**, 15352–15363.

Schlau-Cohen, G. S., Ishizaki, A., Calhoun, T. R., Ginsberg, N. S., Ballottari, M., Bassi, R. & Fleming, G. R. (2012) Elucidation of the timescales and origins of quantum electronic coherence in LHCII. *Nat. Chem.*, **4**, 389–395.

Schrödinger, E. (1943) *What Is Life? The Physical Aspect of the Living Cell and Mind.* Dublin.

Schubert, T., Akopian, A. (2004) Actin filaments regulate voltage-gated ion channels in salamander retinal ganglion cells. *Neuroscience*, **125**, 583–590.

Scott, A. (1995) *Stairway to the Mind: The Controversial New Science of Consciousness.* Springer.

Sekulić, L., Satarić, B. M., Tuszynski, J. A. & Satarić, M. V. (2011) Nonlinear ionic pulses along microtubules. *Eur. Phys. J.*, **34**, 49–60.

Senkowski, D., Schneider, T. R., Foxe, J. J.& Engel A. K. (2008) Crossmodal binding through neural coherence: Implications for multisensory processing. *Trends Neurosci.*, **31**, 401–409.

Shcherbatko, A., Ono, F., Mandel, G., Brehm, P. (1999) Voltage-dependent sodium channel function is regulated through membrane mechanics. *Biophys. J.*, **77**, 1945–1959.

Simon, W., Hapfelmeier, G., Kochs, E., Zieglgänsberger, W. & Rammes, G. (2001) Isoflurane blocks synaptic plasticity in the mouse hippocampus. *Anesthesiology*, **94**, 1058–1065.

Singer, W. (2009) Distributed processing and temporal codes in neuronal networks. *Cogn. Neurodyn.*, **3**, 189–196.

Spencer, K. M., Nestor, P. G., Niznikiewicz, M. A., Salisbury, D. F., Shenton, M. E. & McCarley, R. W. (2003) Abnormal neural synchrony in schizophrenia. *J. Neurosci.*, **23**, 7407–7411.

Stam, C. J. & De Bruin, E. A. (2004) Scale-free dynamics of global functional connectivity in the human brain. *Human Brain Mapping*, **22**, 97–109.

Staxén I., Bergounioux C. & Bornman, J. F. (1993) Effect of ultraviolet radiation on cell division and microtubule organization in Petunia hybrida protoplasts. *Protoplasma*, **173**, 70–76.

Strege, P. R., Holm, A. N., Rich, A., Miller, S. M., Ou, Y., Sarr, M. G. & Farrugia, G. (2003) Cytoskeletal modulation of sodium current in human jejunal circular smooth muscle cells. *Amer. J. Physiol. Cell Physiol.*, **284**, 60–66.

Strümpfer, J., Şener, M. & Schulten, K. (2012) How quantum coherence assists photosynthetic light-harvesting. *J. Phys. Chem. Lett.*, **3**, 536–542.

Sun, H., Hu, X. Q., Emerit, M. B., Schoenebeck, J. C., Kimmel, C. E., Peoples, R. W., Miko, A. & Zhang, L. (2008) Modulation of 5-HT3 receptor desensitization by the light chain of microtubule-associated protein 1B expressed in HEK 293 cells. *J. Physiol (Lond.)* **586**, 751–762.

Sun, Y., Wang, C. H. & Dai, J. (2010) Biophotons as neural communication signals demonstrated by *in situ* biophoton autography. *Photochem. Photobiol. Sci.*, **9**, 315–322.

Tafur, J., Van Wijk, E. P., Van Wijk, R. & Mills, P. J. (2010) Biophoton detection and low-intensity light therapy: A potential clinical partnership. *Photomed. Laser Surg.*, **28**, 23–30.

Takeda, M., Kobayashi, M., Takayama, M., Suzuki, S., Ishida, T., Ohnuki, K., Moriya, T. & Ohuchi, N. (2004) Biophoton detection as a novel technique for cancer imaging, *Cancer Sci.*, **95**, 656–661.

Taylor, K., Mandon, S., Freiwald, W. & Kreiter, A. (2005) Coherent oscillatory activity in monkey area V4 predicts successful allocation of attention. *Cerebral Cortex*, **15**, 1424–1437.

Thar, R. & Kuhl, M. (2004) Propagation of electromagnetic radiation in mitochondria? *J. Theor. Biol.*, **230**, 261–270.

Tononi, G. & Koch, C. (2008) The neural correlates of consciousness. *Ann. NY Acad. Sci.*, **1124**, 239–261.

Tononi, G. (2004) An information integration theory of consciousness. *BMC Neurosci.*, **5**, 42.

Turner, D. B., Dinshaw, R., Lee, K.-K., Belsley, M. S., Wilk, K. E., Curmi, P. M. G. & Scholes, G. D. (2012) Quantitative investigations of quantum coherence for a light-harvesting protein at conditions simulating photosynthesis. *Phys. Chem. Chem. Phys.*, **14**, 4857–4874.

Tuszynski, J. A. & Woolf, N. (2006) The path ahead. In: *The Emerging Physics of Consciousness*. Berlin, Heidelberg: Springer, pp. 1–26.

Tuszynski, J. A. Brown, J. A. & Hawrylak, P. (1998) Dielectric polarization, electrical conduction, information processing and quantum computation in microtubules, are they plausible? *Trans. Roy. Soc. A*, **356**, 1897–1926.

Tuszynski, J. A., Brown, J. A., Crawford, E., Carpenter, E. J., Nip, M. L. A., Dixon J. M. & Sataric, M. V. (2005) Molecular Dynamics simulations of tubulin structure and calculations of electrostatic properties of microtubules, *Math. Comput. Model.*, **41**, 1055–1070.

Tuszynski, J., Portet, S., Dixon, J., Luxford, C., Cantiello, H. (2004) Ionic wave propagation along actin filaments. *Biophys. J.*, **86**, 1890–1903.

Uchida, S., Nakayama, H., Maehara, T., Hirai, N., Arakaki, H., Nakamura, M., Nakabayashi, T. & Shimizu, H. (2000) Suppression of gamma activity in the human medical temporal lobe by sevoflurane anesthesia. *Neuroreport*, **11**, 39–42.

Uhlhaas, P. J. & Singer, W. (2006) Neural synchrony in brain disorders: Relevance for cognitive dysfunctions and pathophysiology. *Neuron*, **52**, 155–168.

Uhlhaas, P. J. & Singer, W. (2010) Abnormal neural oscillations and synchrony in schizophrenia. *Nat. Rev. Neurosci.*, **11**, 100–113.

Uhlhaas, P. J., Pipa, G., Lima, B., Melloni, L., Neuenschwander, S., Nikolić, D. & Singer, W. (2009) Neural synchrony in cortical networks: History, concept and current status. *Front. Integr. Neurosci.*, **3**, 17.

Urban, B. W. (2008) The site of anesthetic action. In *Modern Anesthetics*. Berlin, Heidelberg: Springer, pp. 3–29.

Urban, B. W. & Haydon, D. A. (1987) The actions of halogenated ethers on the ionic currents of the squid giant axon. *Proc. R. Soc. Lond. Ser. B. Biol. Sci.*, **231**, 13–26.

Urban, B. W., Bleckwenn, M. & Barann, M. (2006) Interactions of anesthetics with their targets: Non-specific, specific or both? *Pharmacology & Therapeu.*, **111**, 729–770.

Vale, R. D. (2003) The molecular motor toolbox for intracellular transport. *Cell*, **112**, 467–480.

Van De Ville D., Britz J. & Michel C. M. (2010) EEG microstate sequences in healthy humans at rest reveal scale-free dynamics. *Proc. Natl. Acad. Sci. USA*, **107**, 18179–18184.

Van Wijk, R., Ackerman, J. M. & van Wijk, E. P. (2006) Effects of a color filter used in auriculomedicine on ultraweak photon emission of the human body. *J. Altern. Complement. Med.*, **12**, 955–962.

Vanhooren, A., Devreese, B., Vanhee, K., Van Beeumen, J. & Hanssens, I. (2002) Photoexcitation of tryptophan groups induces reduction of two disulfide bonds in goat α-lactalbumin. *Biochemistry*, **41**, 11035–11043.

Vattay, G., Kauffman, S. & Niiranen, S. (2014) Quantum biology on the edge of quantum chaos. *PloS one*, **9**, e89017.

Vattay, G., Salahub, D. & Csabai, I. (2015) Quantum criticality at the origin of life. arXiv preprint arXiv:1502.06880.

Visser, N. V., Westphal, A. H., van Hoek, A., van Mierlo, C. P. M., Visser, A. J. W. G. & van Amerongen, H., (2008) Tryptophan–tryptophan energy migration as a tool to follow apoflavodoxin folding. *Biophys. J.*, **95**, 2462–2469.

Voeikov, V. (2001) Reactive oxygen species, water, photons and life. *Riv. Biol.*, **94**(2), 237–258.

Wakai, A., Kohno, T., Yamakura, T., Okamoto, M., Ataka, T. & Baba, H. (2005) Action of isoflurane on the substantia gelatinosa neurons of the adult rat spinal cord. *Anesthesiology*, **102**, 379–386.

Walker, E. H. (1970) The nature of consciousness. *Math. Biosci.*, **7**, 131–178.

Wang, N., Butler, J. P., Ingber, D. E. (1993) Mechanotransduction across the cell surface and through the cytoskeleton. *Science*, **260**, 1124–1127.

Wang, N., Tytell, J. D. & Ingber, D. E. (2009) Mechanotransduction at a distance: Mechanically coupling the extracellular matrix with the nucleus. *Nat. Rev. Mol. Cell Biol.*, **10**, 75–82.

Wang, Y., Barakat, A. & Zhou, H. (2010) Electrotonic coupling between pyramidal neurons in the neocortex. *PLoS One*, **5**, e10253.

Wang, Y., Benton, L. A., Singh, V. & Pielak, G. J. (2012) Disordered protein diffusion under crowded conditions. *J. Phys. Chem. Lett.*, **3**, 2703–2706.

Weber, G. & Shinitzky, M. (1970) Failure of energy transfer between identical aromatic molecules on excitation at the long wave edge of the absorption spectrum. *Proc. Natl. Acad. Sci. USA*, **65**, 823–830.

Wegner D. M. (2002) *The Illusion of Conscious Will*. Cambridge, MA: MIT Press.

Weisenborn, P., Meder, H., Egmond, M. R., Visser, T. J. & van Hoek, A. (1996) Photophysics of the single tryptophan residue in Fusarium solani Cutinase: Evidence for the occurrence of conformational substates with unusual fluorescence behaviour. *Biophys. Chem.*, **58**, 281–288.

Wong, G. C. L., Pollack, L. (2010) Electrostatics of strongly charged biological polymers: Ion-mediated interactions and self-organization in nucleic acids and proteins. *Ann. Rev. Phys. Chem.*, **61**, 171–189.

Woolf, N. J., Priel, A. & Tuszynski, J. A. (2009) *Nanoneuroscience: Structural and Functional Roles of the Neuronal Cytoskeleton in Health and Disease*. Heidelberg: Springer-Verlag.

Woolf, N. J., Priel, A. & Tuszynski, J. A. (2010) Nanotechnology, Nanostructure, and Nervous System Disorders. In *Nanoneuroscience*. Berlin, Heidelberg: Springer, pp. 177–226.

Wu, L. Z., Sheng, Y. B., Xie, J. B. & Wang, W. (2008) Photoexcitation of tryptophan groups induced reduction of disulfide bonds in hen egg white lysozyme. *J. Mol. Struct.*, **882**, 101–106.

Xi, J., Liu, R., Asbury, G. R., Eckenhoff, M. F. & Eckenhoff, R. G. (2004) Inhalational anesthetic-binding proteins in rat neuronal membranes. *J. Biol. Chem.*, **279**, 19628–19633.

Yoon, Y. Z., Kim, J., Lee, B. C., Kim, Y. U., Lee, S. K. & Soh, K. S. (2005) Changes in ultraweak photon emission and heart rate variability of epinephrine-injected rats. *Gen. Physiol. Biophys.*, **24**, 147–159.

Zamansky, G. B. & Chou, I. N. (1987) Environmental wavelengths of ultraviolet light induce cytoskeletal damage. *J. Invest. Dermatol.*, **89**, 603–606.

Zaremba, T. G., LeBon, T. R., Millar, D. B., Smejkal, R. M. & Hawley, R. J. (1984) Effects of ultraviolet light on the *in vitro* assembly of microtubules. *Biochemistry*, **23**, 1073–1080.

Biophysics of Consciousness: A Foundational Approach
R. R. Poznanski, J. A. Tuszynski and T. E. Feinberg
Copyright © 2017 World Scientific, Singapore.

# 14

# CONSCIOUSNESS IN THE UNIVERSE AN UPDATED REVIEW OF THE "ORCH OR" THEORY

*Stuart R. Hameroff,* and Roger Penrose*[†]

* *Anesthesiology and Psychology, Center for Consciousness Studies, Banner-University Medical Center*
*The University of Arizona, Tucson, Arizona, USA*

[†]*Mathematical Institute and Wadham College, University of Oxford, Oxford, UK*

## Abstract

The nature of consciousness, the mechanism by which it occurs in the brain, and its ultimate place in the universe are unknown. We proposed in the mid 1990's that consciousness depends on biologically "orchestrated" coherent quantum processes in collections of microtubules within brain neurons, that these quantum processes correlate with, and regulate, neuronal synaptic and membrane activity, and that the continuous Schrödinger evolution of each such process terminates in accordance with the specific Diósi–Penrose (DP) scheme of "objective reduction" ("OR") of the quantum state. This orchestrated OR activity ("Orch OR") is taken to result in moments of conscious awareness and/or choice. The DP form of OR is related to the fundamentals of quantum mechanics and space–time geometry, so Orch OR suggests that there is a connection between the brain's biomolecular processes and the basic structure of the universe. Here we review Orch OR in light of criticisms and developments in quantum biology, neuroscience, physics and cosmology. We also introduce novel suggestions of (1) beat frequencies of faster Orch OR microtubule dynamics (e.g. megahertz) as a possible source of the observed electroencephalographic ("EEG") correlates of consciousness and (2) that OR played a key role in life's evolution. We conclude that consciousness plays an intrinsic role in the universe.

*Keywords*: anthropic principle, consciousness, free will, Orch OR, microtubule, tubulin, objective reduction, OR, neuron, quantum, wavefunction, measurement problem, qualia, pi resonance, general relativity, space-time geometry, universe, electro-encephalography, EEG, anesthesia, memory, evolution, quantum computing.

## 14.1. Introduction: Consciousness in the Universe

Consciousness implies awareness: subjective, phenomenal experience of internal and external worlds. Consciousness also implies a sense of self, feelings, choice, control of voluntary behavior, memory, thought, language, and (e.g., when we close our eyes, or meditate) internally generated images and geometric patterns. But what consciousness actually *is* remains unknown. Our views of reality, of the universe, of ourselves depend on consciousness. Consciousness defines our existence.

Three general possibilities regarding the origin and place of consciousness in the universe have been commonly expressed.

(A) *Consciousness is not an independent quality but arose, in terms of conventional physical processes, as a natural evolutionary consequence of the biological adaptation of brains and nervous systems.* This prevalent scientific view is that consciousness emerged as a property of complex biological computation during the course of evolution. Opinions vary as to when, where and how consciousness appeared, e.g., only recently in humans, or earlier in lower organisms. Consciousness as an evolutionary adaptation is commonly assumed to be epiphenomenal (i.e., a secondary effect without independent influence (Dennett, 1991; Dennett & Kinsbourne, 1991; Wegner, 2002)), and also illusory (largely constructing reality, rather than perceiving it (Chalmers, 2012)). Nonetheless, consciousness is frequently argued to confer beneficial advantages to species (Dennett, 1995). Overall, in this view, consciousness is *not* an intrinsic feature of the universe.

(B) *Consciousness is a separate ("spiritual") quality, distinct from physical actions and not controlled by physical laws, that has always been in the universe*. "Descartes' dualism," religous viewpoints and other spiritual approaches assume consciousness has been in the universe all along, e.g., as the "ground of being," "creator" or component of an omnipresent "God" (Chopra, 2001). In this view, consciousness can causally influence physical matter and human behavior, but has no basis or description in science (Nadaeu & Kafatos, 2001; Kant, 1998). In another approach, panpsychism attributes consciousness to all matter, but without scientific identity *or* causal influence. Idealism contends consciousness is all that exists, the material world (and science) being an illusion (Berkeley, 1975). In all these views, consciousness lies outside science.

(C) *Consciousness results from discrete physical events; such events have always existed in the universe as non-cognitive, proto-conscious events, these acting as part of precise physical laws not yet fully understood. Biology evolved a mechanism to orchestrate such events and to couple them to neuronal activity, resulting in meaningful, cognitive, conscious moments and thence also to causal control of behavior.* These events are proposed specifically to be moments of quantum state reduction (intrinsic quantum "self-measurement"). Such events need not necessarily be taken as part of current theories of the laws of the universe, but should ultimately be scientifically describable. This is basically the type of view put forward, in very general terms, by the philosopher Whitehead (1929, 1933) and also fleshed out in a scientific framework in the Penrose–Hameroff theory of "orchestrated objective reduction" ("Orch OR") (Penrose & Hameroff, 1995, 2011; Hameroff & Penrose, 1996a,b, 2014; Hameroff, 1998a,b). In the Orch OR theory, these conscious events are terminations of quantum computations in brain microtubules reduced by Diósi–Penrose (DP) "objective reduction" ("OR"), and having experiential qualities. In this view, consciousness is an intrinsic feature of the action of the universe.

In summary, we have:

(A) Science/Materialism, with consciousness having no distinctive role.
(B) Dualism/Spirituality, with consciousness (etc.) being outside science.
(C) Science, with consciousness as an essential ingredient of physical laws not yet fully understood.

## 14.2. Consciousness, Computation and Brain Activities

### 14.2.1. *Unexplained features of consciousness*

How does the brain produce consciousness? Most scientists and philosophers view consciousness as an emergent property of complex computation among "integrate-and-fire" brain neurons which interconnect and switch at chemically mediated synapses. However, the mechanism by which such neuronal computation may produce conscious experience remains unknown (Koch, 2004; Chalmers, 1996). Specific unexplained features of consciousness include the following:

*The "hard problem"*: What is the nature of phenomenal experience, and what distinguishes conscious from non-conscious cognition? Perception and behavior may be accompanied or driven by phenomenal conscious awareness, experinces or subjective feelings, composed of what philosophers call "qualia" (Chalmers, 1996). However, perception and behavior may at other times be unaccompanied by consciousness. We could have evolved as full-time non-conscious "zombies" performing complex "autopilot" behaviors without conscious awareness. How and why do we have phenomenal consciousness, an "inner life" of subjective experience?

*'Binding':* Disparate sensory inputs are processed in different brain regions, at slightly different times, and yet are bound together into unified conscious content "binding" (van der Malsburg, 1999). How is conscious content bound together?

*Synchrony:* Neuronal membrane polarization states may be precisely synchronized over large regions of brain (Fries *et al.*, 2002), and also propagate through brain regions as synchronized zones (Hameroff, 2010). Does precise synchrony require electrical synapses ("gap junctions") and/or quantum entanglement? Does synchrony reflect discrete, unified conscious moments?

*'Non-computability' and causal agency:* As shown by Gödel's theorem, Penrose (1989, 1994) described how the mental quality of "understanding" cannot be encapsulated by any computational system and must derive from some "non-computable" effect. Moreover, the neurocomputational approach to volition, where algorithmic computation completely determines all thought processes, appears to preclude any possibility for independent causal agency, or free will. Something else is needed. What non-computable factor may occur in the brain?

*Cognitive behaviors of single cell organisms*: Protozoans like *Physarum* can escape mazes and solve problems, and *Paramecium* can swim, find food and mates, learn, remember and have sex, all without synaptic connections (Nakagaki *et al.*, 2000; Adamatzky 2012). They are not part of a network. How do single cells manifest intelligent behavior?

### 14.2.2. *Conscious moments and computation*

Consciousness has often been argued to be a sequence of discrete moments. James (1890) described the "specious present, the short duration of which we are immediately and incessantly sensible" (though James was vague about duration, and also described a continual "stream of consciousness"). The "perceptual moment" theory of Stroud (1956) described consciousness as a series of discrete events, like sequential frames of a movie (modern film and video present 24 to 72 frames per second, 24 to 72 Hertz, "Hz"). Consciousness is also seen as sequences of discrete events in Buddhism, trained meditators describing distinct "flickerings" in their experience of pure undifferentiated awareness (Tart, 1995). Buddhist texts portray consciousness as "momentary collections of mental phenomena," and as "distinct, unconnected and impermanent moments which perish as soon as they arise." Buddhist writings even quantify the frequency of conscious moments. For example the Sarvaastivaadins (Von Rospatt, 1995) described 6,480,000 'moments' in 24 hours (an average of one 'moment' per 13.3 ms, 75 Hz), and some Chinese Buddhism as one "thought" per 20 ms (50 Hz). The best measurable correlation of consciousness through modern science is gamma synchrony EEG, 30 to 90 Hz coherent neuronal membrane activity occurring across various synchronized brain regions. Slower periods, e.g., 4 to 7 Hz theta frequency, with nested gamma waves could correspond to saccades and visual gestalts (Woolf & Hameroff, 2001; Van Rullen & Koch, 2003). Thus, we may argue that consciousness consists of discrete events at varying frequencies occurring across brain regions, for example 40 conscious moments per second. What are these conscious moments?

The over-arching presumption in modern science and philosophy is that consciousness emerges from complex synaptic computation among brain neurons acting as fundamental information units. In digital computers, discrete voltage levels represent information units (e.g., "bits") in silicon logic gates. McCulloch & Pitts (1943) proposed such gates as integrate-and-fire artificial neurons, leading to "perceptrons" (Rosenblatt, 1962) and other types of "artificial neural networks" capable of learning and self-organized behavior. Similarly, according to the standard "Hodgkin–Huxley" model (Hodgkin & Huxley, 1952), biological neurons are "integrate-and-fire" threshold logic devices in which multiple branched dendrites and a cell body (soma) receive and integrate synaptic inputs as membrane potentials (Fig. 1). According to Hodgkin–Huxley, the integrated potential is then compared to a threshold potential at the axon hillock, or axon initiation segment (AIS). When AIS threshold is reached by the integrated potential, an all-or-none action potential "firing," or "spike" is triggered as output, conveyed along the axon to the next synapse. Cognitive networks of Hodgkin–Huxley neurons connected by variable strength synapses (Hebb, 1949) can self-organize and learn, their axonal firing outputs controlling downstream activity and behavior.

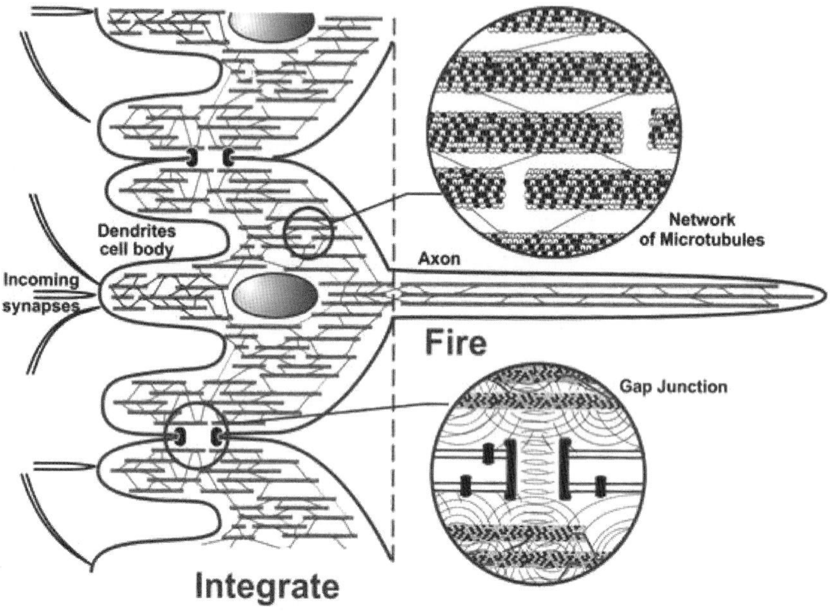


**Fig. 1.** An "integrate-and-fire" brain neuron, and portions of other such neurons are shown schematically with internal microtubules. In dendrites and cell body/ soma (left) involved in integration, microtubules are interrupted and of mixed polarity, interconnected by microtubule-associated proteins (MAPs) in recursive networks (upper circle, right). Dendritic–somatic integration (with contribution from microtubule processes) can trigger axonal firings to the next synapse. Microtubules in axons are unipolar and continuous. Gap junctions synchronize dendritic membranes, and may enable entanglement and collective integration among microtubules in adjacent neurons (lower circle right). In Orch OR, microtubule quantum computations occur during dendritic/somatic integration, and the selected results regulate axonal firings which control behavior.

How does consciousness arise from neurocomputation? Some contend that consciousness emerges from computational complexity due to firings and other brain electrical activity (Scott, 1995; Tononi, 2004). However neither the specific neuronal activities contributing to complexity, nor any predicted complexity threshold for emergence of consciousness have been put forth. Nor is there a sense of how complexity *per se* could give rise to discrete conscious moments. Others contend large scale, cooperative axonal firing outputs, "volleys," or "explosions" produce consciousness (Koch, 2004; Malach, 2007). But coherent axonal firings are in all cases preceded and caused by synchronized dendritic/somatic integrations. Indeed, gamma synchrony EEG, the best correlate of consciousness, is generated not by axonal firings, but by dendritic and somatic integration potentials. Accordingly, some suggest consciousness primarily involves neuronal dendrites and cell bodies/soma, i.e., in integration phases of "integrate-and-fire" sequences (Pribram, 1991; Eccles, 1992; Hameroff, 2012). Integration implies reduction of uncertainty, merging and consolidating multiple possibilities to one, e.g., selecting conscious perceptions and actions.

### 14.2.3. *Consciousness and dendritic integration*

Neuronal integration is commonly approximated as linear summation of dendritic/somatic membrane potentials (Fig. 2(a)). However, actual integration is not passive, actively involving complex processing (Shepherd, 1996; Sourdet & Debanne, 1999; Poirazi & Mel, 2001). Dendritic–somatic membranes generate local field potentials ("LFPs") which give rise to EEG, including coherent gamma synchrony, the best measurable neural correlate of consciousness ("NCC") (Gray & Singer, 1989; Crick, 1990). Anesthetic molecules selectively erase consciousness, acting on post-synaptic dendrites and soma, with little or no effects on axonal firing capabilities. Arguably, dendritic/somatic integration is most closely related to consciousness, with axonal firings serving to convey outputs of conscious (or non-conscious) processes to control behavior. But even complex, active integration in Hodgkin–Huxey neurons would, apart from an entirely probabilistic (random) input, be completely algorithmic and deterministic, leaving no apparent place for consciousness.

However, neurons involved in conscious brain processes apparently deviate from Hodgkin–Huxley. Naundorf *et al.* (2006) showed that firing threshold at the AIS in cortical neurons in brains of awake animals

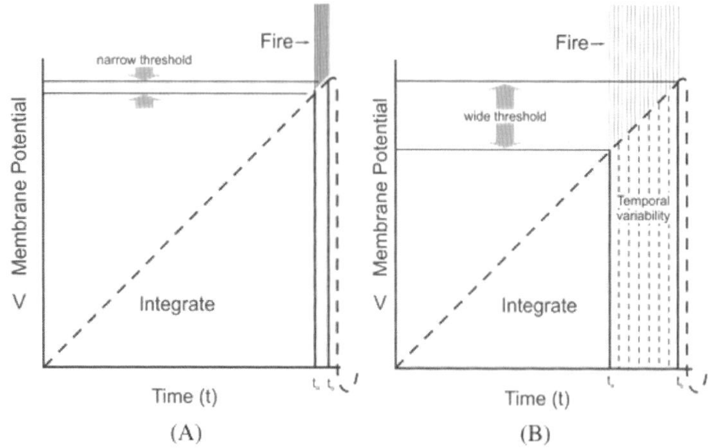


**Fig. 2.** Neuronal "integrate-and-fire" behaviors. (A) The Hodgkin–Huxley model predicts integration by membrane potential in dendrites and soma reach a specific, narrow threshold potential at the proximal axon, and fire with very limited temporal variability (small $t_b - t_a$) for given inputs. (B) Recordings from cortical neurons in awake animals (large $t_b - t_a$) (Naundorf *et al.*, 2006) show a large variability in effective firing threshold and timing. Some unknown "x-factor" (related to consciousness?) modulates integration to exert causal influence on firing and behavior.

(compared to neurons *in vitro*) vary significantly spike-to-spike (Fig. 2(b)). Some factor in addition to inputs, synaptic strengths and the integrated AIS membrane potential apparently contributes to effective integration controlling firing, or not firing, ultimately influencing behavior. This unknown end-integration, pre-firing factor is perfectly positioned for conscious perception and action. What could it involve?

One possible firing-modulating factor comes from lateral connections among neurons via gap junctions, or electrical synapses (Fig. 1). Gap junctions are protein complexes which fuse adjacent neurons and synchronize their membrane polarization states, e.g. in gamma synchrony EEG (Dermietzel, 1998; Draguhn *et al.*, 1998; Galaretta & Hestrin, 2001; Bennett & Zukin, 2004; Fukuda & Kosaka, 2000; Traub *et al.*, 2002). Gap junction-connected cells have fused, synchronized membranes, and also continuous intracellular volumes, as open gap junctions between cells act like doors between adjacent rooms. Neurons connected by dendritic–dendritic gap junctions have synchronized LFPs (giving rise to the EEG) in integration phase, but not necessarily synchronous axonal firing outputs. Gap junction-synchronized dendritic networks can thus *collectively* integrate inputs, enhancing computational capabilities (Hameroff, 2010). However, membrane-based modulations via gap junction connections would be reflected in the integrated membrane potential, and unable to account for threshold variability seen by Naundorf *et al.* (2006). Finer scale processes from within neurons (and conveyed from interiors of adjacent neurons via open gap junctions) could alter firing threshold without changing membrane potentials, and serve as a potential site and mechanism for consciousness.

Finer scale intra-cellular processing, e.g., derived from cytoskeletal structures, are the means by which single-cell organisms perform cognitive functions without synaptic inputs. Observing intelligent actions of unicellular creatures, famed neuroscientist Charles Sherrington said "of nerve there is no trace, but perhaps the cytoskeleton might serve." Neurons have a rich and uniquely organized cytoskeleton, the major components being microtubules (Sherrington, 1957).

## 14.3. A Finer Scale of Neuronal Information Processing

### 14.3.1. *Microtubules*

Interiors of eukaryotic cells are organized and shaped by their cytoskeleton, a scaffolding-like protein network of microtubules, microtubule-associatied proteins (MAPs), actin and intermediate filaments (Tuszynski *et al.*, 1995). Microtubules ("MTs," Fig. 3) are cylindrical polymers 25 nanometers (nm = $10^{-9}$ meter) in diameter, and of variable length, from a few hundred nanometers apparently up to meters in long nerve axons. MTs self-assemble from peanut-shaped "tubulin" proteins, each tubulin being a dimer composed of alpha and beta monomers, with a dipole giving MTs ferroelectric properties. In MTs, tubulins are usually arranged in 13 longitudinal protofilaments whose lateral connections result in two types of hexagonal lattices (A-lattice and B-lattice) (Amos & Klug, 1974), the protofilaments being shifted in relation to their neighbors, slightly differently in each direction, resulting in differing relationships between each tubulin and its six nearest neighbors. Helical pathways following along neighboring tubulin dimers in the A-lattice repeat every five and eight tubulins, respectively, down any protofilament, and following along neighboring tubulin monomers repeat every three monomers, after winding twice

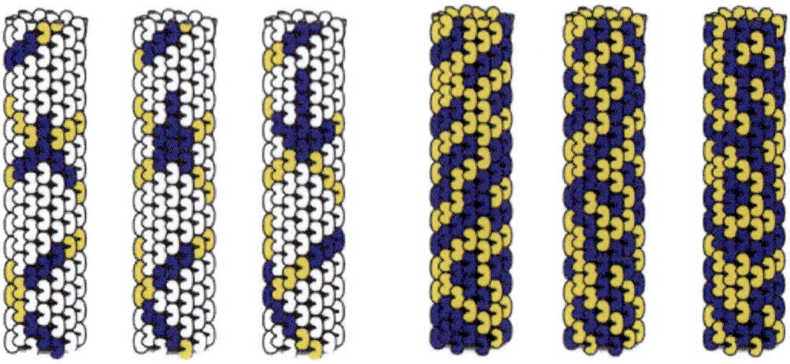


**Fig. 3.** Three timesteps (e.g., at 10 MHz) of a microtubule automaton. Tubulin subunit states due to aligned dipole orientations or spin currents (yellow, blue) represent information. (A) Dipole orientations or spin currents interact and compute along spiral lattice pathways. For example (upper, middle in each microtubule) two upward traveling (blue) dipole/spin waves intersect, generating a new vertical wave (a "glider gun" in cellular automata) (B) A general microtubule automata process (Rasmussen *et al.*, 1990).

around the MT. Thus helical winding pathways in the MT A-lattice follow the Fibonacci series (3, 5, 8...) found widely in nature.

Along with actin and other cytoskeletal structures, MTs self-assemble to establish cell shape, direct growth and organize functions including those of brain neurons. Various types of MAPs bind at specific lattice sites, and bridge to other MTs, defining cell architecture like girders and beams in a building. Another type of MAP is tau, whose displacement from MTs results in neurofibrillary tangles and the cognitive dysfunction of Alzheimer's disease (Brunden *et al.*, 2011; Craddock *et al.*, 2012; Rasmussen *et al.*, 1990). Other MAPs include motor proteins (dynein, kinesin) which move rapidly along MTs, transporting cargo molecules to specific synapses and locations. Tau proteins bound to MTs apparently serve as traffic signals, determining where motor proteins deliver their cargo (Dixit *et al.*, 2008). Thus, specific placement of tau on MT lattices appears to reflect encoded information governing synaptic plasticity.

MTs are particularly prevalent in neurons ($10^9$ tubulins/neuron), and are uniquely stable. Non-neuronal cells undergo repeated cycles of cell division, or mitosis, for which MTs disassemble and re-assemble as mitotic spindles which separate chromosomes, establish cell polarity and architecture, then depolymerize for tubulins and MTs to be re-utilized for cell function. However neurons, once formed, do not divide, and so neuronal MTs can remain assembled indefinitely. Dendritic–somatic MTs are unique in other ways. MTs in axons (and non-neuronal cells) are arrayed radially, extending continuously (with the same polarity) from the centrosome near the nucleus, outward toward the cell membrane. However MTs in dendrites and cell bodies are interrupted, of mixed polarity (Fig. 1), and arranged in local recursive networks suitable for learning and information processing (Dustin, 1985). Finally, MTs in other cells can assemble at one end and dis-assemble at the other ("treadmilling"), or grow and then abruptly dis-assemble ("dynamic instability" or "MT catastrophes" (Guillard *et al.*, 1998). However dendritic–somatic MTs are capped by special MAPs which prevent de-polymerization (Mitchison & Kirschner, 1984), and are thus especially stable and suitable for long-term information encoding and memory (Craddock *et al.*, 2012a)

### 14.3.2. *Microtubule information processing*

After Sherrington's broad observation in 1957 about the cytoskeleton as a cellular nervous system, Atema (1973) proposed that tubulin conformational changes propagate as signals along microtubules. Hameroff & Watt (1982) suggested that distinct tubulin dipoles and conformational states — mechanical changes in protein shape — could represent information, with MT lattices acting as two-dimensional Boolean switching matrices with input/output computation occurring via MAPs. MT information processing has also been viewed in the context of cellular ("molecular") automata ("microtubule automata," Fig. 3) in which tubulin dipole and conformational states interact with neighboring tubulin states in hexagonal MT lattices by dipole couplings, synchronized by biomolecular coherence as proposed by Fröhlich (Fröhlich, 1968, 1970, 1975; Smith *et al.*, 1984; Hameroff, 2006a).

Protein conformational changes occur at multiple scales (Karplus & McCammon, 1983), e.g., $10^{-6}$ sec to $10^{-11}$ sec transitions. Coordinated movements of the protein's atomic nuclei, far more massive than electrons, require energy and generate heat. Early versions of Orch OR portrayed tubulin states as alternate mechanical conformations, coupled to, or driven by London force dipoles in non-polar hydrophobic pockets (Hameroff & Penrose, 1996a,b; Hameroff, 1998a,b; Penrose & Hameroff, 2011). However, recent Orch OR papers do not make use of such large conformational changes, depending instead on tubulin dipole or spin states alone to represent information (Sec. 14.3.3 below).

Within MTs, each tubulin may differ from among its neighbors due to genetic variability, post-translational modifications (Janke & Kneussel, 2010; Hameroff, 2007), phosphorylation states, binding of ligands and MAPs, and moment-to-moment conformational and/or dipole or spin state transitions. Synaptic inputs can register information in dendritic–somatic MTs in brain neurons by metabotropic receptors, MAP2, and CaMKII, a hexagonal holoenzyme able to convey calcium ion influx to MT lattices by phosphorylation (Fig. 4, (Craddock *et al.*, 2012a). Thus, tubulins in MTs can each exist in multiple possible states, perhaps dozens or more. However for simplicity, models of MT automata consider only two alternative tubulin states, i.e., binary "bits."

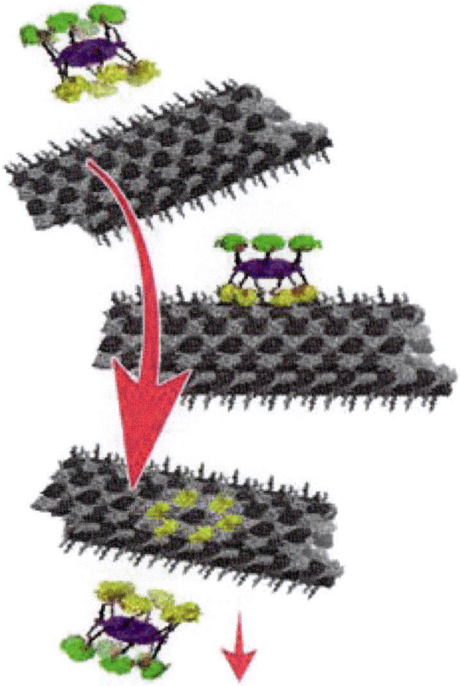

**Fig. 4.** Calcium-calmodulin kinase II ('CaMKII'), a hexagonal holoenzyme activated by synaptic calcium influx extends six leg-like kinase domains above and below an association domain. The six kinase domains precisely match hexagonal size and geometry in both A-lattice and B-lattice microtubules (Craddock, 2012a) with permission from Travis Craddock).

Another potential factor arises from the specific geometry of MT lattices in which helical winding pathways (in the A-lattice) repeat according to the Fibonacci sequence (3, 5, 8…) and may correlate with conduction pathways (Hameroff, *et al.*, 2002). Dipoles or spin states aligned along such pathways may be favored (and coupled to MT mechanical vibrations) thus influencing MT automata computation.

MT automata based on tubulin dipoles in hexagonal lattices show high capacity integration and learning (Rasmussen *et al.*, 1990). Assuming $10^9$ binary tubulins per neuron switching at 10 megahertz ($10^7$) gives a potential MT-based capacity of $10^{16}$ operations per second *per neuron.* Conventional neuronal-level approaches based on axonal firings and synaptic transmissions ($10^{11}$ neurons/brain, $10^3$ synapses/neuron, $10^2$ transmissions/s/synapse) give the same $10^{16}$ operations per second for the entire brain! MT-based information processing offers a huge potential increase in brain capacity (Hameroff, 2007).

How would MT processes be "read out" to influence neuronal and network activities in the brain? First, as previously mentioned, MT processing during dendritic–somatic integration can influence axonal firings to implement behavior. Second, MT processes may directly result in conscious awareness. Third, MT processes can regulate synaptic plasticity, e.g., as tracks and guides for motor proteins (dynein and kinesin) transporting synaptic precursors from cell body to distal synapses. The guidance mechanism in choosing the proper path is unknown, but seems to involve placement of the MAP tau at specific sites on MT lattices. In Alzheimer's disease, tau is hyperphosphorylated and dislodged from destabilized MTs, forming neurofibrillary tangles which correlate with memory loss (Matsuyama & Jarvik, 1989; Brinden *et al.*, 2011; Craddock *et al.*, 2012). Fourth, tubulin states can encode binding sites not only for tau, but also structural MAPs determining cytoskeletal scaffolding and thus directly regulate neuronal structure and synaptic formation. Finally, MT information processing may be directly related to activities at the levels of neurons and neuronal networks through something of the nature of scale-invariant dynamics. Several lines of evidence point to fractal-like (1/f) self-similarity over different spatiotemporal scales in brain dynamics and structure (He *et al.*, 2010; Kitzbichfer *et al.* 2009). Scale-invariance is generally considered at scale levels of neurons and higher-level neuronal networks, but may extend downward in size (and higher in frequency) to intra-neuronal MT dynamics, spanning 4 or 5 scale levels or more, each level separated by several orders of magnitude.

MT information processing depends on interactive states of individual tubulin proteins. What are those states, and how are they governed?

### 14.3.3. *Tubulin dipoles and anesthesia*

Tubulin, like other proteins, is composed of a heterogeneous group of amino acid residues connected to a peptide backbone. The residues include both water-soluble polar, and water-insoluble non-polar groups, the latter including "aromatic" amino acids (phenylalanine, tyrosine and tryptophan) with "π" orbital electron resonance clouds in phenyl and indole rings. π orbital clouds are composed of electrons able to delocalize across a spatial region. Like oil separating from water, non-polar electron clouds coalesce during protein folding to form isolated water-excluding "hydrophobic regions" within proteins with particular ("oily," "lipid-like") solubility. Driving the folding are non-polar, but highly polarizable π orbital electron cloud dipoles which couple by van der Waals London forces (instantaneous dipole-induced dipole attractions between electron clouds) (Voet & Voet, 1995).

Within intra-protein hydrophobic regions, anesthetic gas molecules bind by London force dipole couplings, and thereby (somehow) exert their effects on consciousness (Craddock *et al.*, 2012b, 2015; Hameroff, 1998c, 2006a; Hameroff *et al.*, 1982; Hameroff & Watt, 1983). Historically, views of anesthetic action have focused on neuronal membrane proteins, but actual evidence (e.g., from genomics and proteomics (Xi *et al.*, 2004; Pan *et al.*, 2007) points to anesthetic action in microtubules. In the most definitive anesthetic experiment yet performed, Emerson *et al.* (2013) used fluorescent anthracene as an anesthetic in tadpoles, and showed cessation of tadpole behavior that occurs specifically via anthracene binding in tadpole brain microtubules. Despite prevailing assumptions, actual evidence supports anesthetic action on microtubules.

Tubulin (Fig. 5) contains 32 aromatic (phenyl and indole) amino acid rings with π electron resonance clouds, most within a Forster resonance transfer distance of 1 to 2 nm (Craddock *et al.*, 2012b). Resonance rings align along grooves which traverse tubulin, and appear to meet those in neighbor tubulins along helical lattice pathways (Fig. 6A). Simulation of anesthetic molecules (Fig. 5, red spheres) shows binding in a hydrophobic channel aligned with the five- and eight-start helical winding pathways in the microtubule A-lattice.

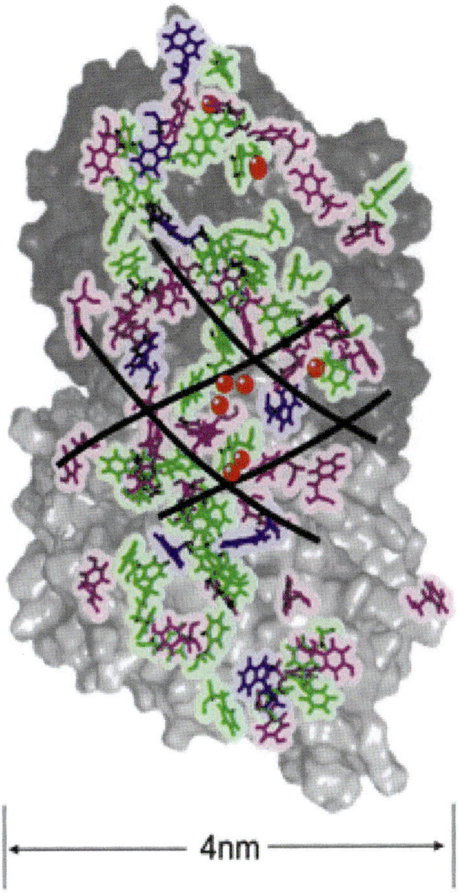


**Fig. 5.** Molecular modeling of tubulin dimer shows aromatic amino acids tryptophan (blue), phenylalanine (purple) and tyrosine (green) in non-polar, hydrophobic regions. Red spheres are anesthetic binding sites (with permission from Craddock *et al.*, (2012b)). Curved lines enclose rings in particular aligned orientation along five and eight-start helical channels, containing anesthetic binding sites.

Figure 6B shows collective dipole couplings in contiguous rings. Quantum superposition of both states is shown in gray. Anesthetics (lower right) appear to disperse dipoles necessary for consciousness, resulting in anesthesia (Hameroff, 1998c, 2006a; Hameroff *et al.*, 1982; Hameroff & Watt, 1983). Electron cloud dipoles may be either charge separation (electric) or electron spin (magnetic). Tubulin dipoles in

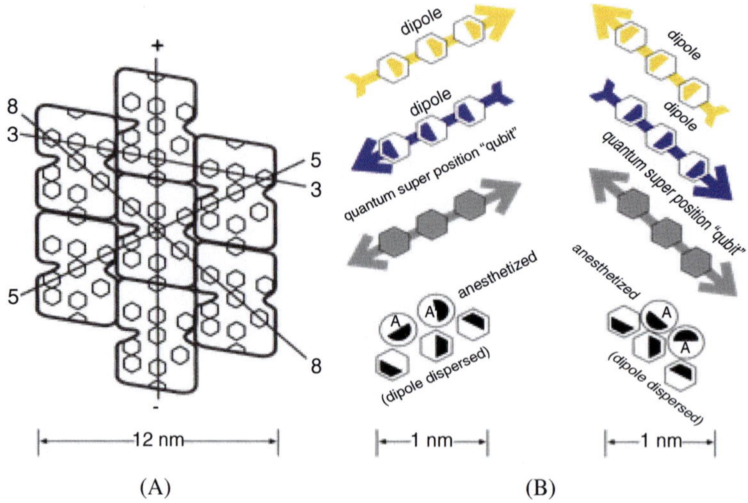


**Fig. 6.** Dipoles in tubulin. (A) Seven tubulin neighborhood in microtubule A-lattice with schematized placement of aromatic rings along three, five and eight-start helical pathways. (B) Five-start (left) and eight-start (right) helical pathways represented by aligned aromatic ring dipole "bits" and superposition of both (quantum bit, or "qubit"). Bottom: anesthetic gas molecules (A) form their own dipole couplings with amino acid rings, dispersing collective dipoles and disrupting quantum and classical computation.

Orch OR were originally described in terms of London-force electric dipoles, involving charge separation. However we now suggest, as an alternative, *magnetic* dipoles, which could be related to electron spin — and possibly related also to nuclear spins (which can remain isolated from their environments for long periods of time). 'Spin-flips' might perhaps relate to alternating currents (ACs) in MTs. Spin is inherently quantum in nature, and quantum spin transfer through aromatic rings is enhanced at warm temperature (Ouyang & Awschalom, 2003). In Figs. 6 and 7, yellow may be considered "spin up," and blue considered "spin down."

It should be made clear, however, that the notions of 'up' and 'down' referred to here need be figurative only. Yet, there are, in fact, directional aspects to the notion of spin; in essence, the spin direction is the direction of the axis of rotation, where conventionally we regard the rotational direction to be *right-handed* about the direction being referred to, and "up"

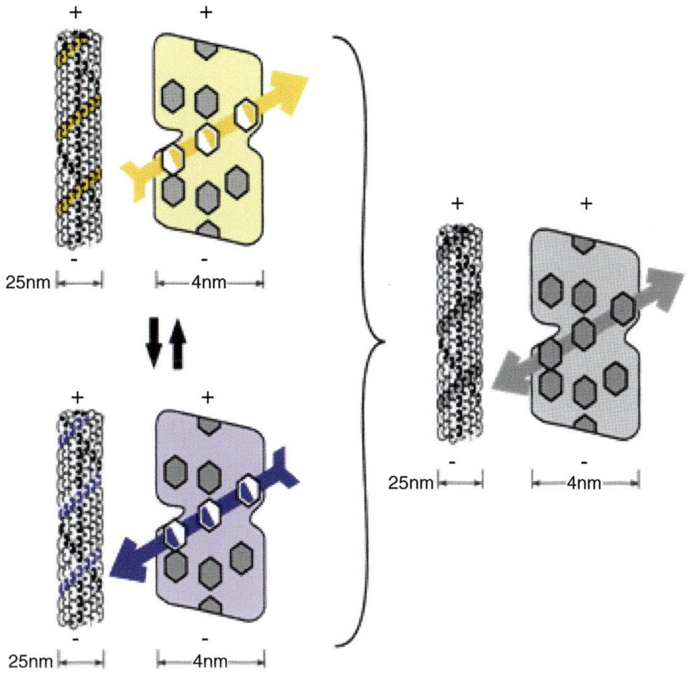


**Fig. 7.** Classical and quantum dipole information states for the "five-start" helical pathway in tubulin and microtubules. Left: The "five-start" helix in microtubule A-lattice aligned with dipoles in intra-tubulin aromatic rings. Top: "upward" dipole (yellow), Bottom, "downward" dipole (blue). Right: Quantum superposition of both upward and downward helical paths coupled to dipole orientations, i.e., "qubits." Dipoles may be electric dipoles due to charge separation, or magnetic dipoles, e.g., related to electronic (and/or nuclear) spin. Similar qubit pathways may occur along eight-start pathways, or other pathways.

would refer to some arbitrarily chosen spatial direction and "down" to the opposite direction. If the particle has a magnetic moment (e.g., electron, proton or neutron), its magnetic moment is aligned (or anti-aligned, according to the type of particle) with its spin. Within a microtubule, we might imagine "up" and "down" are chosen to refer to the two opposite directions along the tube's axis itself, or else some other choice of alignment might be appropriate. However, as indicated earlier, spin is a quintessentially quantum-mechanical quantity, and for a spin-one-half object, like an electron or a nucleon (neutron or proton), all possible directions for the spin rotation axis arise as quantum superpositions of some arbitrarily chosen pair of directions. Indeed the directional features of quantum spin inter-relate with the quantum superposition principle in fundamental ways.

Here, we may speculate that chains of correlated ("up–up–up," "down–down–down") or possibly anti-correlated ("down–up–down," "up–down–up") spin along lattice pathways in microtubules or perhaps something more subtle might provide biologically plausible ways of propagating quantum bit pairs (qubits) along the pathways. If such correlated spin chains make physical sense, one might speculate that periodic spin-flip or spin-precession processes (either electric or magnetic) might occur, and could be correlated with ACs in microtubules at specific frequencies. Electron cloud dipoles can result from either charge separation (electric) or electron spin (magnetic). Tubulin dipoles in Orch OR were originally described in terms of London force electric dipoles, charge separation. However we now favor magnetic dipoles, e.g., related to electron spin, possibly enabling "spin-flip" ACs in MTs.

The group of Anirban Bandyopadhyay at National Institute for Material Sciences in Tsukuba, Japan, has indeed discovered conductive resonances in single microtubules that are observed when there is an applied AC at specific frequencies in gigahertz, megahertz and kilohertz ranges (Sahu *et al.*, 2013a, 2013b; 2014). See Sec. 14.4.5.

Electron dipole shifts do have some tiny effect on nuclear positions via charge movements and Mossbauer recoil (Sataric *et al.*, 1998; Brizhik *et al.*, 2001). A shift of one nanometer in electron position might move a nearby carbon nucleus a few femtometers ("Fermi lengths," i.e. $10^{-15}$ m), roughly its diameter. The effect of electron spin/magnetic dipoles on nuclear location is less clear. Recent Orch OR publications have cast tubulin bits (and quantum bits, or qubits) as coherent entangled dipole states acting collectively among electron clouds of aromatic amino acid rings, with only femtometer conformational change due to nuclear displacement (Penrose & Hameroff 2011; Hameroff, 2012). As it turns out, femotometer displacement might be sufficient for Orch OR (Sec. 14.5.2).

An intra-neuronal finer–scale of MT-based information processing could account for deviation from Hodgkin-Huxley behavior and, one might hope, enhanced computational capabilities. However, like neuronal models, approaches based on MT information processing with classical physics, e.g., those developed by Hameroff and colleagues up through the 1980's, faced a reductionist dead-end in dealing with consciousness. Enhanced computation *per se* fails to address certain aspects of consciousness (Sec. 14.4.1.). Something was missing. Was it some subtle feature of quantum mechanics?

## 14.4. Quantum Physics and Consciousness

### 14.4.1. *Non-computability and OR*

In 1989 Penrose published *The Emperor's New Mind* (Penrose, 1989), which was followed in 1994 by *Shadows of the Mind* (Penrose, 1994). Critical of the viewpoint of "strong artificial intelligence" ("strong AI"), according to which all mental processes are entirely computational, both books argued, by appealing to Gödel's theorem and other considerations, that certain aspects of human consciousness, such as *understanding*, must be beyond the scope of any computational system, i.e., "non-computable." Non-computability is a perfectly well-defined mathematical concept, but it had not previously been considered as a serious possibility for the result of physical actions. The non-computable ingredient required for human consciousness and understanding, Penrose suggested, would have to lie in an area where our current physical theories are fundamentally incomplete, though of important relevance to the scales that are pertinent to the operation of our brains. The only serious possibility was the incompleteness of quantum theory — an incompleteness that both Einstein and Schrödinger (and also Dirac) had recognized, despite quantum theory having frequently been argued to represent the pinnacle of 20th century scientific achievement. This incompleteness is the unresolved issue referred to as the *"measurement problem,"* which we consider in more detail below, in Sec. 4.3. One way to resolve it would be to provide an extension of the standard framework of quantum mechanics by introducing an objective form of quantum state reduction — termed "OR" (objective reduction), an idea which we also describe in detail further in Penrose (1992, 1996, 2000, 2009).

In Penrose (1989), the tentatively suggested OR proposal would have its onset determined by a condition referred to as "the one-graviton" criterion. However, in Penrose (1999, 2009), a much better-founded criterion was used, now frequently referred to as the *Diósi–Penrose proposal* (henceforth "DP;" see Diósi's earlier work, which was a similar gravitational scheme, though not motivated via specific general relativistic principles). The DP proposal gives an objective physical threshold, providing a plausible lifetime for quantum-superposed states. Other gravitational OR proposals have been put forward, from time to time (Karolyhazy 1966; Karolyhazy *et al.*, 1986; Percival, 1995; Ghirardi *et al.*, 1990; Kibble 1981; Pearle, 1989; Pearle & Squires, 1994) as solutions to the measurement problem, suggesting modifications of standard quantum mechanics, but all these differ from DP in important respects. Among these, only the DP proposal (in its role within Orch OR) has been suggested as having anything to do with the consciousness issue. The DP proposal is sometimes referred to as a "quantum gravity" scheme, but it is not part of the normal ideas used in quantum gravity, as will be explained below (Sec. 14.4.4). Moreover, the proposed connection between consciousness and quantum measurement is almost opposite, in the Orch OR scheme, to the kind of idea that had frequently been put forward in the early days of quantum mechanics (see, for example Wigner, (1976) which suggests that a "quantum measurement" is something that occurs only as a result of the conscious intervention of an observer. Rather, the DP proposal suggests each OR event, which is a purely physical process, is itself a primitive kind of "observation," a moment of "proto-conscious experience." This issue, also, will be discussed below.

### 14.4.2. *The nature of quantum mechanics*

The term "quantum" refers to a discrete element of energy in a system, such as the energy $E$ of a particle, or of some other subsystem, this energy being related to a fundamental frequency $\nu$ of its oscillation, according to Max Planck's famous formula (where $h$ is Planck's constant): $E = h\nu$.

This deep relation between discrete energy levels and frequencies of oscillation underlies the wave/particle duality inherent in quantum phenomena. Neither the word "particle" nor the word "wave" adequately conveys the true nature of a basic quantum entity, but both provide useful partial pictures.

The laws governing these submicroscopic quantum entities differ from those governing our everyday classical world. For example, quantum particles can exist in two or more states or locations simultaneously, where such a multiple coexisting superposition of alternatives (each alternative being weighted by a *complex* number) would be described mathematically by a quantum *wavefunction*. The measurement problem (referred to above) is, in effect, the question of why we do not see such superpositions in the consciously perceived macroscopic world; we see objects and particles as material, classical things in specific locations and states.

Another quantum property is "non-local entanglement," in which separated components of a system become unified, the entire collection of components being governed by one common quantum wavefunction. The parts remain somehow connected, even when spatially separated by very significant distances (the present experimental record being 143 km (Xiao *et al.*, 2012)). Quantum superpositions of bit states (quantum bits, or qubits) can be interconnected with one another through entanglement in quantum computers. However, quantum entanglements cannot, by themselves, be used to send a message from one part of an entangled system to another; yet entanglement can be used in conjunction with classical signaling to achieve strange effects — such as the phenomenon referred to as *quantum teleportation* — that classical signaling cannot achieve by itself (Bennett & Wiesner, 1992; Boouwmeester *et al.*, 1997; Macikic *et al.*, 2002).

### 14.4.3. *The measurement problem and OR*

The issue of why we do not directly perceive quantum superpositions is a manifestation of the *measurement problem* mentioned above. Put more precisely, the measurement problem is the conflict between the two fundamental procedures of quantum mechanics. One of these procedures, referred to as *unitary evolution*, denoted here by U, is the continuous deterministic evolution of the quantum state (i.e., of the wavefunction of the entire system) according to the fundamental *Schrödinger equation*. The other is the procedure that is adopted whenever a measurement of the system — or *observation* — is deemed to have taken place, where the quantum state is discontinuously and probabilistically replaced by another quantum state (referred to, technically, as an *eigenstate* of a mathematical operator that is taken to describe the measurement). This discontinuous jumping of the state is referred to as the *reduction* of the state (or the "collapse of the wavefunction"), and will be denoted here by the letter R. This conflict between U and R is what is encapsulated by the term "measurement problem" (but perhaps more accurately it may be referred to as "the measurement *paradox*") and its problematic nature is made manifest when we consider the measuring apparatus itself as a quantum entity, which is part of the entire quantum system consisting of the original system under observation together with this measuring apparatus. The apparatus is, after all, constructed out of the same type of quantum ingredients (electrons, photons, protons, neutrons, etc. — or quarks, gluons, etc.) as is the system under observation, so it ought to be subject also to the same quantum laws, these being described in terms of the continuous and deterministic U. How, then, can the discontinuous and probabilistic R come about as a result of the interaction (measurement) between two parts of the quantum system? This is the paradox faced by the measurement problem.

There are many ways that quantum physicists have attempted to come to terms with this conflict (Bell, 1966; Bohm, 1983; Rae, 2002; Polkinghorne, 2002; Penrose, 2004). In the early 20th century, the Danish physicist Niels Bohr, together with Werner Heisenberg, proposed the pragmatic "Copenhagen interpretation," according to which the wavefunction of a quantum system, evolving according to U, is not assigned any actual physical "reality," but is taken as basically providing the needed "bookkeeping" so that eventually probability values can be assigned to the various possible outcomes of a quantum measurement. The measuring device itself is explicitly taken to behave *classically* and no account is taken of the fact that the device is ultimately built from quantum-level constituents. The probabilities are calculated, once the nature of the measuring device is known, from the state that the wavefunction has U-evolved to at the time of the measurement. The discontinuous "jump" that the wavefunction makes upon measurement, according to R, is attributed to the change in "knowledge" that the result of the measurement has on the observer. Since the wavefunction is not assigned physical reality, but is considered to refer merely to the observer's knowledge of the quantum system, the jumping is considered simply to reflect the jump in the observer's knowledge state, rather than in the quantum system under consideration.

Many physicists remain unhappy with such a point of view, however, and regard it largely as a "stop-gap," in order that progress can be made in applying the quantum formalism, without this progress being held up by a lack of a serious quantum ontology, which might provide a more complete picture of what is actually going on. One may ask, in particular, what it is about a measuring device that allows one to ignore the fact that it is itself made from quantum constituents and is permitted to be treated entirely classically. A good many proponents of the Copenhagen standpoint would take the view that while the physical measuring apparatus ought actually to be treated as a quantum system, and therefore part of an over-riding wavefunction evolving according to U, it would be the *conscious observer*, examining the readings on that device, who actually reduces the state, according to R, thereby assigning a physical reality to the particular observed alternative resulting from the measurement. Accordingly, before the intervention of the observer's consciousness, the various alternatives of the result of the measurement *including* the different states of the measuring apparatus would, in effect, still have to be treated as coexisting in superposition, in accordance with what would be the usual evolution according to U. In this way, the Copenhagen viewpoint puts consciousness outside science, and does not seriously address the ontological nature or physical role of superposition itself nor the question of how large quantum superpositions like Schrödinger's superposed live and dead cat (see below) might *actually* become one thing or another.

A more extreme variant of this approach is the "multiple worlds hypothesis" of Everett (1957) in which each possibility in a superposition evolves to form its own universe, resulting in an infinite multitude of coexisting "parallel" worlds. The stream of consciousness of the observer is supposed somehow to "split," so that there is one in each of the worlds — at least in those worlds for which the observer remains alive and conscious. Each instance of the observer's consciousness experiences a separate independent world, and is not directly aware of any of the other worlds.

A more "down-to-earth" viewpoint is that of *environmental decoherence*, in which interaction of a superposition with its environment "erodes" quantum states, so that instead of a single wavefunction being used to describe the state, a more complicated entity is used, referred to as a *density matrix*. However, decoherence does not provide a consistent ontology for the *reality* of the world, in relation to the density matrix (see, for example, Penrose (1994)), and provides merely a pragmatic procedure. Moreover, it does not address the issue of how R might arise in isolated systems, nor the nature of isolation, in which an external "environment" would not be involved, nor does it tell us which part of a system is to be regarded as the 'environment' part, and it provides no limit to the size of that part which can remain subject to quantum superposition.

Still other approaches include various types of OR in which a specific objective threshold is proposed to cause quantum state reduction (Percival, 1994; Moroz *et al.*, 1998; Ghirardi *et al.*, 1986). The specific OR scheme that is used in Orch OR will be described below.

The quantum pioneer Erwin Schrödinger took pains to point out the difficulties that confront the U-evolution of a quantum system with his still-famous thought experiment called "Schrödinger's cat" (Schrödinger, 1935). Here, the fate of a cat in a box is determined by magnifying a quantum event (say the decay of a radioactive atom, within a specific time period that would provide a 50% probability of decay) to a macroscopic action which would kill the cat, so that according to Schrödinger's own U-evolution the cat would be in a quantum superposition of being both dead and alive at the same time. According to this perspective on the Copenhagen interpretation, if this U-evolution is maintained until the box is opened and the cat observed, then it would have to be the conscious human observing the cat that results in the cat becoming either dead or alive (unless, of course, the cat's own consciousness could be considered to have already served this purpose). Schrödinger intended to illustrate the absurdity of the direct applicability of the rules of quantum mechanics (including his own U-evolution) when applied at the level of a cat. Like Einstein, he regarded quantum mechanics as an incomplete theory, and his 'cat' provided an excellent example for emphasizing this incompleteness. There is a need for something to be done about quantum mechanics, irrespective of the issue of its relevance to consciousness.

### 14.4.4. *OR and quantum gravity*

DP objective reduction is a particular proposal for an extension of current quantum mechanics, taking the bridge between quantum- and classical-level physics as a "quantum-gravitational" phenomenon. This is in contrast with the various conventional viewpoints, whereby this bridge is claimed to result, somehow, from "environmental decoherence," or from "observation by a conscious observer," or from a "choice between alternative worlds," or some other interpretation of how the classical world of one actual alternative may be taken to arise out of fundamentally quantum-superposed ingredients.

The DP version of OR involves a different interpretation of the term "quantum-gravity" from what is usual. Current ideas of quantum-gravity (see, for example Smolin (2002)) normally refer, instead, to some sort of physical scheme that is to be formulated within the bounds of standard quantum field theory — although no particular such theory, among the multitude that has so far been put forward, has gained anything approaching universal acceptance, nor has any of them found a fully consistent, satisfactory formulation. 'OR' here refers to the alternative viewpoint that standard quantum (field) theory is not the final answer, and that the reduction R of the quantum state ("collapse of the wavefunction") that is adopted in standard quantum mechanics is an *actual* physical process which is not part of the conventional unitary formalism U of quantum theory (or quantum field theory). In the DP version of OR, the reduction R of the quantum state does not arise as some kind of convenience or effective consequence of environmental decoherence, etc., as the conventional U formalism would seem to demand, but is instead taken to be one of the consequences of melding together the principles of Einstein's general relativity with those of the conventional unitary quantum formalism U, and this demands a departure from the strict rules of U. According to this OR viewpoint, any quantum measurement — whereby the quantum-superposed alternatives produced in accordance with the U formalism becomes reduced to a single actual occurrence — is a *real* objective physical process, and it is taken to result from the mass displacement between the alternatives being sufficient, in gravitational terms, for the superposition to become unstable.

In the DP scheme for OR, the superposition reduces to one of the alternatives in a timescale $\tau$ that can be estimated (for a superposition of two states each of which is assumed to be taken to be stationary on its own) according to the formula $\tau \approx \hbar/E_G$. An important point to make about $\tau$, however, is that it represents merely a kind of *average* time for the state reduction to take place. It is very much like a *half-life* in a radioactive decay. The actual time of decay in each individual state-reduction event, according to DP (in its current form), is taken to be a random process. Such an event would involve the entire (normally entangled) state, and would stretch across all the superposed material that is involved in the calculation of $E_G$. According to DP (in its current form), the actual time of decay in a particular state-reduction event occurs simultaneously (in effect) over the entire state involved in the superposition, and it is taken to follow the $\tau \approx \hbar/E_G$ formula on the *average* (in a way similar to radioactive decay). Here $\hbar$ $(=h/2\pi)$ is Dirac's form of Planck's constant $h$ and $E_G$ is the *gravitational self-energy* of the *difference* between the two (stationary) mass distributions of the superposition. (For a superposition for which each mass distribution is a rigid translation of the other, $E_G$ is the energy it would cost to displace one component of the superposition in the gravitational field of the other, in moving it from coincidence to the quantum-displaced location (Penrose, 2002).)

It is helpful to have a conceptual picture of quantum superposition in a gravitational context. According to modern accepted physical theories, reality is rooted in three-dimensional space and a one-dimensional time, combined together into a four-dimensional *space–time*. This space–time is slightly curved, in accordance with Einstein's general theory of relativity, in a way which encodes the gravitational fields of all distributions of mass density. Each different choice of mass density effects a space–time curvature in a different, albeit very tiny, way. This is the standard picture according to *classical* physics. On the other hand, when *quantum* systems have been considered by physicists, this mass-induced tiny curvature in the structure of space–time has been almost invariably ignored, gravitational effects having been assumed to be totally insignificant for normal problems in which quantum theory is important. Surprising as it may seem, however, such tiny differences in space–time structure *can* have large effects, for they entail subtle but fundamental influences on the very rules of quantum mechanics (Penrose, 1992, 1996, 2000, 2009).

In the current context, superposed quantum states for which the respective mass distributions differ significantly from one another will have space–time geometries which also correspondingly differ. For illustration, in Fig. 8, we consider a two-dimensional space–time sheet (one space and one time dimension). In Fig. 8, at left, the top and bottom alternative curvatures indicate a mass in two distinct locations. If that mass were in superposition of both locations, we might expect to see both curvatures, i.e. the bifurcating space-time depicted in the right of Fig. 8, this being the union ("glued together version") of the two alternative space–time histories that are depicted on the left. The initial part of each space–time is at the upper left of each individual space–time diagram, and so the bifurcating space–time diagram on right moving downward and rightward illustrates two alternative mass distributions evolving in time, their space–time curvature separation increasing.

Quantum-mechanically (so long as OR has not taken place), the "physical reality" of this situation, as provided by the evolving wavefunction, is being illustrated as an actual superposition of these two slightly differing space–time manifolds, as indicated on the right of Fig. 8. Of course there is additional artistic license involved in drawing the space-time sheets as two-dimensional, whereas the actual space–time constituents are four-dimensional. Moreover, there is no significance to be attached to the imagined "three-dimensional space" within

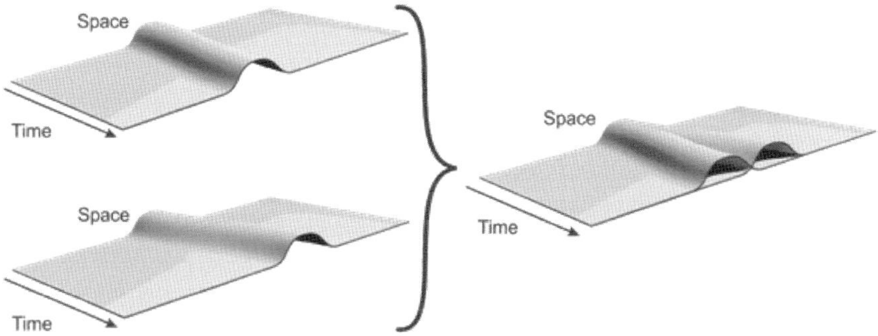


**Fig. 8.** Space–time geometry schematized as one spatial and one temporal dimension in which particle location is represented as curvature. Left: Top and bottom show space–time histories of two alternative particle locations. Right: Quantum superposition of both particle locations as bifurcating space–time depicted as the union ("glued together version") of the two alternative histories. (Adapted from Penrose (1994, p. 338)).

which the space–time sheets seem to be residing. There is no "actual" higher dimensional space there, the "intrinsic geometry" of the bifurcating space–time being all that has physical significance. When the "separation" of the two space–time sheets reaches a critical amount, one of the two sheets "dies" — in accordance with the OR criterion — the other being the one that persists in physical reality. The quantum state thus reduces (OR), by choosing between either the curved or flat space–time in each of the two separations in Fig. 8.

It should be made clear that this measure of superposition separation is only very schematically illustrated as the "distance" between the two sheets in Fig. 8. As remarked above, there is no physically existing "ambient higher dimensional space" inside which the two sheets reside. The degree of separation between the space–time sheets is a more abstract mathematical thing; it would be more appropriately described in terms of a *symplectic measure* on the space of four-dimensional metrics (cf. Penrose (1992); Penrose & Bell (2002)); but the details (and difficulties) of this will not be important for us here. It may be noted, however, that this separation is a space–time separation, not just a spatial one. Thus, the *time* of separation contributes as well as the spatial displacement. It is the product of the temporal separation $T$ with the spatial separation $S$ that measures the overall degree of separation, and OR takes place when this overall separation reaches the critical amount.

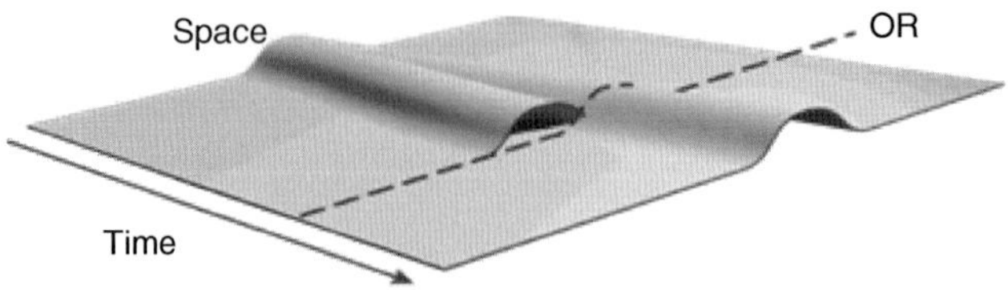


**Fig. 9.** As superposition curvature E reaches threshold (by $E_G=\hbar/\tau$), OR occurs and one particle location/curvature is selected, and becomes classical. The other ceases to exist.

In the absence of a coherent theory of quantum-gravity there is no accepted way of handling such a superposition as a separation (or bifurcation) of space–time geometry, or in any other way. Indeed the basic principles of Einstein's general relativity begin to come into profound conflict with those of quantum mechanics (Penrose, 1996, 2009). Some form of OR is needed.

The OR process is considered to occur when quantum superpositions between such slightly differing space–times take place (Fig. 9), differing from one another by an integrated space–time measure which compares with the fundamental and extremely tiny Planck (4-volume) scale of space–time geometry. As remarked above, this is a 4-volume Planck measure, involving both time and space, so we find that the time measure would be particularly tiny when the space-difference measure is relatively large (as with Schrödinger's hypothetical cat), but for extremely tiny space-difference measures, the time measure might be fairly long. For example, an isolated single electron in a superposed state (very low $E_G$) might reach OR threshold only after thousands of years or more, whereas if Schrödinger's (~1 kg) cat were to be put into a superposition, of life and death, this threshold could be reached in far less than even the Planck time of $10^{-43}$ s.

As already noted, the degree of separation between the space–time sheets is technically a symplectic measure on the space of 4-metrics which is a space–time separation, not just a spatial one, the time of separation contributing as well as spatial displacement. Roughly speaking, it is the product of the temporal separation $T$ with the spatial separation $S$ that measures the overall degree of separation, and (DP) OR takes place when this overall separation reaches a critical amount. This critical amount would be of the order of unity, in *absolute units*, for which the

Planck–Dirac constant $\hbar$, the gravitational constant $G$ and the velocity of light $c$, all take the value unity, (cf. Penrose (1994, pp. 337–339)). For small $S$, the lifetime $\tau \approx T$ of the superposed state will be large; on the other hand, if $S$ is large, then $\tau$ will be small.

To estimate $S$, we compute (in the Newtonian limit of weak gravitational fields) the gravitational self-energy $E_G$ of the difference between the mass distributions of the two superposed states. (That is, one mass distribution counts positively and the other, negatively; (see Penrose (1992, 2002, 2004))). The quantity $S$ is then given by: $S \approx E_G$ and $T \approx \tau$, whence $\tau \approx \hbar / E_G$, i.e., $E_G \approx \hbar / \tau$. Thus, the DP expectation is that OR occurs with the resolving out of one particular space–time geometry from the previous superposition when, on the average, $\tau \approx \hbar/E_G$.

The Orch-OR scheme adopts DP as a physical proposal, but it goes further than this by attempting to relate this particular version of OR to the phenomenon of *consciousness*. Accordingly, the "choice" involved in any quantum state-reduction process would be accompanied by a (miniscule) proto-element of experience, which we refer to as a moment of *proto-consciousness*, but we do not necessarily refer to this as *actual* consciousness for reasons to be described.

### 14.4.5. *OR and Orch OR*

For Orch OR and consciousness to occur, quantum superpositions of gravitational self-energy $E_G$ would need to avoid environmental decoherence long enough to reach time $\tau$ by $\tau \approx \hbar/E_G$. Indeed, it is essential for Orch OR that some degrees of freedom in the system are kept isolated from environmental decoherence, so that OR can be made use of by the system in a controlled way. It should be made clear that in the DP scheme environmental decoherence need not necessarily be playing an important role in any particular instance of state reduction, although in uncontrolled situations the environment may well supply the major contribution to $E_G$. What DP does require is that when state reduction R takes place, this *always* occurs spontaneously, by this gravitational criterion. In nearly all physical situations, there would be much material from the environment that would be entangled with a quantum-superposed state, and it could well be that the major mass displacement — and therefore the major contribution to $E_G$ — would occur in the environment rather than in the system under consideration. Since the environment will be quantum-entangled with the system, the state-reduction in the environment will effect a simultaneous reduction in the system. This could shorten the time for the state reduction R to take place in a superposed system very considerably from what it would have been without the environmental influence. The environment would also introduce an uncontrollable random element into the result of the reduction, so that any non-random (albeit non-computable) element influencing the particular choice of state that is actually resolved out from the superposition would be completely masked by this randomness. In these circumstances, the OR-process would be indistinguishable from the standard R-process of conventional quantum mechanics, which could be considered to be affected by standard environmental decoherence.

If, however, a quantum superposition is (1) "orchestrated," i.e., adequately organized, imbued with cognitive information, and capable of integration and computation, and (2) isolated from non-orchestrated, random environment long enough for the superposition $E_G$ to evolve by the U formalism to reach time $\tau$ by $\tau \approx \hbar/E_G$, then Orch OR will occur and this, according to the scheme, will result in a moment of consciousness. Thus, if the suggested non-computable effects of this OR proposal are to be laid bare, where DP is being adopted and made use of in biological evolution, and ultimately orchestrated for moments of actual consciousness, we indeed need significant isolation from the environment.

As yet, no experiment has been refined enough to determine whether the (DP) OR proposal is actually respected by Nature, but the experimental testing of the scheme is fairly close to the borderline of what can be achieved with present-day technology (see Marshall *et al.*, 2013). For example, one ought to begin to see the effects of this OR scheme if a small object, such as a 10-micron cube of crystalline material could be held in a coherent superposition of two locations, differing by about the diameter of an atomic nucleus, for some seconds, or perhaps minutes to reach threshold by $\tau \approx \hbar/E_G$.

A point of importance, in such proposed experiments, and in estimating requirements for Orch OR, is that in order to calculate $E_G$ it is not enough to base the calculation on an average density of the material in the superposition, since the mass will be concentrated in the atomic nuclei, and for a displacement of the order of the diameter of a nucleus, this inhomogeneity in the density of the material can be crucial, and may well provide a much larger value for $E_G$ than would be obtained if the material is assumed to be homogeneous. The Schrödinger equation (more correctly, in the zero-temperature approximation, the Schrödinger–Newton equation, (see Kibble (1981) and Moz *et al.* (1998)) for the static unsuperposed material would have to be solved, at least approximately, in order to derive the expectation value of the mass distribution in each of the two separate components of the superposition. In the stationary wavefunction of each component, there would be some quantum spread in the locations of the particles constituting the nuclei (i.e., each component's wavefunction would not normally be very sharply peaked at these particle locations, as the locations would be considerably spread out in most materials).

In the situations under consideration here, where we expect a conscious brain to be at far from zero temperature, and because technological quantum computers require zero temperature, it is very reasonable to question quantum brain activities. Nevertheless, it is now well known that superconductivity and other large-scale quantum effects can actually occur at temperatures very far from absolute zero. Indeed, biology appears to have evolved thermal mechanisms to *promote* quantum coherence. In 2003, Ouyang & Awschalom (2003) showed that quantum spin transfer through phenyl ring π orbital resonance clouds (the same as those in protein hydrophobic regions, as illustrated in Figs. 5–7) is enhanced at increasingly warm temperatures. (Spin flip currents through microtubule pathways, as suggested in Sec. 14.3.3. above, may be directly analogous.)

In the past nine years, evidence has accumulated that plants routinely use quantum coherent electron transport at ambient temperatures in photosynthesis (Engel *et al.*, 2007; Hildner *et al.*, 2013). Photons are absorbed in one region of a photosynthetic protein complex, and their energy is conveyed by electronic excitations through the protein to another region to be converted to chemical energy to make food. In this transfer, electrons utilize multiple pathways simultaneously, through π electron clouds in a series of chromophores (analogous to hydrophobic regions) spaced nanometers apart, maximizing efficiency (e.g. via so-called "exciton hopping"). Chromophores in photosynthesis proteins appear to enable electron quantum conductance precisely like aromatic rings are proposed in Orch OR to function in tubulin and microtubules (Figs. 5–7).

Quantum conductance through photosynthesis protein is enhanced by mechanical vibration (Chin *et al.*, 2013), and microtubules appear to have their own set of mechanical vibrations (e.g., in megahertz as suggested by Sahu *et al.* (2013a, 2013b, 2014)). Megahertz mechanical vibrations is ultrasound, and brief, low intensity (sub-thermal) ultrasound administered through the skull to the brain modulates electrophysiology, behavior and affect, e.g., improved mood in patients suffering from chronic pain, perhaps by direct excitation of brain microtubules (Hameroff *et al.*, 2013).

Further research has shown warm quantum effects in bird-brain navigation (Gauger *et al.*, 2011), ion channels (Bernroider & Roy, 2005), sense of smell (Turin, 1996), DNA (Rieper *et al.*, 2011), protein folding (Luo & Lu, 2011) and biological water (Reiter *et al.*, 2011). What about quantum effects in microtubules? In the 1980s and 1990s theoretical models predicted "Fröhlich" gigahertz coherence and ferroelectric effects in microtubules (Rasmussen *et al.*, 1990; Hameroff & Watt, 1982; Smith *et al.*, 1984). In 2001 and 2004, coherent megahertz emissions were detected from living cells and ascribed to microtubule dynamics (powered by mitochondrial electromagnetic fields) by the group of Jiri Pokorny in Prague (Pokorny, 2004; Pokorny *et al.*, 2001).

Beginning in 2009, Anirban Bandyopadhyay and colleagues at the National Institute of Material Sciences in Tsukuba, Japan, were able to use nanotechnology to address electronic and optical properties of individual microtubules (Sahu *et al.*, 2013a, 2013b, 2014). The group has made a series of remarkable discoveries suggesting that quantum effects do occur in microtubules at biological temperatures. First, they found that electronic conductance along microtubules, normally extremely good insulators, becomes exceedingly high, approaching quantum conductance, at certain specific resonance frequencies of applied AC stimulation. These resonances occur in gigahertz, megahertz and kilohertz ranges, and are particularly prominent in low megahertz (e.g. 8.9 MHz). Conductances induced by specific (e.g., megahertz) AC frequencies appear to follow several types of pathways through the microtubule — helical, linear along the microtubule axis, and "blanket-like" along/around the entire microtubule surface. Second, using various techniques, the Bandyopadhyay group also determined that AC conductance through 25nm-wide microtubules is greater than through single 4nm-wide tubulins, indicating cooperative, possibly quantum coherent effects throughout the microtubule, and that the electronic properties of microtubules are programmed within each tubulin. Their results also showed that conductance increased with microtubule length, also indicative of quantum mechanisms.

The resonance conductance ("Bandyopadhyay coherence" — 'BC') through tubulins and microtubules is consistent with the intra-tubulin aromatic ring pathways (see Figs. 5–7) which can support Orch OR quantum dipoles, and in which anesthetics bind, apparently to selectively erase consciousness. Bandyopadhyay's experiments do seem to provide clear evidence for coherent microtubule quantum states at brain temperature.

### 14.4.6. *Beat frequencies*

Quantum-coherent behavior does indeed appear to be relevant, in a way that applies even to biological systems, at surprisingly warm temperatures. Accordingly, we appear to need an extension of the DP proposal that can be used in such "warm" situations. Although such a theory is not yet at hand, it will be of some importance here to indicate certain of the key issues, so that we can get a feeling for the role that we are requiring for DP-related ideas in the suggested proposals put forward in the sections below.

In the first place, it should be pointed out that in standard quantum treatments of systems at non-zero temperature, the description would be in terms of a *density matrix* rather than a simple wavefunction. Such a density-matrix description can be viewed as a *probability mixture* of different wavefunctions — although such an ontology does not reveal the full subtleties involved, since a single density matrix can be interpreted in many different ways as such a probability mixture (see for example Penrose (2004)). As yet, a fully appropriate generalization of the DP scheme to a density-matrix description has not been provided. But in any case it is unlikely that this would be an appropriate thing to do in the present context, and here we shall explore an alternative route to the understanding of quantum effects in warm-temperature systems.

It is important to bear in mind that biological systems are very far from being in thermal equilibrium, so that a crude assignment of an overall 'temperature' to such a system is unlikely to be very revealing. Whenever we are asking for the manifestation of large-scale quantum effects in a warm system, we are not expecting that all the degrees of freedom should be simultaneously involved with these effects and therefore uniformly thermalized. What we really require is that certain of these degrees of freedom can be excited in ways that remain isolated from most of the others, and that these excited degrees can be maintained in some form of quantum oscillation that can preserve its quantum nature for an appreciable time, without dissipation, this time being long enough for the system to reach Orch OR threshold, given by $\tau \approx \hbar/E_G$.

In previous Orch OR publications, the relevant time $\tau$ has been assumed to correlate with physiological parameters of the electro-encephalogram (EEG), i.e., 10 to several hundred milliseconds, which is relatively long for isolated quantum systems. But here we suggest an alternative way in which such oscillation frequencies might come about, namely as *beat* frequencies, arising when OR is applied to superpositions of quantum states of slightly different energies. This makes the task of finding an origin for these observed frequencies far simpler and more plausible.

In order to get some feeling of how the ideas of DP might relate to such situations, let us first address the assumption of *stationarity* that is involved in the DP scheme where, in order to apply DP strictly, we must consider that each of the states in superposition is to be regarded as being stationary, if taken on its own. In standard quantum mechanics, a stationary state is an eigenstate of energy — i.e., a state of definite energy $E$ — which tells us that this quantum state has an (complex) oscillatory nature with a time-dependence that is proportional to $e^{-iEt/\hbar}$ (see, for example, Penrose (2004, Chapter 21)) so that it oscillates with frequency $E/h$ (where we recall that $h=2\pi\hbar$). If we have a state $\Psi$ which is a superposition of two slightly different states $\Psi_1$ and $\Psi_2$, each of which would be stationary on its own, but with very slightly different respective energies $E_1$ and $E_2$, then the superposition would not be quite stationary. Its basic r frequency would be the average $(E_1+E_2)/2h$ of the two, corresponding to the average energy $\frac{1}{2}(E_1+E_2)$, but this would be modulated by a much lower *classical* frequency ("beats") that is the difference between the two, namely $|E_1-E_2|/h$, as follows, very roughly, from the following mathematical identity (where we may take $a=-E_1t/\hbar$ and $b=-E_2t/\hbar$ to represent the quantum wavefunctions for the two energies):

$$e^{ia} + e^{ib} = 2e^{i(a+b)/2} \cos\frac{a-b}{2}.$$

If we imagine the complex oscillatory term $e^{ia}$ to represent one quantum state $\Psi_1$ and $e^{ib}$ to represent the other, then we see that their superposition has a complex quantum oscillation $e^{i(a+b)/2}$, which has a frequency which is the average of the two, but this is modulated by a classical oscillation as given by the cosine term, with a much lower frequency determined by the *difference* between the quantum mechanical frequencies $E_1$ and $E_2$ of the two individual states $\Psi_1$ and $\Psi_2$. This classical "beat" frequency is in fact $|E_1 - E_2|/h$ rather than $|E_1 - E_2|/2h$ because when passing from a quantum amplitude to a classical probability we need to take the squared modulus of the amplitude, and in this case it amounts to taking the squared modulus of half the right-hand side of the above expression, namely $\cos^2\{\frac{1}{2}(a-b)\}=\{1+\cos(a-b)\}/2$ for finding one component of the superposition and $\{1-\cos(a-b)\}/2$ for the other. (This phenomenon is closely related to that found in neutrino oscillations, (see Pontecorvo (1968)).

To be more explicit about how this comes about, it is necessary to appreciate, first, that the eigenstates of energy, $\Psi_1$ and $\Psi_2$, in the superposition — i.e., the two stationary states of which the quantum state is composed, in superposition — will, in the situation under consideration, be different from the two distinguishable *location states* $\Lambda$ and $\Pi$ (taken to be normalized and mutually orthogonal, and without any time-dependence) that would be the states of location arising as a result of the OR process in the original DP proposal (which is concerned with the degenerate case of equal energy eigenvalues) or as the two states between which (as we shall argue) *classical oscillation* takes place (in the case of unequal energy eigenvalues). We consider here the case of unequal energy eigenvalues, so the eigenstates $\Psi_1$ and $\Psi_2$ must be distinct and orthogonal to each other, and we may assume that each is normalized. Accordingly, we can choose phases for the location-state basis $\mathit{\Lambda}$, $\mathit{\Pi}$, so that $\Psi_1$ and $\Psi_2$ when expressed in terms of these location states, take the form

$$\Psi_1 = (\mathit{\Lambda} \cos\theta + \mathit{\Pi} \sin\theta)e^{ia} \text{ and } \Psi_2 = (\mathit{\Lambda} \sin\theta - \mathit{\Pi} \cos\theta)e^{ib}$$

for some angle $\theta$ (measuring the "angle" between the energy basis and the location basis), where the time-dependence of these states is now provided by $a=-E_1t/\hbar$ and $b=-E_2t/\hbar$, as above. The initial quantum state is taken to be a superposition

$$\Psi = \alpha\Psi_1 + \beta\Psi_2,$$

where $\alpha$ and $\beta$ are complex constants satisfying $|\alpha|^2+|\beta|^2=1$. In terms of the location states $\Lambda$ and $\Pi$, we find

$$\Psi = \Lambda(\alpha e^{ia} \cos\theta + \beta e^{ib} \sin\theta) + \Pi(\alpha e^{ia} \sin\theta - \beta e^{ib} \cos\theta).$$

To find the classical oscillation that this ought to reduce to, according to our extended DP proposal, we calculate (in accordance with standard quantum mechanics) the time-dependent probabilities that a measurement to distinguish between the two location states would give us, this being obtained by taking the squared modulus of the coefficients of $\Lambda$ and $\Pi$, namely

$$|\alpha|^2 \cos^2\theta + |\beta|^2 \sin^2\theta + (\alpha\bar{\beta} e^{i(a-b)} + \beta\bar{\alpha}\, e^{i(b-a)}) \cos\theta \sin\theta$$

and

$$|\alpha|^2 \sin^2\theta + |\beta|^2 \cos^2\theta - (\alpha\bar{\beta} e^{i(a-b)} + \beta\bar{\alpha}\, e^{i(b-a)}) \sin\theta \cos\theta,$$

respectively. These two probabilities are seen to sum to 1, as they should, and provide us with a probability value that oscillates between the two locations (though perhaps preferentially with respect to one or the other, depending on the parameters) with a frequency determined by $|a-b|$, namely "beat" difference frequency $|E_1-E_2|/h$, as asserted above. There is also a much higher quantum oscillation frequency which in particular cases (e.g., $|\alpha| = |\beta|$ and $\theta=\pi/4$) we can identify as the average $(E_1+E_2)/2h$ of the two constituent quantum frequencies, but where in general this frequency is not so precisely defined, though can be thought of as being a quantity of this order of size.

According to a (crude) direct application of DP, we might imagine that this "measurement" (i.e., OR action) would be a spontaneous reduction to one or other of these two locations in a timescale of the general order of $\tau \approx \hbar/E_G$ (where $E_G$ is the gravitational self-energy of the difference between the expectation values mass distributions of the two states), but with much apparent randomness as to which of the two locations is taken up upon reduction. However, for an oscillating system like this, where the original quantum state is a superposition of two stationary states of slightly different energies $E_1$ and $E_2$, and which therefore behaves as a state effectively undergoing a *quantum* oscillation with frequency of around $(E_1+E_2)/2h$ and a *classical* "beat" oscillation of frequency $|E_1-E_2|/h$, it seems appropriate that we adopt this suggested *extension* of the original DP proposal, whereby the interfering quantum oscillations reduce spontaneously to a *classical oscillation* whose frequency is the beat frequency, rather than it simply reduces to one location or the other in a seemingly random way that would then not clearly manifest this beat frequency. We take the *time* for the combined quantum oscillation of the state to reduce to be $\tau \approx \hbar/E_G$ (on average), just as in the original DP proposal, but we now take the reduction to be to a classical oscillation (with this beat frequency), rather than to one or the other of the original pair of states. It is the *phase* of this oscillation that becomes definite upon reduction (OR), rather than one or the other of the two locations being singled out. We note that in the limiting situation, where we take $E_1$ and $E_2$ to be identical, the beat period would become infinite, so that in such a situation the reduction simply takes the state to one location or the other, in an average time of the order of $\tau \approx \hbar/E_1 = \hbar/E_2$, just as in the original DP proposal.

We are taking it that $\tau$ is very much larger than the quantum oscillation period $\sim 2h/(E_1+E_2)$, but it could presumably be a lot smaller than the "beats" period $h/|E_1-E_2|$. We must bear in mind that there will be a considerable spread in the actual times at which the reduction will take place (since, as we recall, the role of $\tau$ is really only as a kind of half-life for reduction), but here this only affects the *phase* of the oscillation, the frequency itself being simply the well-defined beat frequency $|E_1-E_2|/h$. Accordingly, if we consider that our system consists of a large number of identical quantum superpositions of the same kind, then this beat frequency would become evident across the system as a whole (as with an orchestra, with many violinists playing the same note, but not phase coherently). Thus, according to this extended DP proposal, we ought to see evidence of this difference frequency $|E_1-E_2|/h$, as a result of the OR process, which would be far lower than the exceedingly high individual frequencies $E_1$ and $E_2$, and the oscillation period $h/|E_1-E_2|$ could be significantly longer than $\tau$.

Thus, we may consider conscious moments to be Orch OR events occurring with beat frequencies $|E_1-E_2|/h$, rather than primary frequencies $E_1/h$ and $E_2/h$. This makes the task far simpler and more plausible than it had been within our earlier scheme. Quantum superpositions need to avoid environmental decoherence only for a time that, while considerably longer than the periods of the primary frequencies, $E_1$ and $E_2$, might nevertheless be short compared with the time period $h/|E_1-E_2|$ of the beat frequencies $|E_1-E_2|/h$. Following Bandyopadhyay's findings, these primary frequencies may be around 10 MHz, with time periods of $\sim 10^{-7}$ s. Decoherence might need be avoided for a mere 10-millionth of a second with consciousness occurring at far slower beat frequencies. For example if $E_1$ and $E_2$ were, 10.000000 MHz and 10.000040 MHz, respectively, a beat frequency of 40 Hz (by $|E_1 - E_2|/h$) could correlate with discrete conscious moments.

These considerations had not been taken into account in our simpler earlier viewpoint that the frequencies of oscillation that appear to be associated with conscious processes are the result of repeated occurrences of OR, and that the periods of oscillation are therefore to be identified with the value of tau itself, e.g., 25 ms for 40 Hz gamma synchrony. It must be borne in mind, in relation to this earlier proposal, that $\tau$ is only a kind of *average* reduction time (like the half-life of a radioactive decay). On that basis, Orch OR events would occur at distinctly irregular intervals, and could be only very roughly related to the required overall ranges such as gamma synchrony (30Hz to 90 Hz) or other EEG frequency bands. It is a little difficult to see how this previous, provisional viewpoint could give rise to a fairly definite characteristic frequency of oscillation, like the 40Hz gamma synchrony EEG.

Nevertheless, for the sake of continuity with our earlier discussions, we shall also refer to this earlier scheme concurrently with our present "beat frequency" point of view, but even this newer perspective must be considered as tentative in various respects. It is to be expected that the actual mechanisms underlying the production of consciousness in a human brain will be very much more sophisticated than any that we can put forward at the present time, and would be likely to differ in many important respects from any that we would be in a position to anticipate in our current proposals. Nevertheless, we do feel that the suggestions that we are putting forward here represent a serious attempt to grapple with the fundamental issues raised by the consciousness phenomenon, and it is in this spirit that we present them here.

## 14.5. Penrose–Hameroff "Orchestrated Objective Reduction"

### 14.5.1. *Orch OR quantum computing in the brain*

Penrose (1989, 1994) suggested that consciousness depends in some way on processes of the general nature of *quantum computations* occurring in the brain, these being terminated by some form of OR. Here the term "quantum computation" is being used in a loose sense, in which information is encoded in some discrete (not necessarily binary) physical form, and where the evolution is determined according to the U process (Schrödinger's equation). In the standard picture of quantum computers

(Benioff, 1982; Deutsch, 1985; Feynman, 1986), information is represented not just as bits of either 1 or 0, but during the U process, also as quantum superposition of both 1 and 0 together (quantum bits or "qubits") where, moreover, large-scale entanglements among many qubits would also be involved. These entangled qubits would compute, in accordance with the Schrödinger equation, in order to enable complex and highly efficient potential parallel processing. As originally conceived, quantum computers would indeed act strictly in accordance with U, but at some point a measurement is made causing a quantum state reduction R (with some randomness normally introduced). Accordingly, the *output* is in the form of a definite state in terms of classical bits.

A proposal was made in Penrose (1989) that something analogous to quantum computing, proceeding by the Schrödinger equation without decoherence, could well be acting in the brain, but where, for *conscious* processes, this would have to terminate in accordance with some threshold for *self*-collapse by a form of non-computable OR. A quantum computation terminating by OR could thus be associated with consciousness. However, no plausible biological candidate for quantum computing in the brain had been available to him, as he was then unfamiliar with microtubules. Penrose and Hameroff teamed up in the early 1990s when, fortunately, the DP form of OR mechanism was then at hand to be applied in extending the microtubule-automata models for consciousness as had been developed by Hameroff and colleagues.

As described in Sec. 14.2.3, the most logical strategic site for coherent microtubule Orch OR and consciousness is in post-synaptic dendrites and soma (in which microtubules are uniquely arrayed and stabilized) during integration phases in integrate-and-fire brain neurons. Synaptic inputs could "orchestrate" tubulin states governed by quantum dipoles, leading to tubulin superposition in vast numbers of microtubules all involved quantum-coherently together in a large-scale quantum state, where entanglement and quantum computation takes place during integration. The termination, by OR, of this orchestrated quantum computation at the end of integration phases would select microtubule states which could then influence and regulate axonal firings, thus controlling conscious behavior. Quantum states in dendrites and soma of a particular neuron could entangle with microtubules in the dendritic tree of that neuron, and also in neighboring neurons via dendritic–dendritic (or dendritic–interneuron–dendritic) gap junctions, enabling quantum entanglement of superposed microtubule tubulins among many neurons (Fig. 1). This allows unity and binding of conscious content, and a large $E_G$ which reaches threshold (by $\tau \approx \hbar/E_G$) quickly, such as at end-integration in EEG-relevant periods of time, e.g., $\tau = 0.5$ s to $\tau = 10^{-2}$ s. In the Orch OR "beat frequency" proposal, we envisage that $\tau$ could be far briefer, e.g., $10^{-7}$ s, a time interval already shown by Bandyopadhyay's group to sustain apparent quantum coherence in microtubules. In either case, or mixture of both, Orch OR provides a possible way to account for frequent moments of conscious awareness and choices governing conscious behavior.

Section 3 described microtubule automata, in which tubulins represent distinct information states interacting with neighbor states according to rules based on dipole couplings which can apply to either London force electric dipoles, or electron spin magnetic dipoles. These dipoles move atomic nuclei slightly (femtometers), and become quantum superpositioned (along with superpositioned atomic nuclei), entangled and perform quantum computation in a U process. In dendrites and soma of brain neurons, synaptic inputs could encode memory in alternating classical phases, thereby avoiding random environmental decoherence to "orchestrate" U quantum processes, enabling them to reach threshold at time $\tau$ for orchestrated objective reduction "Orch OR" by $\tau \approx \hbar/E_G$. At that time, according to this proposal, a moment of conscious experience occurs, and tubulin states are selected which influence axonal firing, encode memory and regulate synaptic plasticity.

An Orch OR moment is shown schematically in Fig. 10. The top panel shows microtubule automata with (gray) superposition $E_G$ increasing over a period up to time $\tau$, evolving deterministically and algorithmically by the Schrödinger equation (U) until threshold for OR by $\tau \approx \hbar/E_G$ is reached, at which time Orch OR occurs, accompanied by a moment of conscious experience. In the "beat frequency" modification of this proposal, these Orch OR events could occur on a faster timescale, for example in megahertz. Their far slower beat frequencies might then constitute conscious moments. The particular selection of conscious perceptions and choices would, according to standard quantum theory, involve an entirely random process, but according to Orch OR, the (objective) reduction could act to select specific states in accordance with some non-computational new physics (in line with suggestions made in Penrose (1989, 1994)).

Figure 10 (middle) depicts alternative superposed space–time curvatures (Figs. 8 and 9) corresponding to the superpositions portrayed in MTs in the top of the figure, reaching threshold at the moment of OR and

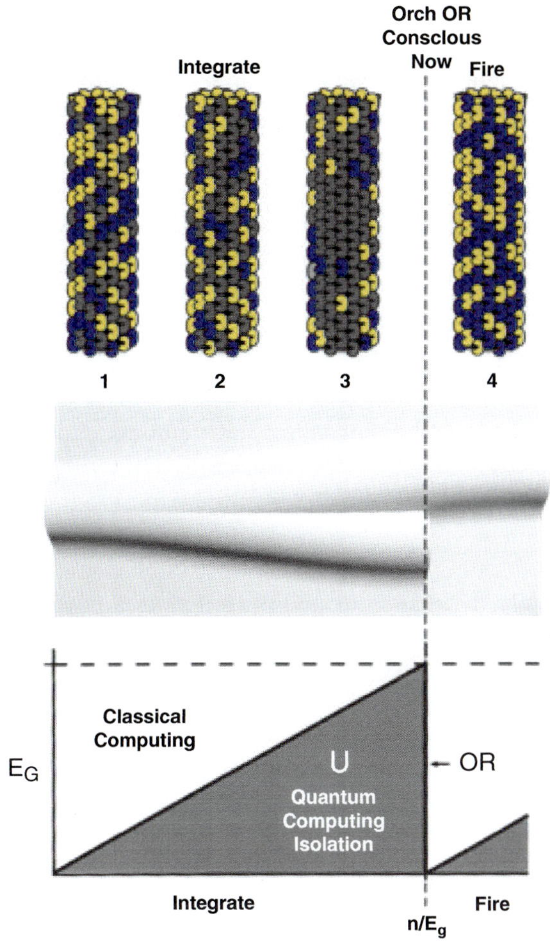


**Fig. 10.** Top: Tentatively proposed picture of a conscious event by quantum computing in one of a vast number of microtubules all acting coherently so that there is sufficient mass displacement for Orch OR to take place. Tubulins are in classical dipole states (yellow or blue), or quantum superposition of both dipole states (gray). Quantum superposition/computation evolves during integration phases (1–3) in integrate-and-fire brain neurons, increasing quantum superposition $E_G$ (gray tubulins) until threshold is met at time $\tau \approx \hbar/E_G$, at which time a conscious moment occurs, and tubulin states are selected which regulate firing and control conscious behavior. Middle: Corresponding alternative superposed space–time curvatures reaching threshold at the moment of OR and selecting one space–time curvature. Bottom: Schematic of a conscious Orch OR event showing U-like evolution of quantum superposition and increasing $E_G$ until OR threshold is met, and a conscious moment occurs by $\tau \approx \hbar/E_G$.

selecting one space–time. Figure 10 (bottom) shows a schematic of the same process.

The idea is that consciousness is associated with this (gravitational) OR process, but (see Sec. 14.4.5) occurs significantly only when (1) the alternatives are part of some highly organized cognitive structure capable of information processing, so that OR occurs in an extremely *orchestrated* form, with vast numbers of microtubules acting coherently, in order that there is sufficient mass displacement overall for the $\tau \approx \hbar/E_G$ criterion to be satisfied. (2) Interaction with environment must be avoided long enough during the U process evolution so strictly orchestrated components of the superposition reach OR threshold without too much randomness, and reflect a significant non-computable influence. Only then does a recognizably conscious Orch OR event takes place. On the other hand, we may consider that any individual occurrence of OR without orchestration would be a moment of random *proto*-consciousness lacking cognition and meaningful content.

We shall be seeing orchestrated OR in more detail shortly, together with its particular relevance to microtubules. In any case, we recognize that the experiential elements of proto-consciousness would be intimately tied in with the most primitive Planck-level ingredients of space–time geometry, these presumed "ingredients" being taken to be at the absurdly tiny level of $10^{-35}$m and $10^{-43}$s, a distance and a time of 20 orders of magnitude smaller than those of normal particle-physics scales and their most rapid processes, and they are smaller by far than biological scales and processes. These scales refer only to the normally extremely tiny differences in *space–time geometry* between different states in superposition, the separated states themselves being enormously larger. OR is deemed to take place when such tiny space–time differences reach the Planck level (roughly speaking). Owing to the extreme weakness of gravitational forces as compared with those of the chemical and electric forces of biology, the energy $E_G$ is liable to be *far* smaller than any energy that arises directly from biological processes.

OR acts effectively instantaneously as a choice between dynamical alternatives (a choice that is an integral part of the relevant quantum dynamics) and $E_G$ is not to be thought of as being in direct competition with any of the usual biological energies, as it plays a completely different role, supplying a needed energy *uncertainty* that then allows a choice to be made between the separated space–time geometries, rather than providing an actual energy that enters into any considerations of energy balance that would be of direct relevance to chemical or normal physical processes. This energy uncertainty is the key ingredient of the computation of the reduction time $\tau$, and it is appropriate that this energy uncertainty is indeed far smaller than the energies that are normally under consideration with regard to chemical energy balance, etc. If it were not so, then there would be a danger of *conflict* with normal considerations of energy balance.

Nevertheless, the extreme weakness of gravity tells us there must be a considerable amount of material involved in the coherent mass displacement between superposed structures in order that $\tau$ can be small enough to be playing its necessary role in the relevant OR processes in the brain. These superposed structures should also process information and regulate neuronal physiology. According to Orch OR, *microtubules* are central to these structures, and some form of biological quantum computation in microtubules (perhaps in the more symmetrical A-lattice microtubules) would have to be involved to provide a subtle yet direct connection to Planck-scale geometry, leading eventually to discrete moments of actual conscious experience and choice. As described above, these are presumed to occur primarily in dendritic–somatic microtubules during integration phases in integrate-and-fire brain neurons, resulting in sequences of Orch OR conscious moments occurring within brain physiology, and able to regulate neuronal firings and behavior.

### 14.5.2. *Tubulin qubits and Orch OR conscious moments*

For Orch OR to be operative in the brain, we would need coherent superpositions of sufficient amounts of (e.g., microtubule) material accounting for $E_G$, undisturbed by environmental entanglement, where this reduction occurs in accordance with the above OR scheme in a timescale of the general order for a conscious experience. For an ordinary type of experience, this might be about $\tau$=0.5s to $\tau$=$10^{-2}$s which concurs with neural correlates of consciousness, such as particular frequencies of EEG, visual gestalts and reported conscious moments.

In order to see whether Orch OR can be implemented for some particular chosen reduction time $\tau$, determined according to $\tau \approx \hbar/E_G$, the gravitational self-energy $E_G$ must be calculated for this $\tau$, which is taken to correspond to the duration of, or perhaps the interval between, conscious moments. We could calculate $E_G$ from the difference between the mass distributions between two states of tubulin in superposition, but as previously mentioned, the use merely of an average density may not be adequate, as the mass is concentrated in the nuclei. There is, however, a large uncertainty about how "smeared out" these nuclei must be considered to be, as referred to above, which is related to how "crystalline" the microtubules may be considered to be. Accordingly, we calculated $E_G$ for tubulin (Hameroff & Penrose, 1996a) separated from itself at three possible levels of separation: (1) the entire smoothed-out protein (partial separation), (2) its atomic nuclei and (3) its nucleons (protons and neutrons). In our picture, the dominant effect is likely to be (2) separation at the level of atomic nuclei, e.g., 2.5 Fermi length for carbon nuclei (2.5 fm; $2.5 \times 10^{-15}$ m). This shift is the same as that predicted to be caused by electron charge separations of one nanometer, e.g., London force dipoles within aromatic amino acid rings.

Using $\tau \approx \hbar / E_G$, where we may choose $\tau$ as 25 ms for "40 Hz" gamma synchrony conscious moments, we calculated the number of required tubulins in superposition, separated by the diameter of their (carbon) atomic nuclei. Because the carbon nucleus displacement is greater than its radius, the gravitational self-energy $E_c$ for superposition separation of one carbon atom is roughly given by: $E_c = Gm^2/a_c$, where $G$ is the gravitational constant, $m$ is the carbon nuclear mass and $a_c$ is the carbon nucleus sphere radius equal to 2.5 fermi distances. We calculated that roughly $2 \times 10^{10}$ tubulins displaced in coherent superposition for 25 ms will, on this basis, self-collapse in that time period, and elicit Orch OR. For a $\tau$ of 500 ms, ~$10^9$ tubulins would be required.

Neurons each contain roughly $10^9$ tubulins, but only a fraction per neuron are likely to be involved in consciousness (e.g., a fraction of those in dendrites and soma). Global macroscopic states such as superconductivity ensue from quantum coherence among only very small fractions of components. If 1% of tubulins within a given set of neurons were coherent for 25 ms, then 20,000 such neurons would be required to elicit OR. In human brain, cognition and consciousness are, at any one time, thought to involve tens of thousands of neurons. Hebb's (1949) "cell assemblies," Eccles's (1992) "modules," and Crick & Koch's (1990) "coherent sets of neurons" are each estimated to contain some 10,000 to 100,000 neurons which may be widely distributed throughout the brain (Scott, 1995). In the "beat frequency" approach, a much smaller time $\tau = 10^{-7}$s might perhaps suffice, but require much larger microtubule superposition $E_G$, involving roughly $10^9$ neurons, or 1% of the brain.

As electron movements may shift atomic nuclei by a distance of the order of a nuclear diameter, we assume that electron-superposition separations of around a nanometer could result in atomic (e.g., carbon) nuclear superposition separations of a few femtometers (Fermi lengths) (Pan *et al.*, 2007), which is about a nuclear diameter, thereby appearing to meet DP requirement for OR.

Assuming that microtubule quantum states occur in a specific brain neuron, how could they involve microtubules in other neurons throughout the brain? Orch OR proposes that quantum states can extend by entanglement between adjacent neurons through gap junctions, primitive electrical connections between adjacent cells (Fig. 1). Structurally, gap junctions are windows which may be open or closed. When open, gap junctions synchronize adjacent cell membrane polarization states, but also allow passage of molecules between cytoplasmic compartments of the two cells. So both membranes and cytoplasmic interiors of gap junction-connected neurons are continuous, essentially one complex "hyper-neuron" or syncytium. (Ironically, before Ramon-y-Cajal showed that neurons were discrete cells, the prevalent model for brain structure was a continuous threaded-together syncytium as proposed by Camille Golgi.) Quantum states in microtubules in one neuron can, we propose, extend by entanglement and tunneling through gap junctions to microtubules in adjacent neurons (including inter-neurons), potentially extending to brain-wide syncytia. Beginning in 1998, evidence began to show that gamma synchrony, the best measureable correlate of consciousness, depended on gap junctions, particularly dendritic–dendritic gap junctions (Dermietzel, 1998; Draguhn *et al.*, 1998; Galaretta & Hestrin, 2001; Bennett & Zukin, 2004; Fukuda & Kosaka, 2000; Traub *et al.*, 2002). To account for the distinction between conscious activities and non-conscious "auto-pilot" activities, and the fact that consciousness can occur in various brain regions, Hameroff (2010) developed the "Conscious pilot" model in which syncytial zones of dendritic gamma synchrony move around the brain, regulated by gap junction openings and closings, in turn regulated by microtubules. The model suggests consciousness literally moves around the brain in a mobile synchronized zone, within which isolated, entangled microtubules carry out quantum computations and Orch OR. Taken together, Orch OR and the conscious pilot distinguish conscious from non-conscious functional processes in the brain.

Applying $\tau \approx \hbar/E_G$ to large numbers of brain neurons, we find that, with this point of view with regard to Orch OR, a spectrum of possible types of conscious events might be able to occur, including those at higher frequency and intensity. It may be noted that Tibetan monk meditators have been found to have 80 Hz gamma synchrony, and perhaps more intense experience (Lutz *et al.*, 2004). Thus, according to the viewpoint proposed above, where we interpret this frequency to be associated with a succession of Orch OR moments, then $E_G \approx \hbar/\tau$ would appear to require that there is twice as much brain involvement for 80 Hz as for consciousness occurring at 40 Hz. More appropriately, it might be $\sqrt{2}$ times as much, since for the calculation of $E_G$, the displacement ought to be entirely coherent, and then the mass enters quadratically in $E_G$. Even higher (frequency), expanded awareness states of consciousness might be expected, according to this scheme, with more neuronal brain involvement. In the beat frequency approach, we might consider that megahertz or higher frequencies might be directly relevant to Orch OR, for which $\tau$ is very low, at $10^{-7}$s, while $E_G$ is large, at roughly $10^9$ neurons, 1% of the brain, or perhaps even faster, larger components with more intense experiences.

There is also the possibility that discernable moments of consciousness are events that normally occur at a much slower pace than is suggested by the considerations above, and that they happen only at rough intervals of the order of, say, several hundreds of milliseconds, rather than ~25ms. One might indeed think of conscious influences as perhaps being rather slow, in contrast with the great deal of vastly faster unconscious computing that might be some form of quantum computing, but without OR. Another possibility is that conscious moments such as visual gestalts may be slower events, e.g., correlating with 4 to 7 Hz theta frequency, with nested gamma waves (Woolf & Hameroff, 2001; VanRullen & Koch, 2003). Yet another possibility, consistent with recent findings of scale-invariant processes in brain function, is that consciousness, according to this version of Orch OR's $\tau \approx \hbar/E_G$ can occur at varying frequencies, moving up and down in scales, with higher frequency events involving more of the brain having greater experiential intensity. At the present stage of uncertainty about such matters it is perhaps best not to be dogmatic about how the ideas of Orch OR are to be applied. In any case, the numerical assignments provided above must be considered to be extremely rough, and at the moment we are far from being in a position to be definitive about the precise way in which the Orch OR is to operate, even according to the particular version of Orch OR that is being described here. Alternative possibilities will need to be considered with an open mind.

### 14.5.3. *Microtubules and environmental "decoherence"*

Technological quantum computers, e.g., those using ion traps as qubits, are plagued by disruption of seemingly delicate quantum states by environmental interactions including thermal vibration. Such technology requires extremely cold temperatures and vacuum to operate. The role of environmental decoherence, according to OR schemes, is that R is effected in a system through its entanglement with its much larger effectively random environment, so that when OR takes place in that environment, the system itself is carried with it and therefore reduces, seemingly randomly, in accordance with a conventional R process. Thus, if we require non-random aspects of OR to play a role in (conscious) brain function, as is required for Orch OR, we need to avoid premature entanglement with the random environment, as this would result in state reduction without non-computable aspects or cognition. For Orch OR, environmental interactions must be avoided during the evolution toward time $\tau$ ($\approx \hbar/E_G$), so that the non-random (non-computable) aspects of OR can be playing their roles. How does quantum computing avoid environmental interaction ("decoherence") in the "warm, wet and noisy" brain?

It was suggested (Hameroff & Penrose, 1996a) that microtubule quantum states avoid decoherence by being pumped, laser-like, by Fröhlich resonance, and shielded by ordered water, C-termini Debye layers, actin gel and strong mitochondrial electric fields. Moreover, quantum states in Orch OR are proposed to originate in non-polar, hydrophobic channels in tubulin interiors, isolated from polar entanglements. These regions are where anesthetic gas molecules bind and act to erase consciousness, are conducive to quantum coherent states, and described collectively as the "quantum underground" in an accompanying chapter in this book (Craddock *et al*, 2016). Further, superpositions in Orch OR are proposed to originate with superposition of only atomic nuclei separation, or spin.

The analogy with high-temperature superconductors may indeed be pertinent. As yet, there is no fully accepted theory of how such superconductors operate, avoiding loss of quantum coherence from the usual processes of environmental decoherence. Yet there are materials which seem to support superconductivity at temperatures roughly halfway between room temperature and absolute zero (He *et al.*, 2011). This is still a long way from body temperature, of course, but increasing evidence suggests functional quantum effects operate in biology.

As described in Sec. 14.4.5, research in the past nine years has clearly shown quantum coherence in warm biological systems. Electronic quantum effects occur at ambient temperatures in proteins involved in photosynthesis (Engel *et al.*, 2007; Hildner *et al.*, 2013), these being thought to be facilitated by coherent protein mechanical vibrations (Chin *et al.*, 2013), very similar to a mechanism proposed by Fröhlich over 40 years ago, and to the mechanism we propose here in tubulin. Evidence for resonance-enhanced quantum conductance along helical pathways in tubulin and microtubules by Bandyopadhyay's group appears to be very supportive of Orch OR. Synthetic systems which function as systems for quantum coherence are chemically close to aromatic ring pathways in tubulin (Sec. 14.3.2, Figs. 5–7) (Hayea *et al.*, 2013). Warm quantum effects have also been discovered or proposed in bird brain navigation (Gauger *et al.*, 2011), ion channels (Bernroider & Roy, 2005), sense of smell (Turin, 1996), DNA (Rieper, 2011), protein folding (Luo & Lu, 2011) and biological water (Reiter *et al.*, 2011). Since Nature has already been found to be able to utilize quantum coherence at biological temperatures in many of the biological systems that have been closely studied, quantum coherence could well be a near-ubiquitous factor in living systems.

If microtubule quantum computations are isolated from the environment, how do they interact with that environment for input and output? Orch OR suggests phases of isolated quantum computing alternate with phases of classical environmental interaction, e.g. at gamma synchrony, roughly 40 times per second. (Computing pioneer Paul Benioff suggested such a scheme of alternating quantum and classical phases in quantum computing robots (Benioff, 1998)). Strictly, according to OR (the DP version or otherwise), it is, in any case precisely the OR procedure that gives rise to the "classical world" that we find in macroscopic systems. All the basic ingredients are, after all, quantum particles of one kind or another, and it is the reduction process (here DP OR) that provides our picture of classicality. According to the DP viewpoint, the classical world actually arises because of continuing OR actions.

### 14.5.4. *Temporal non-locality and free will*

Measurable brain activity correlated with a conscious perception of a stimulus generally occurs several hundred milliseconds after that stimulus. Yet in activities ranging from rapid conversation to competitive athletics, we respond to a stimulus (seemingly consciously) *before* the above activity that would be correlated with that stimulus occurs in the brain. This is interpreted in conventional neuroscience and philosophy (Dennett, 1991; Dennett & Kinsbourne, 1991; Wegner, 2002) to imply that in such cases we respond nonconsciously, on auto-pilot, and subsequently have only an *illusion* of conscious response. The mainstream view is that consciousness is an epiphenomenal illusion, occurring after-the-fact as a false impression of conscious control of behavior. Accordingly, we are merely "helpless spectators" (Huxley, 1986).

Indeed that *might* be the case. However, quantum processes in the brain offer what appear to be loopholes to such implications, where the apparent temporal progression of conscious experience and willed action need not correlate in a clear-cut way with the precise timings of an external clock. In the 1970s, neurophysiologist Benjamin Libet performed experiments on patients having brain surgery while awake, i.e., under local anesthesia (Libet *et al.*, 1997). Able to stimulate and record from conscious human brains, and gather patients' subjective reports with precise timing, Libet determined that conscious perception of a stimulus required up to 500 ms of brain activity post-stimulus, but that conscious awareness occurred at 30 ms post-stimulus. The brain at 30 ms "knew" that activity would continue, or not continue, for several hundred more milliseconds, i.e., that subjective experience was referred "backward in time." Numerous other experiments have also provided strong indications of temporal anomalies in perception and willed choice (Bem, 2012; Bierman & Radin, 1997; Ma *et al.*, 2012).

Bearing such apparent anomalies in mind, Penrose put forward a tentative suggestion (Penrose, 1994) that effects like Libet's backward time referral might be related to the fact that quantum entanglements are not mediated in a normal causal way, so that it might be possible for conscious experience not to follow the normal rules of sequential time progression, so long as this does not lead to contradictions with external causality. In Sec. 14.4.2, it was pointed out that the (experimentally confirmed) phenomenon of "quantum teleportation" (Bennett & Wiesner, 1992; Bouwmeester *et al.*, 1997; Macikic *et al.*, 2002) cannot be explained in terms of ordinary classical information processing, but as a combination of such classical causal influences and the causal effects of quantum entanglement. It indeed turns out that quantum entanglement effects — encompassed by such terms as "quantum information" or "*quanglement*"

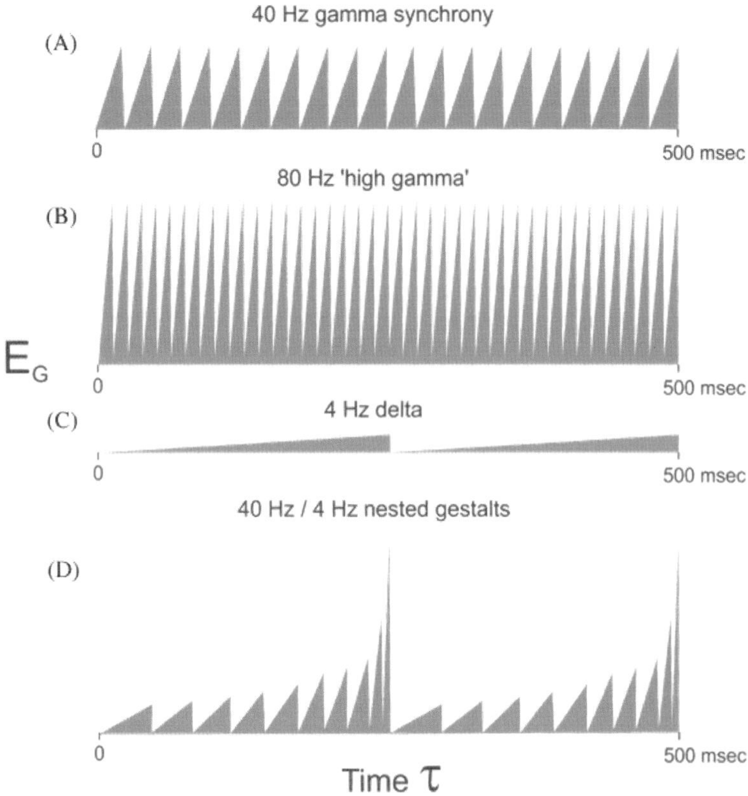


**Fig. 11.** According to a scheme proposed earlier (Fig. 10, bottom), sequences of Orch OR conscious moments occurring by $\tau \approx \hbar/E_G$ with intensity of experience correlated with orchestrated $E_G$. (A) "Normal" Orch OR conscious moments every 25 ms in "40 Hz" gamma synchrony. (B) Heightened, enhanced conscious moments occurring every 12.5 ms in 80 Hz "high gamma" synchrony. (C) Low intensity conscious moments occurring every 250 ms (4 Hz delta wave EEG). (D) Gamma wave conscious moments nested in delta waves in visual gestalts. (Woolf & Hameroff, 2001).

(Penrose, 2004; Percival, 1994) — appear to have to be thought of as being able to propagate in *either* direction in time (into the past or into the future). Such effects, however, cannot by themselves be used to communicate ordinary information into the past. Nevertheless, in conjunction with normal classical future-propagating (i.e., "causal") signaling, these quantum-teleportation influences can achieve certain kinds of "signaling" that cannot be achieved simply by classical future-directed means.

The issue is a subtle one, but if conscious experience is indeed rooted in the OR process, where we take OR to relate the *classical* to the *quantum* world, then apparent anomalies in the sequential aspects of consciousness are perhaps to be expected. The Orch OR scheme allows conscious experience to be *temporally non-local* to a degree, where this temporal non-locality would spread to the kind of timescale that would be involved in the relevant Orch OR process, which might indeed allow this temporal non-locality to spread to a time of Libet's 500 ms or longer. When the "moment" of an internal conscious experience is timed externally, it may well be found that this external timing does not precisely accord with a time progression that would seem to apply to internal conscious experience, owing to this temporal non-locality intrinsic to Orch OR. The effective quantum backward-time referral inherent in the temporal non-locality resulting from the quanglement aspects of Orch OR, as suggested above, enables conscious experience actually to be *temporally non-local*, thus providing a possible means to rescue consciousness from its unfortunate characterization as epiphenomenal illusion. Accordingly, Orch OR could well enable consciousness to have a causal efficacy, despite its apparently anomalous relation to a timing assigned to it in relation to an external clock, thereby allowing conscious action to provide a semblance of free will (Hameroff, 2006b, 2012).

### 14.5.5. *Orch OR and evolution*

In conventional views, the experiential qualities of conscious awareness are assumed to have emerged from complex neuronal computation at some point in evolution, whether recently in human brains, or at some earlier, but unspecified level of development. In these views, consciousness is an emergent property of complex computational activity. On the other hand, Orch OR follows the notion that OR events with primitive "experiential" qualities have been occurring in the universe all along, in the reduction R of quantum superpositions to classical reality. Small superpositions lacking isolation would entangle directly with the random environment, rapidly reaching OR threshold by $\tau \approx \hbar/E_G$, resulting in non-orchestrated OR events. Each such event would lack cognition or any non-computational influence, but *would* be associated with an undifferentiated "proto-conscious" experience, one without information or meaning, but entailing primitive feelings or "qualia," including perhaps rudimentary pleasure. Such undifferentiated experiences are taken, in the Orch OR scheme, to be irreducible, fundamental features of "Planck scale geometry," perhaps ultimately having a role as important to basic physics as those of mass, spin or charge. And undifferentiated, proto-conscious OR events may be significant in biological evolution.

Life on earth is considered to have begun three to four billion years ago in a simmering liquid called the "primordial soup," proposed independently in the 1920's by JBS Haldane and Alexander Oparin. In a famous experiment, Miller & Urey (1959) chemically and energetically simulated a primordial soup and found that it produced "amphipathic" biomolecules having both polar and non-polar components, e.g. non-polar pi electron resonance rings on one end, and charged, polar groups on the other. As "oil and water don't mix," non-polar pi resonance rings of amphipathic biomolecules apparently coalesced, according to Oparin, leading to primitive "micelle" structures with polar ends pointing outward into an aqueous environment (Figure 12), micelles being precursors to more complex biomolecules and cells. For example amphipathic phospholipids coalesce into lipid membranes, amphipathic purines and pyrimidines form DNA and RNA, and amphipathic amino acids form proteins. Non-polar pi resonance groups of amphipathic molecules thus resulted in regions within biomolecules conducive to quantum coherence, entanglement and superposition — the "quantum underground" (Craddock *et al.*, 2016), regions where, in human and animal brains, anesthetics act to selectively erase consciousness. It is perhaps noteworthy that dopamine, an amphipathic biomolecule similar to those apparently

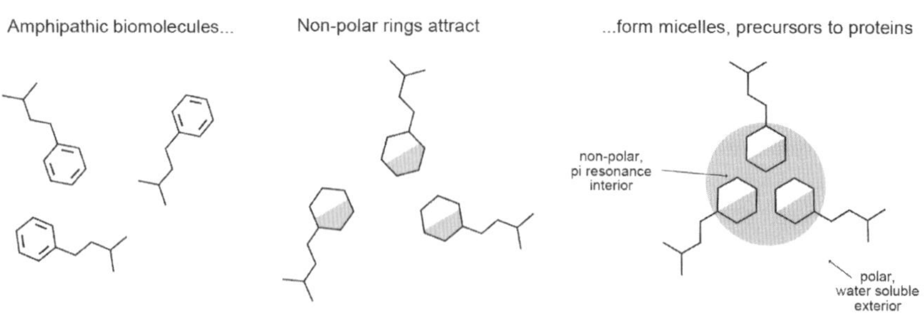


**Fig. 12.** A simplified schematic view of the origin of life. Dopamine-like amphipathic molecules attract and couple by pi electron resonance dipoles forming an Oparin "micelle" with non-polar interior and polar, water-soluble exterior.

present in the primordial soup and early micelle-like biosystems is a brain neurotransmitter which mediates pleasure and reward. We may take the view that OR and primitive pleasure in non-polar, "quantum underground" regions composed of dopamine-like molecules could have played a key role in evolution.

Darwin's theory of natural selection suggests that life evolved by incremental steps and random mutations. Therefore one would not expect the substantial level of coherence across the brain that would be needed for the non-computable Orch OR of human conscious understanding to be reached without something similar, but more primitive, having preceded it. Proto-conscious OR-mediated feelings might themselves have evolved incrementally during evolution, and perhaps even have driven it.

At some stage in biomolecular development, quantum-superposed states of pi resonance groups and their associated atomic nuclei in "quantum underground" regions may have been able to partially avoid polar interactions and premature (i.e. random environment-driven) OR, and therefore might sustain superpositions sufficiently to experience slightly more coherent OR-mediated qualia including, in some cases, primitive feelings of ("proto")-pleasure. As adjacent pi resonance groups arrange in specific geometric relations ("pi stacks"), such primitive feelings might have provided a feedback fitness function to optimize proto-pleasure, or avoid negative feelings ("proto-pain") by pi stack geometry. Primitive proto-pleasure and proto-pain could then have provided positive and negative reinforcement for structure and behavior promoting survival, e.g. in simple biosystems before the advent of genes.

An early, essential role for microtubules and similar structures in living systems is suggested by their deep-rooted genetic origins (Duggin *et al*, 2014). Life on earth is divided into three types of cells, (1) animal (eukaryotic) and plant cells, (2) prokaryotic bacteria, and (3) archaebacteria, the latter existing at extreme temperatures. In (1) animals and plants, microtubules composed of tubulin organize cell activities, movement and behavior. In (2) bacteria, lattice polymers of tubulin analog FtsZ proteins regulate structure and activities, and in (3) archaebacteria, polymers of another tubulin analog, CetZ, control cell shape and function, e.g. converting discoid cells to tube-like shapes able to swim.

Initially, we might presume, rudimentary quantum computing in microtubules or their more primitive counterparts would likely fail to isolate adequately, and consequently the OR threshold would only be reached with considerable randomness of environmental entanglement and therefore no effective qualia. Nevertheless microtubule quantum computing by U evolution could well be advantageous to biological processes, and one may speculate on the presence of positive, if mixed, pleasurable feelings and feedback even without fully reaching threshold for Orch OR and non-computational influence. Indeed, this type of protoconscious but effective processing is likely to be occurring in microtubules, FtsZ and CetZ protein polymers throughout all of biology.

With further evolution, better isolation, resonance with electromechanical vibrations (e.g. at "quantum criticality," Vattay *et al*, 2015), wider entanglement and advent of mixed polarity networks amenable to beat frequencies in dendrites and soma of brain neurons, microtubule quantum superpositions, we argue, led to OR events that became increasingly "orchestrated." Accordingly, our picture is that Orch OR underlies full conscious experience with perceptions and choices influenced by noncomputable Platonic values intrinsic to the structure of the universe.

The origin of eukaryotic animal cells 1.3 billion years ago is suggested to have been a symbiotic event in which motile spirochetes invaded bacterial prokaryotes, spirochetal flagellae being the apparent origin of microtubules which provided movement and internal organization to previously immobile cells (Margulis & Sagan, 1995). As OR events in microtubules became more orchestrated over the course of evolution, the content of conscious experience became more cognitively useful, e.g., representative of the external world, and pleasurable, e.g., food, sex. Pursuit of positive conscious experience would foster survival. Optimization of Orch OR in conscious experience and associated noncomputational effects *per se* may be driving evolution.

As simple nervous systems and arrangements of MTs grew larger and developed isolation mechanisms, quantum cognitive systems would gain selective advantage by avoiding premature OR through environmental decoherence for long enough to be fully orchestrated and reach the OR threshold without involving the random environment. These Orch OR moments can occur across a spectrum defined by $\tau \approx \hbar/E_G$. For small superpositions $E_G$, $\tau$ will be large, requiring prolonged isolation. Larger systems with more frequent conscious moments would be increasingly useful, but more difficult to isolate. In the course of evolution, Orch OR conscious moments (in accordance with $\tau \approx \hbar/E_G$) began in simple organisms involving smaller $E_G$, but requiring longer times $\tau$ during which environmental decoherence is avoided. The scale of $E_G$ would appear also to be related to intensity of experience, so we may anticipate that low $E_G$, with large $\tau$ moments, might be rather dull compared to more intense moments of large $E_G$ and small $\tau$. If this is the case, then such low frequency conscious moments would also be slow and out of step with real-world activities. As systems developed to allow $E_G$ to became larger, the frequency of conscious moments, according to $\tau \approx \hbar/E_G$, could approach present-day biological timescales.

Central nervous systems consisting of approximately 300 neurons, such as those present in tiny worms and urchins at the early Cambrian evolutionary explosion 540 million years ago, theoretically had sufficient microtubules to reach $\tau$ under one minute, and it might thus be just feasible for them to make use of Orch OR (Hameroff, 1998d). Accordingly, one might speculate that the onset of Orch OR and primitive consciousness, albeit exceedingly slow and simple but still with useful conscious moments, precipitated the accelerated evolution of the Cambrian explosion.

Only at a much later evolutionary stage would the selective advantages of a capability for genuine understanding come about, requiring the non-computability of Orch OR that goes beyond mere quantum computation, and depends upon larger scale infrastructure of efficiently functioning MTs, capable of operating quantum-computational processes. Further evolution providing larger sets of MTs (hence larger $E_G$) able to be isolated from decoherence would enable, by $\tau \approx \hbar/E_G$, more frequent and more intense moments of conscious experience, e.g., eventually in human brains every 25 ms in 40 Hz gamma synchrony EEG, or faster. Future evolution might enable brains to accommodate even larger values of $E_G$ and shorter values of $\tau$. At least this is one possibility. Another evolutionary improvement would be to increase the intensity of parallel Orch OR processing, without a requirement that $\tau$ should necessarily become shorter.

### 14.5.6. *Orch OR criticisms and responses*

Orch OR has been criticized repeatedly since its inception. Here we review and summarize major criticisms and responses.

Grush & Churchland (1995) took issue with the Gödel's theorem argument, as well as several biological factors. One objection involved the MT-disabling drug colchicine which treats diseases such as gout by immobilizing neutrophil cells which cause painful inflammation in joints.

Neutrophil mobility requires cycles of MT assembly/disassembly, and colchicine prevents re-assembly, impairing neutrophil mobility and reducing inflammation. Grush and Churchland pointed out that patients given colchicine do not lose consciousness, concluding that microtubules cannot be essential for consciousness. Penrose & Hameroff (1995) responded point-by-point to every objection, e.g., explaining that colchicine does not cross the blood–brain barrier, and so does not reach the brain, and that brain neurons do not disassemble/re-assemble anyway. Colchicine infused directly into the brains of animals does cause severe cognitive impairment and apparent loss of consciousness (Bensimon & Chernat, 1991).

*A-lattice vs. B-lattice microtubules* MTs have two types of hexagonal lattices, A- and B-. Tubulin is a peanut-shaped dimer with alpha and beta monomers. In a 13-protofilament MT A-lattice, tubulin–tubulin sideways interaction occur between alpha monomer on one tubulin, and beta tubulin on the other, i.e. alpha–beta, and beta–alpha interactions (Amos & Klug, 1974). This gives a seamless lattice and Fibonacci geometry which are optimal for quantum computing, and preferred in Orch OR. In the B-lattice, sideways interactions are alpha–alpha and beta–beta, except for a vertical seam of (A-lattice-like) alpha–beta and beta–alpha alignment. Orch OR has predicted A-lattice MTs, but critics point to analysis of MTs from neurons, e.g., from whole mouse brains which are said to show predominantly B-lattice MTs. However, these "B-lattice" (Mandelkow *et al.*, 1992; Kikkawa *et al.*, 1994) brain MTs have multiple seams involving four or more or protofilaments, so A-lattice configuration occurs in a third of so-called B-lattice MTs. Other work shows mixed A and B lattice microtubules (McEwen & Edelstein, 1980).

Orch OR is expected to occur in only a fraction of suitable dendritic and somatic MTs, and perhaps only transiently, and partially. Bandyopadhyay has preliminary evidence MTs may switch between A- and B-lattice configurations. The MT A-lattice configuration may be rare, exist transiently as patches in otherwise B-lattice MTs, and be specifically involved in quantum coherence, Orch OR and consciousness.

Georgiev (2009) questioned Orch OR on the basis of "not enough tubulins." By $\tau \approx \hbar/E_G$, the superposition ($E_G$) required for 25 ms Orch OR events is about $2 \times 10^{10}$ tubulins. Depending on the number of tubulins per neuron, and the percent of tubulin involvement, predictions can be made for the number of neurons, and percent of brain involvement, for Orch OR conscious events. This percentage may be small, as for example superconductors have only a tiny percentage of components in quantum states. Moreover, A-lattice MTs (or A-lattice portions of B-lattice MTs) may be relatively rare, and distributed throughout many neurons. In any case, it might be that many more tubulins are involved (such as in some versions of the beat frequency approach), e.g., $10^{17}$ tubulins, $10^{9}$ neurons, 1% of the brain, or more. It should be noted that Orch OR is the only theory able to meaningfully entertain such quantitative speculation.

Tuszynski *et al.* (1998) questioned how extremely weak gravitational energy in the DP version of OR could influence tubulin protein states. With $2 \times 10^{10}$ tubulins for 25 ms Orch OR, $E_G$ would be roughly $10^{-10}$ eV ($10^{-29}$ joules), seemingly insignificant compared to ambient energy kT at $10^{-4}$ eV. All this serves to illustrate the fact that the energy $E_G$ does not actually play a role in physical processes *as* an energy, in competition with other energies that are driving the physical (chemical, electronic) processes of relevance. As stated in Sec. 14.5.1, $E_G$ is, instead, an *energy uncertainty* — and it is this uncertainty that allows quantum state reduction to take place without violation of energy conservation. The fact that $E_G$ is far smaller than the other energies involved in the relevant physical processes is a necessary feature of the consistency of the OR scheme, particularly with regard to energy conservation. It does not supply the energy to drive the physical processes involved, but it provides the energy uncertainty that allows the freedom for processes having virtually the same energy as each other to be alternative actions. In practice, all that $E_G$ is needed for is to tell us how to calculate the lifetime $\tau$ of the superposition. $E_G$ would enter into issues of energy balance only if gravitational interactions between the parts of the system were important in the processes involved. (The Earth's gravitational field plays no role in this either, because it cancels out in the calculation of $E_G$.) No other forces of nature directly contribute to $E_G$, which is just as well, because if they did, there would be a gross discrepancy with observational physics.

Tegmark (2000) published a critique of Orch OR based on his calculated decoherence times for microtubules of $10^{-13}$ s at biological temperature, far too brief for physiological effects. However Tegmark did not include Orch OR stipulations and in essence created, and then refuted his *own* quantum microtubule model. He assumed superpositions of solitons separated from themselves by a distance of 24 nm along the length of the microtubule. As previously described, superposition separation in Orch OR is at the Fermi length level of atomic nuclei, i.e., seven orders of magnitude smaller than Tegmark's separation value, thus underestimating decoherence time by seven orders of magnitude, i.e., from $10^{-13}$ s to $10^{-6}$ s.

Hagan *et al.* (2001) used Tegmark's same formula and recalculated microtubule decoherence times using Orch OR stipulations, finding $10^{-4}$ to $10^{-3}$ s, or longer. In any case, experimentally, Bandyopadhyay's group has found 10 kHz resonance, i.e., $10^{-4}$ seconds coherence times. Also, as stated earlier, there are versions of the beat-frequency scheme that would require much shorter decoherence times, though at the expense of correspondingly larger bodies of material, more of the brain, being involved in the quantum-coherent states.

Koch & Hepp (2006) challenged Orch OR with a thought experiment, describing a person observing a superposition of a cat both dead and alive with one eye, the other eye distracted by a series of images ("binocular rivalry"). Without explaining how an observable superposition of this kind could be prepared (where according to OR, by $\tau \approx \hbar/E_G$, the cat would already be either dead or alive long before being observed), they asked "Where in the observer's brain would reduction occur?," apparently assuming Orch OR followed the version of the Copenhagen interpretation in which conscious observation, in effect, *causes* quantum state reduction (placing consciousness outside science). This is precisely the opposite of Orch OR in which consciousness *is* the orchestrated quantum state reduction given by OR. But in the straightforward case of conscious observation of an already dead or alive cat, reduction (Orch OR) and consciousness would most likely occur in dendritic-somatic microtubules in gap junction–connected neurons in visual and associative cortex and other brain areas.

Orch OR can (at least in principle) account for the related issue of bistable perceptions (e.g., the famous face/vase illusion, or Necker cube). Non-conscious superpositions of both possibilities (face and vase) during pre-conscious quantum superposition then reduce by OR at time $\tau \approx \hbar/E_G$ to a conscious perception of one or the other, face *or* vase. The reduction could be taken to occur among microtubules within gap junction-connected neurons in various areas of visual and pre-frontal cortex and other brain regions.

Reimers *et al.* (2009) described three types of Fröhlich condensation (weak, strong and coherent, the first classical and the latter two quantum). They validated 8 MHz coherence measured in microtubules by Pokorny (2004) and Pokorny *et al.*, (2001) as weak condensation. Based on simulation of a one-dimensional linear chain of tubulin dimers representing a microtubule, they concluded that only weak Fröhlich condensation occurs in microtubules. Claiming that Orch OR requires strong or coherent Fröhlich condensation, they concluded Orch OR is invalid. However Samsonovich *et al.* (1992) simulated a microtubule as a two-dimensional lattice plane with toroidal boundary conditions and found Fröhlich resonance maxima at discrete locations in super-lattice patterns on the simulated microtubule surface which precisely matched experimentally observed functional attachment sites for MAPs. In any case, these simulations are superseded by experimental evidence for gigahertz, megahertz and kilohertz resonance discovered in single MTs by the Bandyopadhyay group ("Bandyopadhyay coherence," "BC")

McKemmish *et al.* (2009) challenged the Orch OR contention that tubulin switching is mediated by London forces, pointing out that mobile π electrons in a benzene ring (e.g., a phenyl ring without attachments) are completely delocalized, and hence cannot switch between states, nor exist in superposition of both states. Agreed, a single benzene cannot engage in switching. London forces occur between two or more pi electron cloud ring structures, or other non-polar groups. A single benzene ring cannot support London forces. It takes two (or more) to tango. Orch OR has always maintained two or more non-polar groups necessarily, and now invokes contiguous arrays of such groups in quantum channels through tubulin and through microtubules. Moreover, we now add the possibility that magnetic spin dipoles mediate Orch OR.

McKemmish *et al.* (2009) further assert that tubulin switching in Orch OR requires significant conformational structural change, and that the only mechanism for such conformational switching is due to GTP hydrolysis, i.e., conversion of guanosine triphosphate (GTP) to guanosine diphosphate (GDP) with release of phosphate group energy, and tubulin conformational flexing. McKemmish *et al.* (2009) correctly point out that driving synchronized MT oscillations by hydrolysis of GTP to GDP and conformational changes would be prohibitive in terms of energy requirements and heat produced. This is agreed. However, we clarify that tubulin switching in Orch OR need not actually involve significant conformational change, that electron cloud dipoles (London forces), or magnetic spin dipoles are sufficient for bit-like switching, superposition and qubit function (Figs. 5–7). We acknowledge tubulin conformational switching as discussed in early Orch OR publications and illustrations do indicate significant conformational changes. They are admittedly, though unintentionally, misleading. Discovery of gigahertz, megahertz and kilohertz BC in single microtubules supports dipole states providing a favorable signal with regard to the underlying ideas of Orch OR.

The only tubulin conformational factor required in Orch OR is superposition separation at the level of atomic nuclei, e.g., 2.5 Fermi length for carbon nuclei (2.5 femtometers; 2.5 x $10^{-15}$ meters). This shift may be accounted for by electronic cloud dipoles with Mossbauer nuclear recoil and charge effects (Sataric *et al.*, 1998; Brizhik *et al.*, 2001). Tubulin switching in Orch OR requires neither GTP hydrolysis nor significant conformational changes, depending on collective London force dipoles, or magnetic spin dipoles in quantum channels of aromatic rings (Fig. 5–7).

### 14.5.7 *Testable predictions of Orch OR — current status*

Orch OR involves numerous fairly specific and essentially falsifiable hypotheses. In 1998, 20 testable predictions of Orch OR in nine general categories were published (Hameroff, 1998a). They are reviewed here with our comments on their current status in *italics*.

**Neuronal microtubules are directly necessary for cognition and consciousness**

1. Synaptic plasticity correlates with cytoskeletal architecture/activities. *The current status of this is unclear, although microtubule networks do appear to define and regulate synapses.*
2. Actions of psychoactive drugs, including anti-depressants, involve neuronal microtubules. *This indeed appears to be the case. Fluoxitene (Prozac) acts through microtubules* (Bianchi *et al.*, 2009)*; anesthetics also act through MTs* (Emerson *et al.*, 2013).
3. Neuronal microtubule stabilizing/protecting drugs may prove useful in Alzheimer's disease. *There is now some evidence that this may be so; for example, MT-stabilizer epithilone is being tested in this way* (Brunden *et al.*, 2010).

**Microtubules communicate by cooperative dynamics**

4. Coherent gigahertz excitations will be found in microtubules. *Indeed in some remarkable new research, Anirban Bandyopadhyay's group found coherent gigahertz, megahertz and kilohertz excitations in single MTs.* (Sahu *et al.*, 2013a, 2013b; 2014).
5. Dynamic microtubule vibrations correlate with cellular activity. *Evidence on this issue is not yet clear, although mechanical megahertz*

*vibrations (ultrasound) do appear to stimulate neurons and enhance mood* (Hameroff *et al.*,2013).

6. Stable microtubule patterns correlate with memory. *The evidence concerning memory encoding in MTs remains unclear, though synaptic messengers CaMKII and PkMz do act through MTs. Each CaMKII may encode (by phosphorylation) six information bits to six tubulins in a microtubule lattice* (Craddock *et al.*, 2012a, Figure 4).
7. "EPR-like" non-local correlation between separated microtubules. *This is not at all clear, but such things are very hard to establish (or refute) experimentally. Bandyopadhyay's group is testing for "wireless" resonance transfer between separated MTs* (Bandyopadhyay, personal communication).

## Quantum coherence occurs in microtubules

8. Phases of quantum coherence will be detected in microtubules. *There appears to be some striking evidence for effects of this general nature in Bandyopadhyay's recent results* (Sahu *et al.*, 2013a, 2013b, 2014), *differing hugely from classical expectations, where electrical resistance drops dramatically, at certain very specific frequencies, in a largely temperature-independent and length-independent way.*
9. Cortical dendrites contain largely "A-lattice," compared to "B-lattice," microtubules. *Although there is some contrary evidence to this assertion, the actual situation remains unclear. Orch OR has been criticized because mouse brain microtubules are predominantly B lattice MTs. However, these same mouse brain MTs are partially A-lattice configuration, and other research shows mixed A and B lattice MTs* (Amos & Klug, 1974; Mandelkow *et al.*, 1992; Kikkawa *et al.*, 1994; McEwen & Edelstein 1980). *Bandyopadhyay has preliminary evidence that MTs can shift between A- and B- lattice configurations* (Bandyopadhyay, personal communication), *and A-lattices may be specific for quantum processes. Orch OR could also utilize B lattices, although apparently not as efficiently as A-lattice. In any case, A- lattice MTs could well be fairly rare, specific for quantum effects, and sufficient for Orch OR since the A-lattice may be needed only in a fraction of MTs in dendrites and soma, and perhaps only transiently.*
10. Coherent photons will be detected from microtubules. *A positive piece of evidence in this direction is the detection of gigahertz excitations in MTs by Bandyopadhyay's group, which may be interpreted as coherent photons.* (Sahu *et al.*, 2013a, 2013b, 2014)

**Microtubule quantum coherence is protected by actin gelation**

11. Dendritic–somatic microtubules are intermittently surrounded by tight actin gel. *This is perhaps a moot point, now, in view of recent results by Bandyopadhyay's group, as it now appears that coherence occurs at warm temperature without actin gel.*
12. Cycles of actin gelation and solution correlate with electrophysiology, e.g. gamma synchrony EEG *Again, this now appears to be a moot point, for the same reason as above.*
13. Sol–gel cycles are mediated by calcium ion flux from synaptic inputs. *No clear evidence, but again a moot point.*

**Macroscopic quantum coherence occurs among hundreds of thousands of neurons and glia inter-connected by gap junctions**

*Gap junctions between glia and neurons have not been found, but gap junction interneurons would now appear to fill the need, interweaving the entire cortex.*

14. Electrotonic gap junctions synchronize neurons *Gap junction interneurons do appear to mediate gamma synchrony EEG* (Dermietzel,1998; Draguhn *et al.*, 1998; Galaretta & Hestrin, 2001; Bennett & Zukin, 2004; Fukuda & Kosaka 2000; Traub *et al.*, 2002).
15. Quantum tunneling occurs across gap junctions. *As yet untested.*
16. Quantum correlations between microtubules in different neurons occurs via gap junctions. *As yet untested. However Bandyopadhyay has preliminary evidence that spatially separated MTs, perhaps even in different neurons, become entangled in terms of their BC resonances (Bandyopadhyay, personal communication).*

**The amount of neural tissue involved in a conscious event is inversely related to the event time by $\tau \approx \hbar/E_G$**

17. Functional imaging and electrophysiology will show perception and response time shorter with more neural mass involved. *As a "prediction" of Orch OR, the status of this is not very clear; moreover it is very hard to provide any clear exstimate of the neural mass that is involved in a "perception." As a related issue, there does appear to be evidence for some kind of scale-invariance in neurophysiological processes* (He *et al.*, (2010); Kitzbichler *et al.* (2009)).

**An unperturbed isolated quantum state self-collapses (OR) according to $\tau \approx \hbar / E_G$**

18. Technological quantum superpositions will be shown to undergo OR by $\tau \approx \hbar / E_G$ *Various experiments are being developed which should supply an answer to this fundamental question (*Bouwmeester *et al.*, 1997)*, but they appear to remain several years away from being able to achieve firm conclusions.*

**Microtubule-based cilia/centrioles are quantum optical devices**

19. Microtubule-based cilia in retinal rod and cone cells detect photon quantum information. *This appears to be untested, so far.*

**A critical degree of microtubule activity enabled consciousness during evolution**

20. Fossils will show organisms from the early Cambrian period (540 million years ago), had sufficient microtubule capacity for OR by $\tau \approx \hbar / E_G$ of less than a minute, perhaps resulting in rudimentary Orch OR, consciousness and the "Cambrian evolutionary explosion." *It is clearly hard to know an answer to this one, particularly because the level of consciousness in extinct creatures would be almost impossible to determine. However, present day organisms looking remarkably like early Cambrian creatures (actinosphaerum, nematodes) are known to have over $10^9$ tubulins (Dustin, 1985).*

It would appear that the expectations of Orch OR have fared rather well so far, and it gives us a viable scientific proposal aimed at providing an understanding of the phenomenon of consciousness. We believe that the underlying scheme of Orch OR has a good chance of being basically correct in its fundamental conceptions.

## 14.6. Discussion — Consciousness in the Universe

Section 1 described three possibilities regarding the origin and place of consciousness in the universe: (A) as an emergent property of complex brain neuronal computation, (B) as a spiritual quality of the universe, distinct from purely physical actions and (C) as composed of discrete "proto-conscious" quantum events acting in accordance with physical laws not yet fully understood. The Orch OR theory follows (C), and includes aspects of (A) and (B). Orch OR suggests consciousness consists of discrete moments, each an "orchestrated" quantum-computational process terminated by the DP version of OR, an action rooted in quantum aspects of the fine structure of space–time geometry, this being coupled to brain neuronal processes via microtubules.

In standard quantum mechanics, the R procedure is adopted for the action of a *measurement* upon a quantum system, whereby a quantum superposition of two states, these two being distinguishable by that measurement, is probabilistically replaced by one or the other of those states ("reduction of the quantum state" or "collapse of the wavefunction"). But this action is normally taken to be *illusory* in some sense, not being a *real* physical action, but somehow the result of some kind of approximation, or perhaps just as a convenience, or as a shift in the observer's viewpoint, or even as a "split" in the observer's awareness. The hypothesis of OR (objective reduction), on the other hand, asserts that R is a *real objective* physical phenomenon, independent of any observer. Moreover it would be OR that provides the "bridge" between the quantum and classical worlds. This, however, necessitates some kind of modification of the standard U-evolution (i.e. of the Schrödinger equation) for massive-enough systems. The DP version of OR is such a particular scheme, according to which a massive physical body, placed in a quantum superposition of two different stationary locations, would spontaneously find itself located in one or other of these locations in a timescale of order of $\tau \approx \hbar/E_G$, where $E_G$ is the gravitational self-energy of the difference between the expectation values of the two mass distributions in the constituent stationary states. Accordingly, we might say that a quantum-theoretic separation of a material object "from itself" (like Schrödinger's hypothetical dead/alive cat), would be *unstable* and would decay to one or the other of the component states in a timescale that approximates the value $\tau$. The quantity $\tau$ can also be understood as the tiny difference, in fundamental Planck-scale units, between the space–time geometries of the two alternative states. Such superposition/separations tend not to be isolated from their environment, however, and would then entangle with other material in the environment, so that it would be the *entire* entangled system that would evolve until reaching this objective threshold for reduction (OR) at time around $\tau \approx \hbar/E_G$, where $E_G$ is now the gravitational self-energy of the difference between the two superposed mass distributions *including* the relevant entangled environments. At the moment of OR, at an average time of around $\tau$ after the formation of the superposition, the alternative space–time possibilities reduce to just one or the other of the space–time configurations.

So far, this is just the original DP proposal. However, Orch OR goes further than this, and puts forward the suggestion that each action of OR (taken to be in accordance with DP) is accompanied by a moment of *proto-consciousness*. These events would be thought of as the elemental constituents of "subjective experience," but the vast majority of such OR events act without being part of some coherent organized structure, so that the relevant material is normally totally dominated by random behavior in the entangled environment. Accordingly, there would normally be no significant, meaningful experience associated with these ubiquitous proto-conscious events. Yet, these moments of proto-consciousness are taken to be the primitive ingredients of actual full-blown consciousness, when they are appropriately orchestrated together into a coherent whole, somewhat like how sounds, tones and noise (e.g. of an orchestra warming up) become music.

In the version of the DP proposal put forward in (Penrose, 1992, 1996, 2000, 2009) it was, technically speaking, an (not always explicit) assumption that the energies of the two stationary quantum states involved in the superposition were taken to be equal to one another. Here (Sec. 14.4.6), we generalize DP in a novel way, which allows us to consider superposed stationary states of *unequal* energy. We argue that for energies that differ only slightly from one another, the action of OR takes us not just to one or the other of these two constituent states in an average time of about $\tau = \hbar / E_G$, but the result of the OR process is to reduce the superposition to an *oscillation* between the two, whose frequency is given by the *beat* value, given by the *difference* between the two far larger quantum-mechanical frequencies associated with the energies of the two previously superposed states. We suggest that it is these beat frequencies, resulting from the Orch OR processes that involve the reduction of coherently superposed tubulin states with slightly different energies, that result in the characteristic frequencies, such as 40 Hz gamma synchrony which appears to be correlated with conscious states.

In an uncontrolled situation occurring in the physical world, with systems in quantum superposition, OR would normally occur spontaneously when significant environment is entangled with the system, and

$E_G$ can rapidly become relatively large, so $\tau$ is reached quickly, and the choice of particular space–time configuration includes a dominant component of randomness owing to the random nature of the environment. The moment of "subjective experience" that would be associated with this type of OR is an undifferentiated, non-cognitive, insignificantly experiential ("proto-conscious") quality. Due to the random component, such environment-induced OR "experience" would lack information, cognition or meaning, be very brief (low $\tau$ due to high environmental $E_G$) and ubiquitous, playing merely the role of "decoherence" that is familiar in standard interpretations of quantum mechanics.

However, according to Orch OR, biological evolution provided isolated non-polar, quantum-friendly regions, e.g. in structures such as microtubules, within which OR events could be "orchestrated," enabling functional quantum computing in isolated non-polar "aromatic," pi resonance channels (the "quantum underground") within microtubule proteins. With further evolution, orchestrated quantum superpositions in microtubules would have been able to persist for progressively longer times with larger values of $E_G$, with entanglements with other parts of the structure playing meaningful roles, thereby allowing significant "quantum computing" to occur. Yet, with only partial isolation, the OR threshold $\tau \approx \hbar/E_G$ would still only be reached by including unorchestrated environmental entanglement, which introduces randomness in the selection of states. Accordingly, such OR quantum computing would lack fully "orchestrated" cognition, so the claimed non-computable aspects of DP OR would not come into play at this stage. Yet, the advantages of some form of "quantum computation" in these processes could still be of significant relevance, even though the OR action would be only at this "proto-conscious" level.

With even more advanced evolutionary development, biological factors could orchestrate and further isolate microtubule quantum computing so that the OR threshold $\tau \approx \hbar/E_G$ could now be reached by orchestrated microtubule quantum superpositions by themselves, and a relatively large $E_G$ could be achieved without environmental randomness. Such Orch OR moments could provide rich cognitive subjective experience, and control conscious behavior, with a non-computable "willed" influence. Moreover, since the DP version of OR is a *gravitational* proposal, this relates experiential phenomena to the fundamentals of space–time geometry. Evolution may well have favored orchestrated superpositions with larger $E_G$ values, allowing briefer times $\tau$ which are increasingly useful to the organism's cognition. In accordance with our earlier ideas, we might speculate that these eventually reached sufficient $E_G$ for $\tau$ near 25 ms for gamma synchrony with 40Hz or more Orch OR conscious events per second. Alternatively, according to the Orch OR "beat frequency" approach introduced here, natural MT megahertz resonances (perhaps with much larger $E_G$ values) enable much slower beat frequencies in the gamma synchrony range.

Philosophically, Orch OR perhaps aligns most closely with Whitehead (1929, 1933) who viewed mental activity as a process of "occasions," spatio-temporal quanta, each endowed — usually on a very low level, with mentalistic characteristics which were "dull, monotonous, and repetitious." These seem analogous, in the Orch OR context, to "proto-conscious" non-orchestrated OR events. Whitehead viewed high level mentality, consciousness, as being extrapolated from temporal chains of such occasions. In his view, highly organized societies of occasions permit primitive mentality to become intense, coherent and fully conscious. These seem analogous to Orch OR conscious events. Shimony (1993), Stapp (2007) and others recognized that Whitehead's approach was potentially compatible with modern physics, specifically quantum theory, with quantum state reductions — actual events — appearing to represent "occasions," namely Whitehead's high level mentality, composed of "temporal chains...of intense, coherent and fully conscious occasions," these being tantamount to sequences of Orch OR events. These might possibly coincide with gamma synchrony, but with our current "beat frequency" ideas gamma synchrony is more likely to be a beat effect than directly related to the OR reduction time $\tau$. As Orch OR events are indeed quantum state reductions, Orch OR and Whitehead's process philosophy appear to be quite closely compatible.

Whitehead's low-level "dull" occasions of experience would seem to correspond to our non-orchestrated "proto-conscious" OR events. According to the DP scheme, OR processes would be taking place all the time everywhere and, normally involving the random environment, would be providing the effective randomness that is a characteristic of quantum measurement. Quantum superpositions will continually be reaching the DP threshold for OR in non-biological settings as well as in biological ones, and OR would usually take place in the purely random environment such as in a quantum system under measurement. Nonetheless, in the Orch OR scheme, these events are taken to have a rudimentary subjective experience, which is undifferentiated and lacking in cognition, perhaps providing the constitutive ingredients of what philosophers call *qualia*. We term such un-orchestrated, ubiquitous OR events, lacking information and cognition, "proto-conscious." In this regard, Orch OR has some points in common with the viewpoint (B) of Sec. 14.1, which incorporates spiritualist, idealist and panpsychist elements, these being argued to be essential precursors of consciousness that are intrinsic to the universe. It should be stressed, however, that Orch OR is strongly supportive of the *scientific* attitude that is expressed by (A), and it incorporates that viewpoint's picture of neural electrochemical activity, accepting that non-quantum neural network membrane-level functions might provide an adequate explanation of much of the brain's *unconscious* activity. Orch OR in microtubules inside neuronal dendrites and soma adds a deeper level for conscious processes.

Conditions for Orch OR and consciousness are fairly stringent in our scheme: orchestrated superposition must be isolated from the decoherence/OR effects of the random environment for long enough to reach the DP threshold while continuing to perform quantum computation. Small superpositions are easier to isolate for a limited time, but require longer reduction times $\tau$, so that the isolation would need to be correspondingly more perfect.

Large superpositions will reach threshold quickly, but are intrinsically more difficult to isolate. If we consider that the *beat frequency* picture is the appropriate one with regard to the evocation of consciousness, then we may speculate that beat frequencies of faster, e.g. megahertz processes might possibly require only very brief reduction times. These might be even as brief as $10^{-7}$ secs if we take the view that it is actually the case that our extended DP OR proposal allows reduction times to be much briefer than the beat period, while still giving rise to classical beats, as speculated in Sec. 4.6. Accordingly, one suggestion that we can make is that "*Bandyopadhyay coherence*" ("BC") — the megahertz resonance, found by Bandyopadhyay's group, suggesting coherence times of $10^{-7}$ secs, or the tens of kilohertz resonance they found suggesting $10^{-4}$ sec — provide good evidence that such superpositions within sufficiently large collections of microtubules could persist in the brain for reduction times $\tau$ and Orch OR processes that could be relevant to brain function and consciousness.

What about Orch OR in non-biological systems? After all, $\tau \approx \hbar/E_G$ happens everywhere. What kind of role might there be for it in consciousness elsewhere in the universe?

Very large masses can be involved in quantum superpositions, occurring in the universe in quantum-mechanical situations, for example in the cores of neutron stars. One might imagine that $\tau$ would then be rediculously tiny. But $E_G$ could still be relatively small if the mass-displacement remains small owing to the uniformity of the material. But generally, by OR, such large-scale superpositions would reduce extremely quickly, and classically unreasonable superpositions would be rapidly eliminated. Whether such quantum systems could be orchestrated to have meaningful, cognitive Orch OR conscious moments is unknown, but it is certainly conceivable that sentient creatures might have evolved in parts of the universe that would be highly alien to us, for example on neutron-star surfaces, with very large scale superpositions, and presumably very high frequency OR events, an idea that was developed ingeniously and in great detail by Robert Forward in two science-fiction stories (*Dragon's Egg* in 1980, *Starquake* in 1989 (Forward 1980, 1989)). Such creatures (referred to as "cheelas" in the books), with metabolic processes and presumably Orch OR-like events occurring at rates of around a million times that of a human being, could arguably have intense experiences, but whether or not this would be possible in detail is, as of now, a very speculative matter. Nevertheless, the Orch OR proposal offers a possible route to rational argument, as to whether conscious life of a totally alien kind such as this, or some other form of quantum superposition, might be possible, or even probable, somewhere in the universe.

Such speculations also raise the issue of the "anthropic principle," according to which it is sometimes argued that the particular dimensionless constants of Nature that we happen to find in our universe are apparently "fortuitously" favorable to human existence and consciousness. (A *dimensionless* physical constant is a pure number, like the ratio of the electric to the gravitational force between the electron and the proton in a hydrogen atom, which in this case is a number of the general order of $10^{40}$.) The key point is not so much to do with human existence, but the existence of sentient beings of any kind, i.e., the existence of consciousness. Is there anything coincidental about the dimensionless physical constants being of such a nature that conscious life is possible at all? For example, if the mass of the neutron had been slightly less than that of the proton, rather than slightly larger, then neutrons rather than protons would have been stable, and this would be to the detriment of the whole subject of chemistry. These issues are frequently argued about (see Barrow & Tipler (1986)), but the Orch OR proposal provides a little more potential substance to these arguments, since a proposal for the possibility of sentient life is, in principle, provided. A question becomes, why is the universe favorable to consciousness?

The recently proposed cosmological scheme of *conformal cyclic cosmology* (CCC) (Penrose, 2010; Gurzadyan & Penrose, 2010) also has some relevance to these issues. CCC posits that what we presently regard as the entire history of our universe, from its Big-Bang origin (but without inflation) to its indefinitely expanding future, is but one *aeon* in an unending succession of similar such aeons, where the "infinite" future of each matches to the big bang of the next via an infinite change of scale. A question arises whether the dimensionless constants of the aeon prior to ours, in the CCC scheme, are the same as those in our own aeon, and this relates to the question of whether sentient life could exist in that aeon as well as in our own. Could the dimensionless constants change with each successive aeon, might they perhaps "mutate" and evolve to optimize consciousness? Could evolution over aeons thereby account for the anthropic principle? Smolin (1997) has suggested an idea that is somewhat similar to this, but in his scheme, the drive of selective advantage would be for more black holes and baby universes, rather than for consciousness or even for life. Nevertheless, the question of the constancy of these numbers is in principle answerable by observation in CCC, and this issue could have a bearing on the extent or validity of the Orch OR proposal. If Orch OR turns out to be correct, in its essentials, as a physical basis for consciousness, then it opens up the possibility that many questions may become answerable, such as whether life and consciousness could have come about in an aeon prior to our own, that would have previously seemed to be far beyond the reaches of science.

Moreover, Orch OR places the phenomenon of consciousness at a very central place in the physical nature of our universe, whether or not this "universe" includes aeons other than just our own. It is our belief that, quite apart from detailed aspects of the physical mechanisms that are involved in the production of consciousness in human brains, quantum mechanics is an incomplete theory. Some completion is needed, and the DP proposal for an OR scheme underlying quantum theory's R-process would be a definite possibility. If such a scheme as this is indeed respected by Nature, then there is a fundamental additional ingredient to our presently understood laws of Nature which plays an important role at the Planck-scale level of space–time structure. The Orch OR proposal takes advantage of this, suggesting that conscious experience itself plays such a role in the operation of the laws of the universe.

## 14.7. Conclusion

"Orchestrated objective reduction" ("Orch OR") is a theory which proposes that consciousness consists of a sequence of discrete events, each being a moment of OR of a quantum state (according to the DP scheme), where it is taken that these quantum states exist as parts of quantum computations carried on primarily in neuronal microtubules. Such OR events would have to be "orchestrated" in an appropriate way (Orch OR), for genuine consciousness to arise. OR itself is taken to be ubiquitous in physical actions, representing the "bridge" between the quantum and classical worlds, where quantum superpositions between pairs of states get spontaneously resolved into classical alternatives in a timescale $\sim\tau$, calculated from the amount of mass displacement that is between the two states. In our own brains, the OR process that evokes consciousness would be actions that connect brain biology (quantum computations in microtubules) with the fine scale structure of space–time geometry, the most basic level of the universe, where tiny quantum space–time displacements are taken to be responsible for OR. The Orch-OR proposal therefore stretches across a considerable range of areas of science, touching upon the foundations of general relativity and quantum mechanics, in unconventional ways, in addition to the more obviously relevant areas such as neuroscience, cognitive science, molecular biology and philosophy. It is not surprising, therefore, that Orch OR has been persistently criticized from many angles since its introduction in 1994. Nonetheless, the Orch OR scheme has so far stood the test of time better than most other schemes, and it is particularly distinguished from other proposals by the many scientifically tested, and potentially testable, ingredients that it depends upon.

It should be mentioned that various aspects of the Orch OR theory have themselves evolved in response to scientific advances and, in some cases, constructive criticism. We here list some recent adaptations and developments that we have now incorporated into the theory.

## Biological Adaptations and Developments in Orch OR

- For both classical and quantum computing, information states of tubulins within microtubule lattices correlate with dipole orientations, without significant protein conformational changes.
- Tubulin dipoles originate in pi electron resonance clouds in non-polar regions, and may be either electric (London force charge separation), or magnetic (electron spin states and currents). Dipoles oscillate, couple to mechanical movement and nuclear spin at the level of atomic nuclei, and become quantum superpositions of multiple possible dipole orientations. Hence individual tubulins may act as quantum bits, or qubits, as suggested initially in Orch OR (Hameroff & Penrose, 1996a,b) though we now favor pathway qubits.
- Intra-tubulin non-polar pi resonance regions conducive to quantum states appear to align with those in neighboring tubulins, perhaps enabling collective quantum dipoles among many tubulins along helical pathways through microtubule lattices, e.g. along Fibonacci 3, 5 and 8-start helical winding patterns. Collective ("giant dipole," Fröhlich, 1970) pathway dipoles may then form superpositions of possible orientations, now proposed to function as pathway qubits in Orch OR.
- Orch OR qubit pathways may correlate with enhanced electron conductances discovered by Anirban Bandyopadhyay's group (Sahu *et al*, 2013a, 2013b, 2014) in microtubules in warm temperature at specific gigahertz, megahertz and kilohertz frequencies ("Bandyopadhyay coherence," "BC"), strongly supporting Orch OR.
- Non-polar pi resonance channels mediating BC occur not only within microtubules, but lie generally within proteins, lipid membranes, DNA and RNA, apparently pervading all living systems (the "quantum underground," chapter 13).
- Anesthetic gases bind and act in such non-polar pi resonance regions in tubulin, presumably dispersing microtubule quantum dipoles necessary for consciousness, e.g. in mixed polarity microtubules in dendrites and soma of cortical layer 5 pyramidal neurons.
- Scale-invariant (1/f, fractal-like) processes observed at neuronal and network levels might perhaps extend downward to intra-neuronal BC in microtubules, e.g. megahertz excitations. If so, microtubules in mixed-polarity networks vibrating at slightly different megahertz frequencies would interfere, e.g. in layer 5 pyramidal neurons whose apical dendrites rise vertically toward the brain surface, and are primarily responsible for EEG. Interference of microtubule megahertz

vibrations could give rise to much slower "beat frequencies," e.g. between 0 and 100 Hz seen as the electro-encephalogram (EEG).

- As Alzheimer's disease, brain trauma and other disorders are related to microtubule disturbances within brain neurons, therapies including transcranial ultrasound ("TUS," megahertz mechanical vibrations) are in clinical trials to treat such disorders.
- Orch OR with conscious understanding and non-computable Platonic influences, as suggested to occur in human and animal brains, would have been preceded during the course of evolution by simple "proto-conscious" OR events in non-polar pi resonance regions of primitive systems. Behaviors and mutations promoting OR-mediated "proto-pleasure," and avoidance of "proto-pain," may have played significant roles in the evolution of life, and the brain.

The Orch OR proposal suggests conscious experience is intrinsically connected to the fine-scale structure of space–time geometry, and that consciousness could be deeply related to the operation of the laws of the universe.

## Acknowledgments

We thank Dave Cantrell, and Paul Fini, Biomedical Communications, Banner-University Medical Center, The University of Arizona for artwork, Marjan Macphee and Abi Behar-Montefiore, Department of Anesthesiology, University of Arizona for manuscript support, and Travis Craddock PhD, Nova Southeastern University for Figs. 4 and 5.

This article was published in *Physics of Life Reviews*, Vol. 11, Stuart Hameroff and Roger Penrose, Consciousness in the Universe. A review of the "Orch OR" theory, pp. 39–78, Copyright Elsevier (2014). It has been modified and updated slightly from that version.

## References

Adamatzky, A. (2012). Slime mould computes planar shapes. *Int. J. Biol- Inspired Comput.*, **4**, 149–154.

Amos, L. A. & Klug A. (1974) Arrangement of subunits in flagellar microtubules. *J. Cell Sci.*, **14**, 523–549

Atema, J. (1973) Microtubule theory of sensory transduction. *J. Theor. Biol.*, **38**, 181–190.

Bandyopadhyay, A. personal communication.

Barrow, J. D. & Tipler, F. J. (1986) The *Anthropic Cosmological Principle*, Oxford University Press, Oxford.

Bell, J. S. (1966) *Speakable and Unspeakable in Quantum Mechanics*. Cambridge, Massachusetts: Cambridge University Press.

Bem, D. J. (2012) Feeling the future: Experimental evidence for anomalous retroactive influences on cognition and affect. *J. Pers. Soc. Psychol.*, **100**, 407–425.

Benioff, P. (1982) Quantum mechanical Hamiltonian models of Turing machines. *J. Stati. Phys.*, **29**, 515–546.

Benioff, P. (1998) Quantum robots and environments. *Phys. Rev. A*, **58**, 893–904.

Bennett, C. H. & Wiesner, S. J. (1992) Communication via 1- and 2-particle operators on Einstein-Podolsky-Rosen states. *Phys. Rev. Lett.*, **69**, 2881–2884.

Bennett, M. V. & Zukin, R. S. (2004) Electrical coupling and neuronal synchronization in the Mammalian brain. *Neuron*, **41**, 495–511.

Bensimon, G. & Chernat, R. (1991) Microtubule disruption and cognitive defects: effect of colchicine on learning behavior in rats. *Pharmacol. Biochem. Behav.*, **38**, 141–145.

Berkeley, G. (1975) *Philosophical Works*, M. Ayers, ed., London: Dent.

Bernroider, G. & Roy, S. (2005) Quantum entanglement of K ions, multiple channel states and the role of noise in the brain. *Proceedings SPIE*, **5841–5829**, 205–214.

Bianchi, M., Shah, A. J., Fone, K. C., Atkins, A. R., Dawson, L. A., Heidbreder, C. A., Hows, M. E., Hagan, J. J. & Marsden C. A. (2009) Fluoxetine administration modulates the cytoskeletal microtubular system in the rat hippocampus. *Synapse*, **63**, 359–364.

Bierman, D. J. & Radin, D. I. (1997) Anomalous anticipatory response on randomized future conditions. *Percept. Mot. Skills*, **84**, 689–690.

Bohm, D. (1983) Quantum Theory (Prentice–Hall, Englewood-Cliffs.) Chapters 22, section 15–19. Reprinted as: The paradox of Einstein, Rosen and Podolsky. In: J. A. Wheeler and W. H. Zurek, eds. *Quantum Theory and Measurement*. Princeton: Princeton University Press.

Bouwmeester, D., Pan, J. W., Mattle, K., Eibl, M., Weinfurter, H. & Zeilinger, A. (1997) Experimental quantum teleportation. *Nature*, **390**, 575–579.

Brizhik, L., Eremko, A., Piette, B. & Zakrzewski, W. (2001). Electron self-trapping in a discrete two-dimensional lattice. *Physica D*, **159**, 71–90.

Brunden, K. R., Yao, Y., Potuzak, J. S., Ferrer, N. I., Ballatore, C., James, M. J., Hogan, A. M., Trojanowski, J. Q., Smith, A. B. & Lee, V. M. (2011) The characterization of microtubule-stabilizing drugs as possible therapeutic agents for Alzheimer's disease and related taupathies. *Pharm. Res.*, **63**, 341–351.

Brunden, K. R., Zhang, B., Carroll, J., Yao, Y., Poduzak, J. S., Hogan, A. M., Iba, M., James, M. J., Xie, S., Ballatore, C., Smith, A. B., III, Lee, V. M.-Y. & Trojanowski, J. Q. (2000) Epothilone D improves microtubule density, axonal integrity and cognition in a transgenic mouse model of tauopathy. *J. Neurosci.*, **30**, 13861–13866.

Chalmers, D. J. (1996) *The Conscious Mind — In Search of a Fundamental Theory.* New York: Oxford University Press.

Chalmers, D. J. (2012) *Constructing the world*. New York: Oxford University Press.

Chin, A. W., Prior, J., Rosenbach, R., Caycedo-Soler, F., Huelga, S. F. & Plenio, M. B. (2013) The role of non-equilibrium vibrational structures in electronic coherence and recoherence in pigment–protein complexes. *Nat. Phys.* **9**, 113–118.

Chopra, D. (2001) *How to Know God: The Soul's Journey into the Mystery of Mysteries.* New York: Running Press Book Publishers.

Craddock, T. J. A. Tuszynski, J. A. Chopra, D. Casey N. Goldstein, L. E. Hameroff, S. R. & Tanzi, R. E. (2012a) The zinc dyshomeostasis hypothesis of Alzheimer's disease *PLoS One*, **7**, e33552.

Craddock, T., Tuszynski, J. & Hameroff, S. (2012b). Cytoskeletal signaling: Is memory encoded in mictrotubule lattices by CaMKII phosphorylation? *PLoS Comput. Biol.*, **8**, e1002421.

Craddock, T. J. A., St. George, M., Freedman, H., Barakat, K. H., Damaraju, S. Hameroff, S. & Tuszynski, J. A. (2012c) Computational predictions of volatile anesthetic interactions with the microtubule cytoskeleton: Implications for side effects of general anesthesia. *PLoS One*, **7**, e37251.

Craddock, T. J. A., Friesen, D., Mane, J., Hameroff, S. & Tuszynski, J. A. (2014) The feasibility of coherent energy transfer in microtubules. *J. R. Soc. Interface*, 05/11(100).

Craddock, T. J. A., Hameroff, S. R. & Tuszynski, J. A. (2015) Anaesthetics act in quantum channels in brain microtubules to prevent consciousness. *Curr. Top. Med. Chem.*, **15**, 523–533.

Crick, C. K. (1990) Towards a neurobiological theory of consciousness. *Semin. Neurosci.*, **2**, 263–275.

Dennett, D. C. (1991) *Consciousness Explained.* Boston, Massachusetts: Little Brown.

Dennett, D. C. (1995) *Darwin's Dangerous Idea: Evolution and the Meanings of Life.* New York: Simon and Schuster.

Dennett, D. C. & Kinsbourne, M. (1991) Time and the observer: The where and when of consciousness. *Behav. Brain Sci.*, **15**, 183–247.

Dermietzel, R. (1998) Gap junction wiring: A 'new' principle in cell-to-cell communication in the nervous system? *Brain Res. Rev.*, **26**, 176–183.

Deutsch, D. (1985) Quantum theory, the Church–Turing principle and the universal quantum computer, *Proc. R. Soc. Lond.* A, **400**, 97–117.

Diósi, L. (1989) Models for universal reduction of macroscopic quantum fluctuations. *Phys. Rev. A*, **40**, 1165–1174.

Diósi, L. (1987) A universal master equation for the gravitational violation of quantum mechanics. *Phys. Lett. A*, **120**, 377–381.

Draguhn, A., Traub, R. D. & Schmitz, D. (1998) Electrical coupling underlies high-frequency oscillations in the hippocampus *in vitro*. *Nature*, **394**, 189–192.

Duggin, I. G., Aylett, C. H. S., Walsh, J. C., Michie, K. A., Wang, Q., Turnbull, L., Dawson, E. M., Harry, E. J., Whitchurch, C. B., Amos, L. A. & Lowe, J. (2014) CetZ tubulin-like proteins control archaeal cell shape. *Nature*, **519**, 362–365.

Dustin, P. (1985) *Microtubules 2nd edn*. New York: Springer-Verlag.

Eccles, J. C. (1992) Evolution of consciousness. *Proc. Natl. Acad. Sci. USA*, **89**, 7320–7324.

Emerson, D. J., Weiser, B. P., Psonis, J., Liao, Z., Taratula, O., Fiamengo, A., Wang, X., Sugasawa, K., Smith, A. B., Eckenhoff, R. G. & Dmochowski, I. J. (2013) Direct modulation of microtubule stability contributes to anthracene general anesthesia. *J. Am. Chem. Soc.*, **135**, 5389–5398.

Engel, G. S., Calhoun, T. R., Read, E. L., Ahn, T. K., Mancal, T., Cheng, Y. C., Blankenship, R. E. & Fleming, G. R. (2007) Evidence for wavelike energy transfer through quantum coherence in photosynthetic systems. *Nature*, **446**, 782–786.

Everett, H. (1957) Relative state formulation of quantum mechanics. In: J. A. Wheeler, and W. H. Zurek, Eds. *Quantum Theory and Measurement*. Princeton University Press, 1983; *originally in Rev. Mod. Phys.*, **29**, 454–462.

Feynman, R. P. (1986) Quantum mechanical computers. *Found. Phys.*, **16**, 507–531.

Forward, R. (1980) *Dragon's Egg*. Ballentine Books.

Forward, R. (1989) *Starquake*. Ballentine Books.

Fries, P., Schroder, J. H., Roelfsma, P. R., Singer, W. P., Engel, A. K. & Engel, A. K. (2002) Oscillatory neuronal synchronization in primary visual cortex as a correlate of stimulus selection. *J. Neurosci.*, **22**, 3739–3754.

Fröhlich, H. (1968) Long-range coherence and energy storage in biological systems. *Int. J. Quantum Chem.*, **2**, 641–649.

Fröhlich, H. (1970) Long range coherence and the actions of enzymes. *Nature*, **228**, 1093.

Fröhlich, H. (1975) The extraordinary dielectric properties of biological materials and the action of enzymes. *Proc. Natl. Acad. Sci. USA*, **72**, 4211–4215.

Fukuda, T. & Kosaka, T. (2000) Gap junctions linking the dendritic network of GABAergic interneurons in the hippocampus. *J. Neurosci.*, **20**, 1519–1528.

Galaretta, M. & Hestrin, S. (2001) Electrical synapses between GABA-releasing interneurons. *Nat. Rev. Neurosci.*, **2**, 425–433.

Galaretta, M., Galarreta, S. & Hestrin, A. (1999) Network of fast-spiking cells in the neocortex connected by electrical synapses. *Nature*, **402**, 72–75.

Gauger, E., Rieper, E., Morton, J. J. L., Benjamin, S. C. & Vedral, V. (2011). Sustained quantum coherence and entanglement in the avian compass. *Phys. Rev. Letts.*, **106**, 040503.

Georgiev, D. D. (2009) Remarks in the number of tubulin dimers per neuron and implications for Hameroff-Penrose Orch OR. *NeuroQuantology*, **7** 677–679.

Ghirardi, G. C., Grassi, R., Rimini, A. (1990) Continuous-spontaneous reduction model involving gravity. *Phys. Rev.*, **A 42**, 1057–1064.

Ghirardi, G. C., Rimini, A. & Weber, T. (1986) Unified dynamics for microscopic and macroscopic systems. *Phys. Rev., D*, **34**, 470.

Gray, C. M. & Singer, W. (1989) Stimulus-specific neuronal oscillations in orientation columns of cat visual cortex, *Proc. Nat. Acad. Sci. USA.*, **86**, 1698–1702.

Grush, R. & Churchland, P. S. (1995) 'Gaps in Penrose's toilings', *J. Consciousness Studies*, **2**, 10–29.

Guillaud, G. L., Bosc, C., Fourest-Lieuvin, A., Denarier, E., Priollet, F., Lafanachere, L. & Job, D. (1998) STOP proteins are responsible for the high degree of microtubule stabilization observed in neuronal cells. *J. Cell Biol.*, **142**, 167–179.

Gurzadyan, V. G. R. & Penrose R. (2010) Concentric circles in WMAP data may provide evidence of violent pre-Big-Bang activity. arXiv:1011.3706.

Hagan, S., Hameroff, S. & Tuszynski, J. (2001) Quantum computation in brain microtubules? Decoherence and biological feasibility, *Phys. Rev. E.*, **65**, 061901.

Hameroff, S. (1998a) Quantum computation in brain microtubules? The Penrose–Hameroff "Orch OR" model of consciousness. *Philos. Trans. R. Soc. Lond* **A 356**, 1869–1896.

Hameroff, S. (1998b) 'Funda-mentality': Is the conscious mind subtly linked to a basic level of the universe? *Trends Cogn. Sci.*, **2**, 119–127.

Hameroff, S. (1998c) Anesthesia, consciousness and hydrophobic pockets — A unitary quantum hypothesis of anesthetic action. *Toxicol. Lett.*, **100–101**, 31–39.

Hameroff, S. (1998d) Did consciousness cause the Cambrian evolutionary explosion? In: Hameroff, S. R., Kaszniak, A. W. A. C. Scott (eds.), *Toward a Science of Consciousness II: The Second Tucson Discussions and Debates.* Cambridge, Massachusetts, MIT Press.

Hameroff, S. (2006a) The entwined mysteries of anesthesia and consciousness. *Anesthesiology*, **105**, 400–412.

Hameroff, S. (2006b) Consciousness, neurobiology and quantum mechanics: The case for a connection, In J. Tuszynski ed. *The Emerging Physics of Consciousness.* Heidelberg, Germany: Springer, pp.193–252.

Hameroff, S. R. (2007) The brain is both neurocomputer and quantum computer. *Cogn. Sci.*, **31**, 1035–1045.

Hameroff, S. (2010) The "conscious pilot" — dendritic synchrony moves through the brain to mediate consciousness. *J. Biol. Phys.*, **36**, 71–93.

Hameroff, S. (2012) How quantum brain biology can rescue conscious free will. *Front. Integ. Neurosci.*, **6**, 1–17.

Hameroff, S. (2016) The quantum origin of life — How the brain evolved to *feel good*. In: *On Human Nature*, M. Tibayrenc and F. Ayala, eds. Elsevier, in press.

Hameroff, S. R. & Penrose, R. (1996a) Orchestrated reduction of quantum coherence in brain microtubules: A model for consciousness. In: Hameroff, S. R. Kaszniak, A. C. Scott, Eds *Toward a Science of Consciousness; The First Tucson Discussions and Debates*. Cambridge, Massachusetts: MIT Press. pp. 507–540.

Hameroff, S. R. & Penrose, R. (1996b) Conscious events as orchestrated spacetime selections. *J. Consciousness Stud.*, **3**, 36–53.

Hameroff, S. R. Watt, R. C. (1982) Information processing in microtubules. *J. Theor. Biol.*, **98**, 549–61.

Hameroff, S., Nip, A., Porter, M. & Tuszynski, J. (2002) Conduction pathways in microtubules, biological quantum computation and microtubules. *Biosystems*, **64**, 149–168.

Hameroff, S., Trakas, M., Duffield, C., Annabi, E., Gerace, M. B., Boyle, P., Lucas, A. Amos, Q., Buadu, A. & Badal J. J. (2013) Ultrasound (TUS) effects on mental states: A pilot study, *Brain Stimulation*, **3**, 409–415.

Hameroff, S. R., Watt, R. C., Borel, J. D. & Carlson, G. (1982) General anesthetics directly inhibit electron mobility: dipole dispersion theory of anesthetic action. *Physiol. Chem. Phys.*, **14**, 183–187.

Hameroff, S. R. & Watt, R.C. (1983) Do anesthetics act by altering electron mobility? *Anesth. Analg.*, **62**, 936–940.

Hayea, D., Griffin, G. B. & Engel, G. S. (2013) Engineering coherence among excited states in synthetic heterodimer systems, *Science*, **340**, 1431–1434.

He, R.-H., Hashimoto, M., Karapetyan, H., Koralek, J. D., Hinton, J. P., Testaud, J. P., Nathan, V., Yoshida, Y., Yao, H., Tanaka, K., Meevasana, W., Moore, R. G., Lu, D. H. S.-K., Mo, Ishikado, M., Eisaki, H., Hussain, Z., Devereaux, T. P., Kivelson, S. A., Orenstein, K., Kapitulnik, J. A. & Shen, Z.-X. (2011) From a Single-Band metal to a high temperature superconductor via two thermal phase transitions. *Science*, **331**, 1579–1583.

He, B. J., Zemper, J. M., Snyder, A. Z. & Raichle, M. E. (2010) The temporal structures and functional significance of scale-free brain activity. *Neuron*, **66**, 353–369.

Hebb, D. O. (1949) *Organization of Behavior: A Neuropsychological Theory*, New York, Wiley.

Hildner, R., Brinks, D., Nieder, J. B., Cogdell, R. J. & van Hulst. N. F. (2013) Quantum coherent energy transfer over varying pathways in single light harvesting complexes. *Science*, **340**, 1448–1451.

Hodgkin, A. & Huxley, A. (1952) A quantitative description of membrane current and its application to conduction and excitation in nerve. *J. Physiol. — Lond.*, **117**, 500–544.

Huxley, T. H. (1986) Method and Results: *Essays*, Appleton & Co., NY.

James, W. (1899) *The Principles of Psychology*. Vol. 1, Holt & Co., NY.

Janke, C. & Kneussel M. (2010) Tubulin post-translational modifications: Encoding functions on the neuronal microtubule cytoskeleton. *Trends Neurosci.*, **33**, 362–372.

Kant, I. (1998) *Critique of Pure Reason*, Translated by Guyer, P. & Wood, A. W. eds. Cambridge University Press.

Károlyházy, F. (1966) Gravitation and quantum mechanics of macroscopic bodies. *Nuovo. Cim.*, **A42**, 390.

Károlyházy, F., Frenkel, A. & Lukács, B. (1986) On the possible role of gravity on the reduction of the wave function. In: R. Penrose and C. J. Isham, eds. *Quantum Concepts in Space and Time*. Oxford:, Oxford University Press, pp. 109–128.

Karplus, M. & McCammon, J. A. (1983) Protein ion channels, gates, receptors. In J. King (ed.), *Dynamics of Proteins: Elements and Function*. Benjamin/ Cummings, Menlo Park, pp. 263–300.

Kibble, T. W. B. (1981) Is a semi-classical theory of gravity viable? In: Isham, C. J. Penrose, R.& Sciama, D. W. eds. *Quantum Gravity 2: a Second Oxford Symposium Oxford*: Oxford University Press, pp. 63–80.

Kikkawa, M., Ishikawa, T., Nakata, T., Wakabayashi, T. & (1994) Hirokawa, N. Direct visualization of the microtubule lattice seam both *in vitro* and *in vivo J Cell Biol*, **127**, 1965–1971.

Kitzbichler, M. G., Smith, M. L., Christensen, S. R. & Bullmore, E. (2009) Broadband criticality of human brain network synchronization. *PLoS Comput. Biol.*, **5**, e1000314.

Koch, C. (2004) *The Quest for Consciousness: A Neurobiological Approach*. Englewood: CO., Roberts and Co.

Koch, C. & Hepp, K. Quantum mechanics in the brain *Nature*, **440**, 611.

Libet, B., Wright, E. W., Jr., Feinstein, B. & Pearl, D. K. (1979) Subjective referral of the timing for a conscious sensory experience, *Brain*, **102**, 193–224.

Luo, L. & Lu, J. (2011) Temperature dependence of protein folding deduced from quantum transition. arxiv: 1102.3748

Lutz, A., Greischar, A. L., Rawlings, N. B., Ricard, M. & Davidson, R. J. (2004) Long-term meditators self-induce high-amplitude gamma synchrony during mental practice. *Proce. Nat. Acad. Sci. USA*, **101**, 16369–16373.

Ma, X.-S., Zotter, S., Kofler, J., Ursin, R., Jennewein, T., Brukner, C. & Zeilinger, A. (2012) Experimental delayed- choice entanglement swapping. *Nat. Phys.*, **8**, 480–485.

Macikic, I., de Riedmatten, H., Tittel, W., Zbinden, H. & Gisin, N. (2002) Long-distance teleportation of qubits at telecommunication wavelengths *Nature*, **421**, 509–513.

Malach, R. (2007) The measurement problem in human consciousness research. *Behav. Brain Sci.*, **30**, 481–499.

Mandelkow, E., Song Y.-H. & Mandelkow E.-M., (1992) The microtubule lattice — dynamic instability of concepts *Trends Cell Biology*, **5**, 262–266,

Margulis, L. & Sagan, D. (1995) New York: *What is Life*? Simon and Schuster.

Marshall, W., Simon, C., Penrose, R. & Bouwmeester, D. (2003) Towards quantum superpositions of a mirror, *Phys. Rev. Lett.*, **91**, 13–16.

Matsuyama, S. S. & Jarvik, L. F. (1989) Hypothesis: Microtubules, a key to Alzheimer disease. *Proc. Nat. Acad. Sci. USA*, **86**, 8152–8156.

McCulloch, W. & Pitts, W. (1943) A logical calculus of the ideas immanent in nervous activity. *Bull. Math. Biophys.*, **5**, 115–133.

McEwen, B. & Edelstein S. J. Evidence for a mixed lattice in microtubules reassembled *in vitro*. *J. Mol. Biol.*, **139**, 123–145.

McKemmish, L. K., Reimers, J. R., McKenzie, R. H., Mark, A. E. & Hush, N. S. (2009) Penrose-Hameroff orchestrated objective-reduction proposal for human consciousness is not biologically feasible. *Physi. Rev. E.*, **80**, 021912.

Mitchison, T. & Kirschner, M. (1984). Dynamic instability of microtubule growth. *Nature*, **312**, 237–242.

Moroz, I. M., Penrose, R. & Tod, K. P. (1998) Spherically-symmetric solutions of the Schrödinger–Newton equations: *Classical Quant. Grav.*, **15**, 2733–2742.

Nadeau, R. & Kafatos, M. (2001) *The Non-local Universe: The New Physics and Matters of the Mind*, New York: Oxford University Press.

Nakagaki, T., Yamada, H. & Toth, Á. (2000) Maze–solving by an amoeboid organism. *Nature*, **407**, 470.

Naundorf, B., Wolf, F. & Volgushev, M. (2006). Unique features of action potential initiation in cortical neurons. *Nature*, **440**, 1060–1063.

Ouyang, M. & Awschalom, D. D. (2003) Coherent spin transfer between molecularly bridged quantum dots. *Science*, **301**, 1074–1078.

Pan, J. Z., Xi, J., Eckenhoff, M. E. & Eckenhoff, R. G. (2007) Halothane binding proteome in human brain cortex, *J. Proteome Res.* **6**, 582–592.

Pearle, P. (1989) Combining stochastic dynamical state-vector reduction with spontaneous localization, *Phys. Rev., A*, **39** 2277–2789.

Pearle, P. & Squires, E. J. (1994) Bound-state excitation, nucleon decay experiments and models of wave-function collapse, *Phys. Rev. Lett.*, **73**, 1–5.

Penrose, R. (1989) *The Emperor's New Mind: Concerning Computers, Minds, and the Laws of Physics*. Oxford: Oxford University Press.

Penrose, R. (1994) *Shadows of the Mind: An Approach to the Missing Science of Consciousness*. Oxford: Oxford University Press.

Penrose, R. (1992) Gravity and quantum mechanics. In *General Relativity and Gravitation 13*. Part 1: Plenary Lectures Proc. Thirteenth Int. Conf. on General Relativity and Gravitation, Cordoba, Argentina, 28 June–4 July 1992. (Eds.), Gleiser, R. J., Kozameh, C. N. & Moreschi O. M. (1993) (Inst. of Phys. Publ. Bristol and Philadelphia), pp. 179–89.

Penrose, R. (1996) On gravity's role in quantum state reduction, *General Relativity Gravitation*, **28**, 581–600.

Penrose, R. (2000) Wavefunction collapse as a real gravitational effect. In: Fokas, A., Kibble, T. W. B., Grigouriou A. & Zegarlinski, B. eds. *Mathematical Physics*. London: Imperial College Press, pp. 266–282.

Penrose, R. (2004) *The Road to Reality: A Complete Guide to the Laws of the Universe.* London: Jonathan Cape.

Penrose, R. (2009) Black holes, quantum theory and cosmology In: *Fourth Int. Workshop DICE* 2008, *J. Phys. Conf. Seri.*, **174**, 012001.

Penrose, R. (2010) *Cycles of Time: An Extraordinary New View of the Universe.* London: Bodley Head.

Penrose, R. & Hameroff, S. R. (1995) What gaps? Reply to Grush and Churchland. *J. Consciousness Stud.*, **2**, 98–112.

Penrose, R. & Hameroff, S. (2011) Consciousness in the universe: Neuroscience, quantum space–time geometry and Orch OR theory. *J. Cosmology*, **14**, 2011.

Penrose, R. & Bell, J. (2002) State Reduction, and Quanglement. In: R. A. Bertlmann and A. Zeilinger, eds. *Quantum Unspeakables: From Bell to Quantum Information.* Berlin: Springer-Verlag, pp. 319–331.

Percival, I. C. (1994) Primary state diffusion. *Proc. R. Soc. Lond* **A 447**, 189–209.

Percival, I. C. (1995) Quantum spacetime fluctuations and primary state diffusion. *Proc. R. Soc. Lond.*, **A451**, 503–13.

Poirazi, P. F. & Mel, B. W. (2001) Impact of active dendrites and structural plasticity on the memory capacity of neural tissue. *Neuron*, **29**, 779–796.

Pokorný, J. (2004) Excitation of vibration in microtubules in living cells. *Bioelectrochem.*, **63**, 321–326.

Pokorný, J., Hasek, Saroch, J. & Palan, B. (2001) Electromagnetic activity of yeast cells in the M phase. *Electro Magnetobiol.*, **20**, 371–396.

Polkinghorne, J. (2002) *Quantum Theory, A Very Short Introduction.* Oxford: Oxford University Press.

Pontecorvo, B. (1968) Neutrino experiments and the problem of conservation of Leptonic charge, *Soviet Phys. JETP, Vol.* **26**, 984–988.

Pribram, K. H. (1991) *Brain and Perception.* New Jersey: (Lawrence Erlbaum).

Rae, A. I. M. (2002) *Quantum Mechanics.* 4th edn. Institute of Physics Publishing.

Rasmussen, S., Karampurwala, H., Vaidyanath, R., Jensen, K. S. & Hameroff, S. (1990) Computational connectionism within neurons: A model of cytoskeletal automata subserving neural networks. *Physica D*, **42**, 428–449.

Reimers, J. R. McKemmish, L. K. McKenzie, R. H. Mark, A. E. & Hush, N. S. (2009) Weak, strong, and coherent regimes of Frohlich condensation and their applications to terahertz medicine and quantum consciousness *Proc. Nat. Acad. Sci. USA*, **106**, 4219–4224.

Reiter, G. F. Kolesnikov, A. I. Paddison, S. J. Platzman, P. M. Moravsky, A. P. Adams, M. A. Mayers, J. (2011). Evidence of a new quantum state of nano-confined water http://arxiv.org/abs/1101.4994.

Rieper, E., Anders, J. & Vedral, V. (2011) Quantum entanglement between the electron clouds of nucleic acids in DNA, preprint.

Rosenblatt, F. (1962) *Principles of Neurodynamics.* New York: Spartan Books.

Sahu, S., Ghosh, S., Ghosh, B., Aswani, K., Hirata, K., Fujita, D. & Bandyopadhyay, A. (2013a) Atomic water channel controlling remarkable properties of a single brain microtubule: Correlating single protein to its supramolecular assembly. *Biosens. Bioelectron.*, **47**, 141–148.

Sahu, S., Ghosh, S., Hirata, K., Fujita, D. & Bandyopadhyay, A. (2013b) Multi-level memory-switching properties of a single brain microtubule. *Appl. Phys. Lett.*, **102**, 123701.

Sahu, S., Ghosh, S., Fujita, D., Bandyopadhyay, A. (2014) Live visualizations of single isolated tubulin protein self-assembly via tunneling current: effect of electromagnetic pumping during spontaneous growth of microtubule. *Scientific Reports* **3**, 7303.

Samsonovich, A., Scott, A. & Hameroff, S. (1992) Acousto-conformational transitions in cytoskeletal microtubules: Implications for intracellular information processing. *Nanobiology*, **1**, 457–468.

Sataric, M. V., Zekovic, S., Tuszynski, J. A. & Pokorny, J. (1998) The Mossbauer effect as a possible tool in detecting nonlinear excitations in microtubules. *Phys. Rev. E*, **58**, 6333–6339.

Schrödinger, E. (1935) Die gegenwarten situation in der quantenmechanik. Naturwissenschaften, **23**, pp. 807–812, 823–828, 844–849. (Translation by Trimmer, J. T. (1980) in *Proc. Am. Phil. Soc.*, **124**, 323–338.) In: Wheeler, J. A. & Zurek, W. H. ed. *Quantum Theory and Measurement*. Princeton University Press.

Scott, A. C. (1995) *Stairway to the Mind*. New York: Springer-Verlag.

Shepherd, G. M. (1996) The dendritic spine: A multifunctional integrative unit. *J. Neurophysiol.*, **75**, 2197–2210.

Sherrington, C. S. (1957) *Man on His Nature*, 2nd Edn. Cambridge, Massachusetts: Cambridge University Press.

Shimony, A. (1993) *Search for a Naturalistic World View — Volume II. Natural Science and Metaphysics*. Cambridge, UK: Cambridge University Press.

Smith, S., Watt, R. C. & Hameroff, S. R. (1984) Cellular automata in cytoskeletal lattice proteins. *Physica. D*, **10**, 168–174.

Smolin, L. *The Life of the Cosmos*. New York, Oxford University Press.

Smolin, L. (2002) *Three Roads to Quantum Gravity*. New York: Basic Books.

Sourdet, V. & Debanne, D. (1999) The role of dendritic filtering in associative long-term synaptic plasticity. *Learn. Mem.*, **6**, 422–447.

Stapp, H. P. (2007) *Mindful Universe: Quantum Mechanics and the Participating Observer*. Springer, Berlin.

Stroud, J. M. (1956) The fine structure of psychological time. In: *Information Theory in Psychology*, H. Quastler, ed. Free Press, pp. 174–205.

Tart, C. T. (1995) Personal communication and information gathered from "Buddha-1 newsnet".

Tegmark, M. (2000) The importance of quantum decoherence in brain processes. *Phys. Rev., E*, **61**, 4194–4206.

Tononi, G. (2004) An information integration theory of consciousness. *BMC Neurosci.*, **5**, 42.

Traub, R. D., Draguhn, A., Whittington, M. A., Baldeweg, T. Bibbig, A. Buhl, E. H. & Schmitz, D. (2002) Axonal gap junctions between principal neurons: A novel source of network oscillations, and perhaps epileptogenesis. *Rev. Neurosci.*, **13**, 1–30.

Turin, L. (1996) A spectroscopic mechanism for primary olfactory reception. *Chem. Senses*, **21**, 773–791.

Tuszynski, J. A., Brown, J. A., Hawrylak, & P. Marcer, P. (1998) Dielectric polarization, electrical conduction, information processing and quantum computation in microtubules. Are they plausible? *Phil. Trans. R. Soc.*, A, **356**, 1897–1926.

Tuszynski, J. A., Hameroff, S., Sataric, M. V., Trpisova, B. & Nip, M. L. A. (1995) Ferroelectric behavior in microtubule dipole lattices; implications for information processing, signaling and assembly/disassembly, *J. Theor. Biol.*, **174**, 371–380.

van der Malsburg, C. (1999) The what and why of binding: Review — The modeler's perspective. *Neuron*, **24**, 95–104.

VanRullen, R. Koch, C. (2003) Is perception discrete or continuous? *Trends Cogn. Sci.*, **7**, 207–213.

Vattay, G., Salahub, D., Csabai, I., Nassimi, A., Kaufmann, S. A. (2015) Quantum Criticality at the Origin of Life asXiv:1502.06880.

Voet, D. & Voet, J. G. (1995) *Biochemistry*. 2nd edn. Wiley, New York.

von Rospatt, A. (1995) *The Buddhist Doctrine of Momentariness: A survey of the origins and early phase of this doctrine up to Vasubandhu*. Stuttgart: Franz Steiner Verlag.

Wegner, D. M. (2002) *The illusion of Conscious Will*, Cambridge, Massachusetts: MIT Press.

Whitehead, A. N. (1929) *Process and Reality*, New York: MacMillan.

Whitehead, A. N. (1933) *Adventure of Ideas*. London: MacMillan.

Wigner, E. P. Remarks on the mind-body question, in *The Scientist Speculates*, Good, I. J. ed. In: Wheeler, J. A. & Zurek, W. H. eds. Massachusetts *Quantum Theory and Measurement*. Princeton Univsity Press, Princeton; (Reprinted in E. Wigner (1967), Symmetries and Reflections, Indiana University Press, Bloomington).

Woolf, N. J. & Hameroff, S. R. (2001) A quantum approach to visual consciousness. *Trends Cogn. Sci.*, **5**, 472–478.

Xi, J., Liu, R., Asbury, G. R., Eckenhoff, M. F. & Eckenhoff R. G. (2004) Inhalational anesthetic-binding proteins in rat neuronal membranes. *J. Biol. Chem.*, **279**, 19628–19633.

Xiao, S. M., Herbst, T., Scheldt, T., Wang, D., Kropatschek, S., Naylor, W., Wittmann, B., Mech, A., Kofler, J., Anisimova, E., Makarov, V. Jennewein, Y., Ursin, R. & Zeilinger, A. (2012) Quantum teleportation over 143 kilometres using active feed-forward. *Nat. Lett.*, **489**, 269–273.

# APPENDIX A

### Reply to Seven Commentaries on "Consciousness in the Universe: Updated Review of the 'Orch OR' Theory"

Here we respond to seven commentaries (Tuszynski, 2014; Chopra, 2014; Ghosh *et al.*, 2014; Jumper & Scholes, 2014; Lucas, 2014; Tandy, 2014; Pino & Mauro, 2014) which accompanied "Consciousness in the universe: Review of the Orch OR theory" in *Physics of Life Reviews*: The eighth commentary by Reimers *et al.* (2014) is discussed separately in Appendix B. We thank authors of these seven commentaries, and reply to some of their key points.

### A.1. *Jack Tuszynski ("JT") (Tuszynski, 2014)*

First, Orch OR appears to be decoupled from the animate/inanimate divide.... Divorcing consciousness from life forms puts it back into the spiritual/dualistic type of reasoning that Orch OR rejects. How can one account for metabolism taking place in neurons, which would be a minimal requirement for making Orch OR account for life?

#### *Hameroff and Penrose ("H&P")*

Orch OR does not divorce life from consciousness, but marries them within non-polar "quantum channels" inside microtubule cylindrical walls. Although life is commonly equated with some of its properties, e.g., metabolism, reproduction, adaptation, and evolution, these phenomena also occur in inanimate complex systems. They do not define life.

In living cells, metabolism occurs in polar, aqueous environments isolated from nonpolar, water-excluding "quantum channels" (Hameroff & Penrose, 2014; Craddock *et al.*, 2014; 2015). Biological metabolism is no different than inanimate chemical activity ubiquitous in nature. While metabolism, reproduction, adaptation and evolution may be necessary for life and consciousness, they are not sufficient. As suggested by Schrödinger (1944) life, as we know it, depends essentially on quantum processes. In Orch OR, consciousness is a particular manifestation of life's quantum activity, involving "orchestrated" quantum coherent superpositions that persist sufficiently to reach the physical threshold for "objective reduction."

### *JT*

"Second, the emphasis based on the gravitational interactions is difficult to accept within conventional physics. Gravitational interactions are many orders of magnitude weaker than even thermal noise (Tuszynski *et al.* 1998)."

### *H&P*

This is a frequent misunderstanding. The role of gravitation in Orch OR is quite different from that of other forces, and it is not in competition with them. The DP proposal, upon which Orch OR relies, is not even a form of "quantum-gravity" in the normal sense of that term, where the rules of quantum (field) theory would be imposed upon classical gravitational theory. In DP objective reduction, we consider the opposite effect, where principles underlying Einstein's gravitational theory (general relativity) have their effect on quantum mechanics itself. With regard to issues of energy balance, the extremely tiny contribution from gravitation would be merely at the level of basic *uncertainties* in energy conservation. It is good that, in the circumstances under consideration, the gravitational contributions *are* extremely tiny, since they represent only a minute *uncertainty* in the normal energy balance (called upon only when the quantum state reduces according to OR), this balance thereby holding to an enormous precision, *by virtue* of gravitational energy contributions *necessarily* being ignored, as is normal practice.

Indeed, we say in Sec. 14.5.1. of our review: "Owing to the extreme weakness of gravitational forces...the energy $E_G$ is liable to be *far* smaller than any energy that arises directly from biological processes.... $E_G$ is not to be thought of as being in direct competition with any of the usual biological energies, as it plays a completely different role, supplying a needed energy *uncertainty* that then allows a choice to be made between the separated space–time geometries... This energy uncertainty is the key ingredient of the computation of the reduction time $\tau$, and ...it is appropriate that this energy uncertainty is indeed far smaller than the energies that are normally under consideration with regard to chemical energy balance, etc. If it were not so, then there would be a danger of *conflict* with normal considerations of energy balance. Nevertheless, the extreme weakness of gravity tells us there must be a considerable amount of material involved in the coherent mass displacement between superposed structures in order that $\tau$ can be small enough to be playing its necessary role in the relevant OR processes in the brain."

Microtubules may be thought of as "quantum switches," acting within the quantum energy uncertainty that comes about through the influence of Einstein's gravitational principles in a quantum context, thereby — according to Orch OR — enabling weak gravity-related influences to affect conscious thought and behavior.

### *JT*

"I foresee major progress in bridging the gap between nanoscience and consciousness in the area of nanoneuroscience (Woolf *et al.*, 2010) where MT's, actin filaments and motor proteins connect between neurophysiology and molecular biology. Studying the neural phenomena at a nanoscale will lead to monumental breakthroughs in science and medicine and aid in consciousness studies."

### *H&P*

Agreed. Anirban Bandyopadhyay's group has used nanoscience to find quantum resonances at multiple gigahertz, megahertz and kilohertz frequencies in single brain microtubules at warm temperatures (Sahu *et al.*, 2013a, 2013b).

### A.2. *Deepak Chopra ("DC") (Chopra, 2014)*

"The choice is between two non-dual explanations for how mind came into being. [The Eastern spiritual tradition of] Vedanta says that mind is innate in creation. To be viable, this brand of monism must show how mind created matter and energy. The challenge from the Penrose–Hameroff side is to show how matter and energy created mind. Of the two, Vedanta, in my view, has the upper hand. Mind creates matter every time we have thoughts that generate unique electrochemical activity in the brain. But no one has credibly shown how molecules learned to think."

### *H&P*

Dr. Chopra is espousing "View B" in Sec. 1 of our review, that "*Consciousness is a separate quality, distinct from physical actions and not controlled by physical laws, that has always been in the universe.*" This is in marked contradistinction to the conventional "View A" that "*Consciousness is not an independent quality but arose, in terms of conventional physical processes, as a natural evolutionary consequence of the biological adaptation of brains and nervous systems.*"

We agree with Dr. Chopra that "View A" fails to show how matter and energy create mind, how molecules can "think" (i.e., perform cognition accompanied by conscious experience). But we also recognize the shortcomings of his "View B," e.g., how can mind create energy and matter? Orch OR and "View C" bridge these two views, and, in principle, point to a solution for both their problems: (C) *Consciousness results from discrete physical events (objective reductions, OR); such events have always existed in the universe as non-cognitive, proto-conscious events, these acting as part of precise physical laws not yet fully understood.* Thus, in our view, a "proto-conscious" source of mind is omnipresent in the universe as OR events which shape reality (as in "View B"). However, experientially rich, human-like consciousness required biological evolution of a mechanism to "orchestrate" OR events, and couple them to brain neuronal activity (as in "View A").

### *DC*

"Orch OR provides a credible, testable model for how mental activity enters the physical world. I would take its optimism and turn it around: the mind–brain problem is indeed closer to being solved, not because quantum events give rise to mind but because these events indicate that an invisible agency (consciousness) is producing orderly, intelligent, information-infused activity at the very interface where space–time emerges.

### *H&P*

In Orch OR, fundamental Planck scale geometry ("where space–time emerges") is indeed considered to contain Platonic values producing such activity. Orch OR embraces both View A and View B.

### A.3. *Subrata Ghosh, Satyajit Sahu and Anirban Bandyopadhyay ("SSA") (Ghosh et al., 2014)*

The blind faith that Hodgkin–Huxley type neuron bursts explain neural information processing completely will collapse soon, and then brain building projects (Markram, 2006) all over the world will face the danger of banking on an incomplete picture of a neuron.

*H&P*

SSA are alluding to the assumptions that (1) Hodgkin–Huxley "integrate-and-fire" neuron models adequately represent neuronal functions, and (2) mapping and simulating such functions will capture essential brain features including consciousness. We completely agree with SSA that these assumptions are wrong, as they ignore intra-neuronal classical and quantum computation in microtubules, and non-computability necessary for consciousness, memory and synaptic regulation.

*SSA*

The recent finding of microtubule resonant oscillations (Sahu *et al.*, 2013a, 2013b, 2014) that could vibrate (the neuronal) axon (and dendrites/soma) brings Orch-OR into the picture as an extremely essential concept to fill the vacuum.

*H&P*

SSA refers to their own landmark discovery of gigahertz, megahertz and kilohertz quantum vibrations in single microtubules, experimental findings which strongly support theoretical assertions of Orch OR. They imply that such finer scale, deeper order processes are needed for a proper picture of neuronal function. We agree.

*SSA*

Stuart and Roger have rightly argued here (Hameroff & Penrose, 2014) that the wireless communication of axons via resonant vibrations around 100 $\mu$m diameter domain alleviates the biggest criticism of the Orch-OR proposal. The orchestration of resonant vibrations can occur globally among all neurons across the entire brain....This paper therefore closes the series of historical argument/counterargument on the "gap junction" forever.

### *H&P*

SSA refer to the issue in Orch OR of how microtubule-based quantum coherence can extend globally in neurons throughout the brain, enabling sufficiently large $E_G$ to reach OR threshold (by $E_G \approx h/\tau$) in times $\tau$ brief enough to be useful. Orch OR has previously suggested entanglement via window-like gap junctions (avoiding noisy membranes, synapses and neuropil). However SSA have shown "wireless" communication among microtubules in different neurons, possibly obviating the need for gap junctions. This still does not properly explain the necessary *quantum* entanglements that would be needed, between separated microtubules, however. In order for the respective "$E_G$s" of separated parts of a system to add up appropriately, in order to reach the required level for OR to occur, the entire mass displacement needs to be quantum-coherent, i.e., entangled. Quantum entanglements cannot be set up simply by classical communication. One might envisage gap junctions achieving the needed entanglements, but there does not appear to be any evidence, as yet, that SAA's "wireless communication" has thus required quantum-coherent character. This would seem to be an important issue for future research.

As a further point, "wireless" communication detected by the Bandyopadhyay group occurs between microtubule bundles in axons and dendrites/soma of different neurons. Orch OR has emphasized microtubule quantum computation in neuronal dendrites and cell bodies/soma (rather than those in axons) because dendritic–somatic processes correspond with integration in integrate-and-fire neurons, and such processes can trigger (or not trigger) axonal firings, thereby controlling the behavior. Microtubules in neuronal dendrites and soma are uniquely arrayed in mixed polarity networks of short, stabilized microtubules, whereas axonal microtubules extend continuously in unipolar bundles. The functional distinction for the two types of arrays is as yet unknown.

### *SSA*

"...objective reduction also gets a new dimension because resonance frequency bands of brain materials cover a wide range." Therefore, the discovery of resonance and wireless processing lead to a layered architecture of multiple space–time metric stacked one above another (Penrose, 2005). This is exactly what Roger and Stuart have been arguing for as the foundation of brain information processing for decades.... This paper (Hameroff & Penrose, 2014) therefore marks the beginning of developing a comprehensive mathematical modeling of the brain. Hopefully, in the near future, with more experimental understanding of the space–time metric, Orch-OR would evolve to a complete deductive mathematical expression of consciousness, — a dream that entire mankind is eagerly waiting to see.

### *H&P*

SSA refers first to their own discovery of gigahertz, megahertz and kilohertz resonances in microtubules, vibrations which appear to have a fractal-like relationship. Based on their results, we proposed that collective brain oscillations in the range of 10 to 100 Hz seen as the EEG are actually "beat frequencies," interference patterns, of much faster microtubule megahertz vibrations inside neurons. Observed and recorded for over a century, EEG origins have never been understood.

In Orch OR, microtubule quantum events correlate with fluctuations in the structure of space–time geometry ("space–time metric"). SSA are referring to a scale-invariant conscious connection between brain processes and the structure of the universe. One must be careful about drawing too strong of a conclusion of this kind here. The Planck scale of quantum gravity is, indeed, a particular *scale* of size, and scale-invariance for such fluctuations is not to be expected.

### *A.4. Chanelle Jumper and Gregory Scholes (J&S)*

"...A primary issue is decoherence, a process where the phases of quantum wavefunctions are randomized to reduce quantum phenomena to classical processes." Decoherence arises from dynamic disorder, or 'energy noise', and is expected to be extremely rapid in biological environments.

For example, the Orch OR theory requires quantum superposition on a time scale that is not currently accepted for biological, or even chemical, systems (McKemmish *et al.*, 2009).

### *H&P*

As J&S assert, avoidance of decoherence is critical to Orch OR and functional quantum biology. Previous versions of Orch OR called for microtubule coherence times (avoidance of decoherence) for times $\tau$ on the order of EEG frequency intervals, e.g., $10^{-1}$ to $10^{-2}$ s, lengthy in terms of quantum systems, and too long for Bandyopadhyay's measured coherence times (Sahu *et al.*, 2013a, 2013b, 2014) of $10^{-4}$ s (10 kilohertz resonance). However, in the present updated version of Orch OR, EEG markers of consciousness are seen as beat frequencies of much faster microtubule resonances and Orch OR events, e.g., 10 MHz with coherence times of a mere $10^{-7}$ s. Given the Bandyopadhyay coherence times of $10^{-4}$ s, Orch OR seems to now be on firm ground from the standpoint of decoherence.

The DP version of OR essentially replaces decoherence (or, rather, incorporates it within a broader objective scheme), as described in Sec. 14.4.5, but environmental interaction must still be avoided for significant times $\tau$ to have cognitive and meaningful conscious moments.

### *J&S*

"....biological systems are purported to be 'warm, wet and noisy,'... a messy business, incapable of supporting 'delicate' quantum processes...." On the contrary, it appears that living systems are governed by cycles and correlations, requiring massive cooperation across a large range of time and length scales.... Whether it is an acceptance of quantum behavior, or generating a foundation for the study of difficult questions in biology, should we reconsider the premise that living entities are founded on uncorrelated and chaotic machinery?

### *H&P*

Absolutely. Correlations over multiple spatiotemporal scales in living systems may be explained by a largely scale-invariant dynamics derived from quantum processes in non-polar, hydrophobic regions, e.g., quantum channels in microtubules (similar to quantum coherence in $\pi$-conjugated systems as described by Hwang & Scholes (2011)).

Regarding the "warm, wet and noisy" description of the biological environment: warm, yes, biological systems are indeed warm compared to quantum systems developed in laboratory conditions. However as proposed by Fröhlich, biology can utilize thermal energy to drive coherence, in a manner roughly analogous to a laser. "Wet"? No, at least not in non-polar, hydrophobic "quantum channels" within microtubules and other biomolecular interiors isolated from polar, aqueous portions of cell interiors and the extracellular milieu. These "dry," non-polar regions are precisely where anesthetic gases bind to selectively erase consciousness. "Noisy?" Noise-like fluctuations in brain electrical activity appear to be globally correlated, and therefore not noise at all. Of course, J&S are merely pointing out flaws in the common criticisms of schemes like Orch OR that depend upon quantum processes, and with that, we heartily agree.

*J&S*

...one of the next scientific revolutions will involve the evolution of a new framework for unraveling the extraordinary nature of life.

*H&P*

Agreed. And microtubule-based quantum resonance may well be a major part of that framework.

### A.5. *Reply to John Lucas (2014)*

Lucas (1961, 1970, 2013) had suggested in 1961 that Gödel's incompleteness theorem indicates that the human mind is non-computable, that some non-algorithmic factor is necessary. This anti-mechanistic argument was joined and elaborated by Penrose (1989, 1993, 2005) who suggested "objective reduction" as the non-computable factor, bringing quantum mechanics into the brain/mind picture. In his commentary, Lucas eloquently summarizes Orch OR positions, with particular reference to the thorny issue of freedom ("free will").

### A.6. *Reply to Charles Tandy (2014)*

Tandy (2014) continues with Gödel non-computability ("hyper-algorithmicity," as Tandy calls it) a property which Lucas and Penrose have attributed to the conscious mind. Although non-computability has usually been taken as bad news for artificial intelligence ("strong AI") and "Singularity" attempts to emulate conscious brain functions in computers, Tandy concludes that Orch OR and hyper-algorithmicity could in principle lead to such a goal — an Orch OR-based "artificial consciousness."

### A.7. *Reply to Samantha Pino and Ernesto Di Mauro (2014)*

Pino & Di Mauro (2014) put Orch OR in a larger context, relating it, and the problem of consciousness, to issues raised by Popper, Descartes, Darwin, Jung, Pauli, Heisenberg and others. They suggest the failure of the Higgs boson to account for super-symmetry leaves the makeup of the universe as mysterious as ever, perhaps requiring consciousness to complete the picture. Irrespective of the Higgs boson, this is also the essential conclusion of our review.

#### *P&D*

Hameroff & Penrose theory proposes an explanation by quantum physics of an intricately unapproachable phenomenon, of an *unknown*. To do this they feel comforted by the lack of different alternative explanations, possibly overinterpreting the possibilities and the role of quantum physics. However, this might after all not be that negative, potentially opening an experimental field to falsifiable verification.

#### *H&P*

It is worth pointing out that the DP proposal may be fairly close to either experimental confirmation or refutation. Following on from an early proposal (Marshall *et al.*, 2003), Bouwmeester has recently intimated (Penrose, 2013) that within 7 or 8 years we may well have an experimental answer to the viability of DP.

In closing, we refer back to the question of CCC as discussed in Sec. 6 of our review. The reference (Gurzadyan & Penrose, 2010) in the review may now be updated (Gurzadyan & Penrose, 2013). An independent analysis of the WMAP satellite data (Meissner *et al.*, 2013) supports CCC, and groups are now analyzing the more recent (and precise) Planck satellite data (An *et al.*, 2013). Orch OR and consciousness may well have preceded the Big Bang!

### Acknowledgment

We thank these authors for their useful commentaries.

This article was published in *Physics of Life Reviews*, Vol. 11, Stuart Hameroff and Roger Penrose, Reply to seven commentaries on "Consciousness in the universe — A review of the of the 'Orch OR' theory". pp. 94–100, Copyright Elsevier (2014).

### References

An, D., Meissner, K. A. & Nurowski, P. (2013) Structures in the Planck map of the CMB. arXiv:1307.5737.

Chopra, D. (2014) Reality and consciousness: A view from the east — Comment on "Consciousness in the universe — A review of the 'Orch OR' theory". *Phys. Life Rev.*, **11**, 81–82.

Ghosh, S., Sahu, S. & Bandyopadhyay, A. (2014) Evidence of massive global synchronization and the consciousness. Comment on "Consciousness in the universe: A review of the 'Orch OR' theory" by Stuart and Roger. *Phys. Life Rev.*, **11**, 83–84.

Gurzadyan, V. G. & Penrose R. (2013) On CCC-predicted concentric low-variance circles in the CMB sky, *Eur. Phys. J. Plus*, **128**, 22.

Hameroff, S. & Penrose, R. (2014) Consciousness in the universe — A review of the "Orch OR" theory. *Phys. Life Rev.*, **11**, 39–78.

Hameroff, S. & Penrose, R. (2014) Reply to criticism of the 'Orch OR qubit' — 'Orchestrated objective reduction' is scientifically justified. *Phys. Life Rev.*, **11**, 104–112.

Hwang, I. & Scholes, G. (2011) Electronic energy transfer and quantum-coherence in π-conjugated polymers. *Chem. Mater.*, **23**, 610–620.

Jumper, C. & Scholes, G. (2014) Life — Warm, wet and noisy? Comment on "Consciousness in the universe — A review of the 'Orch OR' theory" by Hameroff and Penrose. *Phys. Life Rev.*, **11**, 85–86.

Lucas, J. (2014) The face of freedom. Comment on "Consciousness in the universe — A review of the 'Orch OR' theory". *Phys. Life Rev.*, **11**, 87–88.

Lucas, J. R. (1961) Minds, machines and Gödel. *Philosophy*, **36**, 112–127.

Lucas, J. R. (1970) *The Freedom of the Will*. Oxford University Press.

Markram, H., (2006) The Blue Brain Project. *Nat. Rev. Neurosci.*, **7**, 153–160.

Marshall, W., Simon, C., Penrose, R. & Bouwmeester, D. (2003) Towards quantum superpositions of a mirror. *Phys. Rev. Lett.*, **91**, 13–16.

McKemmish, L. K., Reimers, J. R., McKenzie, R. H., Mark, A. E. & Hush, N. S. (2009) Penrose-Hameroff orchestrated objective-reduction proposal for human consciousness is not biologically feasible. *Phys. Rev. E*, **80**, 021912.

Meissner, K. A., Nurowski, P. & Ruszczycki, B. (2013) Structures in the microwave background radiation, *Proc. R. Soc. Lond. A*, **469**, 20130116.

Penrose R. (1989) *The Emperor's New Mind: Concerning Computers, Minds, and the Laws of Physics*. Oxford: Oxford University Press.

Penrose, R. (1997) On understanding understanding. *Int. Stud. Philos. Sci.*, **11**, 7–20.

Penrose, R. (2005) *The Road to Reality: A Complete Guide to the Laws of the Universe.* CPI group, London Vintage.

Penrose, R. (2013) Personal communication, at a meeting on Galiano Island, BC, Canada, May 2013.

Pino, S. & Di Mauro, E. (2014) How to conciliate Popper with Cartesius. Comment on: "Consciousness in the universe — A review of the 'Orch OR' theory" by S. Hameroff and R. Penrose. *Phys. Life Rev.*, **11**, 91–93.

Reimers, J. R., McKemmish, L. K., McKenzie, R. H., Mark, A. E. & Hush, N. S. (2014) The revised Penrose-Hameroff orchestrated objective-reduction proposal for human consciousness is not scientifically justified: Comment on "Consciousness in the universe — A review of the 'Orch OR' theory" by Hameroff and Penrose. *Phys. Life Rev.*, **11**, 101–103.

Sahu, S., Ghosh, S. Hirata, K., Fujita, D. & Bandyopadhyay, A. (2013a) Multi-level memory-switching properties of a single brain microtubule. *App. Phys. Lett.*, **102**, 123701.

Sahu, S., Ghosh, S., Ghosh, B., Aswani, K., Hirata, K., Fujita, D. & Bandyopadhyay, A. (2013b) Atomic water channel controlling remarkable properties of a single brain microtubule: Correlating single protein to its supramolecular assembly. *Biosens. Bioelectron.*, **47**, 141–148.

Sahu, S., Ghosh, S., Fujita, D., Bandyopadhyay, A. (2014) Live visualizations of single isolated tubulin protein self-assembly via tunneling current: effect of electromagnetic pumping during spontaneous growth of microtuble. *Scientific Reports*, 3, 7303.

Schrödinger, E. (1994) *What is life? The physical aspect of the living cell lectures.* Trinity College, Dublin: Dublin Institute for Advanced Studies.

Tandy, C. (2014) Are You (Almost) a Zombie? (Conscious thoughts about "Consciousness in the universe" by Hameroff and Penrose). *Phys. Life Rev.*, **11**, 89–90.

Tuszynski, J. A., Brown, J. A. & Hawrylak, P. (1998) Dielectric polarization, electrical conduction, information processing and quantum computation in microtubules, are they plausible?, *Trans. R. Soc. Lond. A*, **356**, 1897–1926.

Tuszynski, J. A. (2014) The need for a physical basis of cognitive process. Comment on consciousness in the universe — A review of the "Orch OR" theory by Hameroff and Penrose. *Phys. Life Rev.*, **11**, 79–80.

Woolf, N. J., Priel, A. & Tuszynski, J. A. (2010) *Nanoneuroscience: Structural and Functional Roles of the Neuronal Cytoskeleton in Health and Disease*. Heidelberg: Springer-Verlag.

# APPENDIX B

### Reply to criticism of the "Orch OR Qubit" — "Orchestrated Objective Reduction" *is* Scientifically Justified

The critical commentary by Reimers *et al.* (2014) regarding the Penrose–Hameroff theory of "Orch OR" is largely uninformed and basically incorrect, as they solely criticize non-existent features of Orch OR, and ignore (1) actual Orch OR features, (2) supportive evidence, and (3) previous answers to their objections (See Sec. 14.5.6 above.) Here we respond point-by-point to the issues they raise.

#### *Jeffery Reimers, Laura McKemmish, Ross McKenzie, Alan Mark, Noel Hush (Reimers et al., 2014)*

...For quantum information processing one must have quantum information storage units such as *qubits*.... the involvement of quantum-gravity in the manifestation of consciousness would need to be described in terms of how quantum-gravity affected the operation of these qubits...

#### *H&P*

Basically true. And this is just what we have done. Qubits involve (1) superposition of alternative possible states, and (2) a mechanism by which the possible states reduce, or collapse to definite states.

With regard to (1), any type of superposed state can, in principle, serve as a qubit, e.g., simultaneous alternative electric charge location, spin, dipole orientation, photon polarization, Fock state, topological pathway ("braid"), or current flow direction (e.g., superconducting Josephson junctions).

In Orch OR, the qubit involves electric or magnetic (spin) dipole orientations in "quantum channels" within each tubulin (microtubule subunit proteins), and between such tubulins along helical pathways through microtubule lattices. The dipoles occur due to coupled London force or spin dipole attractions among phenyl and indole rings of aromatic amino acids (tryptophan, phenylalanine, and tyrosine) which comprise "quantum channels," these electron states coupling to position or spin of their atomic nuclei. The coupled dipoles oscillate between alternative orientations, and become superpositions of both states to function as qubits (Figs. B.1(B), and 5–7 in our review).

This helical pathway version of the Orch OR qubit (akin to a "topological qubit" (Collins, 2006)) was developed in 2002 (Hameroff *et al.*, 2002), after the structure of tubulin was elucidated by electron crystallography (Nogales, et al., 1998) to reveal clustered arrays of electron resonance rings ("quantum channels," Fig. 5 in our review) (Craddock, *et al.*, 2012). The helical pathway qubit replaced the earlier notion (1996–1998, Fig. B-1 (A), (Hameroff & Penrose, 1996a,b; Hameroff, 1998a) that oscillating dipoles in individual tubulins were the fundamental qubits. Helical pathway qubits are inherently resistant to decoherence, can couple to natural resonances in microtubule lattices, and associate with alternating current ("AC"). Indeed, Bandyopadhyay's group (Sahu *et al.*, 2013a, 2013b, 2014) has shown remarkable AC conductance at warm temperature through single microtubules at certain resonant frequencies (e.g., gigahertz, megahertz, and kilohertz). This evidence is consistent with oscillating dipoles extending along helical pathways through microtubules, and thus appears to provide considerable support for Orch OR.

How does the "DP" gravitational OR scheme influence the operation of Orch OR qubits? We first point out that DP is not really "quantum-gravity" in the normal sense of that term, as explained in Appendix A in response to Jack Tuszynski ("JT"), and in Sec. 14.4.4. — though Reimers *et al.* continually use this terminology. This relates to (2), a mechanism by which the possible qubit states reduce, or collapse to definite states, the so-called "wavefunction collapse" of the *measurement problem* of quantum mechanics. In Orch OR, "quantum-gravity" (DP objective reduction) causes reduction, or collapse, of superpositions to classical states in accordance with $E_G \approx \hbar/\tau$ (Fig. 10). It may be mentioned that there are current experimental programs aimed at testing the validity of DP, and these ideas have recently become quite popular (Wolchover, 2013).

So in answer to this question raised by Reimers *et al.* ("how does quantum-gravity affect operation of the qubit?"), Orch OR is very clear: it is the gravitational proposal DP that causes microtubule qubits to reduce, or collapse (in a time $\tau \approx \hbar/E_G$), to definite classical microtubule states (which then regulate brain neurons).

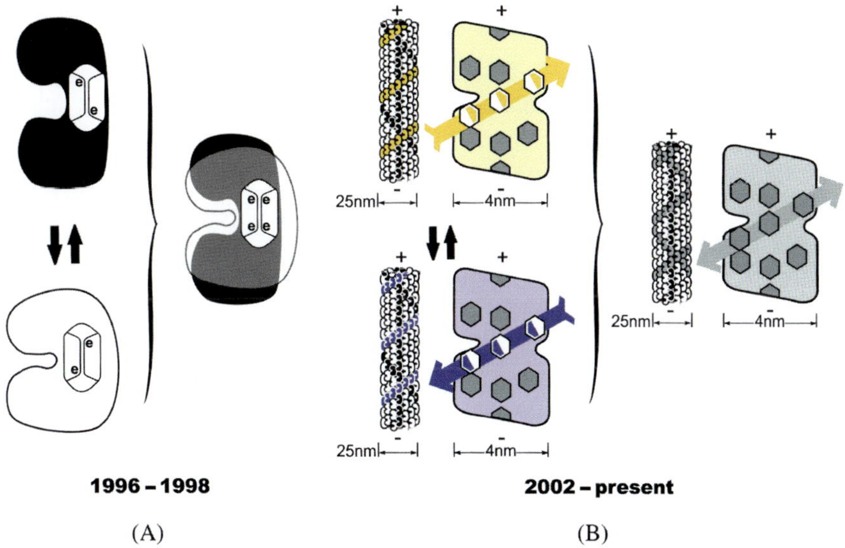


**Fig B.1.** Early, and current, versions of the Orch OR qubit. (A) Schematic cartoon version of Orch OR tubulin protein qubit used in Orch OR publications mainly from 1996 to 1998. On left, tubulin oscillates between 2 states with ~1 nm conformational flexing (~10% tubulin diameter). On right, both states exist in quantum superposition. (Irrespective of the schematic cartoon, the 1 nm displacement has never been implemented in Orch OR calculations.) The states are shown to correlate with electron locations (dipole orientations) in two adjacent phenyl (or indole) resonance rings in a non-polar "hydrophobic pocket." (B) Schematic cartoon version of Orch OR qubit used since 2002 (following identification of tubulin structure by electron crystallography (Hameroff *et al.*, 2002; Nogales *et al.*, 1998). Each tubulin is shown to have 9 rings representing 32 actual phenyl or indole rings per tubulin, with coupled, oscillating London force (or spin) dipole orientations among rings traversing "quantum channels," aligning with rings in adjacent tubulins in helical pathways through microtubule lattices. On the right, superposition of alternative tubulin and helical pathway dipole states. There is no conformational flexing. Mechanical (femtometer) displacement occurs at the level of tubulin atomic nuclei (not shown). Reimers *et al.* (2014) continually, and exclusively, criticize the obsolete, non-implemented cartoon version on left (A), and ignore the actual Orch OR qubit version on right (B).

## Reimers *et al.*

In the current review Hameroff and Penrose suggest that the qubit could be either: (a) "interactive dipole states of individual tubulin proteins"...

*H&P*

"Interactive dipole states" YES, "individual tubulin proteins" NO. The Orch OR qubit is a dipole state extending through mesoscopic helical pathways of many tubulins in microtubule lattices (Figs. B.1(B), 6 and 7 in our review). This is a change from early Orch OR theory in which each tubulin was a qubit (Fig. B.1(A), though without the significant conformational flexing shown there — see below).

**Reimers *et al.***

...interactive dipole states....such as "London-force dipoles" or (b) magnetic dipoles or (c) nuclear spins. "London force electric dipoles" have been discussed in previous publications but the other two options have been introduced for the first time.

*H&P*

True. London force "electric dipoles," magnetic dipoles (electron spin), nuclear spins and AC current flows are possibly synergistic modes. For example microtubule electron dipoles may induce longer-lived nuclear spin states for short-term memory. As we say in Sec. 14.4.6 above: "It is to be expected that the actual mechanisms underlying the production of consciousness in a human brain will be very much more sophisticated than what we can put forward at the present time."

**Reimers *et al.***

Previously, Hameroff and Penrose had also proposed that conformational switching could produce coupled electron-vibration qubits but this claim is withdrawn in the current review.

*H&P*

Not true. There are least two types of "conformational switching" to consider. The type (1) to which Reimers *et al.* refer (see below) was considered and rejected by us in 1996, but implied in cartoon form through 1998, and occasionally thereafter [Fig. B.1(A)]. Reimers *et al.* continually, and exclusively, criticize the obsolete, non-implemented version in Fig. B.1(A), and ignore the actual Orch OR qubit version in Fig. B.1(B) as if they never read our review.

Type (1) conformational switching is shown in Fig. B.1(A). Early Orch OR illustrations showed such conformational "flexing" of entire tubulin proteins (e.g., ~1 nm displacement, ~10% of tubulin diameter) correlating with discrete information states (and superposition of states). It is no longer relevant to Orch OR except historically for the following reason.

In 1996, in attempting to connect brain biology to gravity-induced OR (Hameroff & Penrose, 1996a), we calculated $E_G$ for superposition separation of three different conformational changes in tubulin: (1) a ~10% conformational flexing (~1 nm, $10^{-9}$ m), as shown in Fig. B.1(A), (2) movement and separation at the level of atomic nuclei, e.g., carbon atoms (~5 fm, $5 \times 10^{-15}$ m), and (3) movement and separation at the level of nucleons, i.e., protons and neutrons (~femtometers, $10^{-15}$ m). The dominant effect was found to be separation of intra-tubulin atomic nuclei ($5 \times 10^{-15}$ m), and all Orch OR calculations have used separation at that level. Nanometer flexing was rejected in 1996, not "withdrawn in the current review." However, schematic cartoon illustrations (e.g., Fig. B.1(A) used through 1998 and occasionally thereafter were admittedly, though unintentionally, misleading.

Type (2) conformational switching can occur at the aforementioned much smaller level of atomic nuclei, $5 \times 10^{-15}$ m, femtometers, shown to be optimal for $E_G$. This is also the calculated displacement of one (e.g., carbon) atomic nucleus caused by a one nanometer shift in nearby electrons by charge and Mossbauer recoil (Sataric *et al.*, 1998, Brizhik *et al.*, 2001). Thus, nanometer electron dipoles (London forces) or spin can couple to femtometer nuclear displacements, as needed for $E_G$ in Orch OR.

Slight mechanical, conformational vibrations at the level of atomic nuclear displacement are likely to be associated with (e.g., megahertz) AC electronic resonances discovered by Bandyopadhyay's group in piezoelectric microtubules, e.g., mediated by electric or magnetic dipoles oscillations. London repulsive forces (Pauli exclusion when electron clouds get too close and overlap) are 100 times stronger than attractive forces, likely to promote nonlinear mechanical vibrations in microtubules with displacement at the level of atomic nuclei.

## Reimers *et al.*

The London force is of quantum-mechanical origin. An instantaneous fluctuation of the electronic distribution creates a dipole in one molecule that in turn induces a dipolar response in a neighboring molecule. This leads to a net attractive force.

***H&P***

True. Quantum-mechanical London ("dipole dispersion") force attractions are precisely how anesthetic gas molecules bind (e.g., in brain microtubules; Fig. 6 above) to disperse quantum dipoles and selectively erase consciousness (sparing non-conscious neuronal brain activities (Hameroff *et al.*, 1982; Hameroff & Watt, 1983; Hameroff, 1998; Hameroff, 2006). London forces are weak, but numerous and influential, and able to govern protein function (Voet & Voet, 1995).

**Reimers *et al.***

The key feature is that these electric dipoles are *fluctuations*, not *states*. Individual *states* are needed to construct a qubit, and the review makes no attempt at specifying how qubit states could be associated with these London fluctuations.

***H&P***

Not true, as Fig. 7 in our review demonstrates. London force dipoles are coupled, and "fluctuate" collectively ("oscillate") between two alternative orientation states, with quantum superposition of both states. The London force-mediated shifts themselves are indeed instantaneous, but lifetimes of particular states following each shift are finite, e.g. up to $10^{-4}$ s as shown in microtubule resonances by Bandyopadhyay's group (Sahu *et al.*, 2013a, 2013b, 2014). The evidence appears to be in support of our contention.

It is somewhat gratifying to note that Reimers *et al.* have dropped their previous misguided criticism that electrons within a phenyl or benzene ring are delocalized, and therefore cannot constitute a switch (McKemmish *et al.*, 2009). As we show in Sec. 14.5.6, this would apply only to single phenyl rings, not coupled rings or contiguous arrays of such rings, as exist in tubulin.

**Reimers *et al.***

No model of Orch OR can be treated seriously without the following: (i) a precise description of the quantum states of the qubit.

*H&P*

In an explorative work of this nature, where there remain many unknown factors, it is unreasonable to demand great precision at this stage. Nevertheless, we have provided a plausibly precise description (see above), as dipole states mediated by London forces and/or spin currents in resonance rings aligned in helical pathways through microtubules. Reimers *et al.* ignore this, appearing not to have read or understood our review, criticizing, instead, an irrelavant cartoon. If Reimers *et al.* are asking for, say, a Hamiltonian operator as a precise description for a microtubule quantum state, see our suggestions below.

**Reimers *et al.***

"...(ii) a description of the mechanism through which the wavefunctions representing these states become entangled,"

*H&P*

Orch OR has always linked qubit entanglement to Fröhlich-type coherence, or condensation among pi electron resonance dipoles akin to laser-induced entanglement in technological quantum systems.

**Reimers *et al.***

"...including specification of the basis in which measurements of the qubit's properties are performed *in situ*,..."

*H&P*

"Measurement" of microtubule qubits is taken to occur by the DP proposal for gravitational OR by $E_C = \hbar/\tau$, a "self measurement." Reimers *et al.* appear to have missed the key point.

**Reimers *et al.***

(iii) A means of achieving quantum coherence over the required time scale.

### *H&P*

The non-polar, hydrophobic environment within "quantum channels" shields quantum coherence from polar environmental interactions. This is the solubility phase in which anesthetics act (the "Quantum underground," Chapter 13) and the likely origin of consciousness. Ambient energy, electric fields and mechanical vibrations pump coherence (as occurs in photosynthesis, and as suggested by Fröhlich). Most importantly, apparent quantum coherence up to $10^{-4}$ s has been shown by Bandyopadhyay's group (Sahu *et al.*, 2013a, 2013b, 2014) to occur in single microtubules at warm temperature, which may be sufficient for Orch OR.

### Reimers *et al.*

Hameroff and Penrose provide only a vague set of qubit possibilities. By not specifying the qubits in the current review they fail to provide a means by which the postulated links between quantum-gravity and conscious behavior could be assessed.

### H&P

As remarked upon above, to ask for too much precision in our suggested mechanisms is, at this stage, an unreasonable request. Nevertheless, we have clearly specified our proposal for an Orch OR qubit, as the Orch OR helical pathway that Reimers *et al.* appear to ignore.

Unlike previous proposals for a physical basis for consciousness, Orch OR provides a detailed, testable, falsifiable and moderately rigorous theory, not only for consciousness, but also for microtubule dynamics. It is a broadly based scheme, addressing areas of molecular biology, neuroscience, quantum physics, pharmacology, philosophy, quantum information theory, and even aspects of quantum-gravity. Orch OR has been repeatedly challenged but, in our view, remains to be seriously threatened.

### Reimers *et al.*

In previous versions of Orch OR, they *did* define a qubit that at the time might have been considered a reasonable proposition to advance and test. They proposed that conformational switching produced a coupled electron-vibration qubit that interacted with the cellular environment through associated large changes in microtubule structure and with quantum-gravity via the significant mass displacement associated with vibration.

***H&P***

The electron-vibrational coupling in Orch OR is between (1) pi electron resonance dipoles in aligned tubulin proteins in microtubules, and (2) conformational vibrations at the level of their atomic nuclei, i.e. 5 femtometers.Vattay *et al* (2015) have shown pi resonance rings in proteins are arrayed at "quantum criticality," the precise distance separating quantum and classical behaviors. Conformational vibrations, even slight ones, can thus cause proteins to oscillate between quantum and classical states, e.g. Bandyopadhyay coherence in microtubules.

**Reimers *et al.***

Coupled electron-vibration qubits are indeed considered as possibilities for use in modern quantum information technologies (Ferguson *et al.*, 2002; Hines *et al.*, 2004; McKemmish *et al.*, 2011). Quantum coherence was postulated to be provided by Fröhlich condensation (Fröhlich, 1968a,b, 1970), a predicted but unobserved macroscopic quantum effect. The original proposal thus contained a critical testable hypothesis.

***H&P***

Fröhlich condensation remains in Orch OR, and we consider Bandyopadhyay's gigahertz, megahertz, and kilohertz resonance in microtubules to be clear evidence for Fröhlich condensation, e.g., mediated by oscillating London force dipoles or spin currents (Sec. 14.4.5).

**Reimers *et al.***

We tested this hypothesis and found two fatal shortcomings, resulting in it being withdrawn from Orch OR in this current review.

*H&P*

It is not withdrawn (Sec. 14.4.5). Large scale (nanometer) conformational flexing never was part of Orch OR except in cartoon illustrations (Fig. B-1). The supposed "fatal shortcomings" were in Reimers *et al.* misconceptions. In Orch OR and microtubules, Fröhlich condensation is alive and well as Bandyopadhyay coherence.

**Reimers *et al.***

First, we showed the conformational-switch was not a vibration, as is required for the qubit, but instead involves an irreversible chemical reaction (McKemmish *et al.*, 2009).

*H&P*

There is no requirement for a qubit that it should be a vibration. But Reimers *et al.* again refer to the long-rejected (and non-existent, in terms of underlying calculations) nanometer flexing of tubulin (Fig. B.1(A), instead of femtometer (six orders of magnitude smaller) displacement at the level of atomic nuclei. We agree that coherent tubulin nanometer flexing would require significant GTP hydrolysis, an irreversible chemical reaction, and is not feasible (nor did we ever propose such an idea). Fröhlich condensation and Bandyopadhyay coherence are pumped by ambient energy and mitochondrial electric fields (Hameroff & Penrose, 2014).

**Reimers *et al.***

Second, we examined the postulate that Fröhlich condensation could deliver unprecedented quantum coherence in a qubit involving electronic motion (Reimers *et al.*, 2009). Whilst Fröhlich proposed that the coupled non-linear equations that he solved would show *Bose-Einstein*-like behavior, we found that instead a Fröhlich condensate would be *extremely incoherent*.

Further, we showed that significant *classical* effects of Fröhlich condensation did not manifest unless the system was very far from thermal equilibrium, with component parts needing to be at temperatures in excess of 500 K for room-temperature operation. Fröhlich condensation could not sustain quantum coherence in biological systems and could not support Orch OR.

### *H&P*

The calculations behind these criticisms have nothing to do with Orch OR, microtubules or biology, and are faulty.

In the paper to which they refer (Reimers *et al.*, 2009), Reimers *et al.* applied the Wu-Austin Hamiltonian to "a linear chain of coupled oscillators *as envisaged in the Orch OR proposal*." This is false. Orch OR does ***not*** envisage microtubules as a one-dimensional chain of oscillators. We envisage microtubules as three-dimensional crystalline lattices with Fibonacci geometry and gigahertz, megahertz, and kilohertz resonances.

Moreover, the Wu-Austin Hamiltonian is suspect, as it lacks a lower bound, does not converge, and is therefore considered "unphysical" (Bolterauer, 1999). Moreover it addresses quantum states in rigid polar alpha helical regions in proteins rather than non-polar "quantum channels" as addressed in Orch OR. Using a different Hamiltonian, Samsonovich *et al.* (1992) simulated microtubules as two-dimensional lattice planes with toroidal boundary conditions (approximating three-dimensions). They found strong Fröhlich coherence in super-lattice patterns which precisely match experimentally-observed attachment sites for microtubule-associated-proteins ('MAPs').

### Reimers *et al.*

We also note that the observed decoherence times for quantum processes involving electronic motion are usually in the range of 10 fs to 30 ps. A qubit with dynamics even slightly coupled to electronic motion would not retain quantum coherence on the 25 ms timescale required for Orch OR which Hameroff and Penrose suggest in this current review. This has consequences for all proposed qubits.

### *H&P*

This, at last, is a reasonable objection, but an anticipated one. Bandyopadhyay's findings of gigahertz, megahertz and (as low as) 10 kilohertz resonances indicate microtubule "decoherence times" (duration for which decoherence is avoided, i.e., "coherence times") for microtubules can persist as long as $10^{-4}$ s. This same "coherence time" of $10^{-4}$ s was also calculated based on Orch OR stipulations (Hagan *et al.*, 2001), but is indeed 250 times briefer than the 25 ms we invoked for Orch OR events, e.g., correlating with (40 Hz) gamma synchrony EEG.

In a supplemental modification, we suggest in our review that Orch OR events occur at much higher frequencies (e.g., megahertz) than previously proposed (e.g., at 40 Hz), and that interference between sets of coherent microtubule vibrations (e.g., ~10 MHz) results in much slower "beat frequencies," e.g., at 40 Hz gamma synchrony. Indeed, EEG rhythms (whose origins have never been understood) may actually be "beats" of much faster megahertz Orch OR events in intra-neuronal microtubules.

Moreover, recently-observed very high frequency (kilohertz) EEG (Usui *et al.*, 2010), too fast for membrane depolarizations, may reflect intra-neuronal microtubule dipole oscillations.

Orch OR occurring at ~10 MHz compared to Orch OR occurring at 40 Hz would have greater $E_G$ involving more of the brain (~1% of brain tubulins) and, we suggest, greater experiential intensity. Decoherence, or premature OR, would then need to be avoided for a mere $10^{-7}$ s compared to $10^{-1}$ to $10^{-2}$ s. Bandyopadhyay's group has already shown decoherence time 1000 times longer, at $10^{-4}$ s, so Orch OR is on more solid ground with respect to decoherence.

Orch OR has some support from experimental evidence and it is a scientifically justified proposal. The riddle of how EEG is generated (including kilohertz EEG (Usui *et al.*, 2010) may also perhaps be solved in terms of Orch OR.

### Reimers *et al.*

In an effort to perpetuate their model they now include "electron-cloud dipoles (London forces)," magnetic spin dipoles and nuclear spins in a list of possible qubits, without suggesting how any of these phenomena could in fact be used to make a relevant qubit.

### *H&P*

Decoherence (or premature OR) need be avoided for only $10^{-7}$ s for 10 MHz Orch OR events. Superposition of pi electron resonance dipole states couples to nuclear position (and spin) in specific helical pathways of tubulins in microtubules. The superposition corresponds to separated space–time geometry and gravitational self-energy $E_G$ which undergoes OR at time $\tau \approx \hbar/E_G$. and "self-collapses" to classical states, accompanied by a moment of conscious experience.

The Orch OR proposal is aimed at perpetuating truth and scientific knowledge. If it were to be shown invalid, we would drop the model and acknowledge accordingly. But we believe that, so far, it has survived many extensive criticisms, such as those by Reimers *et al.* (see Sec. 5.6 in our review). Nevertheless, as said earlier, in accordance with scientific principles we have tried to improve the theory when the need arises and to introduce new ideas. Reimers *et al.* do not even address most of these developments.

Indeed we have modified and adapted the original Orch OR proposal as new information has come forth (though "electron cloud London force dipoles" have always been key components). For example quantum channels, helical pathway qubits, and faster (megahertz) Orch OR events with EEG beat frequencies are adaptations based on new knowledge of tubulin structure and Bandyopadhyay coherence.

Magnetic spin dipoles and nuclear spin are indeed also suggested in this review, and for good reasons. Quantum (magnetic) spin transfer through phenyl rings is increased with temperature (Ouyang & Awschalom, 2003), and likely to be important biologically. Oscillating spin currents, or spin-flips, may propagate through quantum channels as easily as electric (London force) dipoles), or together, synergistically, along with nuclear spin and displacement.

**Reimers *et al.***

The review is thus neither self-consistent or scientifically coherent and violates the basic tenants of good scientific practice (van Gunsteren, 2013).

***H&P***

Orch OR *is* self-consistent, scientifically coherent, and increasingly supported by evidence. Reimers *et al.* then accuse us of a lack of good scientific practice. Their reference (van Gunsteren, 2013) describes "Seven deadly sins" of good scientific practice, e.g., "using pretty pictures instead of solid science". We have repeatedly stated that conformational flexing as shown in Fig. B.1(A) (developed and used before tubulin structure was known) is unrelated to the underlying scientific calculations based on atomic nuclear displacement. Figure B1(B) more closely describes the present Orch OR helical pathway qubit, but still is a simplification, e.g., showing 9 rings per tubulin rather than the actual 32 rings per tubulin. These rings within tubulin also are where anesthetic gas molecules bind and act to erase and prevent consciousness.

With regard to the use of "pretty pictures" instead of solid science, in the original criticism of Orch OR (Reimers *et al.*, 2009) by Reimers *et al.*, they depict individual coupled oscillators (presumably tubulins) floating in a bath, the cell as "minestrone soup" in their Fig. B.1. Tubulins and microtubules do not float but exist as solid state crystalline lattices. By their own criteria, Reiner's *et al.*'s Fig. B.1 violates basic tenants of good scientific practice.

However, the "minestrone soup" picture of cell interiors as presented by Reimers *et al.* relates to a critical point raised by Jumper & Scholes (2014) in their constructive commentary:

**Jumper and Scholes (2014)**

"...what do we really know about the fundamental nature and properties of the biological environment? The concept of 'warm, wet and noisy' is put forth when one imagines the complex components and workings of a biological system to be... a messy business, incapable of supporting 'delicate' quantum processes.... Should we reconsider the premise that living entities are founded on uncorrelated and chaotic machinery? [Why] do we expect incoherence in biology?... On the contrary, it appears that living systems are governed by cycles and correlations, requiring massive cooperation across a large range of time and length scales."

**Reimers *et al.***

The specification of the quantum qubit should be the centerpiece of the proposal. All other aspects of the Orch OR proposal are only relevant in terms of how they affect the qubits. Without a viable qubit specification there is no connection between the proposal and the observations of Bandyopadhyay and others. Without a qubit there is no connection to postulated effects of quantum-gravity. Without a qubit there is no testable hypothesis linking together the phenomena of quantum-gravity, elementary biochemical function, and consciousness, and no basis on which "Orch OR theory" can be considered as a proposal worthy of further consideration.

### *H&P*

We agree. That's why we have defined an Orch OR qubit based on oscillating London force dipoles and/or spin dipoles in resonance rings in helical pathways through microtubule lattices. The oscillations occur in kilohertz, megahertz, and gigahertz, and appear to be in line with Bandyopadhyay's findings. Reimers *et al.* ignore our specified qubit and continue to criticize an irrelevant cartoon.

We believe that Orch OR is a detailed, testable, falsifiable, and reasonably rigorous approach to a theory of consciousness, and microtubule function. Supportive evidence for Orch OR (Bandyopadhyay megahertz coherence (Sahu *et al.*, 2013a, 2013b), Eckenhoff anesthetic effects on microtubules (Xi *et al.*, 2004; Pan *et al.*, 2007; Emerson *et al.*, 2013), quantum channels (Craddock *et al.*, 2012)) is of a kind not yet found in other relevant theories. Orch OR has been repeatedly challenged but we do not feel that it has yet been seriously threatened.

## Acknowledgement

This article was published in *Physics of Life Reviews*, Vol. 11, Stuart Hameroff and Roger Penrose, "Reply to criticism of the Orch OR qubit — Orchestrate objective reduction is scientifically justified", pp. 104–112, Copyright Elsevier (2014).

## References

Bolterauer, H. (1999) Elementary arguments that the Wu-Austin Hamiltonian has no finite ground states (the search for a microscopic foundation of Fröhlich's theory). *Bioelectrochem. Bioenerg.*, **48**, 301–304.

Brizhik, L., Eremko, A., Piette, B. & Zakrzewski, W. (2001) Electron self-trapping in a discrete two-dimensional lattice. *Physica D*, **159**, 71–90.

Collins, G. P. (2006) Computing with quantum knots. *Sci. Am.*, 57–63.

Craddock, T., St George, M., Freedman, H., Barakat, K., Damaraju, S., Hameroff, S. & Tuszynski, J. (2012) Computational predictions of volatile anesthetic interactions with the microtubule cytoskeleton: Implications for side effects of general anesthesia. *PLoS One*, **7**, e37251.

Emerson, D., Weiser, B., Psonis, J., Liao, Z., Taratula, O. & Fiamengo, A. Eckenhoff. (2013) Direct modulation of microtubule stability contributes to anthracene general anesthesia. *J. Am. Chem. Soc.*, **135**, 5398.

Ferguson, A. J., Cain, P. A., Williams, D. A. & Briggs, G. A. D. (2002) Ammonia-based quantum computer. *Phys. Rev. A*, **65**, 034303.

Fröhlich H. (1968a) Long-range coherence and energy storage in biological systems. *Int. J. Quantum Chem.* **2**, 641–649.

Fröhlich, H. (1968b) Bose condensation of strongly excited longitudinal electric modes. *Phys. Lett. A*, **26**, 402–403.

Fröhlich, H. (1970) Long range coherence and the action of enzymes. *Nature*, **228**, 1093.

Hagan, S., Hameroff, S. & Tuszynski, J. Quantum computation in brain microtubules? Decoherence and biological feasibility. *Phys. Rev. E*, **65**, 061901.

Hameroff, S. (1998a) Quantum computation in brain microtubules? The Penrose–Hameroff "Orch OR" model of consciousness. Philos. *Trans. R. Soc. Lond Ser. A, Math. Phys. Sci.*, **356**, 1869–1896.

Hameroff, S. (1998b) Anesthesia, consciousness and hydrophobic pockets–a unitary quantum hypothesis of anesthetic action. *Toxicol. Lett.*, **100/101**, 31–39.

Hameroff, S., Nip, A., Porter, M. & Tuszynski, J. (2002) Conduction pathways in microtubules, biological quantum computation and microtubules. *Biosystems*, **64**, 149–168.

Hameroff, S. (2006) The entwined mysteries of anesthesia and consciousness. *Anesthesiology*, **205**, 400–412.

Hameroff, S. R. & Penrose, R. (1996a) Orchestrated reduction of quantum coherence in brain microtubules: A model for consciousness. In: S. R. Hameroff A. W. Kaszniak & A. C. Scott, ed. *Toward a Science of Consciousness; The first Tucson Discussions and Debates*. Cambridge (MA):MIT Press, p. 507. Also published in *Math. Comput. Simul.*, **40**, 453–480.

Hameroff, S. R. & Penrose, R. (1996b) Conscious events as orchestrated space–time selections. *J. Conscious Stud.* **3**, 36–53.

Hameroff, S. & Penrose, R. (2014) Consciousness in the universe — A review of the 'Orch OR' theory. *Phys. Life Rev.*, **11**, 39–78.

Hameroff, S. & Penrose, R. (2014) Reply (1) to seven commentaries on 'Consciousness in the universe — A review of the 'Orch OR' theory. *Phys. Life Rev.*, **11**, 94–100.

Hameroff, S., Watt, R., Borel, J. & Carlson, G. (1982) General anesthetics directly inhibit electron mobility: dipole dispersion theory of anesthetic action. *Physiol. Chem. Phys.*, **14**, 183–187.

Hameroff, S. & Watt, R. (1983) Do anesthetics act by altering electron mobility? *Anesth. Analg.*, **62**, 936–940.

Hines, A. P., Dawson, C. M., McKenzie, R. H. & Milburn, G. J. (2004) Entanglement and bifurcations in Jahn-Teller models. *Phys. Rev. A*, **70**, 022303.

Jumper, C. & Scholes, G. (2014) Life — Warm, wet and noisy? Comment on "Consciousness in the universe: A review of the 'Orch OR' Theory" by Hameroff and Penrose. *Phys. Life Rev.*, **11**, 85–86.

McKemmish, L. K., McKenzie, R. H., Hush, N. S. & Reimers, J. R. (2011) Quantum entanglement between electronic and vibrational degrees of freedom in molecules. *J. Chem. Phys.*, **135**, 244110/1–11.

McKemmish, L. K., Reimers, J. R., McKenzie, R. H, Mark, A. E. & Hush, N. S. (2009) Penrose-Hameroff orchestrated objective-reduction proposal for human consciousness is not biologically feasible. *Phys. Rev. E*, **80**, 021912–021916.

Nogales, E., Wolf, S. G. & Downing K. H. (1998) Structure of the alpha beta tubulin dimer by electron crystallography. *Nature*, **91**, 199–203.

Ouyang, M, & Awschalom, D. D. (2003) Coherent spin transfer between molecularly bridged quantum dots. *Science*, **301**, 1074–1078.

Pan, J., Xi, J., Eckenhoff, M. & Eckenhoff, R. Halothane binding proteome in human brain cortex. *J. Proteome Res.*, **6**, 582–592.

Reimers, J. R., McKemmish, L. K., McKenzie, R. H., Mark, A. E. & Hush N. S. (2014) The revised Penrose-Hameroff orchestrated objective-reduction proposal for human consciousness is not scientifically justified: Comment on "Consciousness in the Universe: A review of the 'Orch OR' theory" by Hameroff and Penrose. *Phys. Life Rev.*, **11**, 101–103.

Reimers, J. R., McKemmish, L. K., McKenzie, R. H., Mark, A. E. & Hush, N. S. (2009) Weak, strong, and coherent regimes of Fröhlich condensation and their applications to terahertz medicine and quantum consciousness. *Proc. Nat. Acad. Sci. USA*, **106**, 4219–4224.

Sahu, S., Ghosh, S., Ghosh, B., Aswani, K., Hirata, K., Fujita, D. & Bandyopadhyay, A. (2013a) Atomic water channel controlling remarkable properties of a single brain microtubule: Correlating single protein to its supramolecular assembly. *Biosens. Bioelectron.*, **47**, 141–148.

Sahu, S., Ghosh, S., Hirata, K., Fujita, D. & Bandyopadhyay, A. (2013b) Multi-level memory-switching properties of a single brain microtubule. *Appl. Phys. Lett.*, **102**, 123701.

Sahu, S., Ghosh, S., Fujita, D., Bandyopadhyay, A. (2014) Live visualizations of single isolated tubulin protein self-assembly via tunneling current: effect of electromagnetic pumping during spontaneous growth of microtuble. *Scientific Reports*, **3**, 7303.

Samsonovich, A., Scott, A. & Hameroff, S. (1992) Acousto-conformational transitions in cytoskeletal microtubules: Implications for intracellular information processing. *Nanobiology*, **1**, 457–468.

Sataric, M. V., Zekovic, S., Tuszynski, J. A. & Pokorny J. (1998) The Mossbauer effect as a possible tool in detecting nonlinear excitations in microtubules. *Phys. Rev. E*, **58**, 6333–6339.

Usui, N., Terada, K., Baba, K., Matsuda, K., Nakamura, F., Usui, K., Tottori, T., Umeoka, S., Fujitana, S., Tottori, T., Umeoka, S., Fujitani, S., Mihara, T. & Inoue, Y. (2010) Very high frequency oscillations (over 1000 Hz) in human epilepsy. *Clin. Neurophysiol.*, **121**, 1825–1831.

van Gunsteren, W. F. ( 2013) The seven sins in academic behavior in the natural sciences. *Angew. Chem. Int. Ed.*, **52**, 118–122.

Vattay, G., Salahub D., Csabai, I., Nassimi, A. & Kauffman, S., (2015) Quantum criticality at the origin of life. *Journal of Physics: Conference Series*. **626**:012023.

Voet, D. & Voet, J. (1995) *Biochemistry*. 2nd ed., Wiley.

Wolchover, N. (2013) Physicists eye quantum-gravity interface. Quanta Magazine, October 31, Simons Foundation.

Xi, J., Liu, R., Asbury, G. R., Eckenhoff, M. F. & Eckenhoff, R. G. (2004) Inhalational anesthetic-binding proteins in rat neuronal membranes. *J. Biol. Chem.*, **279**, 19628–19633.

# INDEX

H



I



J



K



L